Jawaid A. Khan
Jeanne Dijkstra
Editors

Plant Viruses As Molecular Pathogens

Pre-publication
REVIEWS,
COMMENTARIES,
EVALUATIONS . . .

"**T**his book is both up-to-date and very informative for traditional and new generations of plant virologists in both industrialized and developing nations of the world. Moreover, some chapters present very interesting concepts regarding the use of molecular techniques to gain new insight into long-standing pathological issues, such as virus evolution, host adaptation, and epidemiology. I was also pleased to see a good deal of information on plant viruses of importance in the tropics, such as potyviruses, begomoviruses, and some emerging plant viruses transmitted by fungal vectors. Altogether, a very valuable collection of themes on the new art and science of plant virology."

Francisco J. Morales, PhD
Head Virology Research Unit,
International Center for Tropical
Agriculture (CIAT),
Miami, Florida

"**T**his book covers a wide range of subjects, and it is refreshing to have most of the chapters written by people who have not reviewed the specific topics before, which gives new perspectives to their coverage. As a particular example, I refer to the chapter 'Natural Resistance to Viruses' by Jari Valkonen; he covers the field well, and postulates how resistance is engendered. In addition, some subjects, such as the transmission of viruses by nematodes, by fungi, or through the seed, have not been reviewed for many years and have been in need of updating. There are also some chapters dealing with molecular techniques that students and researchers will find useful."

Milton Zaitlin, PhD
Professor Emeritus,
Department of Plant Pathology,
College of Agriculture and Life Sciences,
Cornell University,
Ithaca, New York

NOTES FOR PROFESSIONAL LIBRARIANS AND LIBRARY USERS

This is an original book title published by Food Products Press®, an imprint of The Haworth Press, Inc. Unless otherwise noted in specific chapters with attribution, materials in this book have not been previously published elsewhere in any format or language.

CONSERVATION AND PRESERVATION NOTES

All books published by The Haworth Press, Inc. and its imprints are printed on certified pH neutral, acid free book grade paper. This paper meets the minimum requirements of American National Standard for Information Sciences-Permanence of Paper for Printed Material, ANSI Z39.48-1984.

Plant Viruses
As Molecular Pathogens

FOOD PRODUCTS PRESS
Crop Science
Amarjit S. Basra, PhD
Senior Editor

New, Recent, and Forthcoming Titles of Related Interest:

Dictionary of Plant Genetics and Molecular Biology by Gurbachan S. Miglani

Advances in Hemp Research by Paolo Ranalli

Wheat: Ecology and Physiology of Yield Determination by Emilio H. Satorre and Gustavo A. Slafer

Mineral Nutrition of Crops: Fundamental Mechanisms and Implications by Zdenko Rengel

Conservation Tillage in U.S. Agriculture: Environmental, Economic, and Policy Issues by Noel D. Uri

Cotton Fibers: Developmental Biology, Quality Improvement, and Textile Processing edited by Amarjit S. Basra

Heterosis and Hybrid Seed Production in Agronomic Crops edited by Amarjit S. Basra

Intensive Cropping: Efficient Use of Water, Nutrients, and Tillage by S. S. Prihar, P. R. Gajri, D. K. Benbi, and V. K. Arora

Physiological Bases for Maize Improvement edited by María E. Otegui and Gustavo A. Slafer

Plant Growth Regulators in Agriculture and Horticulture: Their Role and Commercial Uses edited by Amarjit S. Basra

Crop Responses and Adaptations to Temperature Stress edited by Amarjit S. Basra

Plant Viruses As Molecular Pathogens by Jawaid A. Khan and Jeanne Dijkstra

Barley Science: Recent Advances from Molecular Biology to Agronomy of Yield and Quality edited by Gustavo A. Slafer, José Luis Molina-Cano, Roxana Savin, José Luis Araus, and Ignacio Romagosa

In Vitro Plant Breeding by Acram Taji, Prakash Kumar, and Prakash Lakshmanan

Crop Improvement: Challenges in the Twenty-First Century edited by Manjit S. Kang

Tillage for Sustainable Cropping by P. R. Gajri, V. K. Arora, and S. S. Prihar

Plant Viruses
As Molecular Pathogens

Jawaid A. Khan
Jeanne Dijkstra
Editors

CRC Press
Taylor & Francis Group
Boca Raton London New York

CRC Press is an imprint of the
Taylor & Francis Group, an **informa** business

All rights reserved. No part of this work may be reproduced or utilized in any form or by any means, electronic or mechanical, including photocopying, microfilm, and recording, or by any information storage and retrieval system, without permission in writing from the publisher. Printed in the United States of America.

Cover illustrations: (figure on left) symptom induction in *Nicotiana benthamiana* leaves inoculated with cucumovirus recombinants, courtesy of Chikara Masuta; (figure on right) schematic diagram showing a geminivirus infection cycle, courtesy of Rafael F. Rivera-Bustamante.

Cover design by Jennifer M. Gaska.

Library of Congress Cataloging-in-Publication Data

Khan, Jawaid A.
    Plant viruses as molecular pathogens / Jawaid A. Khan, Jeanne Dijkstra.
        p. cm.
    Includes bibliographical references (p.).
    ISBN 1-56022-894-6 (hardcover : alk. paper)—ISBN 1-56022-895-4 (softcover : alk. paper)
    1. Plant viruses. 2. Virus diseases of plants. I. Dijkstra, Jeanne, 1930-II. Title.

QR351 .K476 2001
579.2'8—dc21
                                                                    00-049452

CONTENTS

About the Editors   xii

Contributors   xiii

Preface   xvii

SECTION I: TAXONOMY

**Chapter 1. The Principles and Current Practice
   of Plant Virus Taxonomy**   3
   Mike A. Mayo

   Introduction   3
   The Underlying Principles   4
   The Role of ICTV   7
   The Current System of Plant Virus Classification   8
   Future Prospects   21

Chapter 2. How to Write the Names of Virus Species   25
   Marc H. V. Van Regenmortel

SECTION II: VIRUS TRANSMISSION AND TRANSPORT

**Chapter 3. Genes Involved in Insect-Mediated
   Transmission of Plant Viruses**   31
   Juan José Lopez-Moya

   Introduction   31
   Insect Vectors of Plant Viruses   32
   Classification of Transmission Modes   32
   Nonpersistent and Semipersistent Transmission   34
   Circulative Nonpropagative Transmission   41
   Circulative Propagative Transmission   47
   Conclusions   49

**Chapter 4. Characteristic Features of Virus Transmission
by Nematodes**   **63**
   Jeanne Dijkstra
   Jawaid A. Khan

Introduction   63
Vector Nematodes   63
Viruses Transmitted by Nematodes   64
Mechanism of Virus Transmission   67
Control of Nematode-Transmitted Viruses   70

Chapter 5. Virus Transmission by Fungal Vectors   **77**
   Jeanne Dijkstra
   Jawaid A. Khan

Introduction   77
Fungal Vectors   77
Fungus-Transmitted Viruses   80
Modes of Transmission   93
Mechanism of Virus-Vector Association   94
Epidemiology and Control of Fungus-Borne Viruses   97
Conclusions   98

**Chapter 6. Seed Transmission of Viruses:
Biological and Molecular Insights**   **105**
   Jawaid A. Khan
   Jeanne Dijkstra

Introduction   105
Biological Characteristics   107
Genetic Determinants   111
Conclusions   122

Chapter 7. Molecular Biology of Plant Virus Movement   **127**
   Ayala L. N. Rao
   Yoon Gi Choi

Introduction   127
Movement of Plant Viruses   128
Current Status   138

SECTION III: MOLECULAR BIOLOGY OF RNA VIRUSES

Chapter 8. Mechanism of RNA Synthesis by a Viral
RNA-Dependent RNA Polymerase   **147**
    Kailayapillai Sivakumaran
    Jin-Hua Sun
    C. Cheng Kao

Introduction   147
Initiation of RNA Synthesis   148
Abortive Synthesis   151
Transition from Initiation to Elongation   159
Termination of RNA Synthesis   167
Comparison of RNA-Dependent
    and DNA-Dependent RNA Synthesis   167
Final Comments   170

Chapter 9. Gene Expression Strategies of RNA Viruses   **175**
    Uli Commandeur
    Wolfgang Rohde
    Rainer Fischer
    Dirk Prüfer

Introduction   175
Subgenomic RNAs and Genome Segmentation   176
Initiation of Translation   181
Elongation of Translation   188
Termination of Translation   189
Polyprotein Processing and Host Factors   190
Outlook   191

Chapter 10. Recombination in Plant RNA Viruses   **203**
    Chikara Masuta

Introduction   203
Overview of Research History on the Recombination
    of Plant RNA Viruses   203
Recombination in Cucumoviruses   209
Newly Evolved Recombinant Viruses Between
    CMV and TAV Under Selection Pressure   211
Summary and Conclusion   219

**Chapter 11. Variability and Evolution of *Potato Virus Y,*
the Type Species of the *Potyvirus* Genus**   **225**
    Laurent Glais
    Camille Kerlan
    Christophe Robaglia

Introduction   225
The Family *Potyviridae*   225
Relationships Between Potyviruses Species   230
Biological Diversity of PVY   231
Genetic Variability of PVY   235
Mechanisms Leading to RNA Genome Polymorphisms   239
Conclusions   244

SECTION IV: MOLECULAR BIOLOGY OF DNA VIRUSES

**Chapter 12. Geminivirus Replication
and Gene Expression**   **257**
    Zulma I. Monsalve-Fonnegra
    Gerardo R. Argüello-Astorga
    Rafael F. Rivera-Bustamante

Introduction   257
Geminivirus Classification   258
Genome Structure   258
Geminivirus Infection Cycle: A Brief Overview   261
Geminivirus Replication   263
Control of Viral Gene Expression   267
Concluding Remarks   275

**Chapter 13. The Molecular Epidemiology
of Begomoviruses**   **279**
    Judith K. Brown

Introduction   279
Begomoviruses: Distribution and Characteristics   280
Detection, Identification, and Classification
    of Begomoviruses   284
Begomovirus Suscepts and Hosts in Relation
    to Epidemiology   287

Whitefly Vector Biology That Influences
    Epidemiology   293
The Influence of Vector Biotypes, Races, and Variants
    on Virus Spread   295
New Diseases, Epidemics, and Pandemics   300
Conclusions   307

**Chapter 14. Translational Strategies in Members
of the Family *Caulimoviridae*   317**
    Mikhail M. Pooggin
    Lyubov A. Ryabova
    Thomas Hohn

Introduction   317
Architecture of Viral RNA   320
General Enhancement of Expression   321
Shunting   322
Leaky Scanning (Bacilliform Caulimoviruses)   326
Activated Polycistronic Translation (Icosahedral
    Caulimoviruses)   327
Transactivation and Shunting   329
Biological Significance of Ribosome Shunt   329
Gag-Pol Translation   330
Making Use of Translational Control   331
Concluding Remarks   331

Chapter 15. Recombination in Plant DNA Viruses   339
    Thomas Frischmuth

Introduction   339
Recombination in Plant ssDNA Viruses   347
Recombination in Plant dsDNA Viruses   354
Conclusions   357

SECTION V: RESISTANCE TO VIRAL INFECTION

Chapter 16. Natural Resistance to Viruses   367
    Jari P. T. Valkonen

Introduction   367
Distinguishing a Nonhost from a Host   368
The Terms Describing Virus-Host Interactions
    and Resistance   369

Host Responses Conferring Resistance to Viruses 373
Identification of the Genes Involved in Resistance 377
Recognition of the Virus: The Gene-for-Gene Theory 378
Viral Suppressor of Resistance 385
Future Uses of Natural Virus Resistance Genes 388

Chapter 17. Engineering Virus Resistance in Plants **399**
   Erwin Cardol
   Jan van Lent
   Rob Goldbach
   Marcel Prins

Introduction 399
Structure and Genetic Organization of Plant Viral Genomes 399
Engineered Resistance to Viruses 406
Advantages of RNA-Mediated Resistance 416
Concluding Remarks 417

SECTION VI: METHODS IN MOLECULAR VIROLOGY

Chapter 18. Antibody Expression in Plants **425**
   Thorsten Verch
   Dennis Lewandowski
   Stefan Schillberg
   Vidadi Yusibov
   Hilary Koprowski
   Rainer Fischer

Introduction 425
Antibody-Mediated Viral Resistance in Transgenic Plants 425
Expression of Recombinant Antibodies Using Plant Virus
   Gene Vectors 429
Conclusions 436

**Chapter 19. Nucleic Acid Hybridization for Plant Virus
and Viroid Detection** **443**
   Rudra P Singh
   Xianzhou Nie

Introduction 443
The Principle of the Nucleic Acid Hybridization Method 444
Types of Labels 448

Sample Preparation 450
Prehybridization 454
Hybridization 454
Washing Procedures 455
Detection Procedures 455
Sensitivity of the Detection 456
Application to Viruses and Viroids 461
Conclusions 462

Chapter 20. Application of PCR in Plant Virology **471**
  Ralf Georg Dietzgen

Introduction 471
Aspects of PCR Template Preparation 472
Plant Virus Detection and Differentiation 475
Real-Time, Quantitative PCR 484
Virus Strain Discrimination 485
Viral Genome Characterization 487
Molecular Plant Virology Applications 489
Conclusions 491

Chapter 21. Plant Virus Detection in Animal Vectors **501**
  Rudra P. Singh

Introduction 501
Detection by Earlier Methods 501
Detection by a New Method 506
Detection in Animal Vectors 507
The Presence of Virus in a Vector and Biological
  Transmission to Plants 514
Conclusions 515

Index **523**

ABOUT THE EDITORS

Jawaid A. Khan, PhD, MSc, obtained his MSc in botany with specialization in plant pathology from the Aligarh Muslim University, Aligarh, India. He was awarded a PhD degree in plant virology from the Wageningen Agricultural University, Wageningen, The Netherlands. He carried out research on potyviruses of beans for his PhD thesis. During this period, he described a novel type of interference phenomenon and proposed the taxonomic status of bean common mosaic virus and blackeye cowpea mosaic virus isolates. Dr. Khan has a number of research papers to his credit. His research interests include the molecular characterization of geminiviruses and potyviruses infecting ornamental, horticultural, and other economically important crops and the development of virus detection systems. He is a plant virologist at the National Botanical Research Institute, Lucknow (an organization of the Council of Scientific and Industrial Research, Government of India).

Jeanne Dijkstra, PhD, MSc, obtained her MSc in biology and her PhD in plant virology from the University of Amsterdam. From 1957 to 1995 she was employed as a plant virologist at the Department of Virology, Wageningen Agricultural University in The Netherlands, where she carried out investigations on early events in the infection process, interference between viruses in a plant, and identification and taxonomy of plant viruses. In that period she taught virology to undergraduate and postgraduate students in virology, plant pathology, crop protection, plant breeding, and horticultural and tropical crop science. In addition to producing a number of scientific papers, she is a co-editor of the book *Viruses of Plants* and the author of *Practical Plant Virology: Protocols and Exercises.*

CONTRIBUTORS

Gerardo R. Argüello-Astorga, PhD, Research Associate, Centro de Investigacíon y de Estudios Avanzados-IPN, Unidad Irapuato, Departamento de Ingeniería Genética, Irapuato, México.

Judith K. Brown, PhD, Associate Professor, Department of Plant Sciences, University of Arizona, Tucson.

Erwin Cardol, PhD student, Laboratory of Virology, Wageningen University, Wageningen, The Netherlands.

Yoon Gi Choi, PhD, Postdoctoral Fellow, Department of Plant Pathology, University of California, Riverside, California.

Uli Commandeur, PhD, Assistant Professor, Aachen University of Technology (RWTH), Institute for Biology VII, Molecular Biotechnology, Aachen, Germany.

Ralf Georg Dietzgen, PhD, Principal Scientist, Department of Primary Industries, Queensland Agricultural Biotechnology Centre, The University of Queensland, Australia.

Rainer Fischer, PhD, Professor, Head of Department of Molecular Biotechnology, Aachen University of Technology (RWTH), Institute for Biology VII, Molecular Biotechnology, Aachen, Germany; Professor, Head of Department of Molecular Biotechnology, Fraunhofer-IUCT, Schmallenberg, Germany.

Thomas Frischmuth, PhD, Associate Professor, University of Stuttgart, Biologisches Institut, Department of Plant Molecular Biology and Virology, Stuttgart, Germany.

Laurent Glais, PhD, Researcher, Institut National de la Recherche Agronomique (INRA), Centre de Rennes UMR BiO3P, Domaine de la Motte, Le Rheu, France.

Rob Goldbach, PhD, Professor and Head, Laboratory of Virology, Wageningen University, Wageningen, The Netherlands.

Thomas Hohn, PhD, Professor and Head of Laboratory, Friedrich Miescher Institute, Basel, Switzerland.

C. Cheng Kao, PhD, Associate Professor and Jack Gill Fellow, Department of Biology, Indiana University, Bloomington, Indiana.

Camille Kerlan, PhD, Chairman of the Virology Section of European Association for Potato Research, Institut National de la Recherche Agronomique (INRA), Centre de Rennes UMR BiO3P, Domaine de la Motte, Le Rheu, France.

Hilary Koprowski, MD, Director of the Biotechnology Foundation Laboratories and the Center for Neurovirology, Thomas Jefferson University, Biotechnological Foundation Laboratories, Philadelphia, Pennsylvania.

Dennis Lewandowski, PhD, Assistant in Plant Pathology, University of Florida, Department of Plant Pathology, Citrus Research and Education Center, Lake Alfred, Florida.

Juan José Lopez-Moya, PhD, Titular Scientist, Departamento de Biologia de Plantas, Centro de Investigaciones Biologicas, CIB, Consejo Superior de Investigaciones Cientificas, CSIC, Madrid, Spain.

Chikara Masuta, PhD, Associate Professorr, Graduate School of Agriculture, Hokkaido University, Sapporo, Japan.

Mike A. Mayo, PhD, Senior Principal Scientist, Deputy Head of Virology Unit, Scottish Crop Research Institute, Dundee, Scotland.

Zulma I. Monsalve-Fonnegra, PhD student, Centro de Investigación y de Estudios Avanzados-IPN, Unidad Irapuato, Departamento de Ingeniería Genética, Apartado, Irapuato, México.

Xianzhou Nie, PhD, Research Scientist, Agriculture and Agri-Food Canada, Potato Research Centre, Fredericton, New Brunswick, Canada.

Mikhail M. Pooggin, PhD, Visiting Scientist, Friedrich Miescher Institute, Basel, Switzerland.

Marcel Prins, PhD, Assistant Professor, Laboratory of Virology, Wageningen University, Wageningen, The Netherlands.

Dirk Prüfer, PhD, Group Leader, Fraunhofer-IUCT, Department of Molecular Biotechnology, Applied Genomics and Proteomics, Schmallenberg, Germany.

Ayala L. N. Rao, PhD, Associate Professor, Department of Plant Pathology, University of California, Riverside, California.

Rafael F. Rivera-Bustamante, PhD, Professor and Head, Genetic Engineering Department, Centro de Investigación y de Estudios Avanzados-IPN, Unidad Irapuato, Departamento de Ingeniería Genética, Irapuato, México.

Christophe Robaglia, PhD, Head, Laboratoire du Métabolisme Carboné, CEA, Ecophysiologie Végétale et Microbiologie, Saint Paul Lez Durance, France.

Wolfgang Rohde, PhD, Professor and Head of the Virology Group, Department of Plant Breeding and Yield Physiology, Max-Planck-Institut für Züchtungsforschung, Köln, Germany.

Lyubov A. Ryabova, PhD, Research Assistant, Friedrich Miescher Institute, Basel, Switzerland.

Stefan Schillberg, PhD, Group Leader Molecular Farming, Rheinisch-Westfälische Technische Hochschule (RWTH), Institut für Biologie I (Botanik/Molecular Genetik), Aachen, Germany.

Rudra P. Singh, PhD, Principal Research Scientist, Agriculture and Agri-Food Canada, Potato Research Centre, Fredericton, New Brunswick, Canada.

Kailayapillai Sivakumaran, PhD, Senior Research Associate, Department of Biology, Indiana University, Bloomington, Indiana.

Jin-Hua Sun, MS, Senior Research Associate, Infectious Disease Division, New Haven, Connecticut.

Jari P. T. Valkonen, PhD, Professor of Virology (especially plant viruses) at Swedish University of Agricultural Sciences (SLU), Uppsala, Sweden, and Research Group Leader, Institute of Biotechnology, University of Helsinki, Finland, Department of Plant Biology, Genetic Centre, Swedish University of Agricultural Sciences (SLU), Uppsala, Sweden.

Jan van Lent, PhD, Assistant Professor, Laboratory of Virology, Wageningen University, Wagengen, The Netherlands.

Marc H. V. Van Regenmortel, PhD, Director of Immunochemistry Laboratory, Molecular and Cellular Biology Institute (IBMC), French National Centre for Scientific Reserach (CNRS), Strasbourg, France.

Thorsten Verch, PhD, Research Associate, Thomas Jefferson University, Biotechnological Foundation Laboratories, Philadelphia, Pennsylvania.

Vidadi Yusibov, PhD, Assistant Professor, Thomas Jefferson University, Biotechnological Foundation Laboratories, Philadelphia, Pennsylvania.

Preface

Despite the tremendous amount of knowledge about plant viruses collected in the previous century, these pathogens still constitute a hazard to many economically important crops, and their often unpredictable behavior puzzles virologists. More and more it is realized that for a better understanding of the action of viruses one has to look at their root: the viral genome. The recent successful unraveling of the nucleotide sequences of approximately 97 percent of the human genome has certainly been a great achievement, but it is only a beginning. A far greater challenge still lies ahead: how to correlate those 3 billion letters to the functioning of the human body. Knowing the alphabet is one thing; understanding the language, another. Or, as somebody once remarked, a student who masters 97 percent of the Russian alphabet should have no illusions about understanding Tolstoy.

Similarly, the nucleotide sequences of many plant viruses have been elucidated, but still little is known about the action of the genes and the role of the nontranslated part of the genome. The present book attempts to provide more insight into molecular processes eventually leading to the pathogenic behavior of plant viruses.

This book consists of six sections subdivided into twenty-one chapters. Chapter 1 in Section I gives an overview of the present plant virus classification and provides insight into taxonomic rationalization. The latest rules for writing the names of taxonomic entities are discussed in Chapter 2.

Section II contains five chapters dealing with the transmission and transport of viruses. Chapters 3 through 6 describe the various ways plant viruses are transmitted by insects, nematodes, and fungi as well as through seed. They present current knowledge of genes and gene products involved in their respective transmission strategies. Chapter 7 details molecular mechanisms involved in the movement of a virus in its host plant.

Significant developments in the area of molecular biology of RNA viruses are discussed in Section III. Chapter 8 reviews the advancements made in understanding the mechanism of RNA replication, while Chapter 9 focuses on novel aspects of gene expression strategies. Events and mechanisms of recombinations encountered in nature as well as under experimental conditions are highlighted in Chapter 10. The last chapter in this section,

Chapter 11, considers biological and molecular diversity of a potyvirus and discusses the evolutionary aspects of its variability.

Section IV on DNA viruses, which includes Chapters 12 through 15, covers mechanisms of DNA synthesis and gene expression, molecular epidemiology, translational strategies, and recombination.

How plants develop resistance to viral infection is the focus of Section V. Chapter 16 presents an exhaustive overview of natural resistance, and Chapter 17 describes biotechnological strategies of pathogen-derived resistance, with emphasis on RNA-mediated resistance.

Finally, different molecular methods are presented in Section VI. Chapter 18 describes expression of recombinant proteins, such as antibodies, and their role in engineered resistance. Chapter 19 provides information on the recent developments of nucleic acid-based hybridization methodology, its technological advances, and its potential role in studying plant viruses and vectors. The enormous versatility of PCR (polymerase chain reaction) is extensively reviewed in Chapter 20. This chapter highlights the significant use of PCR and discusses its different forms and technical protocols. Last, in Chapter 21, the early detection of viruses in animal vectors and their impact on epidemiological studies are presented.

This joint venture would not have been successful without the active participation of the contributors who kindly authored different chapters on diverse topics of this fascinating area of science. We would like to thank all of them for their ready and pleasant cooperation in the preparation of this volume. It is hoped that this book will be of help to all those who wish to become acquainted with the current scenario of plant virology.

Jawaid A. Khan
Jeanne Dijkstra

SECTION I:
TAXONOMY

Chapter 1

The Principles and Current Practice of Plant Virus Taxonomy

Mike A. Mayo

INTRODUCTION

To conduct meaningful scientific discussion, the parties involved need to agree about the subject being discussed. Making this agreement is taxonomy. Thus, defining taxa, classifying lower taxa or individuals into higher taxa, and naming them are fundamental to scientific discussion and progress. This need for taxonomy has been recognized in virology, implicitly or explicitly, for many years. In this time, there have been many and various propositions as to how to classify viruses of plants. These have largely been attempts to make order of virus diversity by using as much knowledge of the diversity as was available at the time. Inevitably, any new information prompted revision of these schemes, often in radical ways. This process has been reviewed in Matthews (1983). Current taxonomic systems for viruses date from the years immediately preceding the establishment in 1966 of the International Committee on Nomenclature of Viruses (ICNV). A major background influence on this initiative was the then newly acquired knowledge about particle morphology. These observations led to proposals for a variety of classification schemes, in particular those of Brandes and Wetter (1959) for plant viruses, and that of Lwoff et al. (1962) for all viruses, based on various characters in a highly structured hierarchical scheme. The ICNV, and its later manifestation, the International Committee on Taxonomy of Viruses (ICTV), has since evolved a universal system of virus classification based

As with all taxonomic information, that presented here is a synthesis of information garnered from many virologists, in particular colleagues on the Plant Virus Subcommittee of ICTV. I gratefully acknowledge their contributions. The work was funded in part by the Scottish Executive Rural Affairs Department of the U.K. Government.

on principles and rules agreed upon by all branches of virology. This was reviewed in Murphy et al. (1995).

ICTV is bound by statute to publish reports of its decisions, and the most recent reports illustrate vividly the explosion of detailed information about virus structure and genetics in the past decade or so. In the Fifth ICTV Report (Francki et al., 1991), plant viruses were classified into 33 groups (one of which was in a family) and three genera in two families. In the following several years, many changes were introduced under the stewardship of the then-Chairman of the ICTV Plant Virus Subcommittee, Professor G. P. Martelli. These resulted in the classification of plant viruses into 47 genera, of which 25 were classified among ten families. The remaining 22 were left as unassigned genera (Martelli, 1997). This was summarized in the Sixth ICTV Report (Murphy et al., 1995). Since then, three additional families have been created and a further 23 genera have been recognized. Also, viroids have now been classified among seven genera clustered into two families, and retrotransposons have been classified into two families, with plant-"infecting" retrotransposons classified in one of the two genera in each family. Taxonomic developments since the publication of the Sixth ICTV Report are summarized by Mayo (1999). A more detailed summary is given in the Seventh ICTV Report (Van Regenmortel et al., 2000). This chapter outlines the present (summer 1999) situation and gives some insights into this relatively rapid taxonomic rationalization.

THE UNDERLYING PRINCIPLES

Taxonomy

Despite their not being organisms, it is very obvious that viruses are subject to variation during replication and that selection has operated on populations that contain variants. Thus, viruses have evolved by natural selection. The effects of selection have been to cause discontinuities in the distribution of properties among virus progeny, making it possible to recognize sets of characters that can be used to discriminate between groups of viruses. Virus species are defined following this logic, although the concept of species as applied to viruses has only recently become acceptable (Van Regenmortel, 1990), after a lengthy period of debate and controversy (e.g., Harrison, 1985; Milne, 1985). Virus species is defined as "a polythetic class of viruses that constitutes a replicating lineage and occupies a particular ecological niche" (Van Regenmortel, 1990: 249).

This definition avoids the problem that is posed by some species definitions: how to account for shared gene pools given the asexual nature of viruses. In essence, this species definition formalized what plant virologists had been doing for some time. The series *Descriptions of Plant Viruses* (edited by B. D. Harrison and A. F. Murant; published by the Commonwealth Mycological Institute and Association of Applied Biologists [CMI/AAB], and latterly AAB) is a recognition of the existence of a taxon that includes minor variants but distinguishes agents that differ appreciably in nature. Clearly, a species will contain many variant types that fall within its polythetically defined boundary. "Strains," "serotypes," and "variants" are usually such subspecific groupings. Higher taxa are defined as clusters of lower taxa. Thus, a genus contains one or more species, one of which is the type species to which the genus is formally linked. Families are formed when genera can be clustered. But this is not always so, and some genera are not classified into a higher taxon.

Species Demarcation

The existence of a theoretical definition of a virus species is an important step in rationalizing virus taxonomy and nomenclature. Such a definition, however, must be accompanied by a declaration by the appropriate groups of experts of the several criteria that should be used to make a polythetic appraisal of the status of any particular virus isolate (Van Regenmortel et al., 1997). This criterion list is needed for all genera, though it need not be the same for all genera. This last point is worth emphasizing. The essentially pragmatic way in which virus species and genera are defined means that there is no formal (or indeed realistic) requirement that criteria used to assist in distinguishing between species in one genus be applicable to distinguishing species in another genus. In principle, it is readily conceivable that the forces of selection acting on variation among the different properties of viruses in one genus be more or less powerful than the corresponding forces acting on viruses in another genus. And this difference will in all probability be reflected in the choice of criteria for distinguishing species in each of the genera.

In the Seventh ICTV Report (Van Regenmortel et al., 2000), a considerable effort was made to spell out the criteria lists applicable to viruses in the different genera.

The International Code

Once taxa have been created, it is necessary to name them immediately. In contrast to nomenclatural decisions in other biological fields, virus nomenclature and virus taxonomy are both controlled by one body, ICTV. Decision making is guided by principles and informed by rules that are set out as part of the International Code of Virus Classification and Nomenclature (Mayo and Horzinek, 1998; Van Regenmortel et al., 2000). The Code has evolved from a set of rules through gradual refinement by successive decisions of the ICTV Executive Committee (e.g., Matthews, 1983; Francki et al., 1991; Mayo and Murphy, 1994; Murphy et al., 1995; Mayo, 1996; Mayo and Horzinek, 1998). The current rules are listed in the Code in the form of 49 articles. Table 1.1 is a summary of the key elements from these rules. The main guiding principles of virus nomenclature are intended to maximize convenience for virologists using the Code, in research and in teaching. Thus, name changes should be kept to a minimum, names should only be selected after inspection for potential error or confusion in their use, and names should only be adopted when they are truly necessary (see Table 1.1).

Although a key element in virus taxonomy is that of stability (i.e., there should be a reluctance to make changes, particularly to nomenclature), the rules are subject to modification when necessary, within the broad principles of virus taxonomy. Such modifications are considered as taxonomic proposals. The most notable recent modification has been to the rules concerned with the orthography of species names. These new rules are highlighted in Table 1.1 (section C). They bring procedures for the names of species into line with the way in which other formal taxon names are written; that is, they are to be written in italic script and the first word in the taxon name should begin with a capital letter.

TABLE 1.1. Key Elements of the International Code of Virus Classification and Nomenclature

A. Principles of virus nomenclature
- To aim for stability
- To avoid or reject the use of names that might cause error or confusion
- To avoid the unnecessary creation of names

B. Essentials of rules for naming taxa
- Existing names should be retained when possible. (Rule 3.9)
- Priority rules do not apply. (Rule 3.10)
- Names of persons are not used. (Rule 3.11)
- Names should be easy to use and easy to remember. (Rule 3.12)
- Names may use sigla. (Rule 3.15)
- Names should not mislead. (Rule 3.18)
- Names should not offend. (Rule 3.19)
- Names should have appropriate endings to signify the taxon level. (Rules 3.27, 3.30, 3.32, 3.34, 3.36)

C. Rules for orthography
- Names of virus orders, families, subfamilies, and genera are printed in italics and their first letters are capitalized. (Rule 3.39)
- Species names are printed in italics and the first letter of the first word of the name is capitalized. (Rule 3.40)

THE ROLE OF ICTV

Taxonomic decisions are taken by ICTV, which is authorized by statutes approved by the Virology Division of the International Union of Microbiological Societies (Mayo and Pringle, 1998, and references therein). Thus, decisions are subjected to representative international scrutiny. The ICTV is organized and advised by an Executive Committee that consists of 18 members: four officers, chairs of six subcommittees representative of each major branch of virology, and eight elected members.

Plant virology is represented by the chairman of a subcommittee that itself consists of 19 study group chairs and eight other members. The study groups are concerned with particular taxa or groups of taxa (e.g., the *Potyviridae* Study Group). The subcommittee chair appoints the chairs of the study groups, who then appoint study group members as is appropriate. Figure 1.1 is a diagrammatic representation of the stages in the decision-making process of ICTV.

Ideas for taxonomic change, either creation or modification of plant virus taxa, or decisions about names of these taxa, usually originate in study group deliberations. These ideas are scrutinized by the Plant Virus Subcommittee, largely for their acceptability within the overall taxonomic scheme for plant

FIGURE 1.1. Taxonomic Decision-Making Structure for Plant Viruses

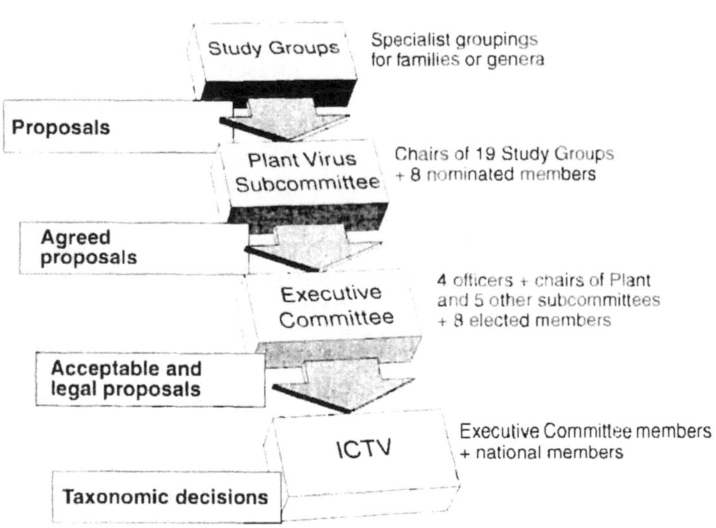

viruses. If approved, the proposals are then submitted to the Executive Committee members who examine their acceptability in the context of all virus taxonomy. Proposals that are again approved are then put to the membership of ICTV for a final vote as to their acceptability. After a favorable vote by ICTV, the proposals become part of the taxonomic scheme for viruses. These decisions are then published in Virology Division News in *Archives of Virology* and/or in the regular ICTV reports (e.g., Van Regenmortel et al., 2000).

THE CURRENT SYSTEM
OF PLANT VIRUS CLASSIFICATION

Broad Outlines

In this section, it is convenient to divide viruses according to the nature of their genomes. However, this division does not imply any taxonomic significance. For example, it is not suggested that all viruses with negative-sense, single-stranded (ss) RNA genomes form a taxonomic cluster at the same hierarchical level as all viruses with double-stranded (ds) RNA or, indeed, dsDNA genomes. Thus, this arbitrary division, which is also used in ICTV reports, is one only of convenience.

Genomes are of two types: (1) those which are reverse transcribed during the expression of their genomes (retroid), and then encapsidate either an RNA copy (retroviruses) or a DNA copy (pararetroviruses) of the genome, and (2) those which are not reverse transcribed. The nonretroid viruses can be divided according to the type of genome nucleic acid, normally that found encapsidated in virus particles. This can be ssDNA, dsRNA, negative/ambisense ssRNA, or positive (i.e., messenger)-sense ssRNA. No genomes of dsDNA that do not encode reverse transcriptase have been found in plant viruses.

Table 1.2 shows the distribution of genome types among viruses of major classes of hosts. The great majority of plant virus genera consist of species of RNA genome viruses. Viruses of vertebrates are found in all genome types, whereas viruses of prokaryotes are very largely dsDNA genome types. The listing detailed in the following pages of virus families and genera grouped according to their genome types and presented largely in the same order as that used in the Seventh ICTV Report (Van Regenmortel et al., 2000). Detail is concentrated on those taxa newly created since the Sixth ICTV Report was published. The listing is summarized in Table 1.3, in which the type species of each of the genera is also given. Further details, together with bibliographic information, are contained in the individual family/genus descriptions by various authors brought together in the Seventh ICTV Report (Van Regenmortel et al., 2000).

TABLE 1.2. Virus Genera by Host and Genome Types

Host	Genome type						
	dsDNA	ssDNA	Retroid DNA	RNA	dsRNA	ssRNA (−)	ssRNA (+)
Plants	–	4	6	2	6	5	49
Vertebrates	25	4	2	7	7	22	19
Invertebrates	10	3	3	–	9	10	8
Fungi/Algae/ Protozoa	5	–	3	–	6	–	3
Prokaryotes	22	6	–	–	1	–	2

Source: Data from Van Regenmortel et al. (2000).

ssDNA Genome Plant Viruses

Family Geminiviridae

Family *Geminiviridae* contains three genera of viruses that have genomes of circular ssDNA in one or two pieces. Geminivirus particles have a geminate appearance and the viruses are transmitted by whiteflies (genus *Begomovirus,* type species *Bean golden mosaic virus*), by leafhoppers (genus *Mastrevirus,* type species *Maize streak virus*), or by planthoppers (genus *Curtovirus,* type species *Beet curly top virus*). These genera were formerly Subgroups III, I, and II, respectively. The classification is based on the sizes and arrangements of genes within the monopartite or bipartite genomes. Sequence similarities are also used as measures of relatedness.

Genus Nanovirus

Other types of ssDNA genome viruses are classified in genus *Nanovirus* (type species *Subterranean clover stunt virus*). These viruses have 20 nm diameter, icosahedral particles that encapsidate circular ssDNA approximately 1 kb in size. They resemble the animal-infecting circoviruses in genome size and genome organization but differ from them in virion size and morphology. The species in the genus all have either six or seven circular single-stranded DNAs, each of which has a single open reading frame (ORF) and is transcribed unilaterally. These definitive nanoviruses are

TABLE 1.3. The Current Classification of Plant Virus Genera

Family	Genus	Type species
ssDNA genomes		
Geminiviridae	**Mastrevirus**	Maize streak virus
	Curtovirus	Beet curly top virus
	Begomovirus	Bean golden mosaic virus
–	**Nanovirus**	Subterranean clover stunt virus
Reverse transcribing		
Caulimoviridae	Caulimovirus	Cauliflower mosaic virus
	"**SbCMV-like**"	Soybean chlorotic mottle virus
	"**CsVMV-like**"	Cassava vein mosaic virus
	"**PVCV-like**"	Petunia vein clearing virus
	Badnavirus	Commelina yellow mottle virus
	"**RTBV-like**"	Rice tungro bacilliform virus
Pseudoviridae	**Pseudovirus**	Saccharomyces cerevisiae Ty1 virus
Metaviridae	**Metavirus**	Saccharomyces cerevisiae Ty3 virus
dsRNA		
Reoviridae	Phytoreovirus	Wound tumor virus
	Fijivirus	Fiji disease virus
	Oryzavirus	Rice ragged stunt virus
Partitiviridae	Alphacryptovirus	White clover cryptic virus 1
	Betacryptovirus	White clover cryptic virus 2
–	**Varicosavirus**	Lettuce big-vein virus
(–)ssRNA		
Rhabdoviridae	Cytorhabdovirus	Lettuce necrotic yellows virus
	Nucleorhabdovirus	Potato yellow dwarf virus
Bunyaviridae	Tospovirus	Tomato spotted wilt virus
–	Tenuivirus	Rice stripe virus
–	**Ophiovirus**	Citrus psorosis virus
(+)ssRNA		
Bromoviridae	Bromovirus	Brome mosaic virus
	Cucumovirus	Cucumber mosaic virus
	Alfamovirus	Alfalfa mosaic virus
	Ilarvirus	Tobacco streak virus
	Oleavirus	Olive latent virus 2
Closteroviridae	Closterovirus	Beet yellows virus
	Crinivirus	Lettuce infectious yellows virus
Comoviridae	Comovirus	Cowpea mosaic virus
	Nepovirus	Tobacco ringspot virus
	Fabavirus	Broad bean wilt virus 1
Luteoviridae	Luteovirus	Barley yellow dwarf virus-PAV

Family	Genus	Type species
	Polerovirus	Potato leafroll virus
	Enamovirus	Pea enation mosaic virus-1
Potyviridae	Potyvirus	Potato virus Y
	Rymovirus	Ryegrass mosaic virus
	Bymovirus	Barley yellow mosaic virus
	Macluravirus	Maclura mosaic virus
	Ipomovirus	Sweet potato mild mottle virus
	Tritimovirus	Wheat streak mosaic virus
Sequiviridae	Sequivirus	Parsnip yellow fleck virus
	Waikavirus	Rice tungro spherical virus
Tombusviridae	Tombusvirus	Tomato bushy stunt virus
	Carmovirus	Carnation mottle virus
	Necrovirus	Tobacco necrosis virus A
	Machlomovirus	Maize chlorotic mottle virus
	Dianthovirus	Carnation ringspot virus
	Avenavirus	Oat chlorotic stunt virus
	Aureusvirus	Pothos latent virus
	Panicovirus	Panicum mosaic virus
–	Tobravirus	Tobacco rattle virus
–	Tobamovirus	Tobacco mosaic virus
–	Hordeivirus	Barley strip mosaic virus
–	Furovirus	Soil-borne wheat mosaic virus
–	**Pomovirus**	Potato mop-top virus
–	**Pecluvirus**	Peanut clump virus
–	**Benyvirus**	Beet necrotic yellow vein virus
–	Sobemovirus	Southern bean mosaic virus
–	Marafivirus	Maize rayado fino virus
–	Umbravirus	Carrot mottle virus
–	Tymovirus	Turnip yellow mosaic virus
–	Idaeovirus	Raspberry bushy dwarf virus
–	**Ourmiavirus**	Ourmia melon virus
–	Potexvirus	Potato virus X
–	Carlavirus	Carnation latent virus
–	**Foveavirus**	Apple stem pitting virus
–	**Allexivirus**	Shallot virus X
–	Capillovirus	Apple stem grooving virus
–	Trichovirus	Apple chlorotic leaf spot virus
–	**Vitivirus**	Grapevine virus A

Viroids

Pospiviroidae	**Pospiviroid**	Potato spindle tuber viroid
	Hostuviroid	Hop stunt viroid

TABLE 1.3 *(continued)*

Family	Genus	Type species
	Cocadviroid	*Coconut cadang-cadang viroid*
	Apscaviroid	*Apple scar skin viroid*
	Coleviroid	*Coleus blumei viroid 1*
Avsunviroidae	**Avsunviroid**	*Avocado sunblotch viroid*
	Pelamoviroid	*Peach latent mosaic viroid*

Source: Data from Van Regenmortel et al. (2000).
Note: Bold text shows taxa and/or names changed since 1995.

transmitted by aphids. The genus also contains a tentative species, Coconut foliar decay virus (CFDV). Only one circular ssDNA of 1.3 kb has been detected so far for CFDV, and the distribution of putative ORFs indicates that the DNA may be transcribed bilaterally. CFDV is transmitted by planthoppers.

Retroid Viruses

Family Caulimoviridae

The main class of plant viruses that replicate in a way that involves reverse transcription includes those related to *Cauliflower mosaic virus*. In the Sixth ICTV Report, such viruses were classified in either genus *Caulimovirus* or genus *Badnavirus*. Recent sequence data have shown that the genomes of these viruses differ appreciably, and the present classification has recognized this diversity by creating new genera (there are now six), and by clustering these genera into the new family *Caulimoviridae*. Viruses with isometric particles are in one of four genera: *Caulimovirus* (type species *Cauliflower mosaic virus*), "SbCMV-like viruses" (type species *Soybean chlorotic mottle virus*), "CsVMV-like viruses" (type species *Cassava vein mosaic virus*), and "PVCV-like viruses" (type species *Petunia vein clearing virus*). Viruses with bacilliform particles, previously badnaviruses, are separated into the genus *Badnavirus* (type species *Commelina yellow mottle virus*) and the genus "RTBV-like viruses" (type species *Rice tungro bacilliform virus*). The main criteria distinguishing the genera are the number and arrangements of the ORFs in the different genomes. Sequence similarities and differences reinforce this division into genera.

Retrotransposons

Recent taxonomic changes have brought retrotransposons into the universal taxonomic scheme. In their genome structure and expression strategies, these agents resemble viruses in the family *Retroviridae* relatively closely. And, except for the encapsidation of an RNA genome, they also resemble viruses in family *Caulimoviridae* that are sometimes referred to as pararetroviruses. So far, two families have been recognized that differ in the possession or not of an *env* gene and in the organization of the *pol* gene. One genus in each family contains plant-"infecting" retrotransposons. For example, genus *Pseudovirus* (type species *Saccharomyces cerevisiae Ty1 virus;* family *Pseudoviridae*) contains *Arabidopsis thaliana Ta1 virus* and genus *Metavirus* (type species *Saccharomyces cerevisiae Ty3 virus;* family *Metaviridae*) contains *Lilium henryi de1 I virus.*

dsRNA Genome Viruses

Family Partitiviridae

The family *Partitiviridae* was formed initially to contain genera of viruses that infect various fungi and that have multipartite dsRNA genomes encapsidated in isometric particles. Cryptic viruses occur in a number of plant species and characteristically induce no symptoms in their hosts. These viruses resemble the fungus-infecting partitiviruses in having isometric particles either 30 or 40 nm in diameter and that contain two or more dsRNA segments. They were added as the genera *Alphacryptovirus* (type species *White clover cryptic virus 1*) and *Betacryptovirus* (type species *White clover cryptic virus 2*).

Family Reoviridae

Plant viruses with isometric particles and dsRNA genomes of ten or more segments are classified, along with similar viruses of animals, in the family *Reoviridae*. Three genera of plant reoviruses differ in particular in particle morphology: *Phytoreovirus* (type species *Wound tumor virus*), *Fijivirus* (type species *Fiji disease virus,* and *Oryzavirus* (type species *Rice ragged stunt virus*).

Genus Varicosavirus

This genus was created to contain viruses typified by *Lettuce big-vein virus* (LBVV). *Tobacco stunt virus* resembles LBVV in a number of respects and has been added to the genus, but as a "tentative species," to allow the possibility of its proving on further analysis to be very closely related to LBVV. These viruses have been known for some time to have dsRNA genomes, probably of two components, that are encapsidated in rod-shaped particles. The viruses are probably transmitted by *Olpidium* spp. In these characters, the viruses are distinct from any others.

Viruses with Negative-Sense RNA Genomes

Family Rhabdoviridae

Plant viruses that form bullet-shaped particles and that have monopartite, negative-sense ssRNA genomes are classified in family *Rhabdoviridae*, either in genus *Cytorhabdovirus* (type species *Lettuce necrotic yellows virus*) or in genus *Nucleorhabdovirus* (type species *Potato yellow dwarf virus*). The genera are distinguishable on the basis of the site at which virus particles accumulate in cells of infected plants. Many structural studies of plant tissues infected with rhabdoviruses have been reported. However, many such viruses are "orphans" in the sense that it is not possible to test their relatedness with well-defined rhabdoviruses, as they are not in culture and no reference complementary (cDNA clones and no antisera) are in existence. These are listed in the Seventh ICTV Report as "unassigned plant rhabdoviruses."

Family Bunyaviridae

Plant viruses with membrane-bound particles that contain tripartite ssRNA genomes made up of negative-sense and/or ambisense components are classified in genus *Tospovirus* (type species *Tomato spotted wilt virus*) in the family *Bunyaviridae*. Viruses in the other four genera of this family infect animals. Tospoviruses multiply both in plant hosts and in the thrips vectors responsible for their transmission.

Genus Tenuivirus

Viruses in genus *Tenuivirus* (type species *Rice stripe virus*) have an unusual particle morphology in that they appear as complex threadlike structures. Tenuiviruses resemble viruses in family *Bunyaviridae* in having genomes of negative-sense and ambisense RNA, but they differ markedly in particle morphology and in having genomes of four to six pieces of ssRNA. They are thought to multiply in their planthopper vectors.

Genus Ophiovirus

Citrus psorosis virus (the type species) and related viruses were recently classified in genus *Ophiovirus*. They resemble tenuiviruses in particle morphology but differ in having only three size classes of ssRNA. Also, particle preparations of ophioviruses are infective, possibly linked to the fact that they are not phloem limited. Also, unlike tenuiviruses, ophioviruses do not infect graminaceous plants.

Viruses with Positive-Sense ssRNA Genomes

Family Sequiviridae

Viruses in this family have been referred to as plant picornaviruses because they have isometric particles consisting of three protein types and a c. 10 kb RNA that is expressed by proteolysis of a large polyprotein translation product. Genome layout and sequences also resemble those of picornaviruses of vertebrates. Sequiviruses are transmitted in the semipersistent manner, and for genus *Sequivirus* (type species *Parsnip yellow fleck virus*), this transmission by aphids is dependent on the gene products of a helper virus. Viruses in genus *Waikavirus* (type species *Rice tungro spherical virus*) are transmitted by leafhoppers or aphids and are known to supply helper factors for the transmission of other viruses.

Family Comoviridae

The genomes of viruses in this family, similar to those of sequiviruses, are expressed by translation into a polyprotein that is cleaved to form functional products. However, the genomes are bipartite and are less reminiscent of picornavirus genomes than are sequivirus genomes. The genera in the family are distinguishable by their vectors. Viruses in genus *Comovirus* (type species *Cowpea mosaic virus*) are transmitted by beetles; those in genus *Fabavirus* (type species; *Broad bean wilt virus 1*) are transmitted by aphids, and viruses in genus *Nepovirus* (type species *Tobacco ringspot virus*) are transmitted by nematodes, in association with pollen, or through seeds.

Family Potyviridae

Potyviruses have filamentous particles and genomes of c. 10 kb that are translated to form a single polyprotein. The genera described by Murphy et al. (1995) were distinguished mainly by their vectors: aphids for genus *Potyvirus* (type species *Potato virus Y*), mites for genus *Rymovirus* (type species *Ryegrass mosaic virus*), and fungi for genus Bymovirus (type species *Barley yellow mosaic virus*). Further molecular characterization has shown that viruses previously classified as "unassigned in the family" are distinct from viruses in recognized genera. These viruses are now classified in the genera *Macluravirus* (type species *Maclura mosaic virus*) and *Ipomovirus* (type species *Sweet potato mild mottle virus*). Macluraviruses are aphid transmitted and ipomoviruses are transmitted by whiteflies. Although classified previously in genus *Rymovirus*, *Wheat streak mosaic virus* (WSMV) differs markedly from *Ryegrass mosaic virus* (the type species of genus *Rymovirus*) in gene sequences and in the species of mite vector. WSMV has now been removed from the genus and is the type species of the new genus *Tritimovirus*.

Family Luteoviridae

The genus *Luteovirus* described in the Sixth ICTV Report (Murphy et al., 1995) contained 16 definitive species. However, analyses of genome sequences made it increasingly implausible that a single genus could contain the genetic diversity in the genomes of these viruses (D'Arcy and Mayo, 1997; Mayo and D'Arcy, 1999). However, a consideration of biological properties (transmission, tissue localization, symptom type), and some molecular properties (e.g., coat protein sequences) suggested that these viruses form a coherent group of pathogens (D'Arcy and Mayo, 1997). To resolve this inconsistency, the family *Luteoviridae* was created so that molecular disparity in replication-related gene sequences could be recognized by separation at the genus level. Thus, genus *Luteovirus* now contains viruses with genomes similar to those of the type species, *Barley yellow dwarf virus-PAV*; that is, they lack a P0 gene, have little overlap between the P1 and P2 genes, and lack a VPg (genome-linked viral protein). Genus *Polerovirus*, on the other hand, contains viruses with genomes akin to those of *Potato leafroll virus* (the type species); that is, they contain a P0 gene, have an extensive overlap between the P1 and P2 genes, and have a 5'-linked VPg.

Pea enation mosaic virus (PEMV) was previously classified as the sole member of a genus because it was unlike other viruses. However, detailed analysis of its apparently bipartite genome has shown that the disease of enation formation in peas is induced by infection with a complex of two viruses (Demler and De Zoeten, 1991). One component resembles viruses of genus *Polerovirus* in gene complement, except that P4 is absent and the capsids formed in cells infected by it cannot move from cell to cell. The other component resembles viruses of genus *Umbravirus* in that it can move systemically and cells infected with it do not form particles. Thus, PEMV is a complex of two viruses. The current classification redefines genus *Enamovirus* (type species *Pea enation mosaic virus-1*) to contain only the larger RNA species that resembles polerovirus genomes and to place it within the family

Luteoviridae. The RNA component that resembles umbraviruses is classified as the species *Pea enation mosaic virus-2* in the genus *Umbravirus.*

Genus Umbravirus

Viruses in the genus *Umbravirus* (type species *Carrot mottle virus*) are peculiar in that their genomes do not encode a coat protein. Each umbravirus is transmitted by aphids when its genome becomes encapsidated in the coat protein of an assistor luteovirus. Umbraviruses exist in nature only as part of a complex that includes a luteovirus because, although they can invade a plant systemically, spread is limited by the distribution of the assistor luteovirus. There are some similarities in sequence between umbravirus genes and those of viruses in the family *Tombusviridae.*

Family Tombusviridae

In the Sixth ICTV Report, the family *Tombusviridae* contained the genera *Tombusvirus* (type species *Tomato bushy stunt virus*) and *Carmovirus* (type species *Carnation mottle virus*). Subsequently, in recognition of a number of molecular similarities, the previously unassigned genera *Necrovirus* (type species *Tobacco necrosis virus A*), *Machlomovirus* (type species *Maize chlorotic mottle virus*), and *Dianthovirus* (type species *Carnation ringspot virus*) were added to the family. The family has recently been further expanded to include genera typified by some newly characterized viruses. A range of molecular characteristics is used to discriminate among these genera. The genera are *Avenavirus* (type species *Oat chlorotic stunt virus*), *Aureusvirus* (type species *Pothos latent virus*), and *Panicovirus* (type species *Panicum mosaic virus*).

Genus Sobemovirus

Viruses in this genus (type species *Southern bean mosaic virus*) have monopartite RNA genomes of c. 4.4 kb linked to a VPg but lacking a poly(A). In some features of the putative RNA-dependent RNA polymerase, sobemoviruses resemble poleroviruses. However, unlike poleroviruses, sobemoviruses are transmitted by beetles and are mechanically transmissible.

Former Genus Furovirus

Viruses previously grouped together on the basis of having rod-shaped particles and being transmitted by soil-inhabiting fungi have proved to be increasingly diverse as more and more molecular characters have become apparent (Torrance and Mayo, 1997). The viruses are markedly diverse in the numbers of genome parts, in possession or not of a triple gene block (TGB), in the expression mechanisms used during genome translation, and in gene sequence. The present classification of these viruses includes four genera. Genus *Furovirus* is, as before (Murphy et al., 1995), typified by *Soil-borne wheat mosaic virus* and thus contains viruses that lack a TGB in their bipartite genomes. Viruses with other characteristics have been reclassified in new genera. Viruses with a tripartite genome that contains a TGB that are transmitted by *Spongospora subterranea* are classified in genus *Pomovirus* (type species *Potato mop-top virus*). Viruses with a bipartite genome and a TGB that are transmitted by *Polymyxa graminis* are classified in genus *Pecluvirus* (type species *Peanut clump virus*). Viruses with a genome of two essential molecules and two or three supplementary, but not minimally obligatory, molecules that are transmitted by *Polymyxa betae* are classified in genus *Benyvirus* (type species *Beet necrotic yellow vein virus*). For the moment, it is believed that clustering any of these genera into higher taxa is not feasible, and they remain as unassigned genera.

Other Genera of Viruses with Rod-Shaped Particles

These viruses are classified in one of three genera: Genus *Tobamovirus* contains viruses with monopartite genomes and is typified by *Tobacco mosaic virus*. Genus *Tobravirus* contains bipartite genome viruses that are transmitted by nematodes and is typified by *Tobacco rattle virus*. Tripartite genome viruses are classified in genus *Hordeivirus* (type species *Barley stripe mosaic virus*).

Family Bromoviridae and Similar Taxa

Family *Bromoviridae* contains virus genera long recognized as meriting clustering into a higher taxon (Van Vloten-Doting et al., 1981). The viruses they contain all have tripartite genomes, a coat protein encoded by a subgenomic RNA, and similar gene product sequences. The viruses differ markedly in particle morphology and in transmission, and these properties distinguish several of the genera. Viruses in genus *Bromovirus* (type species *Brome mosaic virus*) have isometric particles that are transmitted by beetles. Viruses in genus *Cucumovirus* (type species *Cucumber mosaic virus*) have isometric particles that are transmitted by aphids, whereas those in genus *Alfamovirus* (type species *Alfalfa mosaic virus*) have bullet-shaped particles transmitted by aphids. Viruses in genus *Ilarvirus* (type species *To-*

bacco streak virus) have isometric/bacilliform particles that are relatively unstable and that are transmitted in association with pollen. A new genus, *Oleavirus* type species *(Olive latent virus 2)*, has been added to family *Bromoviridae*. Oleaviruses are distinct from other viruses in the family in that the subgenomic coat protein mRNA is not encapsidated.

Another new genus, currently unassigned, is *Ourmiavirus* (type species *Ourmia melon virus*). The ourmiavirus genome is tripartite, but the particle morphology is distinctive in that the particles are bacilliform with conical ends and are markedly resistant to the disruptive effects of a variety of chemicals that would destroy bromovirus particles. No vector is known; the virus is relatively efficiently transmitted through seed.

Genus *Idaeovirus* is also unassigned. It is monotypic, the type (and only) species being *Raspberry bushy dwarf virus*. Although the genome is bipartite, particles are reminiscent of those of ilarviruses and the virus is transmitted in association with pollen. Also, some gene products are similar in sequence to those of viruses in the family *Bromoviridae* (Ziegler et al., 1993).

Genera Tymovirus *and* Marafivirus

Viruses in these well-established genera are superficially very different. Those in genus *Tymovirus* (type species *Turnip yellow mosaic virus*) are transmissible mechanically and by beetles, and they have particles that contain one coat protein and RNA with a 3' terminal tRNA-like structure. Viruses in genus *Marafivirus* (type species *Maize rayado fino virus*) are not mechanically transmissible, have leafhopper vectors, and have particles that contain two coat proteins and RNA that is 3' polyadenylated. However, the RNAs of viruses in either genus are markedly C rich and contain a "tymobox" sequence motif. Also, the RNA-dependent RNA polymerases of viruses in the two genera are much more alike than either is to polymerases of viruses in any other genus. The significance of this similarity is at present unclear.

Family Closteroviridae

Viruses typified by *Beet yellows virus* were previously classified in a single genus, *Closterovirus*. However, the discovery that some viruses of this type have bipartite genomes, and thus two size classes of particles, has led to changes. The cluster has been raised to the level of a family and has been named *Closteroviridae*. Viruses with bipartite genomes have been removed from the genus *Closterovirus* and placed in a new genus, *Crinivirus* (type species *Lettuce infectious yellows virus*). However, the current genus

Closterovirus still includes species or tentative species viruses that have aphids, mealybugs, or whiteflies as vectors, which is an unusual amount of biological diversity to have among species within one genus.

Other Genera of Monopartite Genome Viruses with Filamentous Particles

Viruses with relatively short (<800 nm in length), filamentous particles are classified in the genera *Capillovirus* (type species *Apple stem grooving virus*), *Potexvirus* (type species *Potato virus X*), *Carlavirus* (type species *Carnation latent virus*), *Allexivirus* (type species *Shallot virus X*), *Foveavirus* (type species *Apple stem pitting virus*), *Trichovirus* (type species *Apple chlorotic leaf spot virus*), or *Vitivirus* (type species *Grapevine virus A*). These genera are distinguished largely by genome characteristics. Capillovirus has genomes that are expressed as a polyprotein that is cleaved to release the mature virus proteins and thus differs from viruses in the other six genera. In these genera, of the viruses with a TGB, potexviruses have five ORFs in their genomes and particles of less than 700 nm length; carlaviruses are similar but have six ORFs; allexiviruses have six ORFs, are mite transmitted, and have particles longer than 700 nm; and foveaviruses have five ORFs and no known vector. Viruses that lack a TGB either are insect transmitted and have five ORFs in their genomes (vitiviruses) or have no known vector and fewer than five ORFs (trichoviruses). Sequence comparisons among the various gene products of viruses in these genera confirm the divisions described here.

Viroids

Viroids are not viruses and indeed may not even be linked with them phylogenetically. However, ICTV includes a taxonomic description of these pathogens in its reports under the heading of "subviral agents." Viroids are unencapsidated, small circular ssRNA molecules that do not encode peptide sequences but can replicate autonomously in host plants. Viroids are 350 nucleotides or less in size, and features of the nucleotide sequence are the primary taxonomic characteristics. Recently, by using these discriminatory characteristics, it has been possible to devise a family/genus classification of viroids (Flores et al., 1998). There are two families of viroids. Those in family *Pospiviroidae* contain a central conserved region (CCR), whereas those in family *Avsunviroidae* do not contain a CCR but do undergo self-cleavage. Family *Pospiviroidae* contains the genera *Pospiviroid* (type species *Potato spindle tuber viroid*), *Hostuviroid* (type species *Hop stunt viroid*),

Cocadviroid (type species *Coconut cadang-cadang viroid*), *Apscaviroid* (type species *Apple scar skin viroid*), *and Coleviroid* (type species *Coleus blumei viroid 1*), each characterized by the CCR. Family *Avsunviroidae* contains the genera *Avsunviroid* (type species *Avocado sunblotch viroid*) and *Pelamoviroid* (type species *Peach latent mosaic viroid*); these genera are distinguishable on the basis of RNA size and secondary structure. Within the genera, sequence identities of <90 percent are taken to indicate that the viroids concerned are distinct species rather than strains of one species.

Satellites

Satellites are defined as subviral agents lacking genes that could encode the enzymes needed for their replication; therefore, their multiplication depends on the coinfection of a host cell with a helper virus. A satellite is genetically distinct from its helper virus by virtue of having a nucleotide sequence substantially different from it, although some satellites share short sequences, often at the termini of their RNA, with their helper viruses. Satellites are not classified by species or genera because they are not a homogeneous group of agents and information on their properties (e.g., nucleotide sequence) is insufficient to deduce their evolutionary origins.

For convenience, satellites are divided into two major categories: "satellite viruses" (resembling tobacco necrosis satellite virus) and "satellite nucleic acids." They are then further divided into those with ssDNA, with dsRNA, with large ssRNA, with small linear ssRNA, or with circular ssRNA.

FUTURE PROSPECTS

Two basic principles of virus taxonomy are, in practice, in direct conflict with each other. The first is to strive to bring all viruses into the universal system of virus taxonomy, and the second is to maintain the stability of the existing taxonomy by resisting change (Mayo and Pringle, 1998). However, the first imperative is dominant, and new discoveries can be expected that will expand and complicate the present scheme. As any taxonomy is a man-made attempt to systematize that for which order need not necessarily be achievable, this expanding complexity is inevitable. Some have argued that increasing knowledge of molecular detail will surely lead to an understanding of the evolutionary history of the current set of viruses, and that this will lead to a comprehensive classification scheme that will reflect virus phylogeny accurately. Rather, the reverse seems to be the rule. It is more and more apparent that, at least for many RNA viruses, we cannot deduce the past from the present molecular complexity. Some insights are readily apparent, but there is a limit as to how far back one can perceive evolutionary history.

The current classification contains a number of temporary assignments. Some viruses are classified as "tentative species." These viruses may, as further information becomes available, merit a designation as a species, or they may prove to be more properly classified as "strains" (i.e., subspecies) of an existing species. In some families (e.g., *Luteoviridae*), species have been recognized as being distinctive but, at the same time, because information is lacking on the particular characters needed for assignment to one or another existing genus (or indeed to a wholly new genus), the species remain "unassigned." For both of these categories, hopefully, the coming years will yield the necessary information to formally classify them in the appropriate taxa. However, the categories are unlikely to be left empty for long, as it seems certain that new challenging information about novel or unusual viruses will emerge.

REFERENCES

Brandes J, Wetter C (1959). Classification of elongated plant viruses on the basis of particle morphology. *Virology* 8: 99-115.

D'Arcy CJ, Mayo MA (1997). Proposals for changes in luteovirus taxonomy and nomenclature. *Archives of Virology* 142: 1285-1287.

Demler SA, De Zoeten GA (1991). The nucleotide sequence and luteovirus-like nature of RNA-1 of an aphid non-transmissible strain of pea enation mosaic virus. *Journal of General Virology* 72: 1819-1834.

Flores R, Randles JW, Bar-Joseph M, Diener TO (1998). A proposed scheme for viroid classification and nomenclature. *Archives of Virology* 143: 623-629.

Francki RIB, Fauquet CM, Knudson DL, Brown F (1991). Classification and Nomenclature of viruses. Fifth Report of the International Committee on Taxonomy of Viruses. *Archives of Virology* (Suppl. 2). Springer-Verlag, Wien, New York. 450 pp.

Harrison BD (1985). Usefulness and limitations of the species concept for plant viruses. *Intervirology* 24: 71-78.

Lwoff A, Horne R, Tournier P (1962). A system of viruses. *Cold Spring Harbor Symposia on Quantitative Biology* 27: 51-55.

Martelli GP (1997). Plant virus taxa: Properties and epidemiological characteristics. *Journal of Plant Pathology* 79: 151-171.

Matthews REF (1983). The history of virus taxonomy. In Matthews REF (ed.), *A Critical Appraisal of Viral Taxonomy*. CRC Press Inc., Boca Raton, Florida, pp. 1-35.

Mayo MA (1996). Recent revisions of the rules of virus classification and nomenclature. *Archives of Virology* 141: 2479-2484.

Mayo MA (1999). Developments in plant virus taxonomy since the publication of the Sixth ICTV Report. *Archives of Virology* 144: 1659-1666.

Mayo MA, D' Arcy CJ (1999). Family *Luteoviridae*: A reclassification of luteoviruses. In: Smith HG, Barker H (eds.). *The* Luteoviridae. CABI, Wallingford, Oxon, pp. 15-22.

Mayo MA, Horzinek M (1998). A revised version of the International Code of Virus Classification and Nomenclature. *Archives of Virology* 143: 1645-1654.

Mayo MA, Murphy FA (1994). Modifications to the rules for virus nomenclature. *Archives of Virology* 134: 213-215.

Mayo MA, Pringle CR (1998). Virus taxonomy—1997. *Journal of General Virology* 79: 649-657.

Milne RG (1985). Alternatives to the species concept for virus taxonomy. *Intervirology* 24: 94-98.

Murphy FA, Fauquet CM, Bishop DHL, Ghabrial SA, Jarvis AW, Martelli GP, Mayo MA, Summers MD (eds.) (1995). *Virus Taxonomy—The Classification and Nomenclature of Viruses: Sixth Report of the International Committee on Taxonomy of Viruses*. Springer-Verlag, Vienna, 586 pp.

Torrance L, Mayo MA (1997). Proposed re-classification of furoviruses. *Archives of Virology* 142: 435-439.

Van Regenmortel MHV (1990). Virus species, a much overlooked but essential concept in virus classification. *Intervirology* 31: 241-254.

Van Regenmortel MHV, Bishop DHL, Fauquet CM, Mayo MA, Maniloff J, Calisher CH (1997). Guidelines to the demarcation of virus species. *Archives of Virology* 142: 1505-1518.

Van Regenmortel MHV, Fauquet CM, Bishop DHL, Carstens E, Estes M, Lemon S, Maniloff J, Mayo MA, McGeoch D, Pringle CR, Wickner RB (eds.) (2000). *Virus Taxonomy. Seventh Report of the International Committee on Taxonomy of Viruses*. Academic Press, New York, San Diego, 1162 pp.

Van Vloten-Doting L, Francki RIB, Fulton RW, Kaper JM, Lane LC (1981). Tricornaviridae—A proposed family of plant viruses with a tripartite, single-stranded RNA genome. *Intervirology* 15: 198-203.

Ziegler A, Mayo MA, Murant AF (1993). Proposed classification of the bi-partite genomed raspberry bushy dwarf idaeovirus with tri-partite-genomed viruses in the family *Bromoviridae*. *Archives of Virology* 131: 483-488.

Chapter 2

How to Write the Names
of Virus Species

Marc H. V. Van Regenmortel

In formal virus taxonomy, the names of orders, families, subfamilies, and genera are always printed in italics and the first letters of the names are capitalized. At its meeting in San Diego in March 1998, the Executive Committee of the International Committee on Taxonomy of Viruses (ICTV) decided to extend this practice to the names of species taxa to clearly indicate that the species name had been approved as the official, internationally recognized name (Pringle, 1998).

The new rule 3.40 of the International Code of Virus Classification and Nomenclature is as follows:

> Species names are printed in italics and you have the first letter of the first word capitalized. Other words are not capitalized unless they are proper nouns, or parts of proper nouns. (Mayo and Horzinek, 1998: 1654)

The rule applies when the species name is used to refer to a taxonomic entity, i.e., an abstraction corresponding to a taxon in the virus classification. Examples of correct spelling and typographical style for the corresponding taxonomic entities are *Tobacco mosaic virus*, *Poliovirus*, and *Murray River encephalitis virus* (in this case, river is a proper noun).

It should be stressed that italics and an initial capital letter need to be used only if the species name refers to a taxonomic category. This is the case, for instance, when in the materials and methods section of a paper, the virus used in a study is referred to as a member of a particular species, for example, *Poliovirus*, genus *Enterovirus*, family *Picornaviridae*. However, taxonomic names are not appropriate when referring to physical entities such as the virions found in a preparation or seen in an electron micrograph. It is, indeed, not possible to centrifuge or visualize the family *Picornaviridae*, the genus *Enterovirus*, or the species *Poliovirus*, for abstractions cannot be centrifuged or seen in a microscope.

When referring to concrete viral objects such as virions, italics and initial capital letters are not needed and the names are written in lowercase roman script. This corresponds to informal vernacular usage and is appropriate, for instance, when picornaviruses (not italicized) or poliovirus particles are being centrifuged or are visualized in a microscope. This also applies when the names are used in adjectival form, for instance, tobacco mosaic virus polymerase.

The use of italics when referring to the name of a species as a taxonomic entity will clearly signal that it has the status of an officially recognized species. A complete list of all the virus species that have been recognized so far can be found in the Seventh ICTV Report published in April 2000 (Van Regenmortel et al., 2000). When the taxonomic status of a new putative species is uncertain or its positioning within an established genus has not been clarified, it will be considered a "tentative" species and its name will not be given in italics, although its initial letter will be capitalized. At a later stage, transition to italicization would then signal recognition of full species status.

A uniform italicized spelling of all taxonomic levels from order to species reinforces in a visible manner the status of virus species as taxonomic entities. This new rule also removes some past oddities in orthography of species names. When a virus species name contained a Latin host name, it was customary to use italics and a capital letter for that part of the virus name. However, when the host name was the same in botanical Latin and in English (e.g., iris) it was unclear what the form should be. The new orthography removes such ambiguities.

Scientific names in biology are usually Latin words, and the use of italics indicates their Latin origin. However, the international community of virologists is strongly opposed to the introduction of Latin binomials for naming viruses (Van Regenmortel, 1989), and the nomenclature system developed by ICTV reflects this position.

It is a moot point whether genus names such as *Enterovirus* or *Tobamovirus* should be considered Latin names, and the use of italics in such cases is simply a convenient way to indicate that these terms refer to formal genus taxa. The use of italics for English species names serves the same function.

REFERENCES

Mayo MA, Horzinek MC (1998). A revised version of the international code of virus classification and nomenclature. *Archives of Virology* 143: 1645-1654.

Pringle CR (1998). Virus taxonomy. San Diego 1998. *Archives of Virology* 143: 1449-1459.

Van Regenmortel MHV (1989). Applying the species concept to plant virus. *Archives of Virology* 104: 1-17.

Van Regenmortel MHV, Fauquet CM, Bishop DHL, Carstens E, Estes M, Lemon S, Maniloff J, Mayo M, McGeoch D, Pringle C, Wickner R (2000). *Virus Taxonomy: Seventh Report of the International Committee on Taxonomy of Viruses.* Academic Press, New York, San Diego.

SECTION II:
VIRUS TRANSMISSION
AND TRANSPORT

Chapter 3

Genes Involved in Insect-Mediated Transmission of Plant Viruses

Juan José Lopez-Moya

INTRODUCTION

Plant viruses are important pathogens that challenge the yield and quality of crops worldwide. One of the major obstacles to implementing efficient virus control strategies is our incomplete knowledge of how viruses are spread from plant to plant by vector organisms. Insects form a large group and are the most important vectors of plant viruses.

Plant viruses must go through two stages during their infection cycle. First, they must replicate inside host cells, employing cellular systems complemented with viral functions; to colonize the plant from the initial infection foci, they have to move to adjacent cells (short-distance movement) and, through the vascular system, reach other tissues and organs (long-distance movement). Second, viruses must spread to new hosts; to do that, they have to cross cellular barriers to enter cells. For most plant viruses this process is assisted by vector organisms (Matthews, 1991).

Transmission from plant to plant is an essential process for virus survival. Plant viruses have developed several strategies to perform this task efficiently, in many cases involving the existence of specific viral gene products known to facilitate the transmission process (Hull, 1994; Gray, 1996; Van den Heuvel et al., 1999).

This chapter is intended to update our view of how different plant viruses have adopted strategies to meet this essential requirement for transmission using insect vectors, and of how we envisage the action of genes and gene products involved in the process.

The author gratefully thanks Dr. Thomas P. Pirone for his critical review and valuable suggestions.

INSECT VECTORS OF PLANT VIRUSES

A plant virus vector is an organism that assists transmission of the virus from plant to plant. Among the different groups of vectors, the involvement in virus transmission is known for a few fungi and for some 400 different species of invertebrates, including Nematoda. The majority of vectors belong to the Arthropoda, in the classes Arachnida and Insecta (Harris, 1981). Aphids constitute the most important group of vectors because of their abundance and feeding behavior (piercing-sucking) (Harris, 1991). Leafhoppers and planthoppers also are important vectors of many viruses, and they have a similar feeding mechanism (Nault and Ammar, 1989). Treehoppers, thrips, whiteflies, mealybugs, beetles, and other insects can also act as vectors of different viruses, as can mites (Matthews, 1991).

With regard to all known plant viruses, around 70 percent are insect transmitted, and more than 50 percent of those are transmitted by homopteran vectors (Francki et al., 1991).

Transmission by insects usually does not involve a simple passive transport of virus particles. In some cases, the virus is able to replicate in vector cells. Specificity and selectivity of the transmission process (based on specific features governing interactions among each particular virus, vector, and host plant) influence the epidemic spread of diseases caused by plant viruses (Ferris and Berger, 1993). Therefore, it is of great importance to study the transmission process with the ultimate practical purpose of designing effective means of interfering with the spread of many economically important diseases.

CLASSIFICATION OF TRANSMISSION MODES

Relationships of plant viruses and their insect vectors can be differentiated according to the duration of retention inside the vector. A classification can be outlined based on the properties of the relationship established, including the length of time required for acquisition, latency, and retention. *Acquisition* spans from initiation of probing or feeding in the plant until the vector becomes able to transmit the virus; *latency* is the time required after acquisition before the virus can be readily inoculated; and *retention* is the period for which the vector remains viruliferous. Two major categories of transmission, noncirculative and circulative, can also be recognized based on the sites of retention and the routes of movement through the vector (Matthews, 1991).

Noncirculative viruses are thought to associate temporally with surfaces of the digestive tract of the vector (mouthparts or foregut). These viruses do not demonstrate latency, and they are lost after molting. Noncirculative viruses can be either nonpersistent or semipersistent. *Nonpersistent* viruses are acquired in brief periods (seconds to minutes), they can be inoculated immediately after acquisition, and retention is limited to short periods. Transmission is considered *semipersistent* when its efficiency increases directly with duration of acquisition and inoculation periods, and normally vectors remain viruliferous from hours to days. Transmission of viruses exhibiting both nonpersistent and semipersistent features has been denominated bimodal.

Circulative viruses are those which need translocation inside the vector to be transmitted. Most of these viruses are found in vascular tissues of plants, and some cannot be inoculated mechanically. A common feature is that they need a latent period after acquisition. Circulative transmission can be classified into nonpropagative and propagative. *Nonpropagative transmission* occurs when the virus does not replicate in the vector, although it needs to cross barriers in the digestive tract of the vector to reach the hemolymph, and, from there, the salivary glands to be inoculated during subsequent feeding. In *propagative transmission*, the virus is able to replicate inside cells of the vector during its circulation; thus, the virus is a parasite of both plants and insects. In some cases, the virus can even be passed on transovarially to the vector progeny.

Table 3.1 lists the genera of plant viruses, according to the classification and nomenclature of viruses (Pringle, 1999), that are transmitted by insects, with an indication of their vectors and mode of transmission.

TABLE 3.1. Plant Virus Genera Containing Members Transmitted by Insects, with Indication of Their Vectors and Transmission Manner

Family[1]	Genus[1]	Vector	Transmission Manner
Geminiviridae	Mastrevirus	Leafhoppers	Circulative nonpropagative
	Curtovirus	Leafhoppers/Treehoppers	Circulative nonpropagative
	Begomovirus	Whiteflies	Circulative nonpropagative[2]
Caulimoviridae	Badnavirus	Mealybugs	Semipersistent
	Rice tungro baciliform	Leafhoppers	Semipersistent[1]
	Caulimovirus	Aphids	Nonpersistent/semipersistent
Reoviridae	Fijivirus	Planthoppers	Circulative propagative
	Phytoreovirus	Leafhoppers	Circulative propagative
	Oryzavirus	Planthoppers	Circulative propagative
Rhabdoviridae	Cytorhabdovirus	Leafhoppers/Aphids	Circulative propagative
	Nucleorhabdovirus	Leafhoppers/Aphids	Circulative propagative

TABLE 3.1 *(continued)*

	(Unassigned)	Neuroptera/Leafhoppers/Aphids	Circulative propagative
Bunyaviridae	*Tospovirus*	Thrips	Circulative propagative
Sequiviridae	*Sequivirus*	Aphids	Semipersistent
	Waikavirus	Leafhoppers	Semipersistent
Comoviridae	*Comovirus*	Beetles	Circulative[4]
	Fabavirus	Aphids	Nonpersistent
Potyviridae	*Potyvirus*	Aphids	Nonpersistent
	Macluravirus	Aphids	Nonpersistent
	Ipomovirus	Whiteflies	Nonpersistent
Bromoviridae	*Alfamovirus*	Aphids	Nonpersistent
	Bromovirus	Beetles	Circulative[4]
	Cucumovirus	Aphids	Nonpersistent
Closteroviridae	*Closterovirus*	Aphids	Semipersistent
	Crinivirus	Whiteflies	Semipersistent
Luteoviridae	*Luteovirus*	Aphids	Circulative nonpropagative
	Polerovirus	Aphids	Circulative nonpropagative
	Enamovirus	Aphids	Circulative nonpropagative
	Nanovirus	Aphids	Circulative nonpropagative
	Marafivirus	Leafhoppers	Circulative propagative
	Tenuivirus	Planthoppers	Circulative propagative
	Umbravirus	Aphids	Circulative nonpropagative
	Sobemovirus	Beetles	Circulative[4]
	Tymovirus	Beetles	Circulative[4]
	Carlavirus	Aphids	Nonpersistent

[1]Based on the current ICTV Universal System of Virus Taxonomy (Pringle, 1999).
[2]Members of this genus are transovarially transmitted and could have a circulative propagative relation (see text for details).
[3]RTBV is assisted during transmission for RTSV (see text for details).
[4]Beetle-transmitted viruses have a different circulative relation (see text for details).

NONPERSISTENT AND SEMIPERSISTENT TRANSMISSION

The majority of plant viruses have a noncirculative (nonpersistent and semipersistent) relationship with their vectors. Although the relationship established is of short duration, it can be quite complex. In most cases, the number of actual particles needed for transmission may be rather low (Walker and Pirone, 1972), and extremely sensitive means of detection are needed to identify the presence of virus within the vector (Plumb, 1989; see Chapter 21 in this book). Therefore, for most viruses, the identity of the actual retention sites is still not certain, and significant efforts are being devoted to resolving this point. Although retention time is generally considered short, its duration may depend on specific conditions, and, in practice, nonpersistent viruses have been shown to be retained for sufficient time to travel rather long distances in their vectors (Zeyen and Berger, 1990).

As is typical of piercing-sucking insects, aphids make brief insertions of their stylets to probe the adequacy of the plant as a food source, sucking sap and injecting saliva in the process. As a result, acquisition and inoculation of nonpersistent viruses occur during these probes (Lopez-Abella et al., 1988). Using the electrical penetration graph technique (Tjallingii, 1985), a recent study proposed a mechanism in which the acquisition of noncirculative viruses is related to intracellular ingestion by the vector, and the inoculation of the virus occurs during salivation (Martin et al., 1997). Figure 3.1 represents schematically the interactions of nonpersistent viruses and vector mouthparts.

FIGURE 3.1. Alimentary Tract of a Sucking Insect, with Relevant Parts Indicated to Illustrate the Relevant Sites of Interaction with a Noncirculative Plant Virus

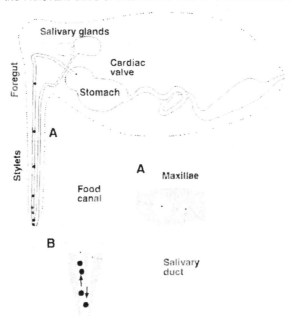

Note: The virus (depicted as a spiked black dot) is acquired and retained in the food canal of the stylets. The cardiac valve is considered the limit beyond which egestion of noncirculative viruses cannot occur. (A) Detail of a cross-section of the interlocked maxillae showing food and salivary ducts. Retention of virus takes place in the cuticle lining the food canal, according with the observations reported in Wang et al., 1996. (B) Detail of a longitudinal section of the stylets' tips, where the particles actually involved in transmission are presumed to be retained (during acquisition) and later released by the saliva flux (during inoculation), according to Martin et al., 1997: 2702.

Two major strategies of vector transmission have been described for noncirculative plant viruses. The transmissibility of viruses belonging to, for instance, the genus *Cucumovirus*, depends on characteristics of only the capsid protein (CP) of the virus particles (capsid protein strategy). For other viruses, such as those belonging to the genera *Potyvirus* and *Caulimovirus*, vector transmission depends on characteristics of both the CP and the virus-encoded nonstructural accessory factors, known as helper components (helper component strategy) (Pirone and Blanc, 1996; Pirone, 1977).

Cucumovirus Transmission

Cucumoviruses have isometric particles, and their genomes are divided into three RNA molecules. They are aphid transmitted in a nonpersistent manner (Palukaitis et al., 1992).

The particles of *Cucumber mosaic virus* (CMV) can be dissociated and re-assembled in vitro. In this way, the involvement of the CP in transmission was established as the primary determinant for aphid transmissibility (Gera et al., 1979; Chen and Francki, 1990; Perry et al., 1994). The CP of CMV, as with many other viral proteins, has multiple functions: in addition to transmission, the CP serves to encapsidate particles and contains systemic movement and host range determinants (Palukaitis et al., 1992).

Sequence comparison and recombination experiments have served to identify which modifications in the CP of CMV affect aphid transmission. Two domains in the CP contain amino acids involved in the process (Shintaku, 1991; Palukaitis et al., 1992). One of the considered domains also plays a role in symptomatology, while the second contains a conserved tyr residue present in all transmissible isolates, substituted by cys or phe in nontransmissible isolates (Owen et al., 1990; Perry et al., 1994). It is interesting that some mutations seem to compensate for the detrimental effect on transmission caused by another. Strains of CMV have different efficiencies of transmission by different species of aphids. The specificity for different vectors, as would be expected in a mechanism involving direct interaction between the CP and the vector, lies in the CP (Perry et al., 1994, 1998). Interestingly, chimeras constructed using CP sequences of two isolates of CMV have shown that amino acid changes that greatly influence transmission rates with a particular aphid vector may have little effect on transmission by another (Perry et al., 1998). Defects in transmission may be attributed either to difficulties in binding and/or release of virions in the mouthparts or foregut of the vector or to instability of virions during the process.

Potyvirus Transmission

The family *Potyviridae,* divided into six genera according to genome composition and vector specificity, contains the largest number of plant viruses (Lopez-Moya and Garcia, 1999). The majority of them are aphid transmitted in a nonpersistent manner, with the involvement of a virus-encoded protein known as a helper component (HC) (Pirone, 1977; Murant et al., 1988). Dependency of helper factor has also been described for other viruses (Lung and Pirone, 1973; Murant et al., 1976; Woolston et al., 1983; Hunt et al., 1988).

Evidence of the involvement of an HC in transmission was shown first with non-aphid-transmissible isolates of potyviruses and with *Potato aucuba mosaic virus* (PAMV, genus *Potexvirus*). These could only be transmitted from infected plants in which a potyvirus was also present (Kassanis and Govier, 1971). Using artificial membrane feeding, the existence of an HC was established in experiments with transmission-deficient isolates of potyviruses and purified virus particles (Govier and Kassanis, 1974a,b).

The potyvirus HC is a virus-encoded protein around 50 kDa (Thornbury et al., 1985) that is implicated in several functions (Maia et al., 1996). The transmission-active form of HC has been recovered from only infected or transgenic plants (Thornbury et al., 1985; Berger et al., 1989; Thornbury et al., 1993). Transmission-active HC is supposed to be a multimeric protein (Pirone, 1977; Thornbury et al., 1985). The HC is specific for viruses belonging to the genus *Potyvirus,* and in some cases it can act in heterologous combinations (Pirone, 1981; Sako and Ogata, 1981; Lecoq and Pitrat, 1985; Lopez-Moya et al., 1995; Flasinski and Cassidy, 1998).

Some potyviral isolates are nontransmissible due to defects in their HCs (Thornbury et al., 1990; Lecoq et al., 1991). By means of mutational analysis in *Tobacco vein mottling virus* (TVMV). some important residues for transmission have been identified in the HC (Atreya et al., 1992; Atreya and Pirone, 1993). Mutations in other potyviruses have confirmed the involvement of certain conserved residues in the transmission process (Granier et al., 1993; Huet et al., 1994; Canto et al., 1995). Repeated mechanical inoculation can result in emergence of poorly aphid-transmissible variants of *Potato virus Y* due to alterations in the HC (Legavre et al., 1996).

Among the different hypotheses postulated to explain how the HC works during transmission, the best-supported one gives the HC a role in the adsorption of particles to unidentified structures in the stylets of the insect. by binding both to virions and to stylets, acting thus as a bridge (bridge hypothesis). Retention sites are located in the cuticular linings of aphid mouthparts (Berger and Pirone, 1986; Ammar et al., 1994). Radioactive labeling of *To-*

bacco etch virus (TEV) particles confirmed this retention site for combinations of active HC plus transmissible virus particles (Wang et al., 1996, 1998). The predicted interaction between the HC and the CP of potyviruses was proved recently by in vitro binding experiments (Blanc et al., 1997). The regions of HC involved in interaction with the CP and with structures in the mouthparts of the vectors have recently been determined. A highly conserved Pro-Thr-Lys (PTK) domain, located in the central region of the protein, seems to be involved in binding to virions (Peng et al., 1998), whereas a conserved Lys-Ile-Thr-Cys (KITC) domain in the N-terminal region is involved in binding to the mouthparts (Blanc et al., 1998). Spontaneous deletion of the latter motif could explain the loss of aphid transmissibility found in TEV mutants (Dolja et al., 1993, 1997). The HC has also been implicated in the vector specificity of a particular potyvirus (Wang et al., 1998).

Existence of nontransmissible isolates with functionally active HCs indicated CP involvement in the process as another determinant of transmission (Pirone and Thornbury, 1983). A highly conserved motif of three amino acids (Asp-Ala-Gly, DAG), located near the N-terminus, was suggested to be involved in the process (Harrison and Robinson, 1988), and this was proved by mutational analysis in TVMV (Atreya et al., 1990, 1991, 1995). Mutations in this motif that impede transmission were correlated with nonretention of virions in the aphid stylets (Wang et al., 1996). It was proposed that the CP N-terminus may bind directly to aphid mouthparts when correctly exposed through a conformational change of the CP mediated by the HC (Salomon and Bernardi, 1995). However, recent works have determined that the DAG motif is involved in the interaction with the HC, further supporting the bridge hypothesis (Blanc et al., 1997). Interestingly, not all transmissible isolates of potyvirus have this DAG motif (Lopez-Moya et al., 1995; Johansen et al., 1996; Flasinski and Cassidy, 1998), and the context in which the motif is found seems to affect greatly the specificity of the process (Lopez-Moya et al., 1999). The DAG motif is also involved in transmission of potexviruses such as PAMV, which can be assisted during transmission by a potyvirus HC, as mentioned before. This phenomenon was explained by the identification of a DAG motif near the N-terminus of the PAMV CP, and when this region was transferred to the potato virus X CP, this virus became aphid transmissible with the aid of a potyvirus HC (Baulcombe et al., 1993).

Since the CP of potyviruses contains determinants of transmission, and the HC might allow transmission of heterologous combinations by complementation, heteroencapsidation in the case of doubly infected plants, or in transgenic plants expressing a functional CP, must be considered as a risk of unwanted spreading of viruses (Bourdin and Lecoq, 1991; Lecoq et al.,

1993). This risk can be minimized using modified versions of CP genes in transgenic plants (Jacquet et al., 1998).

Light microscopic autoradiography of radioactively labeled TEV particles has served to study the mode of action of mineral oil in reducing transmission rate by interference with virion retention in the stylets (Wang and Pirone, 1996a). In contrast, the mechanism by which preacquisition fasting favors transmission appears to be increased retention, probably by reducing the presence of interference substances of plant origin in the stylets (Wang and Pirone, 1996b).

Caulimovirus Transmission

Caulimoviruses are isometric viruses with dsDNA genomes. *Cauliflower mosaic virus* (CaMV, genus *Caulimovirus*), the best-studied caulimovirus, may be transmitted bimodally and presents features of nonpersistent or semipersistent transmission in its association with different aphid vectors (Bouchery et al., 1990).

As in the case of potyviruses, caulimoviruses need the assistance of a virus-encoded helper factor (Lung and Pirone, 1973), which was identified as the 18 kDa product of *gene II* (P18) in the case of CaMV (Armour et al., 1983; Woolston et al., 1983). Naturally occurring variants of CaMV had alterations in the *gene II*, such as large deletions, or a substitution at amino acid position 94 of P18 (Woolston et al., 1987). Helpers from other caulimoviruses can assist transmission of CaMV (Markham and Hull, 1985). Transmission-active-CaMV P18 factor was found associated with viroplasms induced by the virus in host cells (Rodriguez et al., 1987). In addition, P18 forms stable complexes with microtubules (Blanc et al., 1996). Expression of CaMV helper factor in a baculovirus heterologous system allowed sequential acquisition studies using membrane feeding that served to demonstrate a direct involvement of P18 in transmission (Blanc et al., 1993). After acquisition by aphids, baculovirus-expressed P18 assisted transmission of two helper-defective isolates of CaMV acquired from plant or extracts. However, purified virions were not complemented. The baculovirus expression system also served to study in vitro binding of P18 to virus particles (Schmidt et al., 1994). Initially, a bridge mechanism similar to that previously described for potyviruses was considered as the mode of action of P18. However, the failure of baculovirus-expressed P18 to complement transmission of purified particles indicated that existence of an additional factor in aphid transmission could not be totally ruled out. Indeed, a recent report suggests that the mode of action of caulimovirus P18 might differ from the one proposed for potyvirus HC in that a second helper protein, the 15 kDa product

(P15) encoded by *gene III*, has been shown to play a key role in transmission of CaMV (Leh et al., 1999). This protein interacted with P18 in a binding assay and, when included in a membrane feeding assay, allowed a high rate of transmission. This is the first evidence of the involvement of three viral proteins in governing vector transmission of a noncirculative plant virus. Differences in the mode of action of transmission factors between potyviruses and caulimoviruses illustrate how two virus groups have evolved rather different molecular means, both based on the helper strategy, to facilitate vector transmission (Pirone and Blanc, 1996).

Other Noncirculative Viruses

Maize chlorotic dwarf virus, a waikavirus, is transmitted semipersistently by leafhoppers with the involvement of a helper protein (Hunt et al., 1988). Virus particles have been observed in the foregut of the vector (Chidress and Harris, 1989). The feeding behavior of the vector also influences its ability to transmit (Wayadande and Nault, 1993). Another waikavirus, *Rice tungro spherical virus* (RTSV) has a very similar mode of transmission. This virus forms the tungro complex in association with *Rice tungro bacilliform virus* (RTBV), a completely unrelated virus assisted by RTSV for transmission (Hull, 1996). Association of different viruses to facilitate transmission has been described also for the *Parsnip yellow fleck virus* (a sequivirus), depending on the *Anthriscus yellows virus* (AYV), a waikavirus, for its semipersistent transmission by aphids, with the possible aid of a helper component (Murant et al., 1976).

Other noncirculative viruses include members of the family *Closteroviridae*, which includes aphid-, whitefly-, and mealybug-transmitted viruses (Dolja et al., 1994). Whitefly-transmitted closteroviruses are becoming important due to the drastic increase in vector populations throughout the world (Wisler et al., 1997). Some mealybugs are also known vectors of other closteroviruses (Cabaleiro and Segura, 1997; Sether et al., 1998). Aphid-transmitted closteroviruses, such as *Citrus tristeza virus* (CTV), have a semipersistent relationship with their vectors, exhibiting specificity with aphid species (Raccah et al., 1989). The CP of CTV might be involved in the transmission process (Mawassi et al., 1993). The variability observed in CTV populations could be explained in part by aphid transmission and adaptation to hosts (Ayllon et al., 1999). Recently it was demonstrated that *Beet yellows virus* particles have a second CP (encoded by a larger duplicate version of the CP gene) located on a short segment at one end of the particle (rattlesnake structure) (Agranovsky et al., 1995), and, speculatively, this second protein could be involved in the interaction of closteroviruses with their respective vectors. Since purified closterovirus particles cannot be transmitted by their vectors, a putative helper requirement has been suggested (Murant et al., 1988).

Carlavirus, Fabavirus, and *Alfamovirus* are other genera that contain nonpersistently aphid-transmitted plant viruses. Whitefly-transmitted members of carlavirus have been described (Naidu et al., 1998). Carlaviruses and alfamoviruses probably follow a capsid strategy (Pirone and Blanc, 1996).

CIRCULATIVE NONPROPAGATIVE TRANSMISSION

To be transmitted, circulative nonpropagative plant viruses must be actively transported across vector membranes, and they must survive inside the vector during circulation until they are inoculated. The digestive system of insects can be divided into histologically distinct parts with different embryonic origins: foregut, midgut, and hindgut. Entry of circulative plant viruses into the hemolymph may occur during their passage along the digestive tract; in most cases this occurs through the midgut or hindgut. Once in the hemolymph, the virus moves to the salivary glands and passes into the saliva to be excreted later through the salivary duct. Regulation of the process might take place during passage through and out the digestive tract and/or at the stage of entry into the salivary glands (Gray, 1996). Figure 3.2 represents schematically the pathway of circulative viruses inside a piercing-sucking insect vector.

For chewing insects such as beetles, the actual route of circulation could be different, with the viruses being transported across salivary gut membranes to the hemolymph. However, this process might not be totally essential, and the virus might be directly inoculated from the regurgitant (Wang et al., 1992).

Luteovirus Transmission

Based on their genetic organization, three genera of family *Luteoviridae* have been created: *Luteovirus, Polerovirus,* and *Enamovirus.* The PAV and MAV isolates of *barley yellow dwarf virus* (BYDV) have been placed in the genus *Luteovirus,* and *beet western yellows virus* (BWYV), Potato leafroll virus (PLRV), and the RPV isolate of BYDV in the genus *Polerovirus.* Luteoviruses are transmitted by aphids in a circulative nonpropagative manner (Martin et al., 1990). They cannot be transmitted by mechanical inoculation, probably because they are phloem restricted in the plant host, and their relations with vectors have been intensively studied (Gildow, 1987). Evidence exists for three specific barriers during luteovirus circulation: the first

FIGURE 3.2. Alimentary Tract of a Sucking Insect, with Relevant Parts Indicated to Illustrate the Interaction with a Circulative Plant Virus

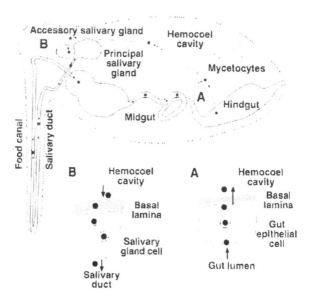

Note: The virus (depicted as a spiked black dot) pathway is indicated by arrows, beginning in the food canal, from where the virus reaches the stomach and the midgut or the hindgut, where it is internalized (see detail in A), as described by Gildow, 1993. Once in the hemocoel cavity (where it can potentially interact with endosymbiotic bacteria-related products), the virus moves through the hemolymph, until it reaches and enters the accessory salivary glands; inoculation occurs when virus-containing saliva is secreted (see detail in B). Two barriers have been identified in the salivary glands, one for attachment to the basal lamina, the other for penetration of the plasmalemma (Peiffer et al., 1997: 498, 502).

regulates entry through endocytosis in the gut (Gildow, 1993); the second regulates selective association of virus with the basal lamina of the accessory salivary glands (ASGs) (Gildow and Gray, 1994); the third has been associated with the penetration of the plasmalemma of the ASGs (Peiffer et al., 1997). Whereas the ASG basal lamina may act as an absolute barrier by preventing virus attachment or impeding penetration, the plasmalemma regulates selectively recognition, attachment, and endocytosis.

Genome expression of luteoviruses includes a readthrough strategy to produce two capsid proteins. The CP gene is separated from the contiguous downstream open reading frame (ORF) by a single stop codon, which can be suppressed by a readthrough mechanism, yielding a fusion protein (Filichkin et al., 1994; Jolly and Mayo, 1994; Brault et al., 1995; see Chapter 14 of this book). This larger protein is present in small amounts and can be further processed, as in the case of transmissible isolates of BYDV that contain a truncated form of this protein after losing a portion from the C-terminus (Filichkin et al., 1994). A relationship between transmissibility and surface-exposed epitopes on PLRV CP has been demonstrated (Van den Heuvel et al., 1993), and changes in the amino acid sequence of the CP readthrough domain of this virus affect its aphid transmission (Jolly and Mayo, 1994). The readthrough extension of BYDV CP is involved in specific recognition of the virus surface during transmission (Wang et al., 1995). As demonstrated for BWYV, the minor capsid readthrough protein is required for aphid transmission, since virions lacking this protein are no longer transmissible (Brault et al., 1995). Other viral products are dispensable for transmission (Ziegler-Graff et al., 1996), and mutational analysis mapped transmission-related functions to the conserved N-terminal half of the readthrough domain (Bruyere et al., 1997). As shown for BYDV-PAV, the readthrough domain is required neither for virus uptake in the aphid hindgut nor for release into the hemocoel cavity, suggesting that it is involved in transport through barriers of the salivary gland (Chay et al., 1996). Expression of capsid plus its extension might constitute an HC-like entity integrated into the particle, with analogous roles to the ones described for separate HCs in promoting interaction with vector receptors (Hull, 1994). Interestingly, a similar readthrough expression strategy is involved in transmission of furoviruses by fungal vectors (Tamada and Kusume, 1991). Other pathogenic agents could take advantage of these mechanisms of insect transmission, for instance, aphid transmission of *Potato spindle tuber viroid* has been demonstrated to occur through heterologous encapsidation in particles of PLRV (Querci et al., 1997).

Another important finding in luteovirus transmission is the involvement of endosymbiont bacteria of the vector in the process of virus transmission. The 60 kDa GroEL homologue known as symbionin (a molecular chaperonin) produced by *Buchnera* sp. played a key role during the process. In vitro binding affinity between PLRV and symbionins from its aphid vector was reported, and the treatment of aphids with antibiotics impeded drastically their ability to transmit (Van den Heuvel et al., 1994). The interaction of BYDV virus particles with chaperonins produced by the bacteria suggests a potential involvement during trafficking through the aphid (Filichkin et al., 1997). Direct injection of BWYV mutants into aphids has provided evidence that virion-GroEL homologue interaction is involved in virus survival (retention) in the hemolymph (Van den Heuvel et al., 1997).

Mutational analysis of recombinant GroEL homologue revealed that the PLRV-binding site was located in the equatorial domain of the protein (Hogenhout et al., 1998).

Enamovirus Transmission

Pea enation mosaic virus (PEMV) constitutes an interesting model of transmission. The genome of PEMV is bipartite, with two autoreplicating unrelated RNAs. The virus is circulatively transmitted by several aphid species, and nontransmissible isolates were identified based in structural features (Adam et al., 1979). Although symptomatology, cytopathological effects, and mode of transmission resemble those of luteoviruses, PEMV is considered separate due to its genomic structure, which can be construed as a stable symbiotic complex that contains two viral components: the RNA1-containing PEMV-1, an enamovirus, and the RNA2 component, both required to establish a systemic infection. The larger RNA1 (similar to the RNAs of genus *Polerovirus*) controls functions related to symptomatology, cytopathology, and aphid transmissibility, while the RNA2 (sharing homology with RNAs of members of the genus *Umbravirus* and other genera) controls functions related with systemic movement and depends on RNA1 for encapsidation and aphid transmission (Demler, Borkhsenious, et al., 1994). RNA1 is also responsible for transmission of a satellite RNA (Demler, Rucker, et al., 1994).

PEMV, in contrast to luteoviruses, can be mechanically transmitted. By successive mechanical passages, variants defective in aphid transmissibility can be generated (Demler et al., 1997). Structural analysis of a 54 kDa minor subunit present only in transmissible isolates demonstrated its structure as a readthrough of the 21 kDa CP followed by the 33 kDa downstream ORF. This domain was not essential for infection, and changes in this region were sufficient to confer or abolish aphid transmissibility (Demler et al., 1997). The same study provided evidence of recombination mechanisms used by the virus to regulate responses to selective pressures. The rapid rise of nontransmissible variants suggests that, in this case (and probably in many others), vector transmission might represent a negative selective pressure for competitiveness, although in natural populations the scenario might be different. PEMV, as happens in the case of luteoviruses, has a high affinity for GroEL homologues (Van den Heuvel et al., 1999).

Geminivirus Transmission

Geminiviruses have single-stranded circular DNA genomes and characteristic isometric paired particles (Harrison, 1985). The family *Geminiviridae* has been divided into three genera, *Mastrevirus, Curtovirus,* and *Begomovirus,* according to genome structure, host range, and vector specificity (Pringle, 1999). Members of the genera *Mastrevirus* and *Curtovirus* are transmitted by leafhoppers and leafhoppers or treehoppers, respectively, while those of the genus *Begomovirus* are transmitted by whiteflies. Although most begomoviruses have genomes composed of two DNA molecules, *Tomato yellow leaf curl virus* (TYLCV) has both monopartite and bipartite isolates, all of them transmitted by *Bemisia tabaci.*

The interaction between geminiviruses and their vectors is considered to be circulative nonpropagative (Harrison, 1985). The virus has to pass through specific cells of the gut to gain access to the hemolymph, from where it moves to the salivary glands. A pathway through the vector, based on localization in the filter chamber, midgut, and salivary gland, has been proposed (Hunter et al., 1998). No evidence of a helper factor has been provided, though the movement protein(s) might play an indirect role in the acquisition process, since both genomic components of a bipartite geminivirus are needed for efficient transmission (Liu et al., 1997). Strains of nontransmissible geminiviruses have been described (Wu et al., 1996). Acquisition and retention have been studied using sensitive DNA-detection-based techniques (Polston et al., 1990; Zeidan and Czosnek, 1991; Azzam et al., 1994). Nonvector whiteflies could acquire virions (Cohen et al., 1989; Polston et al., 1990; Liu et al., 1997). The long-term association of geminiviruses with their vectors affects the longevity and fecundity of the insects, suggesting a possible pathogenic affects on the vector (Rubinstein and Czosnek, 1997). Transovarial transmission of TYLCV has been reported recently (Ghanim et al., 1998). These facts seem to indicate that at least some geminiviruses might have a propagative relationship with their vectors. Observation of viruslike particles in mycetocyte cells could suggest an endosymbiont organism role in transovarial transmission (Costa et al., 1996). It has been reported recently that, as is the case with luteoviruses, GroEL from endosymbiotic bacteria in whiteflies is involved in geminivirus transmission, through a proposed mechanism of virus protection during circulation (Morin et al., 1999).

Geminiviruses of different genera are not serologically related, whereas a strong serological relationship, due to a high conservation of CP amino acid sequence, exists among all whitefly-transmitted geminiviruses, suggesting that the CP is the major determinant of vector specificity (Zeidan and Czosnek, 1991). Furthermore, leafhopper-transmitted geminiviruses have different vector species correlating with rather different CPs. A chimeric virus in which the CP of the whitefly-transmitted *African cassava mosaic virus* (genus *Begomovirus*) was replaced with the CP of *Beet curly top virus* (genus *Curtovirus*) had an altered vector specificity from whiteflies to leafhoppers (Briddon et al., 1990). The construction of mutants with deletions in the CP gene of *Bean golden mosaic virus*, a begomovirus, demonstrated that a functional CP is required for transmission (Azzam et al., 1994). CP replacement was also used to demonstrate whitefly transmission of an insect-nontransmissible isolate of *Abutilon mosaic virus* (Höfer et al., 1997). By studying defective genomic DNAs of the monopartite TYLCV, amino acids essential for insect transmission have been identified, mapping in the region between positions 129 and 134 of the CP (Noris et al., 1998).

A study involving isolates of *Maize streak virus* (genus *Mastrevirus*), selected by alterations of the environmental conditions such as host changes or leafhopper transmission (including single-insect bottlenecks or transmission by large numbers of insects), has confirmed the influence of vector transmission in selection of variants from the quasi-species organization of plant virus populations (Isnard et al., 1998).

Other Nonpropagative Viruses

Four families of beetles (Coleoptera)—Chrysomelidae, Coccinellidae, Curculionidae, and Meloidae, comprise 74 species able to transmit viruses belonging to four genera: *Comovirus, Tymovirus, Bromovirus,* and *Sobemovirus* (Fulton et al., 1987). Generally speaking, all beetle-transmitted viruses have relatively stable isometric particles containing RNA genomes, they are able to be mechanically transmitted, and they are present at high concentrations in infected plant sap. Beetles lack salivary glands in their mouthparts, and transmission is presumed to occur during chewing, while some gut contents are regurgitated. Acquisition of virus occurs rapidly, no latent period has been described, and retention time is highly variable. Although initially beetles were thought to be vectors by a simple contamination mechanism, the high specificity between vectors and viruses suggests a more complex interaction. The viruses can accumulate in the hemolymph, indicating a circulation of the virus inside the vector. However, in several combinations of virus and vector, the retention takes place in the gut (Wang et al., 1992), and the particles present in the regurgitant could come from the hemolymph or from the gut (Wang et al., 1994a,b). The wounds produced by beetles can be reproduced artificially, and in this manner selectivity of transmission has been related to the presence of enzymatic activities (ribonucleases) in the regurgitant (Gergerich et al., 1986). It was suggested that viruses need to avoid inactivation by a translocation mechanism still not well understood (Field et al., 1994). The CP might possess determinants of transmission, although specificity seems to reside in interaction with the host plant.

Groundnut rosette virus (GRV, genus *Umbravirus*) depends on both its satellite RNA (Sat-RNA) and *Groundnut rosette assistor virus* (GRAV, genus *Luteovirus*), for aphid transmission (Murant, 1990). CP of GRAV expressed in transgenic plants served to package sat-RNA efficiently, while GRV genomic RNA was not encapsidated in the absence of the sat-RNA. The characteristics of the sat-RNA that allowed efficient encapsidation of GRV genomic RNA correlated with its capacity to promote aphid transmission (Robinson et al., 1999).

Other viruses with circulative transmission include the members of genus *Nanovirus*, viruses with single-stranded (ssDNA) genomes that are aphid transmitted. A recent report indicates that a helper factor is required for transmission of *Faba bean necrotic yellows virus* of genus *Nanovirus* (Franz et al., 1999).

CIRCULATIVE PROPAGATIVE TRANSMISSION

Some genera consist of plant viruses with complex infection cycles. They can replicate in the cells of their insect vectors, being parasites of both plants and animals. Propagative relationships include a long-term association with the vector that, in many cases, may have adverse effects on the insect host, for instance, in longevity and fecundity. In some cases, propagation includes transovarial transmission of the plant virus to the vector progeny. Interestingly, propagative viruses encode genes that are differentially expressed in their infection cycle, during infection of plants or insects (Falk et al., 1987).

Propagative plant viruses belong to families including viruses that also infect animal hosts (*Bunyaviridae*, *Reoviridae*, and *Rhabdoviridae*), and to the genera *Marafivirus* and *Tenuivirus*.

Tospovirus Transmission

Seven thrips species (out of around 5,000) have been reported to transmit tospoviruses. The most important vector is *Frankliniella occidentalis* (German et al., 1992), due to the economic impact caused by *Tomato spotted wilt virus* (TSWV, genus *Tospovirus*), a virus able to infect more than 800 different plant species (Prins and Goldbach, 1998).

Only larvae can acquire transmissible TSWV from infected plants. Once acquired by larvae, the virus is able to multiply in the vector (Ullman et al., 1993; Wijkamp and Peters, 1993) and is retained during pupation and emergence of the adult stage. Larvae, in spite of their reduced mobility, are better transmitters in experimental conditions when compared to highly mobile adults (Wijkamp and Peters, 1993); for that reason, the importance of larval stages in epidemiology must be carefully considered. The length and factors affecting acquisition, retention, and inoculation periods have been studied (Wijkamp and Peters, 1993). The entry site of virus in larvae of *F. occidentalis* is through the midgut epithelial cells, whereas adults were refractory to passage due to a midgut barrier (Ullman et al., 1993, 1995). Immunohistological techniques have been employed to investigate tissue tropism and the virus pathway inside the insect, revealing that virus accumulation only occurred in midgut, foregut, and salivary glands; the metamorphosis of the vector influenced its transmission ability (Nagata et al., 1999).

Two proteins have been identified in the vector and found to interact with glycoproteins (GPs) of the virus: a 94 kDa protein (Kikkert et al., 1998) and a 50 kDa protein (Bandla et al., 1998). The former is not present in the gut and, speculatively, may represent a receptor involved during circulation of the virus, via recognition of the viral glycoprotein G2; the latter is present in the midgut and binds the viral GPs, representing perhaps a cellular receptor, a hypothesis supported by the use of anti-idiotype antibodies that localize to the plasmalemma of midgut epithelial cells (Bandla et al., 1998).

Other Propagative Viruses

In addition to members of the genus *Tospovirus*, plant viruses in the families *Rhabdoviridae* and *Reoviridae* and in the genera *Tenuivirus* and *Marafivirus* have host ranges that include both plants and their insect vectors.

Members of the family *Rhabdoviridae* infect vertebrates, insects, and plants and have enveloped, bacilliform virions, provided with surface glycoproteins. These proteins of plant rhabdoviruses have been proposed to be involved in interaction with specific cellular receptors in the vector, analogous with animal-infecting viruses, although no experimental evidence is available. Plant rhabdoviruses are aphid or leafhopper transmitted and can be differentiated based on their sites of virus maturation: nucleus or cytoplasm (Jackson et al., 1987).

In the case of plant reoviruses, the three genera *Fijivirus, Phytoreovirus,* and *Oryzavirus* of the family *Reoviridae* are differentiated by particle morphology, number and size of genomic segments, and their specific planthopper or leafhopper transmission. They are transmitted in a circulative propagative mode and can pass transovarially to the vector progeny. These viruses replicate efficiently in many insect tissues, in contrast with their restriction to phloem in the plant (Nuss and Dall, 1990). An interesting case is provided by the *Wound tumor virus* (genus *Phytoreovirus*), which has many isolates from experimental introduction of a vector to several dicotyledonous hosts (Hillman et al., 1991). Other reoviruses have been detected frequently in both vectors and monocotyledonous plants, although accumulation levels tend to be reduced in insects (Falk et al., 1987; Suzuki et al., 1994). Insect transmission has been proposed as a method to separate heterogeneous viruses (Suga et al., 1995). Nonphytopathogenic reoviruses have been found in both plants and insects, and in these cases the plants might serve as reservoirs of virus (Nakashima and Noda, 1995). One of the two outer capsid proteins of *Rice dwarf virus* (RDV, genus *Phytoreovirus*) is considered to be essential for infection of insect cells, as studied using vector cell monolayers (Yan et al., 1996). The loss of P2 protein (due to a point mutation in the segment encoding this protein) results in nontransmissibility of an RDV isolate (Tomaru et al., 1997). When P2-free particles were injected into insects using a glass capillary tube, the insects became infected, suggesting that the ability to infect cells from the hemolymph is retained, while they were unable to interact with and infect cells in the intestinal tract (Omura et al., 1998).

Tenuiviruses have a vector range restricted to delphacid planthoppers, with transovarial passage demonstrated in some cases and high vector specificity (Falk and Tsai, 1998). Proof of propagation of *Maize stripe virus* (MSpV, genus *Tenuivirus*) in *Peregrinus maidis* was achieved by serological analysis of accumulation of the nucleoprotein N (Falk et al., 1987). The virus titer in vectors correlates with transmission efficiency of MSpV isolates (Ammar et al., 1995).

Circulative propagative transmission also occurs in the genus *Marafivirus* (viruses transmitted by leafhoppers) (Rivera and Gamez, 1986).

CONCLUSIONS

In the preceding sections, different relations between plant viruses and insect vectors have been considered. Despite the great diversity of viruses and vector species, as well as the complexity of many of the relationships cited, it can be concluded that transmission is a crucial function during the infection cycle of numerous viruses. During evolution, many plant viruses have found efficient ways to be spread by using insects as their vectors.

Recently, our understanding of the transmission phenomenon has improved, thanks to the application of the powerful tools of molecular biology. The involvement of many virus-encoded genes in transmission has been reported, and the vector counterparts of the interactions involved in transmission are likely to be identified soon. However, we are far from complete comprehension of transmission processes. Molecular analysis has furthered our knowledge, but still new features are being added to each particular situation.

The CP is the viral gene most broadly involved in transmission. In addition to the CP, many noncirculative viruses have developed a strategy that employs accessory helper factors. The helper component strategy might serve the purpose of overcoming more efficiently the bottlenecks introduced by insect transmission. The intimate relationship established in circulative transmission also implicates the involvement of the CP and other viral products, such as readthrough extensions of the CP. Further genetic studies would serve to clarify roles played by these and other viral factors in the transmission process.

A few insect proteins that might be involved in virus transmission have been identified recently. In addition, proteins from insect-endosymbiont organisms have proved to play important roles during retention of circulative viruses. Despite these advances, still little information exists regarding the nature of putative virus receptors in vectors. The characterization of such receptors might serve to elucidate questions regarding the specificity of vectors. Other points, still far from well characterized, that would merit future research efforts include determining how nonpersistent viruses are released after being attached to the unknown sites in the food canal, or how circulative viruses can survive in the hostile environment inside the vectors. New control strategies, based on blockage of the interactions described through genetically engineered manipulation of vectors or host plants, might result from these and future findings. Benefits derived from such an interference justify the efforts dedicated to studying the transmission processes.

REFERENCES

Adam G, Dander E, Shepherd RJ (1979). Structural differences between pea enation mosaic virus strains affecting transmissibility by *Acyrthosiphon pisum* (Harris). *Virology* 92: 1-14.

Agranovsky AA, Lesemann DE, Maiss E, Hull R, Atabekov JG (1995). "Rattlesnake" structure of a filamentous plant RNA virus built of two capsid proteins. *Proceedings of the National Academy of Sciences, USA* 92: 2470-2473.

Ammar ED, Gingery RE, Madden LV (1995). Transmission efficiency of three isolates of maize stripe tenuivirus in relation to virus titre in the planthopper vector. *Plant Pathology* 44: 239-243.

Ammar ED, Järlfors U, Pirone TP (1994). Association of potyvirus helper component protein with virions and the cuticle lining the maxillary food canal and foregut of an aphid vector. *Phytopathology* 84: 1054-1059.

Armour SL, Mercher U, Pirone TP, Lyttle DG, Essemberg RC (1983). Helper component for aphid transmission encoded by region II of cauliflower mosaic virus DNA. *Virology* 129: 25-30.

Atreya CD, Atreya PL, Thornbury DW, Pirone TP (1992). Site-directed mutations in the potyvirus HC-Pro gene affect helper component activity, virus accumulation and symptom expression in infected tobacco plants. *Virology* 191: 106-111.

Atreya CD, Pirone TP (1993). Mutational analysis of the helper component-proteinase gene of a potyvirus: effects of amino acid substitutions, deletions, and gene replacement on virulence and aphid transmissibility. *Proceedings of the National Academy of Sciences, USA* 90: 11919-11923.

Atreya CD, Raccah B, Pirone TP (1990). A point mutation in the coat protein abolishes aphid transmissibility of a potyvirus. *Virology* 178: 161-165.

Atreya PL, Atreya CD, Pirone TP (1991). Amino acid substitutions in the coat protein result in loss of insect transmissibility of a plant virus. *Proceedings of the National Academy of Sciences, USA* 88: 7887-7891.

Atreya PL, Lopez-Moya JJ, Chu M, Atreya CD, Pirone TP (1995). Mutational analysis of the coat protein N-terminal amino acids involved in potyvirus transmission by aphids. *Journal of General Virology* 76: 265-270.

Ayllon MA, Rubio L, Moya A, Guerri J, Moreno P (1999). The haplotype distribution of two genes of citrus tristeza virus is altered after host change or aphid transmission. *Virology* 255: 32-39.

Azzam O, Frazer J, de la Rosa D, Beaver JS, Ahlquist P, Maxwell DP (1994). Whitefly transmission and efficient ssDNA accumulation of bean golden mosaic geminivirus require functional coat protein. *Virology* 204: 289-296.

Bandla MD, Campbell LR, Ullman DE, Sherwood JL (1998). Interaction of tomato spotted wilt tospovirus (TSWV) glycoproteins with a thrips midgut protein, a potential cellular receptor for TSWV. *Phytopathology* 88: 98-104.

Baulcombe DC, Lloyd J, Manoussopoulos IN, Roberts IM, Harrison BD (1993). Signal for potyvirus-dependent aphid transmission of potato aucuba mosaic virus and the effect of its transfer to potato virus X. *Journal of General Virology* 74: 1245-1253.

Berger PH, Hunt AG, Domier LL, Hellmann GM, Stram Y, Thornbury DW, Pirone TP (1989). Expression in transgenic plants of a viral gene product that mediates insect transmission of potyviruses. *Proceedings of the National Academy of Sciences, USA* 86: 8402-8406.

Berger PH, Pirone TP (1986). The effect of helper component on the uptake and localization of potyviruses in *Myzus persicae*. *Virology* 153: 256-261.

Blanc S, Ammar ED, Garcia-Lampasona S, Dolja VV, Llave C, Baker J, Pirone TP (1998). Mutations in the potyvirus helper component protein: Effects on interactions with virions and aphid stylets. *Journal of General Virology* 79: 3119-3122.

Blanc S, Cerutti M, Usmany M, Vlak JM, Hull R (1993). Biological activity of CaMV aphid transmission factor expressed in a heterologous system. *Virology* 192: 643-650.

Blanc S, Lopez-Moya JJ, Wang RY, Garcia-Lampasona S, Thornbury DW, Pirone TP (1997). A specific interaction between coat protein and helper component correlates with aphid transmission of a potyvirus. *Virology* 231: 141-147.

Blanc S, Schmidt I, Vantard M, Scholthof HB, Kuhl G, Esperandieu P, Cerutti M, Louis C (1996). The aphid transmission factor of cauliflower mosaic virus forms a stable complex with microtubules in both insect and plant cells. *Proceedings of the National Academy of Sciences, USA* 93: 15158-15163.

Bouchery Y, Givord L, Monestiez P (1990). Comparison of short- and long-feed transmission of the CaMV Cabb-S strain and S delta hybrid by 2 species of aphid: *Myzus persicae* (Sulzer) and *Brevicoryne brassicae* (L.). *Research in Virology* 141: 677-683.

Bourdin D, Lecoq H (1991). Evidence that heteroencapsidation between two potyviruses is involved in aphid transmission of a non-aphid-transmissible isolate from mixed infections. *Phytopathology* 81: 1459-1464.

Brault V, Van den Heuvel JFJM, Verbeek M, Ziegler-Graff V, Reutenauer A, Herrbach E, Garaud J-C, Guilley H, Richards K, Jonard G (1995). Aphid transmission of beet western yellows luteovirus requires the minor capsid read-through protein P74. *EMBO Journal* 14: 650-659.

Briddon RW, Pinner MS, Stanley J, Markham PG (1990). Geminivirus coat protein gene replacement alters insect specificity. *Virology* 177: 85-94.

Bruyere A, Brault V, Ziegler-Graff V, Simonis M-T, Van den Heuvel JFJM, Richards K, Ghilley H, Jonard G, Herrbach E (1997). Effects of mutations in the beet western yellows virus readthrough protein on its expression and packaging and on virus accumulation, symptoms, and aphid transmission. *Virology* 230: 323-334.

Cabaleiro C, Segura A (1997). Field transmission of grapevine leafroll associated virus 3 (GLRaV-3) by the mealybug *Planococcus citri. Plant Disease* 81: 283-287.

Canto T, Lopez-Moya JJ, Serra-Yoldi MT, Diaz-Ruiz JR, Lopez-Abella D (1995). Different helper component mutations associated with lack of aphid transmissibility of two isolates of potato virus Y. *Phytopathology* 85: 1519-1524.

Chay CA, Gunasinge UB, Dinesh-Kumar SP, Miller WA, Gray ST (1996). Aphid transmission and systemic plant infection determinants of barley yellow dwarf luteovirus-PAV are contained in the coat protein readthrough domain and 17-kDa protein, respectively. *Virology* 219: 57-65.

Chen B, Francki RIB (1990). Cucumovirus transmission by the aphid *Myzus persicae* is determined solely by the viral coat protein. *Journal of General Virology* 71: 939-944.

Chidress SA, Harris KF (1989) Localization of virus-like particles in the foregut of viruliferous *Graminella nigrifrons* leafhoppers carrying the semipersistent maize chlorotic dwarf virus. *Journal of General Virology* 70: 247-251.

Cohen S, Duffus JE, Liu HY (1989). Acquisition, interference, and retention of cucurbit leaf curl viruses in whiteflies. *Phytopathology* 79: 109-113.

Costa HS, Westcot DM, Ullman DE, Rodell DE, Rodell RC, Brown JK, Johnson MW (1996). Virus-like particles in the mycetocyte of the sweetpotato whitefly, *Bemisia tabaci* (Homoptera, Aleyrodidae). *Journal of Invertebrate Pathology* 67: 183-186.

Demler SA, Borkhsenious ON, Rucker DG, De Zoeten GA (1994). Assesment of the autonomy of replicative and structural functions encoded by the luteo-phase of pea enation mosaic virus. *Journal of General Virology* 75: 997-1007.

Demler SA, Rucker DG, Nooruddin L, De Zoeten GA (1994). Replication of the satellite RNA of pea enation mosaic virus is controlled by RNA 2 encoded functions. *Journal of General Virology* 75: 1399-1406.

Demler SA, Rucker-Feeney DG, Skaf JS, De Zoeten GA (1997). Expression and suppression of circulative aphid transmission in pea enation mosaic virus. *Journal of General Virology* 78: 511-523.

Dolja VV, Herndon KL, Pirone TP, Carrington JC (1993). Spontaneous mutagenesis of a plant potyvirus genome after insertion of a foreign gene. *Journal of Virology* 67: 5968-5975.

Dolja VV, Hong J, Keller KE, Martin RR, Peremyslov VV (1997). Suppression of potyvirus infection by coexpressed closterovirus protein. *Virology* 234: 243-252.

Dolja VV, Karasev AV, Koonin EV (1994). Molecular biology and evolution of closteroviruses: Sophisticated build-up of large RNA genomes. *Annual Review of Phytopathology* 32: 261-285.

Falk BW, Tsai JH (1998). Biology and molecular biology of viruses in the genus *Tenuivirus*. *Annual Review of Phytopathology* 36: 139-163.

Falk BW, Tsai JH, Lommel SA (1987). Differences in levels of detection for the maize stripe virus capsid and major non-capsid proteins in plant and insect hosts. *Journal of General Virology* 68: 1801-1811.

Ferris RS, Berger PH (1993). A stochastic simulation model of epidemics of arthropod vectored plant viruses. *Phytopathology* 83: 1269-1278.

Field TK, Patterson CA, Gergerich RC, Kim KS (1994). Fate of viruses in bean leaves after deposition by *Epilachna varivestis*, a beetle vector of plant viruses. *Phytopathology* 84: 1346-1350.

Filichkin SA, Brumfield S, Filichkin TP, Young MJ (1997). In vitro interactions of the aphid endosymbiotic SymL chaperonin with barley yellow dwarf virus. *Journal of Virology* 71: 569-577.

Filichkin SA, Lister RM, McGrath PF, Young MJ (1994). In vivo expression and mutational analysis of the barley yellow dwarf readthrough gene. *Virology* 205: 290-299.

Flasinski S, Cassidy BG (1998). Potyvirus aphid transmission requires helper component and homologous coat protein for maximal efficiency. *Archives of Virology* 143: 2159-2172.

Francki RIB, Fauquet CM, Knudson DL, Brown F (eds.) (1991). Classification and nomenclature of viruses. Fifth Report of the International Committee on Taxonomy of Viruses. *Archives of Virology* (Suppl. 2). Springer-Verlag, Wien, New York, 450 pp.

Franz AW, Van der Wilk F, Verbeek M, Dullemans AM, Van den Heuvel JF (1999). Faba bean necrotic yellows virus (genus *Nanovirus*) requires a helper factor for its aphid transmission. *Virology* 262: 210-219.

Fulton JP, Gergerich RC, Scott HA (1987). Beetle transmission of plant viruses. *Annual Review of Phytopathology* 25: 111-123.

Gera A, Loebenstein G, Raccah B (1979). Protein coats of two strains of cucumber mosaic virus affect transmission by *Aphis gossypii*. *Phytopathology* 69: 396-399.

Gergerich RC, Scott HA. Fulton JP (1986). Evidence that ribonucleases in beetle regurgitant determine the transmission of plant viruses. *Journal of General Virology* 67: 367-370.

German TL, Ullman DE, Moyer JW (1992). Tospoviruses: Diagnosis, molecular biology, phylogeny and vector relationships. *Annual Review of Phytopathology* 30: 315-348.

Ghanim M, Morin S, Zeidan M, Czosnek H (1998). Evidence for transovarial transmission of tomato yellow leaf curl virus by its vector, the whitefly *Bemisia tabaci*. *Virology* 240: 295-303.

Gildow FE (1987). Virus membrane interactions involved in circulative transmission of luteovirus by aphids. *Current Topics in Virus Research* 4: 93-120.

Gildow FE (1993). Evidence for receptor-mediated endocitosis regulating luteovirus acquisition by aphids. *Phytopathology* 83: 270-277.

Gildow FE, Gray SM (1994). The aphid salivary gland basal lamina as a selective barrier associated with vector-specific transmission of barley yellow dwarf luteovirus. *Phytopathology* 83: 1293-1302.

Govier DA, Kassanis B (1974a). Evidence that a component other than the virus particle is needed for aphid transmission of potato virus Y. *Virology* 57: 285-286.

Govier DA, Kassanis B (1974b). A virus-induced component of plant sap needed when aphids acquire potato virus Y from purified preparations. *Virology* 61: 420-426.

Granier F, Durand-Tardiff M, Casse-Delbart F, Lecoq H, Robaglia C (1993). Mutations in zucchini yellow mosaic virus helper component protein associated with loss of aphid transmissibility. *Journal of General Virology* 74: 2737-2742.

Gray SM (1996). Plant virus proteins involved in natural vector transmission. *Trends in Microbiology* 4: 259-264.

Harris KF (1981). Arthropod and nematode vectors of plant viruses. *Annual Review of Phytopathology* 19: 391-426.

Harris KF (1991). Aphid transmission of plant viruses. In Mandahar CL (ed.), *Plant viruses* (Volume 2), CRC Press, Boca Raton, FL, pp. 177-204.

Harrison BD (1985). Advances in geminivirus research. *Annual Review of Phytopathology* 23: 55-82.

Harrison BD, Robinson DJ (1988). Molecular variation in vector-borne plant viruses: Epidemiological significance. *Philosophical Transactions of the Royal Society of London Series B* 321: 447-462.

Hillman BI, Anzola JV, Halpern BT, Cavileer TD, Nuss DL (1991). First field isolation of wound tumor virus from a plant host: Minimal sequence divergence from the type strain isolated from an insect vector. *Virology* 185: 896-900.

Höfer P, Bedford ID, Markham PG, Jeske H, Frischmuth T (1997). Coat protein gene replacement results in whitefly transmission of an insect nontransmissible geminivirus isolate. *Virology* 236: 288-295.

Hogenhout SA, Van der Wilk F, Verbeek M, Goldbach RW, Van den Heuvel JFJM (1998). Potato leafroll virus binds to the equatorial domain of the aphid endosymbiotic GroEL homolog. *Journal of Virology* 72: 358-365.

Huet H, Gal-On A, Meir E, Lecoq H, Raccah B (1994). Mutations in the helper component protease gene of zucchini yellow mosaic virus affect its ability to mediate aphid transmissibility. *Journal of General Virology* 75: 1407-1414.

Hull R (1994). Molecular biology of plant virus vector interactions. In Harris KF (ed.), *Advances in Disease Vector Research* (Volume 10). Springer-Verlag, New York, pp. 361-386.

Hull R (1996). Molecular biology of rice tungro viruses. *Annual Review of Phytopathology* 34: 275-297.

Hunt RE, Nault LR, Gingery RE (1988). Evidence for infectivity of maize chlorotic dwarf virus and for a helper component in its leafhopper transmission. *Phytopathology* 78: 499-504.

Hunter WB, Hiebert E, Webb SE, Tsai JH, Polston JE (1998). Location of geminiviruses in the whitefly *Bemisia tabaci* (Homoptera: Aleyrodidae). *Plant Disease* 82: 1147-1151.

Isnard M, Granier M, Frutos R, Reynaud B, Peterschmitt M (1998). Quasispecies nature of three maize streak virus isolates obtained through different modes of selection from a population used to assess response to infection of maize cultivars. *Journal of General Virology* 79: 3091-3099.

Jackson AO, Francki RIB, Zuidema D (1987). Biology, structure and replication of plant rhabdoviruses. In Wagner RR (ed.), *The Rhabdoviruses*, Plenum Press, New York, pp. 427-508.

Jacquet C, Delecolle B, Raccah B, Lecoq H, Dunez J, Ravelonandro M (1998). Use of modified plum pox virus coat protein genes developed to limit heteroencapsidation-associated risks in transgenic plants. *Journal of General Virology* 79: 1509-1517.

Johansen IE, Keller KE, Dougherty WG, Hampton RO (1996). Biological and molecular properties of a pathotype P-4 isolate of pea seed-borne mosaic virus. *Journal of General Virology* 77: 1329-1333.

Jolly CA, Mayo MA (1994). Changes in the amino acid sequence of the coat protein readthrough domain of potato leafroll luteovirus affect the formation of an epitope and aphid transmission. *Virology* 201: 182-185.

Kassanis B, Govier DA (1971). The role of the helper virus in aphid transmission of potato aucuba mosaic virus and potato virus C. *Journal of General Virology* 13: 221-228.

Kikkert M, Meurs C, Van de Wetering F, Dorfmüller S, Peters D, Kormelink R, Goldbach R (1998). Binding of tomato spotted wilt virus to a 94-kDa thrips protein. *Phytopathology* 88: 63-69.

Lecoq H, Bourdin D, Raccah B, Hiebert E, Purcifull DE (1991). Characterization of a zucchini yellow mosaic virus isolate with a deficient helper component. *Phytopathology* 81: 1087-1091.

Lecoq H, Pitrat M (1985). Specificity of the helper-component-mediated aphid transmission of three potyviruses infecting muskmelon. *Phytopathology* 75: 890-893.

Lecoq H, Ravelonandro M, Wipf-Scheibel C, Monsion M, Raccah B, Dunez J (1993). Aphid transmission of a non-aphid-transmissible strain of zucchini yellow mosaic potyvirus from transgenic plants expressing the capsid protein of plum pox potyvirus. *Molecular Plant-Microbe Interactions* 6: 403-406.

Legavre T. Maia IG, Casse-Delbart F, Bernardi F, Robaglia C (1996). Switches in the mode of transmission select for or against a poorly aphid-transmissible strain of potato virus Y with reduced helper component and virus accumulation. *Journal of General Virology* 77: 1343-1347.

Leh V, Jacquot E, Geldreich A. Herman T. Leclerc D, Cerutti M. Yot P, Keller M. Blanc S (1999). Aphid transmission of cauliflower mosaic virus requires the PIII protein. *EMBO Journal* 18: 7077-7085.

Liu S. Bedford ID, Briddon RW, Markham PG (1997). Efficient whitefly transmission of bipartite geminiviruses requires both genomic components. *Journal of General Virology* 78: 1791-1794.

Lopez-Abella D. Bradley RHE, Harris KF (1988). Correlation between stylet paths made during superficial probing and the ability of aphids to transmit nonpersistent viruses. In Harris KF (ed.), *Advances in Disease Vector Research* (Volume 5). Springer-Verlag. New York, pp. 251-287.

Lopez-Moya JJ, Canto T. Lopez-Abella D. Diaz-Ruiz JR (1995). Transmission by aphids of a naturally non-transmissible plum pox virus isolate with the aid of potato virus Y helper component. *Journal of General Virology* 76: 2293-2297.

Lopez-Moya JJ, Garcia JA (1999). Potyviruses (Potyviridae). In Webster RG, Granoff A (eds.), *Encyclopedia of Virology* (Second Edition. Volume 3). Academic Press, San Diego, pp. 1369-1385.

Lopez-Moya JJ, Wang RY, Pirone TP (1999). Context of the coat protein DAG motif affects potyvirus transmissibility by aphids. *Journal of General Virology* 80: 3281-3288.

Lung MCY, Pirone TP (1973). Studies on the reason for differential transmission of purified cauliflower mosaic virus. *Virology* 60: 260-264.

Maia IG, Haenni A-L, Bernardi F (1996). Potyviral HC-Pro: A multifunctional protein. *Journal of General Virology* 77: 1335-1341.

Markham PG, Hull R (1985). Cauliflower mosaic virus aphid transmission facilitated by transmission factors from other caulimoviruses. *Journal of General Virology* 66: 921-923.

Martin B. Collar JL. Tjallingii WF. Fereres A (1997). Intracellular ingestion and salivation by aphids may cause the acquisition and inoculation of nonpersistently transmitted plant viruses. *Journal of General Virology* 78: 2701-2705.

Martin R. Keese PK, Young MJ, Waterhouse PM, Gerlach WL (1990). Evolution and molecular biology of luteoviruses. *Annual Review of Phytopathology* 28: 341-363.

Matthews REF (1991). *Plant Virology* (Third Edition). Academic Press, San Diego.

Mawassi M, Gafny R, Bar-Joseph M (1993). Nucleotide sequence of the coat protein gene of citrus tristeza virus: Comparison of biologically diverse isolates collected in Israel. *Virus Genes* 7: 265-275.

Morin S, Ghanim M, Zeidan M, Czosnek H, Verbeek M. Van den Heuvel JFJM (1999). A GroEL homologue from endosymbiotic bacteria of the whitefly *Bemisia tabaci* is implicated in the circulative transmission of tomato yellow leaf curl virus. *Virology* 256: 75-84.

Murant AF (1990). Dependence of groundnut rosette virus on its satellite RNA as well as on groundnut rosette assistor luteovirus for transmission by *Aphis craccivora*. *Journal of General Virology* 71: 2163-2166.

Murant AF, Raccah B, Pirone TP (1988). Transmission by vectors. In Milne RG (ed.), *The Plant Viruses, the Filamentous Plant Viruses* (Volume 4), Plenum Press, New York, pp. 237-273.

Murant AF, Roberts IM, Elnager S (1976). Association of virus-like particles with the foregut of the aphid, *Cavariella aegopodii*, transmitting the semipersistent viruses anthriscus yellows and parsnip yellow fleck. *Journal of General Virology* 31: 47-57.

Nagata T, Inoue-Nagata AK, Smid HM, Goldbach R, Peters D (1999). Tissue tropism related to vector competence of *Frankliniella occidentalis* for tomato spotted wilt tospovirus. *Journal of General Virology* 80: 507-515.

Naidu RA, Gowda S, Satyanarayana T, Boyko V, Reddy AS, Downson WO, Reddy DVR (1998). Evidence that whitefly-transmitted cowpea mild mottle belongs to the genus Carlavirus. *Archives of Virology* 143: 769-780.

Nakashima N, Noda H (1995). Nonpathogenic *Nilaparvata lugens* reovirus is transmitted to the brown planthopper through rice plant. *Virology* 207: 303-307.

Nault LR, Ammar ED (1989). Leafhopper and planthopper transmission of plant viruses. *Annual Review of Entomology* 34: 503-529.

Noris E, Vaira AM, Caciagli P, Masenga V, Gronenborn B, Accotto GP (1998). Amino acids in the capsid protein of tomato yellow leaf curl virus that are crucial for systemic infection, particle formation, and insect transmission. *Journal of Virology* 72: 10050-10057.

Nuss DL, Dall DJ (1990). Structural and functional properties of plant reovirus genomes. *Advances in Virus Research* 38: 249-306.

Omura T, Yan J, Zhong B, Wada M, Zhu Y, Tomaru M, Maruyama W, Kikuchi A, Watanabe Y, Kimura I, Hibino H (1998). The P2 protein of rice dwarf phytoreovirus is required for adsorption of the virus to cells of the insect vector. *Journal of Virology* 72: 9370-9373.

Owen J, Shintaku M, Aeschleman P, Tahar SB, Palukaitis P (1990). Nucleotide sequence and evolutionary relationships of cucumber mosaic virus (CMV) strains: CMV RNA3. *Journal of General Virology* 71: 2243-2249.

Palukaitis P, Roosink MJ, Dietzgen RG, Francki RIB (1992). Cucumber mosaic virus. *Advances in Virus Research* 41: 281-348.

Peiffer ML, Gildow FE, Gray SM (1997). Two distinct mechanisms regulate luteovirus transmission efficiency and specificity at the aphid salivary gland. *Journal of General Virology* 78: 495-503.

Peng YH, Kadoury D, Gal-On A, Huet H, Wang Y, Raccah B (1998). Mutations in the HC-Pro gene of zucchini yellow mosaic potyvirus: Effects on aphid transmission and binding to purified virions. *Journal of General Virology* 79: 897-904.

Perry KL, Zhang L, Palukaitis P (1998). Amino acid changes in the coat protein of cucumber mosaic virus differentially affect transmission by the aphids *Myzus persicae* and *Aphis gossypii*. *Virology* 242: 204-210.

Perry KL, Zhang L, Shintaku MH, Palukaitis P (1994). Mapping determinants in CMV for transmission by *Aphis gossypii*. *Virology* 205: 591-595.

Pirone TP (1977). Accessory factors in nonpersistent virus transmission. In Harris KF, Maramorosch K (eds.), *Aphids As Virus Vectors*, Academic Press, London, pp. 221-235.

Pirone TP (1981). Efficiency and selectivity of the helper-component-mediated aphid transmission of purified potyviruses. *Phytopathology* 71: 922-924.

Pirone TP, Blanc S (1996). Helper-dependent vector transmission of plant viruses. *Annual Review of Phytopathology* 34: 227-247.

Pirone TP, Thornbury DW (1983). Role of virion and helper component in regulating aphid transmission of tobacco etch virus. *Phytopathology* 73: 872-875.

Plumb RT (1989). Detecting plant viruses in their vectors. In Harris KF (ed.), *Advances in Disease Vector Research*, (Volume 6), Springer-Verlag, Berlin, pp. 191-209.

Polston JE, Al-Musa A, Perring TM, Dodds JA (1990). Association of the nucleic acid of squash leaf curl virus with the whitefly *Bemisia tabaci*. *Phytopathology* 80: 850-856.

Pringle CR (1999). Virus taxonomy—1999. The universal system of virus taxonomy, updated to include the new proposals ratified by the International Committee on Taxonomy of Viruses during 1998. *Archives of Virology* 144: 421-429.

Prins M, Goldbach R (1998). The emerging problem of tospovirus infection and nonconventional methods of control. *Trends in Microbiology* 6: 31-35.

Querci M, Owens RA, Bartolini I, Lazarte V, Salazar LF (1997). Evidence for heterologous encapsidation of potato spindle tuber viroid in particles of potato leafroll virus. *Journal of General Virology* 78: 1207-1211.

Raccah B, Roistacher CN, Barbagallo S (1989). Semipersistent transmission of viruses by vectors with special emphasis on citrus tristeza virus. In Harris KF (ed.), *Advances in Disease Vector Research* (Volume 6), Springer-Verlag, Berlin, pp. 301-340.

Rivera C, Gamez R (1986). Multiplication of maize rayado fino virus in the leafhopper vector *Dalbulus maidis*. *Intervirology* 25: 76-82.

Robinson DJ, Ryabov EV, Raj SK, Roberts IM, Taliansky ME (1999). Satellite RNA is essential for encapsidation of groundnut rosette umbravirus RNA by groundnut rosette assistor luteovirus coat protein. *Virology* 254: 105-114.

Rodriguez D, Lopez-Abella D, Diaz-Ruiz JR (1987). Viroplasms of an aphid-transmissible isolate of cauliflower mosaic virus contain helper component activity. *Journal of General Virology* 68: 2063-2067.

Rubinstein G, Czosnek H (1997). Long-term association of tomato yellow leaf curl virus with its whitefly vector *Bemisia tabaci*: Effect on the insect transmission capacity, longevity and fecundity. *Journal of General Virology* 78: 2683-2689.

Sako N, Ogata K (1981). Different helper factors associated with aphid transmission of some potyviruses. *Virology* 112: 762-765.

Salomon R, Bernardi F (1995). Inhibition of viral aphid transmission by the N-terminus of the maize dwarf mosic virus coat protein. *Virology* 213: 676-679.

Schmidt I, Blanc S, Esperandieu P, Kuhl G, Devanchelle G, Louis C, Cerutti M (1994). Interaction between the aphid transmission factor and virus particles is a part of the molecular mechanism of CaMV aphid transmission. *Proceedings of the National Academy of Sciences, USA* 91: 8885-8889.

Sether DM, Ullman DE, Hu JS (1998). Transmission of pineapple mealybug wilt-associated virus by two species of mealybug (*Dysmicoccus* spp.). *Phytopathology* 88: 1224-1230.

Shintaku MH (1991). Coat protein gene sequences of two cucumber mosaic virus strains reveal a single amino acid change correlating with chlorosis induction. *Journal of General Virology* 72: 2587-2589.

Suga H, Uyeda I, Yan J, Murao K, Kimura I, Tiongco ER, Cabautan P, Koganezawa H (1995). Heterogeneity of rice ragged stunt oryzavirus genome segment 9 and its segregation by insect vector transmission. *Archives of Virology* 140: 1503-1509.

Suzuki N, Sugawara M, Kusano T, Mori H, Matsuura Y (1994). Immunodetection of rice dwarf phytoreoviral in both insect and plant hosts. *Virology* 202: 41-48.

Tamada T, Kusume T (1991). Evidence that the 75 K readthrough protein of beet necrotic yellow vein virus RNA2 is essential for transmission by the fungus *Polymyxa betae*. *Journal of General Virology* 72: 1497-1504.

Thornbury DW, Hellmann GM, Rhoads RE, Pirone TP (1985). Purification and characterization of potyvirus helper component. *Virology* 144: 260-267.

Thornbury DW, Patterson CA, Dessens JT, Pirone TP (1990). Comparative sequence of the helper component (HC) region of potato virus Y and a HC-defective strain, potato virus C. *Virology* 178: 573-578.

Thornbury DW, Van den Heuvel JFJM, Lesnaw JA, Pirone TP (1993). Expression of potyvirus proteins in insect cells infected with a recombinant baculovirus. *Journal of General Virology* 74: 2731-2735.

Tjallingii WF (1985). Membrane potential as an indication for plant cell penetration by aphid stylets. *Entomologia Experimentalis et Applicata* 38: 187-193.

Tomaru M, Maruyama W, Kikuchi A, Yan J, Zhu Y, Suzuki N, Isogai M, Oguma Y, Kimura I, Omura T (1997). The loss of outer capsid protein P2 results in nontransmissibility by the insect vector of rice dwarf phytoreovirus. *Journal of Virology* 71: 8019-8023.

Ullman DE, German TL, Sherwood JL, Westcot DM, Cantone FA (1993). Tospovirus replication in insect vector cells: Immunocytochemical evidence that the nonstructural protein encoded by the S RNA of tomato spotted wilt tospovirus is present in thrips vector cells. *Phytopathology* 83: 456-463.

Ullman DE, Westcot DM, Chenault KD, Sherwood JL, German TL, Bandla MD, Cantone FA, Duer HL (1995). Compartmentalization, intracellular transport, and autophagy of tomato spotted wilt tospovirus proteins in infected thrips cells. *Phytopathology* 85: 644-654.

Van den Heuvel JFJM, Bruyere A, Hogenhout SA, Ziegler-Graff V, Brault V, Verbeek M, Van der Wilk F, Richards K (1997). The N-terminal region of the luteovirus readthrough domain determines virus binding to *Buchnera* GroEL and is essential for virus persistence in the aphid. *Journal of Virology* 71: 7258-7265.

Van den Heuvel JFJM, Hogenhout SA, Van der Wilk F (1999). Recognition and receptors in virus transmission by arthropods. *Trends in Microbiology* 7: 71-76.

Van den Heuvel JFJM, Verbeek M, Peters D (1993). The relationship between aphid-transmissibility of potato leafroll virus and surface epitopes of the viral capsid. *Phytopathology* 83: 1125-1129.

Van den Heuvel JFJM, Verbeek M, Van der Wilk F (1994). Endosymbiotic bacteria associated with circulative transmission of potato leafroll virus by *Myzus persicae*. *Journal of General Virology* 75: 2559-2565.

Walker HL, Pirone TP (1972). Particle numbers associated with mechanical and aphid transmission of some viruses. *Phytopathology* 82: 1283-1288.

Wang JY, Chay C, Gildow FE, Gray SM (1995). Readthrough protein associated with virions of barley yellow dwarf luteovirus and its potential role in regulating the efficiency of aphid transmission. *Virology* 206: 954-962.

Wang RY, Ammar ED, Thornbury DW, Lopez-Moya JJ, Pirone TP (1996). Loss of potyvirus transmissibility and helper-component activity correlate with non-retention of virions in aphid stylets. *Journal of General Virology* 77: 861-867.

Wang RY, Gergerich RC, Kim KS (1992). Noncirculative transmission of plant viruses by leaf feeding beetles. *Phytopathology* 82: 946-950.

Wang RY, Gergerich RC, Kim KS (1994a). Entry of ingested plant viruses into the hemocoel of the beetle vector *Diabrotica undecipunctata howardii*. *Phytopathology* 84: 147-152.

Wang RY, Gergerich RC, Kim KS (1994b). The relationship between feeding and virus retention time in beetle transmission of plant viruses. *Phytopathology* 84: 995-998.

Wang RY, Pirone TP (1996a). Mineral oil interferes with retention of tobacco etch potyvirus in the stylets of *Myzus persicae*. *Phytopathology* 86: 820-823.

Wang RY, Pirone TP (1996b). Potyvirus transmission is not increased by pre-acquisition fasting of aphids reared on artificial diet. *Journal of General Virology* 77: 3145-3148.

Wang RY, Powell G, Hardie J, Pirone TP (1998). Role of the helper component in vector-specific transmission of potyviruses. *Journal of General Virology* 79: 1519-1524.

Wayadande AC, Nault LR (1993). Leafhopper probing behavior associated with maize chlorotic dwarf virus transmission to maize. *Phytopathology* 83: 522-526.

Wijkamp I, Peters D (1993). Determination of the median latent period of two tospoviruses in *Frankliniella occidentalis*, using a novel leaf disk assay. *Phytopathology* 83: 986-991.

Wisler GC, Duffus JE, Liu H-Y, Li RH (1997). Ecology and epidemiology of whitefly-transmitted closteroviruses. *Plant Disease* 82: 270-280.

Woolston CJ, Covey SN, Penswick JR, Davies JW (1983). Aphid transmission and a polypeptide are specified by a defined region of the cauliflower mosaic virus genome. *Gene* 23: 15-23.

Woolston CJ, Czaplewski LG, Markham PG, Goad AS, Hull R, Davies JW (1987). Location and sequence of cauliflower mosaic virus gene 2 responsible for aphid transmissibility. *Virology* 160: 246-251.

Wu ZC, Hu JS, Polston JE, Ullman DE, Hiebert, E (1996). Complete nucleotide sequence of a nonvector-transmissible strain of Abutilon mosaic geminivirus in Hawaii. *Phytopathology* 86: 608-613.

Yan J, Tomaru M, Takahashi A, Kimura I, Hibino H, Omura T (1996). P2 protein encoded by genome segment S2 of rice dwarf phytoreovirus is essential for virus infection. *Virology* 224: 539-541.

Zeidan M, Czosnek H (1991). Acquisition of tomato yellow leaf curl virus by the whitefly *Bemisia tabaci*. *Journal of General Virology* 72: 2607-2614.

Zeyen RJ, Berger PH (1990). Is the concept of short retention times for aphid-borne nonpersistent plant viruses sound? *Phytopathology* 80: 769-771.

Ziegler-Graff V, Brault V, Mutterer JD, Simonis M-T, Herrbach E, Guilley H, Richards KE, Jonard G (1996). The coat protein of beet western yellows luteovirus is essential for systemic infection but the viral gene products P29 and P19 are dispensable for systemic infection and aphid transmission. *Molecular Plant-Microbe Interactions* 9: 501-510.

Chapter 4

Characteristic Features
of Virus Transmission by Nematodes

Jeanne Dijkstra
Jawaid A. Khan

INTRODUCTION

Since the discovery of a nematode as the vector of *Grapevine fanleaf virus* in California (Hewitt et al., 1958), a large number of economically important viruses have been shown to be transmitted by ectoparasitic soil-inhabiting nematodes (Brown et al., 1995).

Nematode-borne virus diseases in a crop are usually characterized by patches of diseased plants in the field that slowly increase in size.

In general, a high degree of specificity of association exists between vector nematodes and their viruses (Brown et al., 1995, 1996; Taylor and Brown, 1997).

Very often no correlation is found between the presence of virus in a field and numbers of potential vector nematodes (Brown et al., 1990; Ploeg et al., 1996). This lack of correlation may be due to great variation in the proportion of viruliferous nematodes within a population (Ploeg et al., 1992b); soil factors, such as moisture and texture, that may affect the mobility of the nematodes through soil (Van Hoof, 1976); the availability of virus sources; and variability among virus isolates.

With a few exceptions, all instances of virus transmission by nematodes have been described from Europe and North America.

VECTOR NEMATODES

The virus-transmitting nematodes belong to two families, Trichodoridae (order Triplonchida) and Longidoridae (order Dorylaimida), with species in

The authors wish to thank Professor John F. Bol for critical review of the manuscript.

the genera *Paratrichodorus,* and *Trichodorus* (trichodorids), and *Longidorus, Paralongidorus,* and *Xiphinema* (longidorids), respectively.

Longidorids are slender nematodes, 2 to 12 mm long, with relatively long (60 to 250 μm) hollow stylets (Brown et al., 1995). The stylets consist of an odontostyle, surrounded by a stylet guide sheath, for penetration of root tip cells as deep as the vascular cylinder, and a stylet extension, the odontophore, with nerve tissues and protractor muscles. The odontophore passes into the esophagus and the esophageal bulb, containing large gland cells that secrete saliva (Harrison, Murant, et al., 1974; Brown et al., 1995). In the feeding process, the stylet is inserted and after salivation the cytoplasm of penetrated cells is ingested.

Trichodorids are smaller nematodes (0.1 to 1 mm long) with a slightly curved stylet, the onchiostyle, 20 to 80 μm long, implanted on the dorsal wall of the pharynx. Trichodorids usually feed on epidermal cells by pressing their lips against the cell wall that is torn by the stylet so that the cell contents can be sucked in. Subsequently, the food passes through the pharynx and esophagus into the gut.

In general, nematodes have four larval stages, with *X. americanum* being the exception in having only three (Brown et al, 1995). At molting, the cuticular lining of the alimentary canal is shed and replaced.

The natural distribution of *Longidorus* and *Trichodorus* spp. depends mainly on climate. Most *Xiphinema* spp. are found in the tropics and the Mediterranean, with their numbers decreasing toward the north. In contrast, the number of *Trichodorus* and *Paratrichodorus* spp. tends to decrease from north to south (Dijkstra and De Jager, 1998). Soil type is another important factor that plays a role in the distribution of some longidorids and trichodorids (the latter prefer light sandy or light loamy soils). The vertical distribution of *Longidorus* and *Trichodorus* spp. shows great variation. *Longidorus* spp. prefer surface-rooted hosts; hence, most of them live in the upper soil layers, about 20 cm deep (Taylor, 1967). In contrast, *Xiphinema* spp. are present in large numbers around deep-rooted host plants at depths varying from 20 cm to a couple of meters, depending on the type of soil (Taylor, 1972). Usually, nematodes move to deeper layers in the soil during dry or very cold periods.

VIRUSES TRANSMITTED BY NEMATODES

Nematode-transmitted viruses belong to two genera, *Nepovirus* (family *Comoviridae*) and *Tobravirus.* In contrast to tobraviruses, not all nepoviruses are transmitted by nematodes. Nepoviruses have longidorids as their vectors, whereas tobraviruses are transmitted by trichodorids.

Nepoviruses and their vectors have worldwide distribution and a wide host range. Viruses belonging to this group have polyhedral particles and a bipartite single-stranded positive-sense RNA genome. The larger genome segment, RNA1, contains the genes for replication, host range, symptom expression, and seed transmissibility. The smaller segment, RNA2, contains the coat protein (CP) gene and genes for symptom expression, cell-to-cell transport, and nematode transmissibility (Harrison, Murant, et al., 1974; Harrison, Robertson, and Taylor, 1974; Murant, 1981; Matthews, 1991). Both RNA1 and RNA2 encode a polyprotein that is proteolytically processed into functional proteins.

Tobraviruses and their vector trichodorids are predominant in Europe, North America, Brazil, and Japan. Viruses in the genus *Tobravirus, Tobacco rattle virus* (TRV), *Pea early browning virus* (PEBV), and *Pepper ringspot virus* (PRV), are transmitted by nematodes (Brown et al., 1989b). Tobraviruses have tubular particles and, similar to nepoviruses, a bipartite single-stranded positive-sense RNA genome (see Figure 4.1). The RNA1, which is highly similar between isolates belonging to the same species, encodes four nonstructural proteins: two proteins (134 to 141 kDa and 194 to 201 kDa) involved in RNA replication, one protein (29 to 0 kDa) involved in cell-to-cell movement of the viral RNA, and one protein (12 to 16 kDa) that, in the case of PEBV, has been implicated in seed transmission (Hamilton et al., 1987; MacFarlane et al., 1989; Visser and Bol, 1999; see Chapter 6 in this book). The last two proteins are translated from subgenomic RNAs. As expected, RNA1 can replicate and spread on its own, but such infections are unstable because no virions are formed (so-called nonmultiplying [NM] viruses, as opposed to normal multiplying [M] ones).

The RNA2, which is highly variable in length and nucleotide sequence, encodes the CP (Bergh et al., 1985; Angenent et al., 1986, 1989) and additional proteins with a putative role in nematode transmission, their number being dependent on the length of RNA2 (Hernandez et al., 1995). All proteins are translated from subgenomic RNAs.

Tobacco rattle virus has a very extensive host range, and many serological and symptomatological variants have been described (see, for example, Brown et al., 1989a,b).

Pseudorecombinants obtained by mixing RNA2 from one isolate with RNA1 from another isolate were infectious, provided the isolates were strains of the same virus (Robinson and Harrison, 1985). Besides these laboratory-produced pseudorecombinants, evidence suggests that some isolates belonging to two different viruses can undergo recombination in doubly infected plants. Such recombination in nature has been described for TRV isolates in which part of the RNA2 segment was replaced by sequences from

FIGURE 4.1. Organization and Expression of the Genome of *Tobacco Rattle Virus*

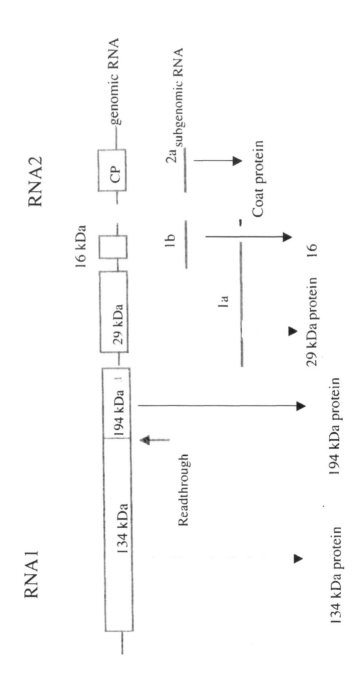

PEBV (Robinson et al., 1987). Symptoms caused by the recombinant virus were those of TRV but the virus was serologically related to PEBV.

MECHANISM OF VIRUS TRANSMISSION

For successful virus transmission by vector nematodes, virus particles must be ingested from an infected plant. After ingestion, the particles must first be retained in the vector and later dissociate from the retention site to be introduced into a susceptible plant cell. Nematode transmission resembles semipersistent transmission by insects.

Ingestion of virus is not a specific process, since both nematode-transmissible and nematode-nontransmissible viruses have been found within members of the same nematode species (Harrison, Robertson, and Taylor, 1974). However, to acquire sufficient virus for transmission, the nematode must have access to the source plant for a minimum time. This acquisition access period varies from some minutes to many hours (Das and Raski, 1968). Most of the virus particles thus ingested pass into the gut and are not retained.

Unlike ingestion, retention of virus particles has been found for only virus-vector associations (Harrison, Robertson, and Taylor, 1974). Electron microscopy of ultrathin sections through the alimentary tract of *Longidorus, Paratrichodorus, Trichodorus,* and *Xiphinema* spp. feeding on virus-infected plants revealed retention sites for virus particles. Based on deductions from the size and packing of TRV particles and on results from TaqMan RT-PCR (reverse transcription–polymerase chain reaction) assays for detection of TRV in nematodes, the number of virus particles in individual nematodes was estimated at 100,000 (F. C. Zoon, personal communication). Depending on the nematode genus or, rarely, the species, virus particles were found to be adsorbed to the inner surface of the odontostyle of *L. macrosoma* (Brown et al., 1995), between the odontostyle and the guide sheath of *L. elongatus* (Taylor and Robertson, 1969, 1975), to the cuticular lining of the lumen of the odontophore and the esophagus of *Xiphinema* spp. (McGuire et al., 1970; Taylor and Robertson, 1969, 1975; Wang and Gergerich, 1998), or to the cuticular lining of the lumina of the pharynx and esophagus in *P. pachydermus* and *T. similis* (Brown et al., 1995).

The genetic makeup of the vector nematode determines the retention and subsequent transmission of virus. Surface structures on the cuticle at the site of retention may play an important role in the interaction between the vector, on the one hand, and the virus with its genetic determinants for nematode transmissibility, on the other (Harrison, Robertson, and Taylor, 1974; Brown et al., 1995).

Observations that serologically distinct serotypes of viruses are transmitted by different species of nematodes (Brown et al., 1989a; Ploeg et al., 1991, 1992a,b) and the established role of RNA2 in nematode transmission as revealed by experiments with pseudorecombinants suggest that determinants on the CP are involved in the specific virus-nematode association (Harrison, Robertson, and Taylor, 1974; Ploeg, Robinson, and Brown, 1993).

However, the CP need not be the only protein responsible for nematode transmissibility, as RNA2 of TRV and PEBV encodes also a number of nonstructural proteins (Angenent et al., 1986; Goulden et al., 1990; Hernandez et al., 1995; MacFarlane and Brown, 1995; MacFarlane et al., 1995, 1996).

The RNA2 of TRV isolate PpK20 transmitted by *Paratrichodorus* spp. contains, besides the CP gene, genes for 40 and 32.8 kDa proteins (Hernandez et al., 1995), whereas RNA2 of PEBV isolate TpA56 encodes 29.6, 23, and 9 kDa proteins (MacFarlane and Brown, 1995).

To find out which of these nonstructural proteins might play a role in nematode transmission, the nucleotide compositions of the nematode-transmissible complementary (c) DNA clone of RNA2 of PEBV isolate TpA56 and of the nontransmissible clone SP5 of this virus were compared (MacFarlane and Brown, 1995). It was demonstrated that a difference of two amino acids (aa) in the 29.6 kDa protein is responsible for the lack of nematode transmission.

Results from experiments in which mutations were introduced into infectious cDNA clones of RNA2 of PEBV isolate TpA56 showed that most genes in this RNA2 are, to some extent, involved in vector transmission (MacFarlane et al., 1995,1996; Schmitt et al., 1998). Deletions in the 23 kDa gene and some mutations in the 9 kDa open reading frame (ORF) led to a reduction in the frequency of transmission but not to its complete loss.

When an infectious cDNA clone of RNA2 of TRV isolate PpK20 was used for construction of deletion mutants, the 40 kDa, but not the 32.8 kDa, protein was shown to be required for transmission by *P. pachydermus* (Hernandez et al., 1997). It was suggested that the latter protein might specifically interact with other species of trichodorids.

By comparing the genome organization of a new isolate of TRV (TpO1) with that of TRV isolate PpK20 and PEBV isolate TpA56, it was concluded that nonstructural proteins are important for the specificity of nematode transmission (MacFarlane et al., 1999).

Nothing is known about the actual mechanism of virus retention. Specific staining for carbohydrates at the site of virus retention in the alimentary canal of the vector nematode suggests involvement of these compounds in linking virus particles to surface structures on the cuticle of the esophagus (Robertson and Hendry, 1986; Brown et al., 1995).

Studies on the architecture of tobraviruses have revealed the presence of long, protruding C-terminal peptides consisting of 29 aa in PEBV-TpA56, 22 aa in TRV-PpK20, and 17 aa in TRV-TpO1 (Mayo et al., 1993; MacFarlane et al., 1999; Visser and Bol, 1999). Such surface-located peptides form mobile structures that may be involved in retention of tobraviruses (Mayo et al., 1994; Legorburu et al., 1995). To investigate a possible role of these peptides in vector transmission, mutations were made in the C-terminal flexible domain of the PEBV-TpA56 CP gene (MacFarlane et al., 1996). Mutant CP1 lacking the first 13 aa of the 29 aa mobile segment was less readily transmitted by the vector than wild-type virus, showing that this part is not indispensable for transmission, although it has some effect on the latter. In contrast, mutant CP2 lacking 15 aa from the distal part of the flexible domain was not transmitted by the vector, indicating that this part plays an essential role in vector transmission. Recently, the C-terminal domain of the CP of TRV-PpK20 has been shown to interact with the 40 and 32.8 kDa proteins of this virus (Visser and Bol, 1999). No interaction between the CP and the 40 kDa protein was observed after deletion of the C-terminal 19 aa of the CP.

Nonstructural viral proteins essential for transmission by vectors have already been described for many viruses, e.g., the helper component proteinase (HC-Pro) of potyviruses that is required for their nonpersistent transmission by aphids (e.g., Pirone and Blanc, 1996; Van den Heuvel et al., 1999). These helper proteins are thought to possess two functional domains: one for interaction with the CP and the other for linking the protein to a receptor site in the food canal of the maxillary stylets or in the foregut of the aphid.

The nonstructural RNA2-encoded 40 kDa protein of PpK20 and, possibly, the three proteins of PEBV-TpA56 may play a similar helper role in nematode transmission by forming a bridge between the virus particle and the lining of the esophagus (Brown et al., 1995; Hernandez et al., 1997; Visser and Bol, 1999). Such a protein complex might then combine with the aforementioned carbohydrates, which are probably of nematode origin (Robertson and Hendry, 1986). Reactions between proteins and carbohydrates are known to occur with antibodies and lectins (plant proteins with combining groups that react with carbohydrates). The helper proteins may have properties comparable to those of lectins.

In contrast to tobraviruses, much less is known about factors that play a role in vector transmission of nepoviruses. On the basis of amino acid sequence similarities in the region of the RNA2 polyprotein situated near the N-terminal side of the CP, four nepoviruses have been placed into two groups: *Raspberry ringspot virus* and *Tomato black ring virus* in one group, and *Grapevine fanleaf virus* and *Tomato ringspot virus* in another (Blok et al.,

1992). Remarkably, viruses of the former group are transmitted by *Longidorus* spp., whereas those of the latter group have *Xiphinema* spp. as their vectors. The polypeptides near the N-terminal side of the CP may therefore play a role in vector transmission. However, the possibility cannot be excluded that RNA2-encoded nonstructural proteins are also involved.

After feeding on the infected plant, the nematode withdraws its stylet. When such a viruliferous nematode starts feeding on another plant, virus particles dissociate from the site of retention through saliva, and are introduced into the punctured cells. In contrast to the CP and the 29.6 kDa protein of PEBV, which are thought to be essential for retention of virions in the vector, the 23 kDa protein, on the other hand, may have a role in releasing virions from the mouthparts of the nematode during feeding (Schmitt et al., 1998). Carbohydrates may also be involved in dissociation of particles from the site of retention (Brown et al., 1995).

To introduce sufficient virus into the plant, the nematode must have access to the plant for a minimum time (the inoculation access period). Inoculation access periods ranging from 15 min to 1 h have been reported (Das and Raski, 1968).

CONTROL OF NEMATODE-TRANSMITTED VIRUSES

The usual practice for control of nematode-transmitted viruses, in particular tobraviruses, is the application of toxic chemicals to the soil. Such chemicals may either immobilize the nematodes temporarily (nematostats, for instance, oxime-carbamates) or kill them (nematicides, for instance, dichloropropane-dichloropropene [D-D]). However, it is now realized that those chemicals constitute a great hazard to the environment; thus, alternative control measures are needed.

Generally applied control measures such as crop rotation, leaving land fallow, and breeding for resistance have not been successful for the following reasons: (1) many nematode-borne viruses possess extensive host ranges (including many weed species); (2) these viruses are very persistent within their vectors; (3) usually the nematode can survive in the soil for a long time without food; (4) the virus isolates are highly variable.

Currently one possibility being explored is to check nematode transmission by using transgenic plants. Efforts to introduce engineered resistance by transforming *Nicotiana tabacum* plants with the CP gene of TRV failed, as the transgenic plants became infected when the virus was introduced by its natural vector, *P. pachydermus*, even though the plants were resistant upon mechanical inoculation with the virus (Ploeg, Mathis, et al., 1993). Transgenic plants expressing the CP of some nepoviruses, however, did not become infected when nematodes carrying these viruses fed on those plants (Brown et al., 1995).

In other efforts to achieve engineered resistance, plants were transformed with deletion mutants of RNA2 of TRV. It is well known that virus maintenance by mechanical transmission leads to loss of the ability of the virus to be transmitted by vectors. Isolates of TRV transmitted in this manner not only lost their vector transmissibility but also showed a reduction in the length of their short particles (Lister and Bracker, 1969), possibly due to deletions in their RNA2. Analyses of the RNA2 of deletion mutants that had arisen in the TRV isolate PpK20 during repeated mechanical transmission showed a plethora of defective RNA2s and even some recombinants possessing sequences of both RNA2 and RNA1 (Hernandez et al., 1996). Among the deletion mutants was one that encoded a C-terminally truncated CP with encapsidating capacity but none of the nonstructural proteins. This mutant RNA2 showed characteristics of a defective interfering (DI) RNA, as it strongly interfered with the replication of wild-type RNA2. Since such DI RNAs were thought to be useful tools for obtaining engineered resistance, tobacco plants were transformed with cDNA of a DI RNA (DI7) (Visser et al., 1999). When leaves of the transgenic tobacco plants were inoculated with the nematode-transmissible TRV isolate PpK20, the majority of RNA2 in the roots consisted of DI7 RNA. Nematodes that fed on the roots transmitted the virus, but only the one with wild-type RNA2. Most likely, the virus particles encapsidated in truncated CP from DI7 RNA lacked the recognition site needed for retention by nematodes. From these results it was clear that control of TRV transmission by nematodes under field conditions could not be achieved by transformation of plants with DI7.

Effective control requires still more knowledge of the mechanisms governing virus-vector-plant associations.

REFERENCES

Angenent GC, Linthorst HJM, Van Belkum AF, Cornelissen BJC, Bol JF (1986). RNA2 of tobacco rattle virus strain TCM encodes an unexpected gene. *Nucleic Acids Research* 14: 4673-4682.

Angenent GC, Posthumus E, Brederode FT, Bol JF (1989). Genome structure of tobacco rattle virus strain PLB: Further evidence on the occurrence of RNA recombination among tobraviruses. *Virology* 171: 271-274.

Bergh ST, Koziel MG, Huang S-C, Thomas RA, Gilley DP, Siegel A (1985). The nucleotide sequence of tobacco rattle virus RNA2 (CAM strain). *Nucleic Acids Research* 13: 8507-8518.

Blok VC, Wardell J, Jolly CA, Manoukian A, Robinson DJ, Edwards ML, Mayo MA (1992). The nucleotide sequence of RNA-2 of raspberry ringspot nepovirus. *Journal of General Virology* 73: 2189-2194.

Brown DJF, Boag B, Jones AT, Topham PB (1990). An assessment of the soil-sampling density and spatial distribution required to detect viruliferous nematodes (Nematoda: Longidoridae and Trichodoridae) in fields. *Nematologia Mediterranea* 18: 153-160.

Brown DJF, Ploeg AT, Robinson DJ (1989a). The association between serotypes of tobraviruses and *Trichodorus* and *Paratrichodorus* species. *OEPP/EPPO Bulletin* 19: 611-617.

Brown DJF, Ploeg AT, Robinson DJ (1989b). A review of reported associations between *Trichodorus* and *Paratrichodorus* species (Nematoda, Trichodoridae) and tobraviruses with a description of laboratory methods for examining virus transmission by nematodes. *Revue de Nématologie* 12: 235-241.

Brown DJF, Robertson WM, Trudgill DL (1995). Transmission of viruses by plant nematodes. *Annual Review of Phytopathology* 33: 223-249.

Brown DJF, Trudgill DL, Robertson WM (1996). Nepoviruses: Transmission by nematodes. In Harrison BD, Murant AF (eds.). *The Plant Viruses* (Volume 5), *Polyhedral Virions and Bipartite RNA Genomes*, Plenum Press, New York, pp. 187-209.

Das S, Raski DJ (1968). Vector efficiency of *Xiphinema index* in the transmission of grapevine fanleaf virus. *Nematologica* 14: 55-62.

Dijkstra J, De Jager CP (1998). *Practical Plant Virology: Protocols and Exercises*. Springer-Verlag, Berlin, Heidelberg, New York, 459 pp.

Goulden MG, Lomonosoff CP, Davies JW, Wood KR (1990). The complete nucleotide sequence of PEBV RNA2 reveals the presence of a novel open reading frame and provides insight into the structure of tobraviral subgenomic promoters. *Nucleic Acids Research* 18: 4507-4512.

Hamilton WDO, Boccara M, Robinson DJ, Baulcombe DC (1987). The complete nucleotide sequence of tobacco rattle virus RNA-1. *Journal of General Virology* 68: 2563-2575.

Harrison BD, Murant AF, Mayo MA, Roberts IM (1974). Distribution and determinants for symptom production, host range and nematode transmissibility between the two RNA components of raspberry ringspot virus. *Journal of General Virology* 22: 233-247.

Harrison BD, Robertson WM, Taylor CE (1974). Specificity of retention and transmission by nematodes. *Journal of Nematology* 6: 155-164.

Hernandez C, Carette JE, Brown DJF, Bol JF (1996). Serial passage of tobacco rattle virus under different selection conditions results in deletion of structural and nonstructural genes in RNA2. *Journal of Virology* 70: 4933-4940.

Hernandez C, Mathis A, Brown DJF, Bol JF (1995). Sequence of RNA2 of a nematode-transmissible isolate of tobacco rattle virus. *Journal of General Virology* 76: 2847-2851.

Hernandez C, Visser PB, Brown DJF, Bol JF (1997). Transmission of tobacco rattle virus isolate PpK20 by its nematode vector requires one of the two non-structural genes in the viral RNA2. *Journal of General Virology* 78: 465-467.

Hewitt WB, Raski DJ, Goheen AC (1958). Nematode vector of soil-borne fan leaf virus of grapevine. *Phytopathology* 48: 586-595.

Legorburu FJ, Robinson DJ, Torrance L, Duncan GH (1995). Antigenic analysis of nematode-transmissible and non-transmissible isolates of tobacco rattle tobravirus using monoclonal antibodies. *Journal of General Virology* 76: 1497-1501.

Lister RM, Bracker CE (1969). Defectiveness and dependence in three related strains of tobacco rattle virus. *Virology* 37: 262-271.

MacFarlane SA, Brown DJF (1995) Sequence comparison of RNA2 of nematode-transmissible and nematode-non-transmissible isolates of pea early-browning virus suggests that the gene encoding the 29 kDa protein may be involved in nematode transmission. *Journal of General Virology* 76: 1299-1304.

MacFarlane SA, Brown DJF, Bol JF (1995). The transmission by nematodes of tobraviruses is not determined exclusively by the virus coat protein. *European Journal of Plant Pathology* 101: 535-539.

MacFarlane SA, Taylor SC, King DI, Hugues G, Davies JW (1989). Pea early browning virus RNA11 encodes four polypeptides including a putative zinc-finger protein. *Nucleic Acids Research* 17: 2245-2260

MacFarlane SA, Vassilakos N, Brown DJF (1999). Similarities in the genome organization of tobacco rattle virus and pea early-browning virus isolates that are transmitted by the same vector nematode. *Journal of General Virology* 80: 273-276.

MacFarlane SA, Wallis CV, Brown DJF (1996). Multiple genes involved in the nematode transmission of pea early browning virus. *Virology* 219: 417-422.

Matthews REF (1991). *Plant Virology* (Third Edition). Academic Press, San Diego, London. 835 pp.

Mayo MA, Brierly KM, Goodman BA (1993). Developments in the understanding of the particle structure of tobraviruses. *Biochemie* 75: 639-644.

Mayo M, Robertson WM, Legorburu FJ, Brierley KM (1994). Molecular approaches to an understanding of the transmission of plant viruses by nematodes. In Lamberti F, De Giorgi C, Bird DMcK (eds.), *Advances in Molecular Nematology*. Plenum Press, New York, pp. 277-293.

McGuire JM, Kim KS, Douthit LB (1970). Tobacco ringspot virus in the nematode *Xiphinema americanum*. *Virology* 42: 212-216.

Murant AF (1981). Nepoviruses. In Kurstak E (ed.). *Handbook of Plant Virus Infections and Comparative Diagnosis*, Elsevier/North Holland, Amsterdam, pp. 197-238.

Pirone TP, Blanc S (1996). Helper-dependent vector transmission of plant viruses. *Annual Review of Phytopathology* 34: 227-247.

Ploeg AT, Asjes CJ, Brown DJF (1991). Tobacco rattle virus sero types and associated nematode vector species of Trichodoridae in the bulb-growing areas in the Netherlands. *Netherlands Journal of Plant Pathology* 97: 311-319.

Ploeg AT, Brown DJF, Robinson DJ (1992a). Acquisition and subsequent transmission of tobacco rattle virus isolates by *Paratrichodorus* and *Trichodorus* nematode species. *Netherlands Journal of Plant Pathology* 98: 291-300.

Ploeg AT, Brown DJF, Robinson DJ (1992b). The association between species of *Trichodorus* and *Paratrichodorus* vector nematodes and serotypes of tobacco rattle tobravirus. *Annals of Applied Biology* 121: 619-630.

Ploeg AT, Mathis A, Bol JF, Brown DJF, Robinson DJ (1993). Susceptibility of transgenic tobacco plants expressing tobacco rattle virus coat protein to nematode-transmitted and mechanically inoculated tobacco rattle virus. *Journal of General Virology* 74: 2709-2715.

Ploeg AT, Robinson DJ, Brown DJF (1993). RNA-2 of tobacco rattle virus encodes the determinants of transmissibility by trichodorid vector nematodes. *Journal of General Virology* 74: 1463-1466.

Ploeg AT, Zoon FC, De Bree J, Asjes CJ (1996). Analysis of the occurrence and distribution of tobacco rattle virus in field soil and disease in a subsequent tulip crop. *Annals of Applied Biology* 129: 461-469.

Robertson WM, Hendry CE (1986). An association of carbohydrate with particles of arabis mosaic virus retained within *Xiphinema diversicaudatum*. *Annals of Applied Biology* 109: 299-305.

Robinson DJ, Hamilton WDO, Harrison BD, Baulcombe DC (1987). Two anomalous tobravirus isolates. Evidence for RNA recombination in nature. *Journal of General Virology* 68: 2551-2561.

Robinson DJ, Harrison BD (1985). Unequal variation in the two genome parts of tobraviruses and evidence for the existence of three separate viruses. *Journal of General Virology* 66: 171-176.

Schmitt C, Mueller A-M, Mooney A, Brown D, MacFarlane S (1998). Immunological detection and mutational analysis of the RNA2-encoded nematode transmission protein of pea early browning virus. *Journal of General Virology* 79: 1281-1288.

Taylor CE (1967). The multiplication of *Longidorus elongatus* (de Man) on different host plants with reference to virus transmission. *Annals of Applied Biology* 59: 275-281.

Taylor CE (1972). Transmission of viruses by nematodes. In Kado EI, Agrawal HO (eds.), *Principles and Techniques in Plant Virology*. Van Nostrand Reinhold. New York, pp. 226-247.

Taylor CE, Brown DJF (1997). *Nematode Vectors of Plant Viruses*. CAB International, London, 286 pp.

Taylor CE, Robertson WM (1969). The location of raspberry ringspot virus and tomato black ring viruses in the nematode vector, *Longidorus elongatus* (de Man). *Annals of Applied Biology* 64: 233-237.

Taylor CE, Robertson WM (1975). Acquisition, retention and transmission by nematodes. In Lamberti F, Taylor CE, Seinhorst JW (eds.), *Nematode Vectors of Plant Viruses, NATO Advanced Study Institutes Series A: Life Sciences* (Volume 2) Plenum Press, New York, pp. 253-276.

Van den Heuvel JFJM, Franz AWE, Van der Wilk F (1999). Molecular basis of virus transmission. In Mandahar CL (ed.), *Molecular Biology of Plant Viruses*, Kluwer Academic Publishers, Boston, Dordrecht, London, pp. 183-200.

Van Hoof HA (1976). The effect of soil moisture on the activity of trichodorid nematodes. *Nematologica* 22: 260-264.

Visser PB, Bol JF (1999). Nonstructural proteins of *Tobacco rattle virus* which have a role in nematode-transmission: Expression pattern and interaction with viral coat protein. *Journal of General Virology* 80: 3273-3280.

Visser PB, Brown DJF, Brederode FTh, Bol JF (1999). Nematode transmission of tobacco rattle virus serves as a bottleneck to clear the virus population from defective interfering RNAs. *Virology* 263: 155-165.

Wang S, Gergerich RC (1998). Immunofluorescent localization of tobacco ringspot nepovirus in the vector nematode *Xiphinema americanum*. *Phytopathology* 88: 885-889.

Chapter 5

Virus Transmission by Fungal Vectors

Jeanne Dijkstra
Jawaid A. Khan

INTRODUCTION

The existence of a disease agent with a soil-inhabiting fungus as its vector was demonstrated in 1958, when Fry gave evidence that the fungus *Olpidium brassicae* was involved in transmission of the big-vein disease of lettuce. Since that year, several soil-borne viruses have been shown to be transmitted by fungal vectors (for reviews, see Grogan and Campbell, 1966; Maraite, 1991; Campbell, 1996).

With a few exceptions, the vectors themselves are not serious pathogens, but they derive their importance from being able to transmit economically important viruses, particularly those of cereals and root and tuber crops.

For a long time, not much information was available about the transmission of these viruses by their vectors. The main obstacles to studies of virus-fungus associations are the difficult mechanical transmission of many of these viruses and the fact that the fungal vectors are obligate parasites.

This chapter deals with the association between fungus-borne viruses and their vectors and presents new insights into possible mechanisms of transmission.

FUNGAL VECTORS

The fungi known to be vectors of viruses are obligate endoparasites of plants. All of them form zoospores. They belong to the Chytridiomycota (*Olpidium*

The authors would like to thank Dr. G. Bollen for his critical review, helpful comments, and valuable additions to the mycological part of the manuscript; Dr. S. Bouzoubaa for valuable suggestions, corrections, and additional information on beny- and pecluviruses; and Mr. Nazimuddin (artist, NBRI) for drawing the life cycle of *Polymyxa betae*.

spp.) or the Plasmodiophoromycota (*Polymyxa* spp. and *Spongospora* spp.). Two species of *Olpidium (O. bornovanus = O. radicale = O. cucurbitacearum* [Campbell and Sim, 1994] and *O. brassicae),* two species of *Polymyxa (P. betae* and *P. graminis),* and one species of *Spongospora (S. subterranea)* have been shown to be natural vectors of viruses (Campbell, 1996) (see Table 5.1).

The life cycles of the two categories of fungal vectors have much in common (Adams, 1991; Campbell, 1996) (see Figure 5.1). Thick-walled resting spores are formed inside roots or young tubers of the host plant. With the plasmodiophorids the resting spores are formed in clusters (spore balls or cystosori), whereas the chytrids have single resting spores. When the infected roots or tubers decay in the soil, the spores are released. Depending on the conditions in the soil, resting spores germinate and release motile primary zoospores that move to roots. Zoospores of chytrids are uniflagel-

TABLE 5.1. Viruses with Their Fungal Vectors—*Olpidium (O.), Polymyxa (P.)* and *Spongospora (S.)* Species—and Viral Proteins Involved in Vector Transmission

Virus/sat	Morphology	Vector	Viral Protein
		chytrid	
CNV	icosahedral	*O. bornovanus*	CP
TNV	icosahedral	*O. brassicae*	CP
TNSV	icosahedral	*O. brassicae*	–
		plasmodiophorid	
BMMV	filamentous	*P. graminis*	73 kDa
BYMV	filamentous	*P. graminis*	70 kDa
BNYVV	rod-shaped	*P. betae*	RT (75 kDa)
PCV	rod-shaped	*P. graminis*	39 kDa
PMTV	rod-shaped	*S. subterranea*	RT (67 kDa)
SBWMV	rod-shaped	*P. graminis*	RT (84 kDa)

Note: BMMV = *Barley mild mosaic virus;* BYMV = *Barley yellow mosaic virus;* BNYVV = *Beet necrotic yellow vein virus;* CNV = *Cucumber necrosis virus;* PCV = *Peanut clump virus;* PMTV = *Potato mop-top virus;* SBWMV = *Soil-borne wheat mosaic virus;* TNSV = tobacco necrosis satellite virus; TNV = *Tobacco necrosis virus;* CP = coat protein; RT = readthrough protein; sat = satellite.

FIGURE 5.1. Life Cycle of *Polymyxa betae* and Acquisition of *Beet Necrotic Yellow Vein Virus* (BNYVV) by the Fungus

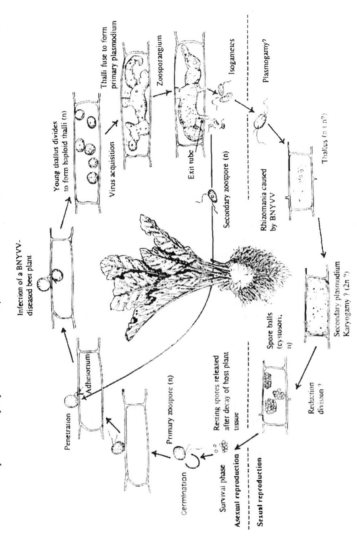

Source: Courtesy of G. J. Bollen. Department of Phytopathology, Wageningen University, The Netherlands.

late and those of plasmodiophorids are biflagellate. The zoospores attach to the root hairs or epidermal cells, often in the zone of elongation (Campbell and Fry, 1966; Temmink, 1971), and enter into the encystment phase. In this process, the flagella are withdrawn and a cyst wall is secreted. Detailed studies of the ultrastructure of zoospores of *O. brassicae* and the life cycle of this fungus have revealed that, upon encystment of the zoospore, the axonema (the fibrillar component of a flagellum) with its axonemal sheath is withdrawn inside the zoospore body (Temmink and Campbell, 1969a,b; Temmink, 1971).

The two types of fungal vectors use different mechanisms for penetration of the host cell. With *Olpidium* spp., belonging to the chytrids, the protoplast of the cyst enters the host through a minute pore dissolved in the wall of the host cell. With the plasmodiophorid fungi, *Polymyxa* spp. and *Spongospora* spp., the wall of the host cell is penetrated by a stylet (*Stachel*, a German term used in the first description of the infection process (see Figure 5.1). As soon as the cyst has settled down on root hairs or epidermal cells of the roots it forms a tube (Ger. *Rohr*), the end of it being pointed at the surface of the host. The tube contains the stachel. Infection proceeds rapidly by evagination of the tube, resulting in a firm attachment to the host with an adhesorium and, subsequently, in puncturing the host wall with the stachel. The stachel is released into the host cell, whereafter the protoplast of the cyst follows. With both types of vectors the protoplast of the fungus moves into the cytoplasm of the host, where the young thallus evolves into a multinucleate primary plasmodium that is enveloped in a thin thallus membrane (immature thallus). The thallus develops into zoosporangia from which the secondary zoospores are released into soil water. With *Olpidium* spp. the zoospores escape from the sporangia through a distinct exit tube penetrating the outer wall of the host cell. In the later part of the cycle, the thallus, now enveloped in a thicker membrane, develops into resting spores or resting sporangia that may remain viable in root debris for a long time (in the case of *P. betae,* even for 20 years). The fungal vectors exhibit considerable host specificity.

Methods and techniques of isolation, culture, maintenance, and use of fungal vectors have been presented by Jones (1993) for chytrids and plasmodiophorids, in general; by Campbell (1988) for *Olpidium* spp.; by Paul et al. (1993) and Dijkstra and De Jager (1998) for *P. betae;* by Adams et al. (1986) for *P. graminis;* and by Arif et al. (1995) for *S. subterranea.*

FUNGUS-TRANSMITTED VIRUSES

By now, transmission by fungi has been established or suggested for about 30 viruses and virus-like agents (for a review, see Campbell, 1996).

According to the current classification of viruses (Pringle, 1999; see Chapter 1 of this book), fungus-borne viruses are found in the genera *Tombusvirus, Carmovirus, Necrovirus,* and *Dianthovirus* of the family *Tombusviridae;* the former genus *Furovirus,* now divided into the four genera *Furovirus, Pomovirus, Pecluvirus,* and *Benyvirus;* the genus *Bymovirus* of the family *Potyviridae;* and a group of unassigned viruses, namely, *Tobacco stunt virus* and, possibly, *Lettuce big-vein virus,* with rod-shaped particles and a double-stranded RNA genome (Mayo, 1995).

The following material presents a selection of important fungus-transmitted viruses, with their characteristics (see Table 5.1).

Cucumber necrosis virus, *Member of the Genus* Tombusvirus

Cucumber necrosis virus (CNV) has icosahedral particles (30 nm in diameter) containing a monopartite positive-sense single-stranded RNA genome.

The virus is transmitted by *O. bornovanus* (Dias, 1970; Campbell and Sim, 1994) to cucumber, one of the few plants in which it becomes systemic (Dias and McKeen, 1972).

The coat protein (CP) of CNV consists of 180 identical subunits. By analogy to the CP of *Tomato bushy stunt virus,* type species of the genus *Tombusvirus,* the CP of CNV is thought to possess three distinct structural domains: an N-terminal RNA-binding (R) domain, a tightly packed shell (S) domain, and an outward-facing protruding (P) domain (Robbins et al., 1997). The R and S domains are connected by an arm (a). The P domain has been shown to be essential for particle assembly and/or stability (Sit et al., 1995).

The genome of CNV has a size of about 4.7 kb and consists of four open reading frames (ORFs) (see Figure 5.2). ORF 1 encodes a 33 kDa protein and, together with ORF 2, a 92 kDa protein resulting from readthrough (RT) of the amber termination codon of ORF 1. ORF 3 codes for a CP of 41 kDa and is expressed through subgenomic (sg) RNA3 of 2.1 kb. The sgRNAs 4 and 5, each of 0.9 kb, code for two smaller polypeptides of 19 and 22 kDa, respectively.

Tobacco necrosis virus, *Type Species of the Genus* Necrovirus

Tobacco necrosis virus (TNV) has icosahedral particles (26 to 30 nm in diameter) containing a monopartite positive-sense single-stranded RNA genome.

Transmission of the virus by *O. brassicae* (Teakle, 1960) was established during the same period as that of lettuce big-vein virus by the same fungus (Grogan et al., 1958).

FIGURE 5.2. Organization and Expression of the Genome of a Tombusvirus

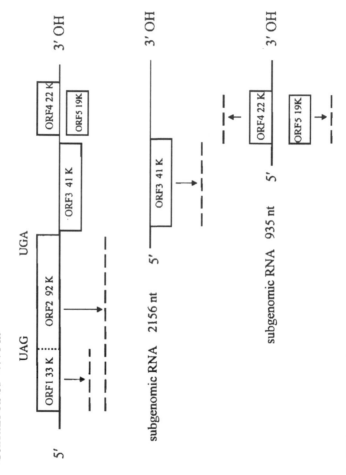

Note: ORF = open reading frame.

Although TNV has a very wide experimental host range, it usually causes necrotic local lesions only. Exceptions are tulip and bean, which are systemically infected and show severe symptoms (tulip necrosis or Augusta disease of tulip [Kassanis, 1949]; stipple-streak of bean [Bawden and Van der Want, 1949]). Although these diseases are of some economic importance, TNV is mainly of academic interest because of its transmission characteristics and its association with tobacco necrosis satellite virus.

The genome of TNV is 3,759 nucleotides (nt) in size and consists of five ORFs (in some isolates a sixth ORF is present) (Lommel, 1995) (see Figure 5.3). ORF 1 encodes a 23 kDa polypeptide, and after RT of its amber termination codon, it expresses, together with ORF 2, a polypeptide of 82 kDa (most likely an RNA-dependent RNA polymerase, RdRp). ORFs 3 and 4, translated from sgRNAs, encode polypeptides of 7.9 and 6.2 kDa, respectively. At present, their functions are not known. The CP, also expressed from an sg messenger, is encoded by ORF 5.

Soil-borne wheat mosaic virus, *Type Species of the Genus* Furovirus

Soil-borne wheat mosaic virus (SBWMV) has rod-shaped virions, about 20 nm in width, with predominant lengths of 92 to 160 nm and 250 to 300 nm (Brunt, 1995). The virions possess a bipartite positive-sense single-stranded RNA genome.

The virus is transmitted by *P. graminis* to wheat and barley, in which it may cause severe diseases, especially in winter wheat (Estes and Brakke, 1966; Brakke, 1971).

In wild-type (wt) isolates of SBWMV, the RNA 1, present in the larger particle, is about 7.0 kb in size, and the RNA2 about 3.6 kb (Brunt, 1995) (see Figure 5.4). The 3'-termini of the RNAs have a transfer (t) RNA-like structure.

RNA1 contains three ORFs. ORF 1 encodes a 150 kDa polypeptide and, together with ORF 2, an RT product of 209 kDa, possibly with RdRp function. ORF 3 expresses a 37 kDa protein that may play a role in cell-to-cell transport.

RNA2 contains two ORFs. The 5'-proximal ORF encodes the capsid protein of 19 kDa, and, as a result of RT translation, an 84 kDa protein is produced. A 3'-proximal ORF encodes a 19 kDa protein of unknown function (Chen et al., 1995).

Potato mop-top virus, *Type Species of the Genus* Pomovirus

Potato mop-top virus (PMTV) has labile rod-shaped particles, 18 to 20 nm in width, and 100 to 150 or 250 to 300 nm in length (Savenkov et al., 1999). The virions contain a tripartite positive-sense single-stranded RNA genome (Harrison and Jones, 1970).

FIGURE 5.3. Organization and Expression of the Genome of *Tobacco necrosis virus*

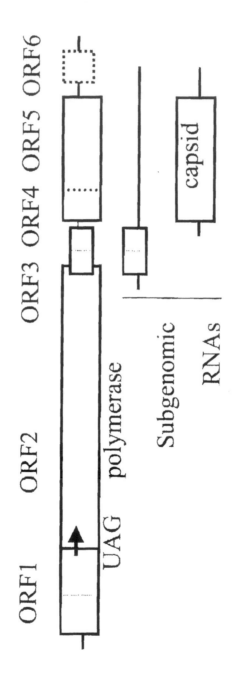

Note: Arrow indicates readthrough of ORF1 amber termination; ORF = open reading frame.

FIGURE 5.4. Organization and Expression of the Genome of *Soil-borne wheat mosaic virus*

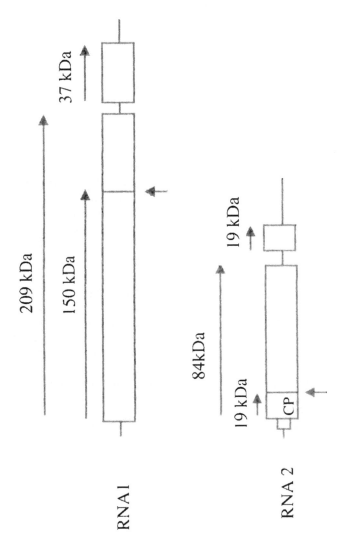

Note: Rectangles = open reading frames; arrows = corresponding translation products; CP = coat protein.

The virus is transmitted by *S. subterranea* (Jones and Harrison, 1969) to potato and has a narrow host range. In contrast to other fungal vectors, *S. subterranea* itself is an important pathogen, as it causes powdery scab of potato. The virus has been reported from potato-growing regions of northern and central Europe, the Andean region, China, Japan, and Canada (Arif et al., 1995).

The distribution of PMTV in potato plants is rather erratic, as presence of the virus can be demonstrated in some stems of the plants but not in others (Torrance et al., 1992). Such an erratic distribution is also found in potato stem-mottle caused by *Tobacco rattle virus*, a tobravirus, and in root and tuber crops infected by some other viruses.

The complete sequences of all three RNAs are now available (Savenkov et al., 1999) (see Figure 5.5).

RNA1 consisting of 6,043 nt has a 5'-terminal ORF (ORF 1) that encodes a polypeptide of 148 kDa and, together with ORF 2, an RT product of 206 kDa, probably with RdRp function. The 3'-untranslated region of RNA1 contains a tRNA-like structure (Savenkov et al., 1999).

RNA2 of 3 kb encodes four polypeptides of 51, 13, 21, and 8 kDa, respectively. The first three are encoded by three overlapping ORFs (triple gene block, TGB). These proteins are found in many virus genera and are involved in cell-to-cell movement of the viruses. The function of the fourth protein is unknown (Scott et al., 1994).

RNA3 of 2.5 kb contains an ORF that encodes the virus CP of 20 kDa and is terminated by an amber codon. A second ORF encodes an RT domain of 47 kDa, resulting in an RT protein of 67 kDa (Kashiwazaki et al., 1995).

Peanut clump virus, *Type Species of the Genus* Pecluvirus

Peanut clump virus (PCV) has rod-shaped virions of two predominant lengths, 190 and 245 nm, with a width of about 21 nm. The virus particles contain a bipartite positive-sense single-stranded RNA genome (Thouvenel and Fauquet, 1981b).

The virus is transmitted by *P. graminis* (Thouvenel and Fauquet, 1981b) and has a wide host range, including monocots and dicots. However, peanut is a noncompatible host for the fungal vector. Consequently, the fungus is unable to acquire virus in a diseased peanut plant. By contrast, *Sorghum* spp. are compatible hosts for both fungus and virus (Thouvenel and Fauquet, 1981a).

RNA1 contains three ORFs encoding proteins of 131, 191 (an RT product of ORF1), and 15 kDa, respectively (see Figure 5.6). The first two proteins constitute the replication RdRp (Herzog et al., 1994), and the last protein is involved in the regulation of replication (Herzog et al., 1998).

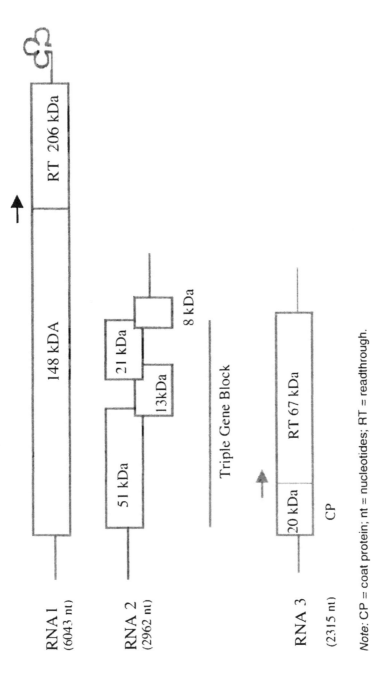

FIGURE 5.5. Organization and Expression of the Genome of *Potato mop-top virus*

RNA 1
(6043 nt)

148 kDA   RT 206 kDa

RNA 2
(2962 nt)

51 kDa   13kDa   21 kDa   8 kDa

Triple Gene Block

RNA 3
(2315 nt)

20 kDa   RT 67 kDa

CP

Note: CP = coat protein; nt = nucleotides; RT = readthrough.

FIGURE 5.6. Organization and Expression of the Genome of *Peanut clump virus*

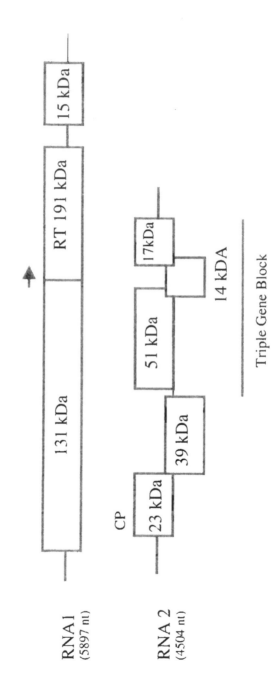

Note: CP = coat protein; nt = nucleotides; RT = readthrough.

RNA2 contains five ORFs. The 5'-proximal ORF encodes the 23 kDa CP, and the second ORF encodes the 39 kDa protein whose translation initiation on RNA2 occurs by a leaky-scanning mechanism (Herzog et al., 1995). This protein is not required for infection of host plant leaves by mechanical inoculation but probably plays a part in vector transmission (Manohar et al., 1993). The closely spaced ORFs 3, 4, and 5, coding for proteins of 51, 14, and 17 kDa, respectively, constitute the TGB whose gene products govern cell-to-cell movement (Herzog et al., 1998). Results from complementation experiments between BNYVV and PCV for cell-to-cell movement have indicated the existence of specific interactions between the three TGB proteins (Lauber, Bleykasten-Grosshans, et al., 1998; Erhardt et al., 1999).

Beet necrotic yellow vein virus,
Type Species of the Genus Benyvirus

Beet necrotic yellow vein virus (BNYVV) has labile rigid rod-shaped particles, about 20 nm in width and with predominant lengths of 390, 265, 100, and 85 nm (Scholten, 1997). The virus possesses a positive-sense single-stranded RNA genome, and each virion contains a single RNA molecule (Richards and Tamada, 1992; Brunt, 1995).

BNYVV has a narrow host range and is transmitted by *P. betae* (Tamada, 1975). In *Beta vulgaris* the virus causes rhizomania ("root madness"), an important disease characterized by abnormal proliferation of rootlets from the tap root and lateral roots (Brunt and Richards, 1989). The disease leads to serious reduction in root yield and sugar content (Scholten, 1997).

In the field, the virus usually remains confined to the roots. Only occasionally, BNYVV is found to be present in the aerial parts of a beet plant, where it may cause yellowing of veins and some necrosis, after which the virus has been named (Tamada, 1975). Since its first reported occurrences in Italy (Canova, 1959) and Japan in 1969 (Scholten, 1997). BNYVV has been found in most sugar beet-growing areas in the world (Brunt and Richards, 1989).

Depending on the virus isolate, four or five distinct RNA species have been described (Bouzoubaa et al., 1985; Richards and Tamada, 1992) (see Figure 5.7). The size of the quadripartite genome correlates with the length distribution of the virions (Putz et al., 1988). All five viral RNAs have 3'-poly(A)-tails.

RNA1 of 6.8 kb contains one ORF that codes for a polypeptide of about 237 kDa with RdRp function. It can replicate alone in protoplasts.

RNA2 of 4.7 kb has a 5'-proximal ORF (ORF 1) that encodes the 21 kDa CP and, together with an ORF 2 that codes for a 54 kDa protein, produces an RT protein of a predicted 75 kDda. This 75 kDa protein plays a vital role in virus assembly (Richards and Tamada, 1992) and is also necessary for transmission of BNYVV by its vector (Tamada and Kusume, 1991). ORF 2

FIGURE 5.7. Organization and Expression of the Genome of *Beet necrotic yellow vein virus*

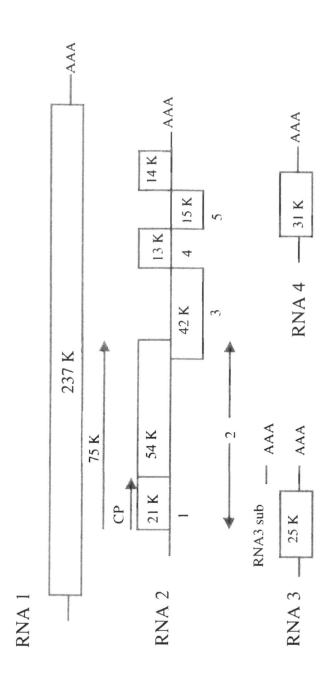

Note: sub = subgenomic RNA; rectangles = open reading frames; arrows = corresponding translation.

is followed by three overlapping ORFs (TGB) coding for polypeptides of 42, 13, and 15 kDa. These proteins have been shown to be involved in cell-to-cell movement of the virus (Gilmer et al., 1992). A 14 kDa protein encoded by the last ORF of RNA2 is involved in regulation of replication and translation of viral proteins (Hehn et al., 1995).

Both RNA1 and RNA2 are needed for stable infections and for vascular transport of the virus in *Spinacia oleracea* (Lauber, Guilley, et al., 1998).

RNA3 of 1.8 kb is expressed from an sg messenger and produces a 25 kDa protein that has an effect on leaf symptoms (Richards and Tamada, 1992) and probably enhances vascular movement of virus in the plant (Brunt, 1995). Recently it has been shown to be essential for the development of rhizomania symptoms (Tamada et al., 1999).

RNA4 of 1.5 kb encodes a 31 kDa polypeptide. This RNA is required for transmission by the fungal vector, but transmission is more efficient in combination with either RNA3 or RNA5 (Richards and Tamada, 1992).

RNA5 of 1.4 kb, which is present in some isolates from Japan, China, and France (Tamada et al., 1999), codes for a 19 kDa polypeptide. This RNA species may facilitate systemic infection of the roots (Richards and Tamada, 1992), and it is associated with severity of symptoms in the latter (Tamada et al., 1999).

Barley mild mosaic virus, *Member of the Genus* Bymoviruse

Barley mild mosaic virus (BMMV) has slightly flexuous filamentous particles, about 13 nm in width and with predominant lengths of 250 to 300 and 500 to 600 nm (Barnett et al., 1995). The virus particles contain a bipartite positive-sense single-stranded RNA genome.

The only known host of BMMV is barley, to which the virus is transmitted by *P. graminis*. The virus causes a serious disease of winter barley in Europe and Asia and is often found in mixed infections with *Barley yellow mosaic virus* (BYMV) (Inouye and Saito, 1975), type species of the genus *Bymovirus*, which is transmitted by the same vector (Adams et al., 1988). Both viruses produce similar symptoms in some barley cultivars, but they are serologically unrelated, with hardly any nucleic acid homology between them (Jacobi et al., 1995).

Both RNA1 and RNA2 encode a polyprotein that is proteolytically processed into functional proteins (see Figure 5.8).

RNA1 of 7,262 nt encodes a putative polyprotein of 2,258 amino acids (aa) that is most likely processed into six to eight proteins (Meyer and Dessens,

FIGURE 5.8. Organization and Expression of the Genome of *Barley mild mosaic virus*

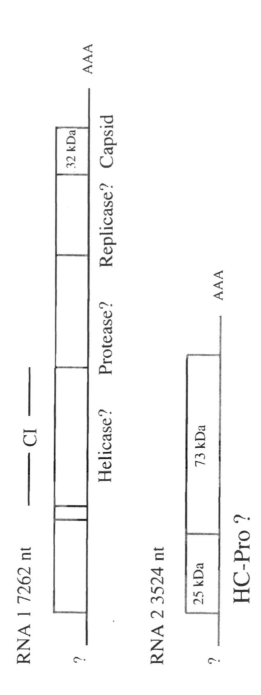

RNA 1 7262 nt

? CI

Helicase? Protease? Replicase? Capsid

32 kDa

AAA

RNA 2 3524 nt

? 25 kDa 73 kDa AAA

HC-Pro ?

Note: CI = cylindrical inclusions; HC-Pro = helper component proteinase. Possible functions of the gene products are presented by analogy with genus *Potyvirus*.

1996; Peerenboom et al., 1997). One of these proteins is the CP of about 32 kDa. Other products of processing are putative proteins with possible replicase, proteinase, or helicase functions.

RNA2 of 3,524 nt encodes a putative polyprotein of 894 aa with an Mr of 98 kDa. The polyprotein could produce a 25 kDa N-terminal protein (the putative helper component proteinase of potyviruses) and a 73 kDa C-terminal protein with a possible role in vector transmission (Jacobi et al., 1995). Comparisons made between the sequences of RNA2 of both BMMV and BYMV (Kashiwazaki et al., 1990, 1991) showed some regions of consistent homology or similarity (Peerenboom et al., 1996). The RNA2 molecule has been found to be very unstable (Jacobi et al., 1995).

MODES OF TRANSMISSION

Two different types of relationship between virus and fungal vector have been described. The distinction is based on the way the virus is acquired by the fungus and its location within the fungal resting spore.

In the cases of CNV and TNV, virions are adsorbed to the zoospores outside the host plant and the virus does not enter the resting spore. This mode of transmission is called *in vitro transmission* (Campbell, 1979) or *nonpersistent transmission* (Teakle, 1983).

In contrast, other viruses, such as SBWMV, PMTV, BNYVV, and BMMV, are acquired inside the host plant and virions are located within the zoospores and resting spores. For this type of transmission, the terms *in vivo transmission* (Campbell, 1979) and *persistent transmission* (Teakle, 1983) are used.

The present authors, however, are not in favor of any of these terms for the following reasons: The terms *in vitro* and *in vivo* for types of natural transmission are confusing and should be used only in the context of their original meaning, i.e., for experiments carried out in the laboratory (in vitro, literally "in glass") or for those performed in the living organism (in vivo). The terms *nonpersistent* and *persistent* for transmission by fungal vectors should also be avoided, as the comparison with insect transmission does not hold well. Campbell (1996) rightly pointed out that these terms refer to virus retention by feeding vectors, whereas fungus transmission is based on virus survival in dormant resting spores.

In this chapter, the two modes of transmission are referred to as "virus acquired by adsorption to zoospores outside the host plant," or externally borne virus, and "virus acquired by zoospores and resting spores inside the host plant," or internally borne virus, respectively.

MECHANISM OF VIRUS-VECTOR ASSOCIATION

Virus Acquired by Adsorption to Zoospores Outside the Host Plant

Most of the available information on this type of transmission has been supplied by research carried out in the 1960s and 1970s. As the viruses studied (TNV, CNV) were of minor economic importance, investigations of their transmission were discontinued.

After decay of infected roots or other plant material in the soil, both the virus and the virus-free fungal zoospores are released in soil water, where association between virus and vector will take place. Particles of TNV have been found to be tightly adsorbed to the plasmalemma of zoospores (Campbell and Fry, 1966). Electron microscopic observations of TNV and zoospores of a lettuce isolate of *O. brassicae* led to the following conclusions (Temmink and Campbell, 1969a,b): Virus particles are adsorbed to the body and axonemal sheath of zoospores. When the zoospore enters into the encystment phase, the axonema with its axonemal sheath is retracted inside the zoospore body. During this process, virus particles adsorbed to the axonemal sheath are taken into the cytoplasm of the zoospore. After penetration of the viruliferous protoplast of the cyst into the host and formation of the young thallus, virus particles are released from the thallus into the cytoplasm of the root epidermal cell.

Viral transmission depends on the ability of the virus to bind to the zoospores and is highly specific, as shown by experiments with TNV. Particles of TNV were adsorbed to zoospores of a lettuce isolate of *O. brassicae*, but there was no adsorption of this virus to zoospores of a mustard isolate of the same vector (Temmink et al., 1970). Moreover, zoospores of *O. bornovanus*, in vitro mixed with CNV and TNV, adsorbed CNV but not TNV (Temmink et al., 1970). Specificity of fungus transmission is thought to be due to interaction between specific receptors in the zoospore plasmalemma and the CP of the virus (Campbell, 1996).

In experiments with CNV and the cherry strain (Ch) of *Tomato bushy stunt virus* (TBSV), type species of the genus *Tombusvirus*, zoospores of O. bornovanus acquired and transmitted only CNV, not TBSV-Ch. By using full-length infectious CNV/TBSV chimeric complementary (c) DNA clones in which the CP genes had been exchanged, the role of CP in specificity of transmission was demonstrated (McLean et al., 1994). Virus particles containing a modified TBSV-Ch genome with the CP gene of CNV were transmitted, but there was no transmission of virus particles possessing a modified CNV genome with the CP of TBSV-Ch.

Further evidence of the role of CP in transmission was obtained from experiments with a mutant (LL5) of CNV, deficient in vector transmission (Robbins et al., 1997). This mutant had been isolated from long-term mechanically transmitted CNV. Prolonged mechanical transmission of viruses is known to lead to development of mutants that have lost their vector transmissibility (Campbell, 1996; Gray, 1996). Analysis of the CP of LL5 and comparisons with wt CP revealed that the mutant contained two amino acid substitutions: one (Phe to Cys) in the a region (between R and S domains) and another (Glu to Lys) in the S domain. In vitro mutagenesis produced the CNV CP mutants $LL5_s$ and $LL5_a$ that contained the single nucleotide substitution found in the S and a region domains, respectively. The $LL5_s$ mutant had lost most of its vector transmissibility, whereas the transmissibility of $LL5_a$ resembled that of wt CNV. From these results, it is clear that the substitution of Glu to Lys in the S domain of CP leads to impaired vector transmission. In-vitro binding assays showed that LL5 and $LL5_s$ particles bind less efficiently to zoospore plasmalemma than wt CNV. As less adsorption leads to less efficient transmission of virus (Temmink et al., 1970), specific bonding between domains on the virus CP and putative receptors in the zoospore plasmalemma may determine the degree of vector transmission (Campbell, 1996; Robbins et al., 1997).

Viruses Acquired by Zoospores and Resting Spores Inside the Host Plant

In contrast to the previous category, internally borne viruses are already present in zoospores released into the soil from viruliferous resting spores or cystosori. Acquisition of virus by the fungus is thought to take place as follows (see Figure 5.1).

Virus-free zoospores penetrate the rootlets or epidermal cells of a virus-infected host plant and form a plasmodium-like thallus (Barr and Asher, 1996). At that stage, the virus most likely enters the fungus through the thin membrane of the young thallus (Rysanek et al., 1992) and accumulates in it. There is no evidence that the virus multiplies in the thallus, as in many instances; for example, BMMV, BNYVV, LBVV, and SBWMV fungal cultures gradually lose the virus (Campbell, 1996). During the formation of zoosporangia or resting spores from the mature thallus, the virus is incorporated in these structures and is spread when newly formed viruliferous zoospores are released into the soil.

The virus can survive inside the resting spores in dry root material or dry infested soil for very long periods. Survival times of fifteen years (BNYVV), ten years (BYMV), two years (PMTV), and one year (SBWMV) have been reported (Adams, 1991).

The virus-vector relationship is highly specific, and molecular studies have shed light upon factors that regulate virus transmission.

Two spontaneous mutants of BNYVV with deletions in their RNA2 produced CP RT polypeptides of 67 kDa and 58 kDa, respectively, instead of 75 kDa proteins of wt RNA2, and could no longer be transmitted by *P. betae* (Tamada and Kusume, 1991). The RT domain of the RT protein contains a KTER (Lys-Thr-Glu-Arg) motif in a hydrophilic region flanked by hydrophobic regions, and this sequence has been reported to be a determinant of vector transmission (Tamada et al., 1996; Reavy et al., 1998). Isolates of BNYVV have been shown to possess small deletions in their KTER-encoding domain after prolonged mechanical transmission (Koenig, 2000).

Also, in other virus-vector combinations, CP RT protein seems to play an important role in vector transmission. In contrast to a field isolate (S) of PMTV, the T isolate of this virus, which had been mechanically transmitted to *Nicotiana debneyi* or *N. benthamiana* for 30 years, could no longer be transmitted by *S. subterranea* (Reavy et al., 1998). Analyses of the RT protein-coding region of RNA3 of PMTV-S showed the presence of 543 nt in the 3'-half of the coding region that were lacking in the corresponding region in the PMTV-T isolate. A hydrophilic region flanked by hydrophobic regions, but no KTER motif, was present in the additional sequences of PMTV-S RT protein. It has therefore been postulated that the RT domain encoded by RNA3 possesses determinants for interaction with the vector (Reavy et al., 1998).

An isolate of BMMV that had been manually transmitted for more than 15 years lost its fungus transmissibility (Adams et al., 1988). This isolate contained a deletion of 364 aa in the 73 kDa C-terminal protein of its RNA2. The deletion with a calculated Mr of 39 kDa resulted in the production of a protein of 34 kDa instead of a wt protein of 73 kDa (Jacobi et al., 1995). It is suggested that such a truncated protein is responsible for the lack of vector transmission of BMMV, and also for that of an isolate of BYMV shown to possess a comparable truncated protein of about 45 kDa, instead of the wt protein of 70 kDa (Jacobi et al., 1995).

Conserved regions have been found in the 73 kDa protein of BMMV and the 70 kDa protein of BYMV, and in the RT proteins of BNYVV, PMTV-S, and SBWMV, all viruses known to have plasmodiophorid fungi as their vectors. In BMMV, BYMV, PMTV, and SBWMV, the amino acid combinations ER (Glu-Arg) or QR (Gln-Arg) were consistently present in regions of homology, whereas in BNYVV such a region contained a KTER motif. These sequences occurred within a hydrophilic region flanked by hydrophobic regions. The positions of these amino acid combinations were predicted to be on the outside of the protein, which makes them suitable sites for contact with the vector (Peerenboom et al., 1996; Reavy et al., 1998).

Deletion mutations resulting from repeated mechanical transmission have been found to occur also in other parts of the viral genomes, for instance, in RNAs 3 and 4 of BNYVV (Lemaire et al., 1988). RNA4 has been shown to be more important for vector transmission than RNA3 (Tamada and Abe, 1989). Even complete loss of RNAs 3 and 4 of BNYVV isolates, which had been transmitted mechanically for over 15 years, has been reported (Koenig, 2000). It has not yet been established whether these smaller RNAs have direct effects on vector transmission or whether they act indirectly, for instance, by enhancing accumulation of virus or its systemic transport (Richards and Tamada, 1992).

EPIDEMIOLOGY AND CONTROL
OF FUNGUS-BORNE VIRUSES

The different transmission characteristics of internally borne and externally borne viruses also have implications for their epidemiology. As has been pointed out earlier, virus present in resting spores of the fungus may remain viable in the soil for a long time even though these viruses are labile. Consequently, fields become more or less permanently infested.

By contrast, the externally borne viruses are not present in resting spores and have to be acquired by zoospores in the soil. These viruses are released from decaying plant material into the soil, where they have to survive for some time. However, unlike the previous category, externally borne viruses such as TNV are very stable; they occur in high concentrations in the plant and have a wide host range. These characteristics increase their chances of being transmitted. Spread of such viruses is especially promoted in protected cultures with recirculating nutrient solution systems, providing an ideal environment for both virus and vector. This has often led to serious epidemics in such cultures (Campbell, 1996).

Compatibility between fungus and host plant does not affect virus transmission very much, provided that the fungus is able to penetrate the plant, i.e., when its zoospores encyst on the roots (Temmink et al., 1970; Barr et al., 1995). Once the virus has been introduced into the root cell of a nonhost, impaired development of thalli or their disintegration does not affect virus multiplication but is only an impediment for further spread of internally borne viruses. Isolates of *P. betae* could infect *B. patellaris* and *B. procumbens*, but after penetration of the zoospores there was no development of thalli in these resistant *Beta* spp. (Scholten, 1997). Although infection by BNYVV had taken place, the concentration of the virus in rootlets was reduced as compared to that in susceptible plants.

In light of the high persistence of internally borne viruses and their vectors during the resting stage and the great longevity of externally borne viruses in the soil and their wide host ranges, crop rotation is ineffective in controlling these viruses.

Efforts have been made to eliminate the fungus by chemical treatment, e.g., fumigation of the soil with methyl bromide. However, these control measures, apart from not being very effective under field conditions (Scholten, 1997), also constitute a serious hazard to the environment.

Spread of the fungal vectors can be reduced to some extent, by preventive measures, such as, in the case of rhizomania, stricter rules for transportation of propagation material with adhering soil. This is especially important in areas where potatoes are grown in rotation with beet.

However, a more promising solution may be breeding for resistance. Most breeding programs aim for resistance to the virus, but in some cases resistance to the vector has also been considered (Scholten, 1997). For rhizomania, progress has been made through introgression of natural resistance genes into breeding stocks (Scholten, 1997).

Another possibility to obtain resistance is through transformation of plants with viral genes. Such pathogen-derived resistance has been achieved by transforming *N. benthamiana* plants with the CP gene of PMTV. The transformed plants displayed immunity not only after mechanical inoculation with PMTV but also after inoculation with viruliferous zoospores (Reavy et al., 1995).

Isolation and combination of plant resistance genes effective against both virus and fungus through transformation of plants might lead to more durable resistance.

CONCLUSIONS

Although progress has been made in elucidating the mechanisms of virus transmission by fungal vectors, the exact nature of the association between virus and fungus is still not clear. The essential role of virus-coded proteins has been demonstrated unequivocally. However, nothing is known about the putative fungal receptors. Therefore, besides further investigations of structures present on the outside of virus particles, study is required of specific domains on the zoospore body where bonding between virus and vector takes place.

Another still unexploited area that demands more attention is the way virus particles go in and out of the protoplasm of zoospores or thalli.

REFERENCES

Adams MJ (1991). Transmission of plant viruses by fungi. *Annals of Applied Biology* 118: 479-492.

Adams MJ, Swaby AG, Jones P (1988). Confirmation of the transmission of barley yellow mosaic virus (BaYMV) by the fungus *Polymyxa graminis*. *Annals of Applied Biology* 112: 133-141.

Adams MJ, Swaby AG, MacFarlane I (1986). The susceptibility of barley cultivars to barley yellow mosaic virus (BaYMV) and its fungal vector, *Polymyxa graminis*. *Annals of Applied Biology* 109: 561-572.

Arif M, Torrance L, Reavy B (1995). Acquisition and transmission of potato mop-top furovirus by a culture of *Spongospora subterranea* f. sp. *subterranea* derived from single cystosorus. *Annals of Applied Biology* 126: 493-503.

Barnett OW, Adam G, Brunt AA, Dijkstra J, Dougherty WG, Edwardson JR, Goldbach R, Hammond J, Hill JH, Jordan RL, et al. (1995). Family *Potyviridae*. In Murphy FA, Fauquet CM, Bishop DHL, Ghabrial SA, Jarvis AW, Martelli GP, Mayo MA, Summers MD (eds.), *Virus Taxonomy*, Springer-Verlag, Wien, New York, pp. 348-351.

Barr KJ, Asher MJC (1996). Studies on the life-cycle of *Polymyxa betae* in sugar beet roots. *Mycological Research* 100: 203-208.

Barr KJ, Asher MJC, Lewis BG (1995). Resistance to *Polymyxa betae* in wild *Beta* species. *Plant Pathology* 44: 301-307.

Bawden FC, Van der Want JPH (1949). Bean stipple-streak caused by a tobacco necrosis virus. *Tijdschrift over Plantenziekten* 55: 142-150.

Bouzoubaa S, Guilley H, Jonard G, Richards K, Putz C (1985). Nucleotide sequence analysis of RNA-3 and -4 of beet necrotic yellow vein virus, isolates F2 and G1. *Journal of General Virology* 66: 1553-1564.

Brakke MK (1971). Soil-borne wheat mosaic virus. *CMI/AAB Descriptions of Plant Viruses* No. 77, CMI/AAB, Kew, Surrey, England, 4pp.

Brunt AA (1995). Genus *Furovirus*. In Murphy FA, Fauquet CM, Bishop DHL, Ghabrial SA, Jarvis AW, Martelli GP, Mayo MA, Summers MD (eds.), *Virus Taxonomy*, Springer-Verlag, Wien, New York, pp. 445-449.

Brunt AA, Richards KE (1989). Biology and molecular biology of furoviruses. *Advances in Virus Research* 36: 1-32.

Campbell RN (1979). Fungal vectors of plant viruses. In Molitoris HP, Hollings M, Wood HA (eds.), *Fungal Viruses*, Springer-Verlag, Berlin, pp. 8-24.

Campbell RN (1988). Cultural characteristics and manipulative methods. In Cooper JI, Asher MJC (eds.), *Viruses with Fungal Vectors*, Association of Applied Biologists, Wellesbourne, UK, pp. 153-165.

Campbell RN (1996). Fungal transmission of plant viruses. *Annual Review of Phytopathology* 34: 87-108.

Campbell RN, Fry PR (1966). The nature of the associations between *Olpidium brassicae* and lettuce big-vein and tobacco necrosis viruses. *Virology* 29: 222-233.

Campbell RN, Sim ST (1994). Host specificity and nomenclature of *Olpidium bornovanus* (= *Olpidium radicale*) and comparisons to *Olpidium brassicae*. *Canadian Journal of Botany* 72: 1136-1143.

Canova A (1959). Appunti di patologia della barbabietola. *Informatore Fitopatologico* 9: 390-396.

Chen J, MacFarlane SA, Wilson TMA (1995). An analysis of spontaneous deletion sites in soil-borne wheat mosaic virus RNA2. *Virology* 209: 213-217.

Dias HF (1970). The relationship between cucumber necrosis virus and its vector *Olpidium cucurbitacearum*. *Virology* 42: 204-211.

Dias HF, McKeen CD (1972). Cucumber necrosis virus. *CMI/AAB Descriptions of Plant Viruses* No. 82, CMI/AAB, Kew, Surrey, England, 4 pp.

Dijkstra J, De Jager CP (1998). *Practical Plant Virology*, Springer-Verlag, Berlin, Heidelberg, Germany, 459 pp.

Erhardt M, Herzog E, Lauber E, Fritsch C, Guilley H, Jonard G, Richards K, Bouzoubaa S (1999). Transgenic plants expressing the TGB1 protein of peanut clump virus complement movement of TGB1-defective peanut clump virus but not of TGB1-defective beet necrotic yellow vein virus. *Plant Cell Reports* 18: 614-619.

Estes AP, Brakke MK (1966). Correlation of *Polymyxa graminis* with transmission of soil-borne wheat mosaic virus. *Virology* 28: 772-774.

Fry PR (1958). The relationship of *Olpidium brassicae* (Wor.) Dang to the big-vein disease of lettuce. *New Zealand Journal of Agricultural Research* 1: 301-304.

Gilmer D, Bouzoubaa S, Hehn A, Guilley H, Richards K, Jonard G (1992). Efficient cell-to-cell movement of beet necrotic yellow vein virus requires 3' proximal genes located on RNA2. *Virology* 189: 40-47.

Gray SM (1996). Plant virus proteins involved in natural vector transmission. *Trends in Microbiology* 4: 259-264.

Grogan RG, Campbell RN (1966). Fungi as vectors and hosts of viruses. *Annual Review of Phytopathology* 4: 29-52.

Grogan RG, Zink FW, Hewitt WB, Kimble KA (1958). The association of *Olpidium* with the big-vein disease of lettuce. *Phytopathology* 48: 292-297.

Harrison BD, Jones RAC (1970). Host range and some properties of potato mop-top virus. *Annals of Applied Biology* 65: 393-402.

Hehn A, Bouzoubaa S, Bate A, Twell D, Marbach J, Richards K, Guilley H, Jonard G (1995). The small cysteine-rich protein P14 of beet necrotic yellow vein virus regulates accumulation of RNA2 in *cis* and coat protein in *trans*. *Virology* 210: 73-81.

Herzog E, Guilley H, Fritsch C (1995). Translation of the second gene of peanut clump virus RNA2 occurs by leaky scanning in vitro. *Virology* 208: 215-225.

Herzog E, Guilley H, Manohar SK, Dollet M, Richards K, Fritsch C, Jonard G (1994). Complete nucleotide sequence of peanut clump virus RNA1 and relationships with other fungus-transmitted rod-shaped viruses. *Journal of General Virology* 75: 3147-3155.

Herzog E, Hemmer O, Hauser S, Meyer G, Bouzoubaa S, Fritsch C (1998). Identification of genes involved in replication and movement of peanut clump virus. *Virology* 248: 312-322.

Inouye T, Saito Y (1975). Barley yellow mosaic virus. *CMI/AAB Descriptions of Plant Viruses* No. 143, CMI/AAB, Kew, Surrey, England, 4 pp.

Jacobi V, Peerenboom E, Schenk PM, Antoniw JF, Steinbiss H-H, Adams MJ (1995). Cloning and sequence analysis of RNA-2 of a mechanically transmitted UK isolate of barley mild mosaic bymovirus (BaMMV). *Virus Research* 37: 99-111.

Jones AT (1993). Virus transmission through soil and by soil-inhabiting organisms in diagnosis. In Matthews RF (ed.). *Diagnosis of Plant Virus Diseases*. CRC Press, Inc., Boca Raton, Florida, pp. 77-99.

Jones RAC, Harrison BD (1969). The behaviour of potato mop-top virus in soil, and evidence for its transmission by *Spongospora subterranea* (Wallr) Lagerh. *Annals of Applied Biology* 63: 1-17.

Kashiwazaki S, Minobe Y, Hibino H (1991). Nucleotide sequence of barley yellow mosaic virus RNA2. *Journal of General Virology* 72: 995-999.

Kashiwazaki S, Minobe Y, Omura T, Hibino H (1990). Nucleotide sequence of barley yellow mosaic virus RNA1: A close evolutionary relationship with potyviruses. *Journal of General Virology* 71: 2781-2790.

Kashiwazaki S, Scott KP, Reavy B, Harrison BD (1995). Sequence analysis and gene content of potato mop-top virus RNA-3: Further evidence of heterogeneity in the genome organization of furoviruses. *Virology* 206: 701-706.

Kassanis B (1949). A necrotic disease of forced tulips caused by tobacco necrosis viruses. *Annals of Applied Biology* 36: 14-17.

Koenig R (2000). Deletions in the KTER-encoding domain, which is needed for *Polymyxa* transmission, in manually transmitted isolates of *Beet necrotic yellow vein benyvirus*. *Archives of Virology* 145: 165-170.

Lauber E, Bleykasten-Grosshans C, Erhardt M, Bouzoubaa S, Jonard G, Richards K, Guilley H (1998). Cell-to-cell movement of beet necrotic yellow vein virus. I. Heterologous complementation experiments provide evidence for specific interactions among the triple gene block proteins. *Molecular Plant-Microbe Interactions* 11: 618-625.

Lauber E, Guilley H, Tamada T, Richards KE, Jonard G (1998). Vascular movement of beet necrotic yellow vein virus in Beta macrocarpa is probably dependent on an RNA 3 sequence domain rather than a gene product. *Journal of General Virology* 79: 385-393.

Lemaire O, Merdinoglu D, Valentin P, Putz C, Ziegler-Graff V, Guilley H, Jonard G, Richards K (1988). Effect of beet necrotic yellow vein virus RNA composition on transmission by *Polymyxa betae*. *Virology* 162: 232-235.

Lommel SA (1995). Genus *Necrovirus*. In Murphy FA, Fauquet CM, Bishop DHL, Ghabrial SA, Jarvis AW, Martelli GP, Mayo MA, Summers DM (eds.), *Virus Taxonomy*, Springer-Verlag, Wien, New York, pp. 398-400.

Manohar SK, Guilley H, Dollet M, Richards K, Jonard G (1993). Nucleotide sequence and genetic organization of peanut clump virus RNA2 and partial characterization of deleted forms. *Virology* 195: 33-41.

Maraite H (1991). Transmission of viruses by soil fungi. In Beemster ABR, Bollen GJ, Gerlagh M, Ruissen MA, Schippers B, Tempel A (eds.), *Biotic Interactions and Soil-Borne Diseases*, Elsevier Science Publishers, Amsterdam, pp. 67-82.

Mayo MA (1995). Unassigned viruses. In Murphy FA, Fauquet CM, Bishop DHL, Ghabrial SA, Jarvis AW, Martelli GP, Mayo MA, Summers MD (eds.), *Virus Taxonomy*, Springer-Verlag, Wien, New York, pp. 504-507.

McLean MA, Campbell RN, Hamilton RI, Rochon DM (1994). Involvement of the cucumber necrosis virus coat protein in the specificity of fungus transmission by *Olpidium bornovanus*. *Virology* 204: 840-842.

Meyer M, Dessens JT (1996). The complete nucleotide sequence of barley mild mosaic virus RNA 1 and its relationship with other members of the *Potyviridae*. *Virology* 219: 268-273.

Paul H, Henken B, Scholten OE, Lange W (1993). Use of zoospores of *Polymyxa betae* in screening beet seedlings for resistance to beet necrotic yellow vein virus. *Netherlands Journal of Plant Pathology* 99 (Suppl. 3): 151-160.

Peerenboom E, Cartwright EJ, Foulds I, Adams MJ, Stratford R, Rosner A, Steinbiss H-H, Antoniw JF (1997). Complete RNA 1 sequences of two UK isolates of barley mild mosaic virus: A wild-type fungus-transmissible isolate and a non-fungus-transmissible derivative. *Virus Research* 50: 175-183.

Peerenboom E, Jacobi V, Antoniw JF, Schlichter U, Cartwright EJ, Steinbiss H-H, Adams MJ (1996). The complete nucleotide sequence of RNA-2 of a fungally-transmitted UK isolate of barley mild mosaic bymovirus and identification of amino acid combinations possibly involved in fungus transmission. *Virus Research* 40: 149-159.

Pringle CR (1999). Virus taxonomy. The Universal System of Virus Taxonomy, updated to include the new proposals ratified by the International Committee on Taxonomy of Viruses during 1998. *Archives of Virology* 144: 421-429.

Putz C, Wurtz M, Merdinoglu D, Lemaire O, Valentin P (1988). Physical and biological properties of beet necrotic yellow vein virus isolates. In Cooper JI, Asher MJC (eds.), *Viruses with Fungal Vectors*, Association of Applied Biologists, Wellesbourne, UK, pp. 83-97.

Reavy B, Arif M, Cowan GH, Torrance L (1998). Association of sequences in the coat protein/readthrough domain of potato mop-top virus with transmission by *Spongospora subterranea*. *Journal of General Virology* 79: 2343-2347.

Reavy B, Arif M, Kashiwazaki S, Webster KD, Barker H (1995). Immunity to potato mop-top virus in *Nicotiana benthamiana* plants expressing the coat protein gene is effective against fungal inoculation of the virus. *Molecular Plant-Microbe Interactions* 8: 286-291.

Richards KE, Tamada T (1992). Mapping functions on the multipartite genome of beet necrotic yellow vein virus. *Annual Review of Phytopathology* 30: 291-313.

Robbins MA, Reade RD, Rochon DM (1997). A cucumber necrosis virus variant deficient in fungal transmissibility contains an altered coat protein shell domain. *Virology* 234: 138-146.

Rysanek P, Stocky G, Haeberlé AM, Putz C (1992). Immunogoldlabelling of beet necrotic yellow vein virus particles inside its fungal vector, *Polymyxa betae* K. *Agronomie* 12: 651-659.

Savenkov EI, Sandgren M, Valkonen JPT (1999). Complete sequence of RNA 1 and the presence of tRNA-like structures in all RNAs of *Potato mop-top virus*, genus *Pomovirus*. *Journal of General Virology* 80: 2779-2784.

Scholten OE (1997). Characterisation and inheritance of resistance to beet necrotic yellow vein virus in *Beta*. PhD Thesis, Wageningen Agricultural University, Wageningen, The Netherlands.

Scott KP, Kashiwazaki S, Reavy B, Harrison BD (1994). The nucleotide sequence of potato mop-top virus RNA-2: A novel type of genome organization for a furovirus. *Journal of General Virology* 75: 3561-3568.

Sit TL, Johnston JC, TerBorg MG, Frison E, McLean MA, Rochon DM (1995). Mutational analysis of the cucumber necrosis virus coat protein gene. *Virology* 206: 38-48.

Tamada T (1975). Beet necrotic yellow vein virus. *CMI/AAB Descriptions of Plant Viruses* No. 144, CMI/AAB, Kew, Surrey, England, 4 pp.

Tamada T, Abe H (1989). Evidence that beet necrotic yellow vein virus RNA-4 is essential for efficient transmission by the fungus *Polymyxa betae*. *Journal of General Virology* 72: 1497-1504.

Tamada T, Kusume T (1991). Evidence that the 75 K readthrough protein of beet necrotic yellow vein virus RNA-2 is essential for transmission by the fungus *Polymyxa betae*. *Journal of General Virology* 72: 1497-1504.

Tamada T, Schmitt C, Saito M, Guilley H, Richards K, Jonard G (1996). High resolution analysis of the readthrough domain of beet necrotic yellow vein virus readthrough protein: A KTER motif is important for efficient transmission of the virus by *Polymyxa betae*. *Journal of General Virology* 77: 1359-1367.

Tamada T, Uchino H, Kusume T, Saito M (1999). RNA 3 deletion mutants of beet necrotic yellow vein virus do not cause rhizomania disease in sugar beets. *Phytopathology* 89: 1000-1006.

Teakle DS (1960). Association of *Olpidium brassicae* and tobacco necrosis virus. *Nature* 188: 431-432.

Teakle DS (1983). Zoosporic fungi and viruses: Double trouble. In Buczaki ST (ed.), *Zoosporic Plant Pathogens: A Modern Perspective*, Academic Press, London, pp. 233-248.

Temmink JHM (1971). An ultrastructural study of *Olpidium brassicae* and its transmission of tobacco necrosis virus. *Mededelingen Landbouwhogeschool Wageningen* 71: 1-135.

Temmink JHM, Campbell RN (1969a). The ultrastructure of *Olpidium brassicae*. II. Zoospores. *Canadian Journal of Botany* 47: 227-231.

Temmink JHM, Campbell RN (1969b). The ultrastructure of *Olpidium brassicae*. III. Infection of host roots. *Canadian Journal of Botany* 47: 421-424.

Temmink JHM, Campbell RN, Smith PR (1970). Specificity and site of in vitro acquisition of tobacco necrosis virus by zoospores of *Olpidium brassicae*. *Journal of General Virology* 9: 201-213.

Thouvenel J-C, Fauquet C (1981a). Further properties of peanut clump virus and studies on its natural transmission. *Annals of Applied Biology* 97: 99-107.

Thouvenel J-C, Fauquet C (1981b). Peanut clump virus. *CMI/AAB Descriptions of Plant Viruses* No. 235. CMI/AAB, Slough, England, 4 pp.

Torrance L, Cowan GH, Scott KP, Pereira LG, Roberts IM, Reavy B, Harrison BD (1992). Detection and diagnosis of potato mop-top virus. *Annual Report of Scottish Crop Research Institute for 1991*, pp. 80-82.

Chapter 6

Seed Transmission of Viruses: Biological and Molecular Insights

Jawaid A. Khan
Jeanne Dijkstra

INTRODUCTION

Seeds provide an ideal medium for the survival of viruses, especially when suitable host plants are in short supply. Irrespective of their transmission frequency, which may vary from less than 1 percent to 100 percent, seed-transmitted viruses have significant impact on disease epidemics. A low rate of virus transmission coupled with secondary spread by animal vectors may lead to disease epidemics in a new area, while a high rate of transmission may cause virus self-extinction as a result of poor development of flowers and seeds. More than 100 plant viruses are reported to have seed-based transmission (Mink, 1993; Van den Heuvel et al., 1999). The recent classification of viruses; their proper division into families, genera, and species; and the advent of sensitive virus detection assays will, definitely, continue to change the existing list (see Table 6.1). During the past decade much attention was focused on elucidation of the genetic (by and large viral) determinants involved in the transmission of viruses through seeds. Although many viruses have been studied extensively, the mechanism involved in transmission is still not well understood. This chapter aims to apprise readers of the present status of the subject. The biological and molecular characteristics of some seed-transmitted viruses are described. To understand the molecular basis of virus transmission, genetic determinants of few viruses are discussed in detail. For further information, readers are referred to extensive reviews on this subject (Mink, 1993; Johansen et al., 1994; Maule and Wang, 1996).

105

TABLE 6.1. Seed-Transmitted Viruses

Family*	Genus*	Virus species*
Bromoviridae	Alfamovirus	Alfalfa mosaic virus
	Cucumovirus	Cucumber mosaic virus
		Peanut stunt virus
		Tomato aspermy virus
	Ilarvirus	Asparagus virus 2
		Elm mottle virus
		Prune dwarf virus
		Prunus necrotic ringspot virus
		Spinach latent virus
		Tobacco streak virus
Bunyaviridae	Tospovirus	Tomato spotted wilt virus
Comoviridae	Comovirus	Broad bean true mosaic virus
		Broad bean stain virus
		Cowpea mosaic virus
		Cowpea severe mosaic virus
		Squash mosaic virus
	Nepovirus	Arabis mosaic virus
		Arracacha virus A
		Arracacha virus B
		Artichoke yellow ringspot virus
		Blueberry leaf mottle virus
		Cherry rasp leaf virus
		Chicory yellow mottle virus
		Cacao necrosis virus
		Crimson clover latent virus
		Grapevine fanleaf virus
		Grapevine bulgarian latent virus
		Lucerne australian latent virus
		Peach rosette mosaic virus
		Raspberry ringspot virus
		Strawberry latent ringspot virus
		Tobacco ringspot virus
		Tomato black ring virus
		Tomato ringspot virus
Potyviridae	Potyvirus	Bean common mosaic virus
		Bean yellow mosaic virus
		Cowpea aphid-borne mosaic virus
		Cowpea green vein banding virus
		Desmodium mosaic virus

Family*	Genus*	Virus species*
		Guar symptomless virus
		Lettuce mosaic virus
		Onion yellow dwarf virus
		Pea seed-borne mosaic virus
		Peanut mottle virus
		Plum pox virus
		Potato virus Y
		Soybean mosaic virus
		Sunflower mosaic virus
	Tritimovirus	*Wheat streak mosaic virus*
Tombusviridae	*Tombusvirus*	*Tomato bushy stunt virus*
	Carmovirus	Blackgram mottle virus
		Cowpea mottle virus
		Melon necrotic spot virus
–	*Carlavirus*	*Blueberry scorch virus*
		Cowpea mild mottle virus
		Red clover vein mosaic virus
–	*Pecluvirus*	*Peanut clump virus*
–	*Hordeivirus*	*Barley stripe mosaic virus*
		Lychnis ringspot virus
–	*Sobemovirus*	*Subterranean clover mottle virus*
		Sowbane mosaic virus
		Southern bean mosaic virus
–	*Trichovirus*	*Potato virus T*
–	*Tobamovirus*	*Tobacco mosaic virus*
		Sunn-hemp mosaic virus
–	*Tobravirus*	*Tobacco rattle virus*
		Pepper ringspot virus
		Pea early-browning virus
–	*Tymovirus*	*Turnip yellow mosaic virus*
		Eggplant mosaic virus
		Dulcamara mottle virus

Source: Modified from Mink, 1993: 388-390.

*Based on the current ICTV system; see Pringle, 1999.

BIOLOGICAL CHARACTERISTICS

Seed-borne virus transmission, a complex phenomenon, involves an intimate host-virus interaction, a floral-infection stage, and the influence of the environment. Although a few general characteristics of virus transmission through seeds are mentioned in this section, the involved mechanisms of some specific virus-host combinations will be discussed separately. Infection of an embryo with a virus is the most important factor of plant virus transmission through seed, with the few exceptions being viruses that do not favor seed transmission. *Tobacco mosaic virus* (TMV, genus *Tobamovirus*) is a very stable virus that remains infectious on the surface of the seed coat. During germination or planting, seedlings get infected with TMV as a result of mechanical injury (Taylor et al., 1961; Broadbent, 1965). *Southern bean mosaic virus* (genus *Sobemovirus*) is found to occur in the seed coat. The transmission frequency is, however, very low, and the virus is inactivated during the process of seed transmission (Crowley, 1959; McDonald and Hamilton; 1972; Uyemoto and Grogan, 1977). *Melon necrotic spot virus* (genus *Carmovirus*) is seed transmitted, but no infection occurs when seeds containing the virus are sown in soil without the fungal vector *Olpidium bornovanus*. However, virus-infected seeds sown in soil containing the virus-free fungus gave rise to infection of the seedlings (vector-mediated seed transmission; Hibi and Furuki, 1985).

In general, plants infected after or shortly before the onset of flowering escape virus transmission. Seed transmission depends upon the ability of the virus to infect micro- and megagametophyte tissues that give rise to infected pollen and ovaries, respectively, or on its ability to infect the embryo during embryogenesis. Ovule-based virus transmission is quite common, and few seed-transmissible viruses infect their progeny through pollen (Carroll and Mayhew, 1976a,b; Carroll, 1981; Hunter and Bowyer, 1997). In ovule-based transmission the virus infects floral parts early in their development. In pollen transmission, on the other hand, the virus should be able to infect the floral meristems and pollen mother cells at an early stage, before the appearance of the callose layer (Hunter and Bowyer, 1997). The virus-host combination plays a significant role in determining the frequency of seed transmission. Different isolates of the same virus show differences in frequency in the same or different cultivars of the same host (Timian, 1974; Wang et al., 1993; Johansen, Dougherty, et al., 1996, details in later part). As observed with other modes of virus transmission, age of plant and environmental factors such as temperature also affect transmission rate (Hanada and Harrison, 1977; Xu et al., 1991, Wang and Maule, 1997).

Pathways of Virus Transmission

Embryo infection, a prerequisite for seed transmission, is accomplished as a result of infection of gametes during gametophyte formation (indirect invasion) or by invasion of the virus after fertilization (direct invasion). Although the former is most commonly reported, examples of the virus entering the embryo directly after fertilization do exist. As seed-transmissible virus does not follow the same route, different host factors are suggested to be involved in distinct pathways.

Indirect Embryo Infection

The virus is present in the embryo as a result of the union of infected gametes at fertilization. Examples of this type of transmission are *Tobacco rattle virus* (TRV, genus *Tobravirus*), *Barley stripe mosaic virus* (BSMV, genus *Hordeivirus*), and *Cucumber mosaic virus* (CMV, genus *Cucumovirus*). In a BSMV/barley combination, a seed-transmissible isolate (MI-1) showed its presence in both the megaspore and pollen mother cells and later in mature egg and pollen cells. However, a non-seed-transmissible isolate (NSP) was absent in both female and male gametes. It was concluded that seed transmission of BSMV is linked to the ability of BSMV to infect megaspore mother cells or pollen mother cells at an early stage of development (Carroll and Mayhew, 1976a,b). Similarly, *Tobacco ringspot virus* (TRSV, genus *Nepovirus*) was shown to be present in the megagametophyte and in pollen, but it was predominantly transmitted through the megagametophyte. Due to the poor quality of infected pollen, TRSV is mainly transmitted to soybean through infection of the megaspore mother cells (Yang and Hamilton, 1974). In *Lettuce mosaic virus* (LMV, genus *Potyvirus*), the virus should be able to infect pollen mother cells before the callose layer surrounds them (Hunter and Bowyer, 1997).

Studying the mechanism of CMV transmission in *Spinacia oleracea*, electron microscopic and reverse transcriptase–polymerase chain reaction (RT-PCR) observations confirmed the presence of CMV in almost all of the reproductive tissues, such as ovary wall cells, ovule integuments and nucellus, anther and seed coat walls. Further, its multiplication was demonstrated by the presence of amorphous bodies and other inclusions in anther parenchyma and tapetum cells. It was demonstrated that the virus infected the embryo indirectly; however, the possibility of direct infection was not ruled out. The observed lower frequency of CMV transmission in the progeny plants produced by infected female and healthy male plants as compared to that resulting from healthy female and infected male plants, was attributed to the lower viability of infected egg cells as compared to that of pollen (Yang et al., 1997). Therefore, in the CMV/*S. oleracea* combination, infection of zygote from infected pollen was suggested to be the major determinant of seed transmission. It is not known whether CMV is present on the surface or inside the pollen. For a better comparison, a nontransmissible CMV isolate should also have been included in this study. In *Alfalfa mosaic virus* (genus *Alfamovirus*; Frosheiser, 1974) and *Bean common mosaic virus* (genus *Potyvirus*; Morales and Castaño, 1987), pollen transmission was also found to be more effective than ovule transmission.

Direct Embryo Infection

In the case of direct infection, the embryo is infected after fertilization. Viruses are transmitted directly into the embryos early in development. This phenomenon has been well studied in *Pea seed-borne mosaic virus* (PSbMV, genus *Potyvirus;* Wang and Maule, 1992, 1994). PSbMV has a range of transmission in cultivars of *Pisum sativum*. Cultivar (cv.) Vedette shows 60 to 80 percent seed transmission, while there was no seed transmission in cv. Progreta. Detailed enzyme-linked immunosorbent assays (ELISA) and electron microscopic studies showed that before fertilization PSbMV was detectable in the floral tissues of sepals, petals, anthers, and carpels of both cultivars. However, it was absent in ovules, pollen, and integument of infected cv. Progreta or cv. Vedette (funicle tissues adjacent to the ovule showed some PSbMV-infected cells). The fertilization process pushed the virus into the ovule along the vascular strand and surrounding tissues in both the cultivars. In late embryonic stage, the virus continued moving from the vascular-associated tissues into the ovule until it reached the micropylar region of cv. Vedette. Whereas cv. Progreta saw hardly any invasion into ovule tissues surrounding the vascular strand, and the amount of virus in the earlier invaded tissues was significantly reduced. After fertilization, PSbMV infected embryo directly, multiplied in the embryonic tissues, and persisted during seed maturation in cv. Vedette. However, it was absent in the embryo of cv. Progreta, preventing virus transmission at this stage. It was demonstrated that in cv. Vedette, PSbMV reached the micropylar region of testa and entered into the embryo, at an early stage, through the suspensor (a transient structure responsible for supplying nutritional and positional support to the developing embryo). Transport of virus via the suspensor is possible because plasmodesmatal connections are present among the suspensor cells and between them and the embryo at an early developmental stage (Maule and Wang, 1996). However, inability of the virus to reach the suspensor before its degeneration stops the seed transmission process. In cv. Progreta, virus invasion of nonvascular tissues is stopped by either blockage of PSbMV movement or its replication, consequently keeping the virus from crossing the boundary between the maternal and progeny tissues. As during the process of meiosis, plasmodesmatal connections between megaspore mother cells and nucellar tissues and between pollen mother cells and the tapetal cells are lost, egg and embryo sac and pollen are separated from their respective parental tissues. The underlying transmission mechanism is also believed to be controlled by multiple maternal host genes that prevent the virus from reaching the suspensor. They control or regulate the ability of PSbMV to spread and multiply in the nonvascular tissues and finally halt the seed transmission in cv. Vedette (Wang and Maule, 1994). This is a complex process in which a specific virus-host interaction is believed to control seed transmissibility by preventing embryo invasion as a result of regulating replication and movement of the virus to reproductive tissues.

Possible roles of viral determinants—e.g., the coat protein (CP), non-structural viral gene products such as the N-terminal region of the helper component protinease (HC-Pro^N), and the 5'-untranslated region (5'-UTR) of PSbMV—in replication and movement will be discussed separately in the following section.

GENETIC DETERMINANTS

Identification of genetic determinants may facilitate the understanding of mechanisms involved in transmission of viruses through seeds. Availability of full-length complementary (c) DNA clones of several transmissible and nontransmissible viral isolates has made it possible to understand the molecular basis of virus transmission through seeds. In this section, based on pseudo-true recombinant studies, viral determinants of a few selected viruses are defined (see also reviews by Johansen et al., 1994; Maule and Wang, 1996).

Pea seed-borne mosaic virus

Complete nucleotide sequences of the seed-transmitted isolate P-1 (31 percent transmission) and practically non-seed-transmissible isolate P-4 (0.2 percent transmission) of PSbMV reveal so many differences that it is difficult to determine the possible roles of genomic parts involved in seed-based transmission. To understand the molecular basis of PSbMV seed transmission, chimeric viruses—namely, vP-1114, vP-4111, vP-1144, and vP-4144—were generated by exchanging various genomic parts between full-length cDNA clones of isolates P-1 and P-4. It was demonstrated that an increased transfer of P-1 nucleotide sequences into the P-4 genome proportionally increased transmissibility of isolate P-4. The chimeras vP-4111, vP-1144, and vP-1114 showed 4.6, 7, and 18 percent transmission, respectively, with regard to vP-1 (wild type). Particularly, the 5'-genomic end of the P-1 isolate, containing 5'-UTR, P1-Pro, HC-Pro^N, was shown to play a major role in seed transmission (Johansen, Doughtery, et al., 1996; see Figures 6.1. and 6.2).

FIGURE 6.1. Schematic Diagram Showing Chimeric Viruses Generated by Exchanging 5'- Genomic Parts of *Pea seed-borne mosaic virus* Isolates P-1 and P-4

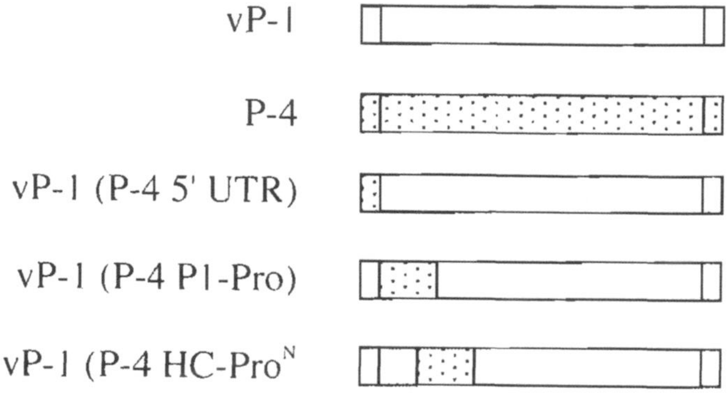

Source: Modified from Johansen, Dougherty, et al., 1996: 3151.

Note: vP-1 = infectious transcript of full-length cDNA clone of isolate P1; vP-1 (P-4 5'-UTR) = vP-1 representing 5'-untranslated region (UTR) of P-4 genome; vP-1 (P-4 P1-Pro) = vP-1 with P1-protease (P1-Pro) of P-4; vP-1 (P-4 HC-Pro^N) = vP-1 containing N-terminal region of helper component protinease (HC-Pro^N) of P-4.

For further analysis, 5'-UTR, P1-Pro, HC-Pro^N regions of isolate P-1 were exchanged with those of isolate P-4, and chimeric viruses vP-1 (P-4 5'-UTR), vP-1 (P-4 P1-Pro), and vP-1 (P-4 HC-Pro^N) were made (see Figure 6.1). The chimeric viruses represent full-length cDNA clones of isolate P-1, with 5'-UTR |vP-1(P-4 5'-UTR)|; with P1-protease |vP-1(P-4 P1-Pro)|; with HC-Pro^N |(vP-1(P-4 HC-Pro^N)| of isolate P- 4. *Pisum sativum* plants inoculated with chimeras vP-1(P-4 5'-UTR) and vP-1(P-4 HC-Pro^N) relatively reduced the seed transmissibility to 50 percent and 20 percent, respectively, in comparison with the wild-type transcript vP-1. In chimeric vP-1(P-4 P1-Pro), the frequency remained the same as in vP-1. Notably, the 5'-UTR and the HC-Pro^N regions of both isolates differ in size as well as in the amino acid level (Johansen et al., 1991; Johansen, Dougherty, et al., 1996). HC-Pro^N has significant roles in aphid transmission, replication, and long-distance movement of the virus (Atreya et al., 1992; Maia and Bernardi, 1996; Guo et al., 1999), while the 5'-UTR is involved in replication and transmission of potyviruses (Riechman et al., 1992, Niepel and Gallie, 1999). As discussed earlier, PSbMV is presumed to enter the embryo through a functional suspensor. If so, PSbMV should be available at the micropylar region before the disintegration of the suspensor. In this reference, it may be

FIGURE 6.2. *Pea seed-borne mosaic virus* (PSbMV) Chimeras (vP-1114, vP-1144, and vP-4144) Generated by Exchanging Genomic Parts of Full-Length cDNA Clones of Isolates DPD1 (P-1 Pathotype) and NY (P-4 Pathotype) of PSbMV

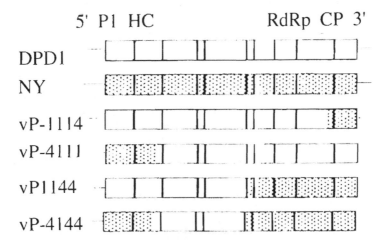

Source: Modified from Andersen and Johansen, 1998: 307.

Note: DPD1-derived sequences are in white and those of NY in black dots.

speculated that HC-Pro^N influences transmission efficiency by regulating replication and/or long-distance movement in reproductive tissues, thus influencing infection of embryos. In *Tobacco etch virus*, a well-characterized potyvirus, mutation in amino acids of CCCE of HC-Pro to APAN abolished the systemic movement (Cronin et al., 1995). In the isolates DPD1 and NY (identified as pathotypes P-1 and P-4, respectively), the corresponding region is, however, CSCV (Johansen, Keller, et al., 1996), which overrules the possible involvement of HC-Pro in the movement of PSbMV. Seed-based transmission of PSbMV actually involves several viral determinants and their expected influence on replication and movement of the virus, particularly in the reproductive tissues of infected hosts.

Besides HC-Pro, the CP gene significantly contributes to the long-distance movement of potyviruses. Isolates DPD1 and NY both infect *P. sativum* systemically. Interestingly, in *Chenopodium quinoa*, DPD1 causes systemic infection while NY is restricted to the inoculated leaf. The chimeric viruses vP-1114, vP-4111, vP-1144, and vP-4144 were generated from full-length cDNA clones of DPD1 and NY isolates by exchanging various genomic parts, as shown in Figure 6.2. The CP was detectable in leaves inoculated with all the chimeras. However, CP was present in upper noninoculated leaves of *C. quinoa* infected with chimera vP-4111. The chimeras vP-1114 and vP-4144 contain sequences from the 3'-genomic region of isolate NY, whereas vP-4111 does not contain sequences from 3'-NY, suggesting that the determinant of long-distance movement of PSbMV in *C. quinoa* is located in this genomic part. To analyze the 3'- region further, full-length clones exchanging CP genes of isolates NY and DPD1 were constructed (see Figure 6.3). The CP of chimeric NY(CP-DPD1) (full-length cDNA clone of NY with the CP of DPD1) was detectable in both inoculated and noninoculated leaves, whereas that of DPD1(CP-NY) (full-length cDNA clone of DPD1 with the CP of NY) was detectable in the inoculated leaf only. These recombinant studies on DPD1 and NY isolates identified the CP gene as the main determinant of the long-distance movement of isolate DPD1 in *C. quinoa* (Andersen and Johansen, 1998; see Figure 6.3).

In fact, N-terminal regions of the CP genes of isolates DPD1 and NY differ by 12 amino acids. Based on functional and structural significance, three potential amino acids—and i.e., threonine (position 11), serine (position 47), serine (position 130)—were selected. NY mutant full-length clones, namely, NY(T11R), NY(S47P), and NY(S130N), with changes from threonine to arginine, serine to proline, and serine to asparagine, respectively, were gen-

FIGURE 6.3. Schematic Representation of the *Pea seed-borne mosaic virus Coat Protein (CP) Chimeras*

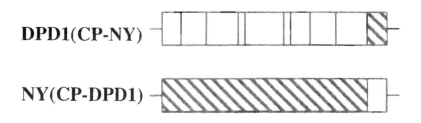

DPD1(CP-NY)

NY(CP-DPD1)

Source: Modified from Andersen and Johansen, 1998: 307.

Note: DPD1(CP-NY) represents a full-length cDNA clone of isolate DPDI (P-1 pathotype) and the CP gene of isolate NY (P-4 pathotype); NY(CP-DPD1) contains a full-length cDNA clone of isolate NY, with the CP gene from isolate DPD1.

erated. All three mutant viruses were detectable in inoculated leaves, while NYS47P) could be detected in noninoculated leaves. Further, mutational studies of the CP gene demonstrated that changing serine to proline at position 47 in the CP of NY allowed systemic spread of the NY(S47P) mutant. However, when proline was replaced with serine, the mutant DPD1(P47S) also lost its ability to systemically infect *C. quinoa* (Andersen and Johansen, 1998; see Figure 6.4).

To study further the specific role of the CP gene in the systemic spread of DPD1 and NY, a comparison was made among the CP gene sequences of ten isolates of PSbMV. Six isolates that were restricted to the inoculated leaves of *C. quinoa*, had serine at position 47. O the remaining four isolates, all systemically infecting, two (SL-25 and L-1) had proline at that position, while the other two (S6 and NEP-1), interestingly, had serine. Although these studies demonstrate proline/serine differences to be crucial for the systemic spread of DPD1 and NY isolates in *C. quinoa*, these amino acids are not solely responsible for the systemic spread of the virus (Andersen and Johansen, 1998). On the other hand, biological characteristics of these two isolates show that a specific interaction between virus and host is necessary for systemic movement. Both pathotypes P-1 and P-4 differ in their biologi-

FIGURE 6.4. Schematic Representation of the *Pea seed-borne mosaic virus* Coat Protein (CP) Mutants

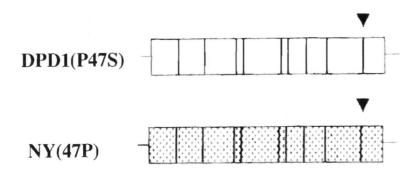

Source: Modified from Andersen and Johansen, 1998: 307.

Note: DPD1(P47S) = mutant full-length cDNA clone of isolate DPD1 in which amino acid proline at position 47 (▼) of the CP gene was replaced by serine; in NY(S47P) = mutant full-length cDNA clone of isolate NY in which the amino acid serine present at position 47 (▼) of the CP gene was replaced by proline.

cal reaction to *P. sativum* and *C. quinoa* genotypes. Moreover, *C. quinoa* supports systemic infection of DPD1, and not NY, which is restricted to the inoculated leaves only (Johansen, Keller, et al., 1996). PSbMV resistance genes *sbm-1, sbm-3, sbm-4,* which give resistance to pathotypes P-1, P-2, and P-4, respectively, are located on chromosome 6 of the pea genotypes (Gritton and Hagedorn, 1975; Provvidenti and Alconero, 1988a,b). Whereas P-1 is unable to replicate in homozygous recessive *sbm-1* pea genotypes, P-4 is infectious in the *sbm-1/sbm-1* genotype. Studies of the *sbm-1/sbm-1* protoplast transfected with P-1 showed that resistance occurs at the cellular level and that blockage of cell-to-cell movement is not the actual cause of resistance. Further, recombinant studies exchanging the full-length clones of P-1 and P-4 demonstrated that the infectious recombinant clones containing the VPg (genome-linked viral protein) of P-4 were able to overcome the *sbm-1* resistance, while recombinants with VPg of P-1 were noninfectious to pea genotypes with *sbm-1/sbm-1* genes. Since VPg is involved in replication, it was assumed that a disruption of P-1 replication by the *sbm-1* genes occurred at an early stage of infection (Keller et al., 1998). Furthermore, maternal genes also determine resistance to seed transmission of PSbMV. Cross-pollination experiments involving cv. Vedette (seed transmissible) and cv. Progreta (non-seed transmissible) demonstrated that seed transmission resistance genes were functional in the maternal tissues of the seed, and that genotype of the progeny embryo did not influence transmission. It was shown that a few maternal genes segregated as quantitative trait loci determined seed transmission of PSbMV (Wang and Maule, 1994).

Replication capacity and the systemic movement of PSbMV are supposed to be the two main factors responsible for virus transmission through seed. In most cases, however, virus movement plays a more important role than its replication behavior. This was shown by experiments in which virus movement, particularly at the point of unloading the virus from phloem into mesophyll tissue, was blocked. In contrast, at an early stage of infection of *sbm-1* pea genotypes with P-1 pathotypes, replication was a more important factor. Different isolates behave differently in the same or different cultivars of the same host. Thus, several viral/host determinants are involved in the seed transmission process of PSbMV.

Cucumber mosaic virus

Cucumber mosaic virus (CMV) contains three single-stranded RNAs, namely, RNA1, RNA2, and RNA3. Strain Pg of CMV is seed transmissible in *Phaseolus vulgaris,* whereas strain Le is not seed transmissible. To study the role of genomic RNA(s) in seed transmissibility, pseudorecombinants were produced from these two strains. Pseudorecombinant studies showed that seed transmissibility was associated with RNA1 of strain Pg (Hampton and Francki, 1992). RNA1, involved in viral replication and movement, was shown to regulate seed transmissibility of pseudorecombinants produced from the genomic RNAs of CMV-Le and CMV-Pg strains. Regarding the involved mechanism, it was suggested that CMV replication influences seed transmission after viral entry into the embryo, contrary to the postulated before-entry influence of PSbMV (Hampton and Francki, 1992; Wang and Maule, 1994).

Similar pseudorecombinant studies with *Raspberry ringspot virus* (RRSV, genus *Nepovirus*) and *Tomato black ring virus* (TBRV) showed RNA1 as the determinant of seed transmissibility (Hanada and Harrison, 1977). In pseudorecombinants from CMV, RRSV, and TBRV, however, seed transmissibility was lower than in their respective wild types. Incompatibility between heterologous RNAs (as observed in pseudorecombinant studies) is suspected to alter the viral function, thereby reducing transmission. However, the influence of other minor determinants on seed transmission may not be ruled out (Hampton and Francki, 1992). Although slow virus movement and its restricted replication limit the accumulation of virus in reproductive and embryo meristems necessary for seed transmission, so far no plant genotype is known to resist seed transmission by restricting either virus movement or virus replication.

Pea early-browning virus

Pea early-browning virus (PEBV, genus *Tobravirus*) has a bipartite single-stranded RNA genome, consisting of RNA1 and RNA2. The RNA2 encodes the CP and three nonstructural genes that are involved in nematode transmission (MacFarlane and Brown, 1995; MacFarlane et al., 1996; see Chapter 4 of this book). Despite the different size of RNA2 and the absence of nonstructural genes in *Pepper ringspot virus* (genus *Tobravirus*), its seed transmissibility characteristics show that RNA2 is not involved in seed transmission (Costa and Kitajima, 1968, as cited in Wang et al., 1997; Bergh et al., 1985; MacFarlane and Brown, 1995). Mutational studies of nonstructural genes of PEBV also favor the hypothesis that RNA1 is involved in seed transmission. Removal or disruption of the nonstructural gene coding sequence of RNA2 did not affect or abolish seed transmissibility of PEBV.

The RNA1 encodes four polypeptides of 201, 141, 30, and 12 kDa, respectively (201K, 141K, 30K, and 12K genes; MacFarlane et al., 1989). No defined function of the 12K gene and its resemblance to HC-Pro and the RNAγ gene (involved in PSbMV and BSMV seed transmission, respectively), prompted a study of the possible role of the 12K gene in seed transmissibility. Two mutants (△12K-deletion mutant, 12FS frame shift) were produced from the 12K gene of the full-length cDNA clone of RNA1 (Wang et al., 1997; see Figure 6.5). After inoculating *Nicotiana benthamiana* with mutants △12K or 12FS, no symptoms were seen in 12FS-inoculated *N. benthamiana*. Moreover, 12FS was not detectable in either inoculated or noninoculated leaves. On the other hand, the mutant △12K caused necrotic symptoms on pea pods and leaves. It was absent in pollen grains, eggs, or ovules, with poor detection in anthers and carpels. It was detectable in the vascular tissues of anthers and carpels, with a higher concentration in the vegetative tissues and pods than that of the wild type. If transmitted through seed, the transmission frequency was lower than 1 percent. In contrast, the wild type was present in all the male as well as female reproductive tissues (Wang and Maule, 1997). Under normal circumstances, infection of the gametes is a prerequisite for seed transmission of PEBV. As there was no infection of gametes in the △12K-infected plants, a possible role of △12K protein in the infection of gametes is suggested. Further, increased virus concentration in

FIGURE 6.5. *Pea early-browning virus* mutants (12FS, ▲12K) Derived from the RNA1 of the 12K Gene

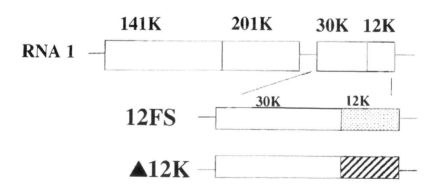

Source: Modified from Wang et al., 1997: 114.

Note: Mutation in PEBV RNA1 was introduced as a result of frame shift (12FS) or deletion (12K) of the 12K gene.

the vegetative tissues (Δ12K-infected plants) showed that the Δ12K gene behaves in a tissue/organ-specific manner, regulating the movement of the virus and, consequently, seed transmission. Moreover, the Δ12K gene has cysteine-rich domains similar to HC-Pro (PSbMV) and γ genes (BSMV), the potential viral determinants of seed transmission (discussed later).

Barley stripe mosaic virus

Barley stripe mosaic virus (BSMV) contains three single-stranded RNAs, namely, α, β, and γ. RNAα has a single open reading frame (ORF)-translating putative replicase; RNAβ encodes the CP and three nonstructural proteins; RNAγ encodes a polymerase protein (γa) and a 3'-nonstructural gene product (γb) that is expressed by a subgenomic RNA.

To understand the role of viral determinants involved in BSMV seed transmission, recombinants were generated between RNAγs of strains CV17 and ND18 by a progressive substitution of CV17 RNAγ sequences into ND18 RNAγ and vice versa. Strain ND18 shows 64 percent seed transmission in cv. Dickson of *Hordeum vulgare*, whereas strain CV17 has practically no seed transmission (less than 1 percent) in the same cultivar. Deletion of a 369 nucleotide (nt) repeat present in the CV17 RNAγ increased seed transmission frequency as compared to wild-type CV17 RNAγ. Coinoculation of ND18 RNAs α and β did not show any increase in seed transmission (Edwards, 1995; see Figure 6.6).

Seed transmissibility was observed to be increased by substitution of ND18 RNAγ leader for CV17 (18/17X versus 17X; see Figure 6.6). However, reciprocal substitution proved deleterious to seed transmission (17/18X versus ND18γ; see Figure 6.7). Interestingly, increasing participation of the ND18 RNAγ gene did not increase the seed transmission frequency proportionally (18/17A; 18/17K; see Figure 6.6), irrespective of parental source of RNAs α and β. Two small ORFs are present in the leader of strain CV17 but absent in strain ND18. The mutation in these ORFs did not influence the frequency of transmission.

On the other hand, seed transmission frequency was reduced by substituting CV17 γa sequence for ND18 γa sequence (17/18A or K versus 17/18X; see Figure 6.7). This substitution, however, added the 369 nt repeat to the RNAγ construct. To check for a possible role of this repeat in seed transmission, the repeat was deleted from the 17/18A,K,M,H RNAγ recombinants and 17/18AX,KX,MX,HX RNAγ recombinants were generated. Notably, no reduction in the seed transmission frequency occurred (Edwards, 1995; see Figure 6.8).

FIGURE 6.6. Schematic Diagram Showing RNAγ Recombinants Generated by Deleting the 369 Nucleotide Repeat (17X) and Progressive Substitution of RNAγ Sequences of Strain CV17 of *Barley stripe mosaic virus* into RNAγ of Strain ND 18 (18/17X,A,K,M,H)

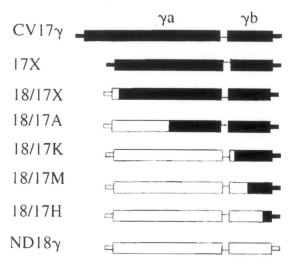

Source: Modified from Edwards, 1995: 911.

The sequence exchanges between the γ genes of strains ND18 and CV17 affect the frequency of seed transmission (18/17M,H and 17/18M,H; 17/18MX,HX; see Figures 6.6, 6.7, and 6.8). Although seed transmission frequency increased when ND18 γb sequences were substituted for those of CV17, the reciprocal substitution decreased the frequency. Although the obtained results were independent of the parental source of RNAs α and β, it was most striking when γb recombinants were coinoculated with RNAs α and β of strain CV17. An increased participation of the ND18 γb gene remarkably increased seed transmission frequency (18/17K versus 18/17M,H; see Figure 6.6), whereas increased participation of the CV17 γb gene resulted in a significant decrease in seed transmission (17/18K versus 17/18M,H; see Figure 6.7; Edwards, 1995).

These results indicate that, in BSMV, the 5'-untranslated leader and a 369 nt repeat in the γa gene and the γb gene are major viral determinants of seed transmission. They affect seed transmission frequency by regulating the replication and movement of BSMV. For instance, the 5'-leader is an untranslated region involved in replication and gene expression. It appears that mutation in the

FIGURE 6.7. Schematic Diagram of RNAγ Recombinants Generated by Substituting RNA γa Sequence of Strain CV17 of *Barley stripe mosaic virus* for That of Strain ND18, Resulting in the Addition of a 369 Nucleotide Repeat (17/18A,K, M,H)

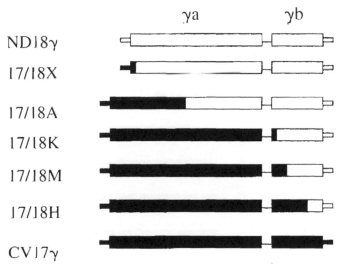

Source: Modified from Edwards, 1995: 911.

5'-leader changes its secondary structure, which probably regulates replication and translation. Whereas the presence of a 369 nt repeat in the γa gene blocked seed transmission, its absence increased the frequency, most likely by enhancing the γa-encoded replicase, resulting in improved replication and movement. The γb gene, a cysteine-rich product, is believed to play a key role in gene expression and virulence. Its mutation affects both transmission and symptom phenotype. Although the exact role of γb in replication is not yet fully explained, the protein contains two zinc finger-like cysteine-rich domains having RNA-binding capacity (Donald and Jackson, 1994, 1996). Similarly, the 12K gene of PEBV, HC-Pro of PShMV, and the γb gene of BSMV, representing three different genera of plant viruses, have cysteine-rich products that influence seed transmissibility by affecting replication and the movement of viruses. Although virus replication and/or movement may strongly influence seed transmissibility, the mechanism of BSMV transmission influenced by these viral determinants is not yet fully understood. Moreover, resistance to BSMV seed transmission in barley cv. Modjo is also controlled by a single recessive host gene (Carroll et al., 1979).

FIGURE 6.8. Schematic Diagram of RNAγ Recombinants (17/18AX,KX,MX,HX) Generated As a Result of Deleting the 369 Nucleotide Repeat from the RNAγ Recombinants 17/18 A,K,M,H (See Figure 6.7)

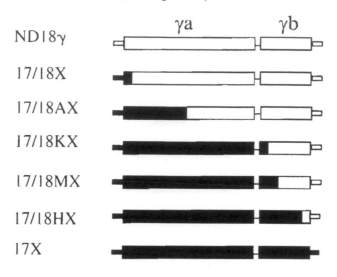

Source: Modified from Edwards, 1995: 911.

CONCLUSIONS

Although some progress has been made in understanding the biological characteristics and transmission pathways of a few seed-borne viruses during the past decade, much remains to be studied. Identification of common factors or processes implicated in transmission pathways of different viruses will greatly contribute to a better understanding of the involved mechanisms. Regarding host genetic determinants, only a few host genes are known to be involved in seed transmission, and molecular mapping may facilitate the identification of such genes in other virus-host combinations.

Results from experiments with full-length cDNA clones have shown, among others, the HC-Pro (PSbMV), 12K (PEBV), and γb (BSMV) genes are the major viral genetic determinants. These genes of viruses belonging to different virus genera have potential roles in movement and replication. Interestingly, all of them have common cysteine-rich zinc finger-like domains involved in RNA-binding activities. As replication and movement of viruses have been suggested to be the main factors in transmission, it is likely that cysteine-rich domains influence transmissibility by affecting these factors. However, substitution of cysteine and histidine to disrupt the fin-ger-like motifs of BSMV γb failed to have appreciable effects on nucleic acid binding (Donald and Jackson, 1996). Similarly, potyvirus HC-Pro gene prod-ucts have conserved cysteine and histidine and possess RNA-binding activ-ity. The HC-Pro gene is involved in genome amplification and long-distance movement. Substitution of conserved cysteine-rich motifs showed moder-ate effect on genome amplification, no effect on cell-to-cell movement, and a very weak effect on long-distance movement (Maia and Bernardi, 1996; Kasschau et al., 1997). The significance of cysteine-rich domains in seed transmission remains speculative, and much more research is needed to elu-cidate their putative role in seed transmission.

REFERENCES

Andersen K, Johansen E (1998). A single conserved amino acid in the coat protein gene of pea seed-borne mosaic potyvirus modulates the ability of the virus to move systemically in *Chenopodium quinoa. Virology* 241: 304-311.

Atreya CD, Atreya PL, Thornbury DW, Pirone TP (1992). Site-directed mutations in the potyvirus HC-Pro gene affect helper component activity, virus accumula-tion, and symptom expression in infected tobacco plants. *Virology* 191: 106-111.

Bergh ST, Koziel MG, Huang S-C, Thomas RA, Gilley DP, Siegel A (1985). The nucleotide sequence of tobacco rattle virus RNA2 (CAM strain). *Nucleic Acids Research* 13: 8507-8518.

Broadbent L (1965). The epidemology of tomato mosaic. XI. Seed transmission of TMV. *Annals of Applied Biology* 56: 177-205.

Carroll TW (1981). Seedborne viruses virus-host interactions. In Maramorosch K, Harris KF (eds.). *Plant Disease and Vectors: Ecology and Epidemiology*. Aca-demic Press, New York. pp. 293-317.

Carroll TW, Gossel PL, Hockel EA (1979). Inheritance of resistance to seed trans-mission of barley stripe mosaic virus in barley. *Phytopathology* 69: 431-433.

Carroll TW, Mayhew DE (1976a). Anther and pollen infection in relation to the pol-len and seed transmissibility of two strains of barley stripe mosaic virus in bar-ley. *Canadian Journal of Botany* 54: 1604-1621.

Carroll TW, Mayhew DE (1976b). Occurrence of virions in developing ovules and embryo sacs in relation to the seed transmissibility of barley stripe mosaic virus. *Canadian Journal of Botany* 54: 2497-2512.

Cronin S, Verchot J, Haldeman-Cahill R, Schaad MC, Carrington JC (1995). Long distance movement factor: A transport function of the potyvirus helper compo-nent protease. *Plant Cell* 7: 549-559.

Crowley NC (1959). Studies on the time of embryo infection by seed-transmitted viruses. *Virology* 8: 116-123.

Donald RGK, Jackson AO (1994). The barley stripe mosaic virus γb gene encodes a multifunctional cysteine-rich protein that affects pathogenesis. *Plant Cell* 6: 1593-1606.

Donald RGK, Jackson AO (1996). RNA-binding activities of barley stripe mosaic virus γb fusion proteins. *Journal of General Virology* 77: 879-888.

Edwards MC (1995). Mapping of the seed transmission determinants of barley stripe mosaic virus. *Molecular Plant Microbe-Interactions* 8: 906-915.

Frosheiser FI (1974). Alfalfa mosaic virus transmission seed through alfalfa gametes and longevity in alfalfa seed. *Phytopathology* 64: 102-105.

Gritton ET, Hagedorn DJ (1975). Linkage of the genes *sbm* and *wlo* in peas. *Crop Science* 15: 447-448.

Guo D, Merits A, Saarma M (1999). Self association and mapping of interaction domains of helper component-proteinase of potato A potyvirus. *Journal of General Virology* 80: 1127-1131.

Hampton RO, Francki RIB (1992). RNA-1 dependent seed transmissibility of cucumber mosaic virus in *Phaseolus vulgaris*. *Phytopathology* 82: 127-130.

Hanada K, Harrison BD (1977). Effects of virus genotype and temperature on seed transmission of nepoviruses. *Annals of Applied Biology* 85: 79-92.

Hibi, T and Furuki I (1985). Melon necrotic spot virus. In *AAB Description of Plant Viruses*, No. 302, AAB, Wellesbourne, Warwick, UK. 4pp.

Hunter DG, Bowyer JW (1997). Cytopathology of developing anthers and pollen mother cells from lettuce plants infected by lettuce mosaic potyvirus. *Journal of Phytopathology* 1 145: 521-524.

Johansen E, Edwards MC, Hampton RO (1994). Seed transmission of viruses: Current perspectives. *Annual Review of Phytopathology* 32: 363-386.

Johansen E, Rasmussen OF, Heide M, Borkhardt B (1991). The complete nucleotide sequence of pea seed-borne mosaic virus RNA. *Journal of General Virology* 72: 2625-2632.

Johansen IE, Dougherty WG, Keller KE, Wang D, Hampton RO (1996). Multiple viral determinants affect seed transmission of pea seedborne mosaic virus in *Pisum sativum*. *Journal of General Virology* 77: 3194-3154.

Johansen IE, Keller KE, Dougherty WG, Hampton RO (1996). Biological and molecular properties of a pathotype P-1 and a pathotype P-4 isolate of pea seed-borne mosaic virus. *Journal of General Virology 77: 1329-1333*.

Kasschau KD, Cronin S, Carrington JC (1997). Genome amplification and long-distance movement functions associated with the central domain of tobacco etch potyvirus helper-component-proteinase. *Virology* 228: 251-262.

Keller KE, Johansen E, Martin RR, Hampton RO (1998). Potyvirus genome-linked protein (VPg) determines pea seedborne mosaic virus pathotype-specific virulence in *Pisum sativum*. *Molecular Plant-Microbe Interactions* 11: 124-130.

MacFarlane SA, Brown DJF (1995). Sequence comparison of RNA2 of nematode-transmissible and nematode-nontransmissible isolates of pea early-browning virus suggests that the gene encoding the 29 kDa protein may be involved in nematode transmission. *Journal of General Virology* 76: 1299-1304.

MacFarlane SA, Taylor SC, King DI, Hughes G, Davies JW (1989). Pea early browning virus RNA 1 encodes four polypeptides including a putative zinc-finger protein. *Nucleic Acids Research* 17: 2245-2260.

MacFarlane SA, Wallis CV, Brown DJF (1996). Multiple virus genes involved in the nematode transmission of pea early browning virus. *Virology* 219: 417-422.

Maia IG. Bernardi F (1996). Nucleic acid-binding properties of a bacterially expressed potato virus Y helper component-proteinase. *Journal of General Virology* 77: 869-877.

Maule AJ, Wang D (1996). Seed transmission of plant viruses: A lesson in biological complexity. *Trends in Microbiology* 4: 153-158.

McDonald JG, Hamilton RJ (1972). Distribution of southern bean mosaic virus in the seed of *Phaseolus vulgaris*. *Phytopathology* 62: 387-389.

Mink GI (1993). Pollen and seed transmitted viruses and viroids. *Annual Review of Phytopathology* 31: 375-402.

Morales FJ, Castaño M (1987). Seed transmission characteristics of selected bean common mosaic virus strains in differential bean cultivars. *Plant Disease* 71: 51-53.

Niepel M, Gallie DR (1999). Identification and characterization of the functional elements within the tobacco etch virus 5' leader required for cap-independent translation. *Journal of Virology* 73: 9080-9088.

Pringle CR (1999) Virus taxonomy. The Universal System of Virus Taxonomy, updated to include the new proposals ratified by the International Committee on Taxonomy of Viruses during 1998. *Archives of Virology* 144: 421-429.

Provvidenti R, Alconero R (1988a). Inheritance of resistance to a lentil strain of pea seed-borne mosaic virus in *Pisum sativum*. *Journal of Heredity* 79: 45-47.

Provvidenti R, Alconero R (1988b). Inheritance of resistance to a third pathotype of pea seed-borne mosaic virus in *Pisum sativum*. *Journal of Heredity* 79: 76-77.

Riechman JL, Lain S, García JA (1992). Highlights and prospects of potyvirus molecular biology. *Journal of General Virology* 73: 1-16.

Taylor RH, Grogan RG, Kimble KA (1961). Transmission of tobacco mosaic virus in tomato seed. *Phytopathology* 51: 837-842.

Timian RG (1974). The range of symbiosis of barley and barley stripe mosaic virus. *Phytopathology* 64: 342-345.

Uyemoto JK, Grogan RG (1977). Southern bean mosaic virus: Evidence for seed transmission in bean embryos. *Phytopathology* 67: 1190-1196.

Van den Heuvel JFJM, Franz AWE, Van der Wilk F (1999). Molecular basis of virus transmission. In Mandahar CL (ed.), *Molecular Biology of Plant Viruses*. Kluwer Academic Publishers, Boston, Dordrecht, London, pp. 183-200.

Wang D, MacFarlane SA, Maule AJ (1997). Viral determinants of pea early browning virus seed transmission in pea. *Virology* 234: 112-117.

Wang D, Maule AJ (1992). Early embryo invasion as a determinant in pea of the seed transmission of pea seed-borne mosaic virus. *Journal of General Virology* 73: 1615-1620.

Wang D, Maule AJ (1994). A model for seed transmission of a plant virus: Genetic and structural analysis of pea. *Plant Cell* 6: 777-787.

Wang D, Maule AJ (1997). Contrasting patterns in the spread of two seed-borne viruses in pea embryos. *Plant Journal* 11: 1333-1340.

Wang D, Woods RD, Cockbain AJ, Maule AJ, Biddle AJ (1993). The susceptibility of pea cultivars to pea seed-borne mosaic virus infection and virus transmission in the UK. *Plant Pathology* 42: 42-47.

Xu Z, Chen K. Zhang Z, Chen J (1991). Seed transmission of peanut stripe virus in peanut. *Plant Disease* 75: 723-726.

Yang AF, Hamilton RI (1974). The mechanism of seed transmission of tobacco ring spot virus in soybean. *Virology* 62: 26-37.

Yang Y, Kim KS. Andersen EJ (1997). Seed transmission of cucumber mosaic virus in spinach. *Phytopathology* 87: 924-931.

Chapter 7

Molecular Biology
of Plant Virus Movement

Ayala L. N. Rao
Yoon Gi Choi

INTRODUCTION

One of the most important prerequisites for a plant virus to establish an infection is its ability to enter and be successfully transported from the point of entry to other parts of a given host plant. The cell wall is a formidable barrier to a plant virus. Not only does this rigid component offer structural integrity to a plant, it is one of the first lines of defense against a viral infection. Unlike animal viruses, which enter cells through receptor-mediated endocytosis, most plant viruses must physically penetrate the cell wall barrier to infect a host. No known receptors mediate the initial transfer of a plant virus into a plant cell. Plant viruses enter a plant cell by mechanical inoculation carried out by humans, animals, insects, nematodes, or fungi. Once inside a cell, viruses may undergo replication, and their movement between cells is less hindered by the structure of the cell wall. The replicated progeny can move and accumulate in a plant until the virus is vectored to another host. Outside the plant, the virus is dependent upon vectors for its transport. By contrast, inside the plant, the virus can facilitate its own movement. Comprehending this process of virus-mediated intercellular transport entails an understanding of the similarities between the cellular transport mechanisms of heterologous viruses, the interactions between a virus and its host, and the plant defense responses that can inhibit further viral movement after the cell wall barrier has been overcome.

Lately, much attention has been focused on elucidating various molecular mechanisms underlying the plant virus movement process. The purpose of this chapter is to inform plant virologists of the current status of knowledge on virus movement and not to elaborate on the literature concerning this subject. Readers are advised to refer to some excellent reviews that have appeared in recent years (Lucas and Gilbertson, 1994; Carrington et al., 1996; Lazarowitz and Beachy, 1999).

MOVEMENT OF PLANT VIRUSES

Plant virus movement is divided into two phases: (1) cell to cell, or short distance, and (2) long distance. While describing short-distance movement, two terms are normally used: *cell-to-cell movement* and *subliminal infection*. In a susceptible host plant, when an invading virus is transported sequentially from initially infected epidermal cells through the mesophyll, bundle sheath, and phloem parenchyma, it is said to have moved *cell to cell* (Carrington et al., 1996). In the absence of such cell-to-cell movement, the infection is confined to the initially infected cell and said to be *subliminal* (Cheo, 1970; Schmitz and Rao, 1996). The advent of recombinant DNA technology together with our ability to transcribe in vitro the desired RNA sequences and test their infectivity in whole plants allowed precise identification of viral genes involved in the movement process. If not all, the majority of plant viruses encode a nonstructural protein, referred to as a movement protein (MP), exclusively dedicated to promoting viral movement between cells. However, it is becoming increasingly evident that in some viral systems, in addition to MP, the structural or coat protein (CP) is also required to mediate this active process. Thus, the overall movement process can either be coat protein independent or coat protein dependent.

Coat Protein-Independent Movement

In those viral systems which do not require the CP for cell-to-cell movement, the MP alone is sufficient (see Figure 7.1A). The best-understood example is *Tobacco mosaic virus* (TMV, genus *Tobamovirus*). TMV is a rod-shaped, single-stranded RNA virus. Its 6.4 kb genome encodes four open reading frames (ORFs). The first two genes encode replicase proteins and the fourth encodes the structural CP (Dawson and Lehto, 1990). The third gene specifies the production of a 30 kDa protein that is not required for replication or encapsidation. A TMV mutant with deletions in this gene replicates and encapsidates in protoplasts but does not move systemically in plants (Meshi et al., 1987). This suggests that the 30 kDa protein is involved in viral spread. The Ls1 mutant strain of TMV does not infect tobacco at 32°C, whereas the parental L strain remains infectious (Nishiguchi et al., 1978). Later, it was shown that Ls1 would infect tobacco in the presence of L at 32°C. This implies that L can complement the movement function of Ls1

FIGURE 7.1. Schematic Diagram Showing Coat Protein-Independent (A) and Coat Protein-Dependent (B) Cell-to-Cell Movement in Selected Plant Virus Groups

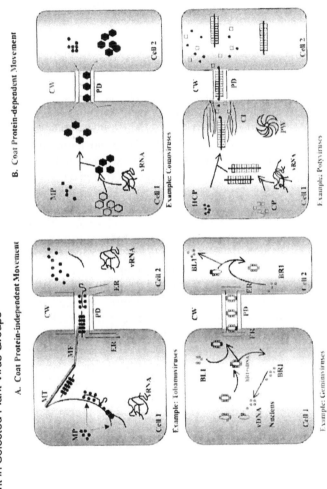

Note: BL1 and BR1 = two genes on DNA B of Bean dwarf mosaic virus; CI = cytoplasmic inclusion; CW = cell wall; ER = endoplasmic reticulum; HCP = helper component proteinase; MF = microfilament; MP = movement protein; MT = microtubules; PD = plasmodesmata; PW = pinwheels; vDNA = viral DNA; vRNA = viral RNA.

(Taliansky, Atabekova, et al., 1982). Sequence analyses showed that L and Lsl differ by a single amino acid change of a serine to a proline in the 30 kDa protein, and this finding supported the hypothesis that this protein was responsible for viral movement (Ohno et al., 1983). Deom et al. (1987) regenerated transgenic plants expressing the 30 kDa gene of TMV Ul which would support the systemic spread of Lsl at 32°C. This conclusive evidence linked the 30 kDa gene product to functions associated with systemic spread and identified the 30 kDa protein as an MP.

The virus moves from cell to cell via plasmodesmata, which are, however, too small to allow free passage of virions or viral genomes (the gateway capacity or size exclusion limit [SEL] is not sufficient). Since a virus must overcome this limitation, the MP might have increased the SEL of the plasmodesmata in the aforementioned transgenic plants. To test this, fluorescent molecules of difference sizes were injected into mesophyll cells of transgenic and nontransgenic plants. Molecules no larger than 0.7 kDa moved from cell to cell in nontransgenic plants, whereas 9.4 kDa molecules moved from cell to cell in the transgenic plants that accumulate the TMV MP (Wolf et al., 1989). Although the plasmodesmata could accommodate the passage of these large molecules, which were predicted to have diameters between 2.4 and 3.1 nm, the modified plasmodesmatal SEL was still not large enough for the passage of virions or free-folded viral RNA. However, the modified plasmodesmata could allow the passage of viral RNA as a single-strand complex. Since TMV mutants unable to encapsidate can move from cell to cell (Saito et al., 1990), the virus must be able to move from cell to cell either as a naked RNA or as a virus-specific ribonucleoprotein complex (vRNP; Dorokhov et al., 1983). In either case, the viral RNA could traverse the plasmodesmata in the presence of an MP.

A number of other biochemical properties of the TMV MP have been determined. Ultrastructural studies have demonstrated the TMV MP in transgenic plants (Tomenius et al., 1987; Atkins et al., 1991; Ding et al., 1992), and TMV MP is found in purified subcellular cell wall and organelle fractions from infected plants (Moser et al., 1988). The MP accumulates in areas of cytoplasm outside the nuclei of infected protoplasts (Meshi et al., 1992). The MP can be found at high levels in the cell wall fraction derived from older leaves of transgenic plants (Deom et al., 1990). Injected 9.4 kDa molecules move through old leaves of transgenic plants but not through developed but unexpanded young leaves. The lack of movement of 9.4 kDa molecules in young leaves that do accumulate MP suggests that young plasmodesmata resist having their SELs modified. This contradicts the observation that young leaves are susceptible to infection and, thus, movement. Therefore, either the TMV MP must reach a threshold concentration level to modify SELs or young plasmodesmata can

regulate their SELs. Fluorescent molecules of 20 kDa, coinjected with purified TMV MP, move into adjacent and distal mesophyll cells. Thus, the MP
can quickly modify plasmodesmata and is able to move from cell to cell, or it
sends a signal to modify the SELs of neighboring plasmodesmata (Waigmann
et al., 1994).

Functional domains of the TMV MP have been determined. Nonviral RNA
binds to the amino acids 65 through 86 of the MP (Citovsky et al., 1990). These
TMV MP/RNA complexes form long thin strands that could be transported
across modified plasmodesmata as vRNPs (Citovsky et al., 1992; Ivanov et al.,
1994). The deletion of up to 55 amino acids from the carboxyl-terminus (amino
acids 214 through 268) of the TMV MP does not drastically affect the movement of TMV. These amino acids are not essential for accumulation of MP on
cell walls, plasmodesmata modification, or viral movement complementation.
Deletion of as few as three amino acids from the amino-terminus disrupts
viral movement altogether (Gafny et al., 1992).

Viral MPs Similar in Function to the TMV MP

In recent years, several other plant viral MPs have been characterized.
The 3a gene of *Cucumber mosaic virus* (CMV, genus *Cucumovirus*) is dispensable for replication and encapsidation (Boccard and Baulcombe,
1993). Kaplan et al. (1995) identified the 3a protein as an MP by showing
that transgenic plants accumulating a CMV 3a protein would support the
movement of CMV with a defective 3a gene. CMV MP can be found in cell
wall subcellular fractions isolated from transgenic plants, and it potentiates
the cell-to-cell movement of 10 kDa molecules (Vaquero et al., 1994).
Since the CMV virions are 29 nm in diameter, the virus likely moves as a
vRNP. CMV MP will move from cell to cell and help move CMV RNA
(Ding et al., 1995). Unlike TMV, CMV needs its CP for efficient cell-to-cell
movement (Suzuki et al., 1991). Amino acids 209 through 236 of the CMV
MP are critical for viral movement and are analogous to a similar region of
amino acids in the TMV MP (Kaplan et al., 1995).

An MP gene in *Alfalfa mosaic virus* (AMV, genus *Alfamovirus*) has been
identified (Van der Kuyl et al., 1991). This AMV MP moderately modifies
the SELs of plasmodesmata (Poirson et al., 1993), and it localizes to the cell
walls of infected tissue and is found in cell wall subcellular fractions likely
to contain plasmodesmata (Godefroy-Colburn et al., 1986; Stussi-Garaud
et al., 1987). Similar to CMV, AMV needs its CP and its MP for efficient
cell-to-cell movement (Van der Kuyl et al., 1991). Amino acids 36 through
81 of the AMV MP comprise an RNA-binding domain (Schoumacher et al.,
1994) and amino acids 13 through 77 are necessary for cell wall association

(Erny et al., 1992). The positions of these functional domains in the AMV MP, however, do not correspond with the positions of the analogous functional domains in the TMV MP.

The genome of *Red clover necrotic mosaic virus* (RCNMV, genus *Dianthovirus*) is divided, and RNA1 can replicate in the absence of RNA2 (Osman and Buck, 1987). RNA2 encodes a single 35 kDa MP that has been detected in the cell wall subcellular fractions from infected plants (Osman and Buck, 1991). This MP potentiates its own cell-to-cell movement in RCNMV RNA. Since the RCNMV MP does not unfold RNA by itself, host factors may be necessary to unfold the vRNP complex (Fujiwara et al., 1993).

Geminiviruses are single-stranded DNA viruses that also encode MPs. Whitefly-transmitted geminiviruses with bipartite genomes include *Tobacco golden mosaic virus*, *Bean dwarf mosaic virus* (BDMV) and *African cassava mosaic virus* (all three belonging to the genus *Begomovirus*), whereas the leafhopper-transmitted geminiviruses with undivided genomes include *Maize streak virus* (genus *Mastrevirus*) and *Beet curly top virus* (genus *Curtovirus*) (Lazarowitz et al., 1989). DNA A of geminiviruses with bipartite genomes encodes genes for encapsidation and replication, and DNA B encodes genes essential for viral movement (Harrison, 1985). Mutations in *BR1* or *BL1*, the two genes on DNA B of BDMV, abolish systemic movement but not replication of the virus (Noueiry et al., 1994). BL1 proteins have been found in cell wall subcellular fractions from systemic hosts (Von Arnim et al., 1993). BL1 also potentiates its own cell-to-cell movement, modifies plasmodesmata SELs, and moves double-stranded DNA from cell to cell. On the other hand, BR1 transports single- and double-stranded DNA from the nucleus (Noueiry et al., 1994). Since geminiviruses replicate in the cell nucleus, BR1 possibly moves replicated DNA from the nucleus to the cytoplasm so that BL1 can transport the viral DNA to the neighboring cell. Geminiviruses are phloem limited in certain hosts. Consequently, the MPs may fail to facilitate mesophyll cell invasion in these hosts, or there may be other MP roles for geminiviruses when they are phloem limited (Harrison, 1985). Since CPs are not needed for efficient geminivirus movement (Gardiner et al., 1988), the MPs might be responsible for transport from companion cells to sieve elements. Geminiviruses with undivided genomes are also thought to encode similar proteins (Lazarowitz et al., 1989).

Other proteins that provide viral movement are less similar to the TMV MP. *Potato virus X* (PVX, genus *Potexvirus*), *Barley stripe mosaic virus* (BSMV, genus *Hordeivirus*), *Beet necrotic yellow vein virus* (genus *Benyvirus*) and *Potato virus M* (genus *Carlavirus*) have three potentially overlapping internal ORFs known as a triple gene block (TGB) that is involved in viral transport (Davies et al., 1993). Although no significant homology has been found between the TGB protein and TMV MP, there are similarities in function and location. For instance, nucleic acid- and NTP-binding domains present in the TMV MP have also been found in the 25 kDa protein of potexviruses, a nonstructural protein encoded by TGB (Saito et al., 1988; Citovsky et al., 1990). Moreover, viruses containing TMV-like MPs can be complemented by an SEL increase brought about by a 25 kDa protein, although the latter has not been found associated with plasmodesmata (Angell et al., 1996). The 25 kDa protein has been detected in cell wall subcellular fractions of infected plants, although ultrastructural examination localizes this protein to the inclusion bodies instead (Davies et al., 1993). TGB mutants of *White clover mosaic virus*, a potexvirus, replicate in protoplasts but fail to move systemically in infected plants (Beck et al., 1991).

Coat Protein-Dependent Movement

Tubule-Guided Mechanism

The movement mechanism of *Cowpea mosaic virus* (CPMV, genus *Comovirus*) seems to be quite different from the previously discussed mechanisms (see Figure 7.1B). Cells infected with CPMV have distinct tubules that penetrate the plasmodesmata (Van Lent et al., 1990). When penetrated by tubules, plasmodesmata lose their characteristic desmotubules. Since the tubules penetrate the plasma membranes of protoplasts, the tubules are not modified desmotubules (Van Lent et al., 1991). Such tubular structures are involved in cell-to-cell movement of CPMV (Kasteel et al., 1996). Wellink and Van Kammen (1989) demonstrated that two overlapping genes that can produce peptides 58 kDa/48 kDa in size are needed along with the viral CP gene to establish a successful CPMV infection. The 58 kDa/48 kDa proteins are not necessary for replication, but they do localize to the tubular structures (Van Lent et al., 1990). Recent studies have demonstrated that the 48 kDa protein is involved in tubule formation (Kasteel, 1999).

Nepovirus infection also induces the formation of movement-associated tubules. An antibody raised against the 45 kDa protein of *Tomato ringspot virus* (genus *Nepovirus*), analogous to the CPMV 48 kDa protein, recognizes the tubules (Wieczorek and Sanfacon, 1993). Spherical objects appear to move through the tubules induced by both nepo- and comoviruses (Deom et al., 1992). It has not yet been determined if these objects are virions or virus-like particles. Likewise, the functional roles that the MPs might play in tubule formation or viral movement are unknown.

The MP of *Cauliflower mosaic virus* (CaMV, genus *Caulimovirus*) resembles the TMV MP, but CaMV may move similar to a comovirus. CaMV is a double-stranded DNA virus with movement functions attributed to *gene 1*. As with the TMV MP, the gene 1 protein can be localized to the cell wall in infected cells (Linstead et al., 1988) and has a nucleic acid-binding domain (Citovsky et al., 1991). However, tubules are formed in CaMV-infected protoplasts (Perbal et al., 1993).

Members of bromo- and cucumoviruses, having tripartite genomes, are also transported between cells in a CP-dependent manner, but the required form of CP is distinct for each member of these groups. For example, molecular and genetic analysis of the bromovirus MP revealed that this gene is critical for cell-to-cell movement and also dictates host specificity (Mise et al., 1993; Mise and Ahlquist, 1995; Rao and Grantham, 1995b). However, in situ analyses of movement characteristics of several CP-defective variants of *Brome mosaic virus* (BMV), a monocot-adapted bromovirus, demonstrated that synthesis of a functional encapsidation competent CP, in addition to MP, is also essential for cell-to-cell spread (Rao and Grantham, 1995a; Schmitz and Rao, 1996; Rao, 1997). The cell-to-cell movement of another member of genus *Bromovirus*, dicot-adapted *Cowpea chlorotic mottle virus* (CCMV), appears to be different from that of BMV. The CPs of BMV and CCMV share 70 percent identity at the amino acid level (Speir et al., 1995). Two hybrid viruses constructed between BMV and CCMV and capable of expressing heterologous CPs exhibited neutral effects on the overall movement process and host range (Osman et al., 1998). However, when the cell-to-cell movement of CP-defective CCMV was compared to that of BMV, it appeared that CCMV can move between cells independent of CP (Rao, 1997), indicating that these two viruses utilize different mechanisms to cross different cell types. Available information suggests that, similar to como- and nepoviruses, BMV also utilizes a tubule-guided mechanism (Kasteel et al., 1997), although this needs to be verified at the whole-plant level. At present it is not known whether CCMV induces tubules or is transported between cells in nonvirion form (i.e., TMV-like).

Non-Tubule-Guided Mechanism

As with BMV, the cell-to-cell movement of CMV is also dependent on both the MP and the CP (Taliansky and Garcia-Arenal, 1995; Canto et al., 1997). However, the form of CP required to support CMV movement is distinct from that of BMV CP. For example, similar to BMV, CMV variants lacking a CP failed to move from cell to cell (Canto et al., 1997). Unlike BMV, virion assembly is not a prerequisite for CMV movement, since assembly-defective CMV variants were able to induce local lesions due to efficient cell-to-cell spread (Kaplan et al., 1998; Schmitz and Rao, 1998). Recent experiments with CMV-P (*Physalis* strain) revealed that, similar to BMV, CMV also induces tubules in transfected protoplasts. However, tubules do not appear to contribute to viral movement, since mutant CMV RNA3 defective in tubule production is competent for cell-to-cell and systemic spread (Canto and Palukaitis, 1999). These authors speculate that tubules promote viral movement through plasmodesmata interconnecting tobacco epidermal cells but not through plasmodesmata either connecting epidermal cells with mesophyll cells or between mesophyll cells.

Movement Complementation by Heterologous Movement Proteins

A virus normally unable to move from cell to cell in a particular plant may be able to move with the help of a second virus of heterologous origin. Despite extensive variation in morphology, host range, and genome organization, many taxonomically distinct plant viruses exhibit complementary movement functions that may be a result of MP cross-compatibility (Atabekov et al., 1990). For example, whereas TMV-L can complement the movement of TMV-Lsl under high temperatures, PVX can complement the movement of TMV in Tm-2 gene tomato plants that normally resist TMV infection (Taliansky, Malyshenko, et. al. 1982). Likewise, TMV and RCNMV are functionally homologous, since the cell-to-cell spread of movement-defective variants of TMV and RCNMV can be complemented in transgenic *Nicotiana benthamiana* plants expressing heterologous MPs (Giesman-Cookmeyer et al., 1995). Similarly, while examining the cross-compatibility of MPs of CCMV and cowpea-adapted *Sunn-hemp mosaic virus* (SHMV, genus *Tobamovirus*), it was demonstrated that, when the MP of CCMV was replaced with that of SHMV, the hybrid was competent for systemic infection in cowpea (De Jong and Ahlquist, 1992). However, while examining the crosscompatibility between MPs of TMV and CMV, it was observed that transgenic *N. tabacum* cv. Xanthi (tobacco) plants expressing the TMV-MP gene supported cell-to-cell movement, but not the systemic movement, of a movement-defective CMV (Cooper et al., 1996). These studies support the hypothesis that the MP of one virus can facilitate the movement of another virus, although this has not been proven directly in all cases. Other viral factors could be involved in the phenomenon.

Not all MPs are capable of mobilizing a virus that is transport defective in a particular host. Transgenic plants accumulating CMV MP can complement the movement of a movement-defective CMV and a wild-type (wt) BMV in inoculated leaves but cannot support the movement of TMV-Lsl, RCNMV, or *Potato leafroll virus* (genus *Polerovirus;* Kaplan et al., 1995). Apparently, other factors are required for viral movement that cannot be entirely supplied by a heterologous MP. Despite supplying such a universal function as viral movement, MPs share only a few identical amino acids (Melcher, 1990). Based on amino acid and structural similarities in a nontaxonomic sense, an attempt was made to group the 30 kDa MPs. Eighteen groups are identified as "30K" superfamilies: the MPs of alfamo-/ilar-, badna-, bromo-, capillo-/tricho-, caulimo-, cucumo-, diantho-, furo-, gemini-, idaeo-, nepo(A)-, nepo(B)-, tobamo-, tobra-, tombus-, and umbraviruses. Five groups of possible candidates are the MPs of clostero-, rhabdo-, tenui-, and waikaviruses, and the phloem proteins. These groups can be subgrouped into four different subsuperfamilies, but boundaries between groups are not clear, except for the tubule-forming MPs (Melcher, 2000). Atabekov and Taliansky (1990) describe many examples in which viruses from one movement family complement the movement of viruses from a different family. If an MP is responsible for the complementation, then the cross-compatibility of these MPs is broad indeed.

Unidentified Movement Proteins

Transport loci have been found in viruses not related to the viruses already mentioned. *Tobacco rattle virus* (genus *Tobravirus*) can move from cell to cell without RNA2, which encodes the CP (Harrison and Robinson, 1986). TRV RNA1, unable to move from cell to cell with a mutated 29 kDa gene, will move systemically in the transgenic plants that accumulate the TMV MP (Ziegler-Graff et al., 1991). Thus, the TRV 29 kDa protein could be an MP. Interestingly, the NSm protein gene of *Tomato spotted wilt virus* (genus *Tospovirus,* family *Bunyaviridae*) is not present in any of the animal members of this family. However, it can be detected in cell wall and organelle subcellular fractions from infected plants, and ultrastructural studies have shown the presence of NSm in the cell wall (Kormelink et al., 1994). The protein is not a component of the viral membrane. Hence, NSm resembles other MPs that are associated with cell walls but are not viral structural components. Likewise, *Tobacco etch virus* (TEV, genus *Potyvirus*) has no known protein with a singular dedicated movement function. Although the TEV CP functions in cell-to-cell movement as well as in systemic transport (Dolja et al., 1995), other protein components may also be involved in the movement of TEV (see Figure 7.1B).

Role of Other Viral Genes in Movement

It is well documented that, depending on the virus group, movement is regulated by either the MP alone or the MP in combination with the CP.

However, other gene products, such as replicase, also appear to influence the movement process. For example, several BMV replicase mutants capable of efficient replication and packaging in protoplasts failed to systemically infect barley plants (Traynor et al., 1991). Likewise, replicase genes of BSMV (Weiland and Edwards, 1994). CMV (Gal-On et al., 1995), and TMV (Nelson et al., 1993), as well as nonstructural protein p19 of *Tomato bushy stunt virus* (genus *Tombusvirus*; Scholthof et al., 1995) and a helper component proteinase of potyviruses (Cronin et al., 1995) have demonstrated specific roles in movement (see Figure 7.1.B).

Role of Host Plant in Viral Movement

In addition to the virus-encoded genes such as MP, CP, and replicase genes, viral movement in a given host plant is largely regulated by the type of host itself. The reason why an MP would allow the long-distance movement of one virus in a particular host but not in another may be found at the molecular level. It has been assumed that, in addition to MP, an unidentified host factor(s) is also involved in potentiating the cell-to-cell movement of progeny viruses (Deom et al., 1992). It is well documented in the literature that *N. benthamiana* is susceptible to many viruses. For example, BMV is a monocot-adapted host and has a very narrow host range. However, *N. benthamiana* is susceptible to BMV infection (although symptomless), and the virus accumulates to very high concentrations (Rao and Grantham, 1995a). *Nicotiana benthamiana* could contain a host factor that is absent in other host plants. It is well established that, following viral infections such as TMV, the MP increases the plasmodesmatal SEL in the previous plant species, permitting cell-to-cell movement of progeny virus (Lucas and Gilbertson, 1994). At present, the plasmodesmatal SEL in *N. benthamiana* is not known. It is possible that the plasmodesmatal SEL at the bundle sheath/phloem parenchyma cell barrier is inherently higher in *N. benthamiana* than in *N. tabacum*. This perhaps would explain why *N. benthamiana* is susceptible to a heterologous MP-mediated systemic infection by CMVFnyδMP-δKPN and also to monocot-adapted viruses such as BMV (Rao et al., 1998). Likewise, the behavior in several hosts of a hybrid virus constructed between BSMV and RCNMV suggests that host-specific factors are involved in virus transport function (Solovyev et al., 1997).

Another important host-related aspect that regulates movement is the defense response elicited following a viral infection. Unlike humans, plants do not have an immune system to defend themselves against a given pathogen (fungal, bacterial, or viral). However, an incompatible interaction between a given pathogen and the host is manifested by a hypersensitive response

(HR), which is often considered a defense mechanism (see also Chapter 16 of this book). While analyzing the effects of several N-terminal mutations introduced into the BMV CP, an interesting phenomenon was observed. In *Chenopodium quinoa,* wt BMV induces chlorotic local lesions followed by systemic mottling. Although the N-terminal basic arm has been shown to be essential for interaction with viral RNA during encapsidation (Sgro et al., 1986), several BMV RNA3 variants harboring a deletion of the first seven N-proximal amino acids were infectious and formed particles but, unlike wt, induced an HR in *C. quinoa* (Rao and Grantham, 1995a). Based on these observations, it was hypothesized that, in addition to RNA binding, the N-terminal basic arm of BMV CP must be encoding additional functions, and BMV evolutionarily could have sustained the entire basic arm to overcome resistance and increase its host range. This conjecture is further reinforced by the observation that either a single amino acid substitution or a five amino acid deletion in the N-terminal basic arm of AMV CP (a member of family *Bromoviridae*) also induced HRs in tobacco plants (Neeleman et al., 1991; Van der Vossen et al., 1994).

CURRENT STATUS

At present, virus movement is an active field of study. With the development of sensitive cell biology-related techniques, such as application of green fluorescent protein in combination with confocal microscopy, our knowledge of how plant viruses are transported between cells is advancing at an astonishing rate. The majority of current research has focused on delineating viral genes involved in the movement process because of the ease with which viral genomes can be manipulated using recombinant DNA techniques. As discussed earlier, since the host plant plays a major role in regulating viral movement, there is a compelling need to isolate and characterize host factors involved in this important phase of virus infection. We are all optimistic that use of *Arabidopsis* as a model system is likely to provide many answers to understanding the contribution of the host plant in virus infection.

REFERENCES

Angell SM, Davies C, Baulcombe DC (1996). Cell-to-cell movement of potato virus X is associated with a change in the size exclusion limit of plasmodesmata in trichome cells of *Nicotiana clevelandii. Virology* 216: 197-201.

Atabekov JG, Taliansky ME (1990). Expression of a plant virus-encoded transport function by different viral genomes. *Advances in Virus Research* 38: 201-248.

Atabekov JG, Taliansky ME, Malyshenko SI, Mushegian AR, Kondakova OA (1990). The cell to cell movement of viruses in plants. In Pirone TP and Shaw JG (eds.), *Viral Genes and Plant Pathogenesis*, Springer-Verlag, New York, pp. 53-55.

Atkins D, Hull R, Wells B, Roberts K, Moore P, Beachy RN (1991). Tobacco mosaic 30K movement protein in transgenic tobacco plants is localized to plasmodesmata. *Journal of General Virology* 72: 209-211.

Beck DL, Guilford PJ, Voot DM, Andersen MT, Forster RLS (1991). Triple gene block of white clover mosaic potexvirus are required for transport. *Virology* 183: 695-702.

Boccard F, Baulcombe DC (1993). Mutational analysis of *cis*-acting sequences and gene function in RNA 3 of cucumber mosaic virus. *Virology* 193: 563-578.

Canto T, Prior DAM, Hellwald KH, Oparka KJ, Palukaitis P (1997). Characterization of cucumber mosaic virus. IV. Movement protein and coat protein are both essential for cell-to-cell movement of cucumber mosaic virus. *Virology* 237: 237-248.

Canto T, Palukaitis P (1999). Are tubules generated by the 3a protein for cucumber mosaic virus movement? *Molecular Plant-Microbe Interactions* 12: 985-993.

Carrington JC, Kasschau KD, Mahajan SK, Schaad MC (1996). Cell-to-cell and long-distance transport of viruses in plants. *Plant Cell* 8: 1669-1681.

Cheo PC (1970). Subliminal infection of cotton by tobacco mosaic virus. *Phytopathology* 60: 41-46.

Citovsky V, Knorr D, Schuster G, Zambryski P (1990). The P30 movement protein of tobacco mosaic virus is a single-stranded nucleic acid binding protein. *Cell* 60: 637-647.

Citovsky V, Knorr D, Zambryski P (1991). Gene I, a potential cell-to-cell movement locus of cauliflower mosaic virus, encodes an RNA-binding protein. *Proceedings of the National Academy of Sciences, USA* 88: 2476-2480.

Citovsky V, Wong ML, Shaw AL, Prasad BMV, Zambrysky P (1992). Visualization and characterization of tobacco mosaic virus movement protein binding to single-stranded nucleic acids. *Plant Cell* 4: 397-411.

Cooper B, Schmitz I, Rao ALN, Beachy RN, Dodds JA (1996). Cell-to-cell transport of movement-defective cucumber mosaic and tobacco mosaic viruses in transgenic plants expressing heterologous movement protein genes. *Virology* 216: 208-213.

Cronin S, Verchot J, Haldeman-Cahill R, Schaad MC, Carrington JC (1995). Long-distance movement factor: A transport function of the potyvirus helper component proteinase. *Plant Cell* 7: 549-559.

Davies C, Hills G, Baulcombe DC (1993). Sub-cellular localization of the 25-kDa protein encoded in the triple gene block of potato virus X. *Virology* 197: 166-175.

Dawson WO, Lehto KM (1990). Regulation of tobamovirus gene expression. *Advances in Virus Research* 38: 307-342.

De Jong W, Ahlquist P (1992). A hybrid plant RNA virus made by transferring the noncapsid movement protein from a rod-shaped to an icosahedral virus is competent for systemic infection. *Proceedings of the National Academy of Sciences, USA* 89: 6808-6812.

Deom CM, Lapidot M, Beachy RN (1992). Plant virus movement proteins. *Cell* 69: 221-224.

Deom CM, Oliver MJ, Beachy RN (1987). The 30-kilodalton gene product of tobacco mosaic virus potentiates virus movement. *Science* 237: 389-394.

Deom CM, Schubert KR, Wolf S, Holt CA, Lucas WJ, Beachy RN (1990). Molecular characterization and biological function of the movement protein of tobacco mosaic virus in transgenic plants. *Proceedings of the National Academy of Sciences, USA* 87: 3284-3288.

Ding B, Li Q, Nguyen L, Palukaitis P, Lucas WJ (1995). Cucumber mosaic virus 3a protein potentiates cell-to-cell trafficking of CMV RNA in tobacco plants. *Virology* 207: 345-353.

Ding SW, Haudenshield JS, Hull RJ, Wolf S, Beachy RN, Lucas WJ (1992). Secondary plasmodesmata are specific sites of localization of the tobacco mosaic virus movement protein in transgenic tobacco plants. *Plant Cell* 4: 915-928.

Dolja VV, Haldeman-Cahill R, Montgomery AE, Vanden Bosch KA, Carrington JC (1995). Capsid protein determinants involved in cell-to-cell and long-distance movement of tobacco etch potyvirus. *Virology* 207: 1007-1016.

Dorokhov YL, Alexandrova NM, Miroschnichenko NA, Atabekov JG (1983). Isolation and analysis of virus-specific ribonucleoprotein of tobacco mosaic virus-infected tobacco. *Virology* 127: 237-252.

Erny C, Schoumacher F, Godefroy-Colburn T, Stussi-Garaud C (1992). Nucleic acid binding properties of the 92-kDa replicase subunit of alfalfa mosaic virus produced in yeast. *European Journal of Biochemistry* 203: 459-465.

Fujiwara T, Giesman-Cookmeyer D, Ding B, Lommel SA, Lucas WJ (1993). Cell-to-cell trafficking of macromolecules through plasmodesmata potentiated by the red clover necrotic mosaic virus movement protein. *Plant Cell* 5: 1783-1794.

Gafny R, Lapidot M, Berna A, Holt CA, Deom C M, Beachy RN (1992). Effects of terminal deletion mutations on function of the movement protein of tobacco mosaic virus. *Virology* 187: 499-507.

Gal-On A, Kaplan I, Palukaitis P (1995). Differential effects of satellite RNA on the accumulation of cucumber mosaic virus RNAs and their encoded proteins in tobacco versus zucchini squash with two strains of CMV helper virus. *Virology* 208: 58-66.

Gardiner WE, Sunter G, Brand L, Elmer JS, Rogers SG, Bisaro DM (1988). Genetic analysis of tomato golden mosaic virus: The coat protein is not required for systemic spread or symptom development. *European Molecular Biology Journal* 7: 899-904.

Giesman-Cookmeyer D, Silver S, Vaewhongs AA, Lommel SA, Deom CM (1995). Tobamovirus and dianthovirus movement proteins are functionally homologous. *Virology* 13: 38-45.

Godefroy-Colburn T, Gagey M-J, Berna A, Stussi-Garaud C (1986). A nonstructural protein of alfalfa mosaic virus in the walls of infected tobacco cells. *Journal of General Virology* 67: 549-552.

Harrison BD (1985). Advances in geminivirus research. *Annual Review of Phytopathology* 23: 55-82.

Harrison BD, Robinson DJ (1986). Tobraviruses. In Van Regenmortel MVH, Fraenkel-Conrat H (eds.), *The Plant Viruses*, Volume 2. New York, Plenum, pp. 339-369.

Ivanov KI, Inovov PI, Timofeeva EK, Dorokov YL, Atabekov JG (1994). The immobilized movement proteins of two tobamoviruses form stable ribonucleoprotein complexes with full-length viral genomic RNA. *FEBS Letters* 346: 217-220.

Kaplan JM, Shintaku MH, Li Q, Zhang L, Marsh LE, Palukaitis P (1995). Complementation of movement defective mutants in transgenic tobacco expressing cucumber mosaic virus movement gene. *Virology* 209: 188-199.

Kaplan JM, Zhang L, Palukaitis P (1998). Characterization of cucumber mosaic virus. V. Cell-to-cell movement requires capsid protein but not virions. *Virology* 246: 221-231.

Kasteel DTG (1999). Structure, morphogenesis and function of tubular structures induced by cowpea mosaic virus. PhD thesis, Wageningen Agricultural University, Wageningen, The Netherlands, 71 pp.

Kasteel DTJ, Perbal C-M, Boyer J-C, Wellink J, Goldbach RW, Maule AJ, Van Lent JWM (1996). The movement proteins of cowpea mosaic virus and cauliflower mosaic virus induce tubular structures in plant and insect cells. *Journal of General Virology* 77: 2857-2864.

Kasteel DTJ, Van der Wal NN, Jansen KAJ, Goldbach RW, Van Lent JWM (1997). Tubule-forming capacity of the movement proteins of alfalfa mosaic virus and brome mosaic virus. *Journal of General Virology* 78: 2089-2093.

Kormelink R, Storms M, Van Lent JWM, Peters D, Goldbach R (1994). Expression and subcellular location of the NSM protein of tomato spotted wilt virus (TSMV), a putative viral movement protein. *Virology* 200: 56-65.

Lazarowitz SG, Beachy RN (1999). Viral movement proteins as probes for intracellular and intercellular trafficking in plants. *Plant Cell* 11: 535-548.

Lazarowitz SG, Pinder AJ, Damsteegt VD, Rogers SG (1989). Maize streak virus genes essential for systemic spread and symptom development. *European Molecular Biology Journal* 8: 1023-1032.

Linstead PJ, Hills GJ, Plaskitt KA, Wilson IG, Harker CL, Maule AJ (1988). The subcellular location of the gene 1 product of cauliflower mosaic virus is consistent with a function associated with virus spread. *Journal of General Virology* 69: 1089-1818.

Lucas WJ, Gilbertson RL (1994). Plasmodesmata in relation to viral movement within leaf tissues. *Annual Review of Phytopathology* 32: 387-411.

Melcher U (1990). Similarities between putative transport proteins of plant viruses. *Journal of General Virology* 71: 1009-1018.

Melcher U (2000). The "30" superfamily of viral movement proteins. *Journal of General Virology* 81: 257-266.

Meshi T, Hosokawa D, Kawagishi M, Watanabe Y, Okada Y (1992). Reinvestigation of intracellular localization of the 30K protein in tobacco protoplasts infected with tobacco mosaic virus RNA. *Virology* 187: 809-813.

Meshi T, Watanabe Y, Saito T, Sugimoto A, Maeda T, Okada Y (1987). Function of the 30kd protein of tobacco mosaic virus: Involvement in cell-to-cell movement and dispensability for replication. *European Molecular Biology Journal* 6: 2557-5047.

Mise K, Ahlquist P (1995). Host specificity restriction by bromovirus cell-to-cell movement protein occurs after initial cell-to-cell spread of infection in non-host plants. *Virology* 206: 276-286.

Mise K, Allison RF, Janda M, Ahlquist P (1993). Bromovirus movement protein genes play a crucial role in host specificity. *Journal of Virology* 67: 2815-2823.

Moser O, Gagey M-J, Godefroy-Colburn T, Stussi-Garaud C, Ellwart-Tschürtz M, Nitschko H, Mundry KW (1988). The fate of the transport protein of tobacco mosaic virus in systemic and hypersensitive hosts. *Journal of General Virology* 69: 1367-1373.

Neeleman L, Van der Kuyl AC, Bol JF (1991). Role of alfalfa mosaic virus coat protein gene in symptom formation. *Virology* 181: 687-693.

Nelson RS, Li G, Hodgson RAJ, Beachy RN, Shintaku M (1993). Impeded phloem-dependent accumulation of the masked strain of tobacco mosaic virus. *Molecular Plant-Microbe Interactions* 6: 45-54.

Nishiguchi M, Motoyoshi F, Oshima N (1978). Behaviour of a temperature sensitive strains of tobacco mosaic virus in tomato leaves and protoplasts. *Journal of General Virology* 39: 53-61.

Noueiry AO, Lucas WJ, Gilbertson RL (1994). Two proteins of a plant DNA virus coordinate nuclear and plasmodesmal transport. *Cell* 76: 925-932.

Ohno T, Takamatsu N, Meshi T, Okada Y, Santo T, Shikata E (1983). Single amino acid substitution in 30K protein of TMV defective in virus transport function. *Virology* 131: 255-258.

Osman F, Choi YG, Grantham GL, Rao ALN (1998). Molecular studies on bromovirus capsid protein. V. Evidence for the specificity of brome mosaic virus encapsidation using RNA3 chimera of brome mosaic and cucumber mosaic viruses expressing heterologous coat proteins. *Virology* 251: 438-448.

Osman TAM, Buck KW (1987). Replication of red clover necrotic mosaic virus RNA in cowpea protoplasts: RNA1 replicates independently of RNA2. *Journal of General Virology* 68: 289-296.

Osman TAM, Buck KW (1991). Detection of the movement protein of red clover mosaic virus in a cell wall fraction from infected *Nicotiana clevelandii* plants. *Journal of General Virology* 72: 2853-2856.

Perbal MC, Thomas CL, Maule AJ (1993). Cauliflower mosaic virus gene-1 product (P1) forms tubular structures which extend from the surface of infected protoplasts. *Virology* 195: 281-285.

Poirson A, Turner AP, Giovana C, Berna A, Roberts K, Godefroy-Colburn T (1993). Effect of the alfalfa mosaic virus movement protein expressed in transgenic plants on the permeability of plasmodesmata. *Journal of General Virology* 74: 2459-2461.

Rao ALN (1997). Molecular studies on bromovirus capsid protein. III. Analysis of cell-to-cell movement competence of coat protein defective variants of cowpea chlorotic mottle virus. *Virology* 232: 385-395.

Rao ALN, Cooper B, Deom CM (1998). Defective movement of viruses in the family *Bromoviridae* is differently complemented in *Nicotiana benthamiana* expressing tobamovirus movement proteins. *Virology* 88: 666-672.

Rao ALN, Grantham GL (1995a). Biological significance of the seven amino-terminal basic residues of brome mosaic virus coat protein. *Virology* 211: 42-52.

Rao ALN, Grantham GL (1995b). A spontaneous mutation in the movement protein gene of brome mosaic virus modulates symptom phenotype in *Nicotiana benthamiana*. *Journal of Virology* 69: 2689-2691.

Saito T, Imai Y, Meshi T, Okada Y (1988). Interviral homologies of the 30 K proteins of tobamoviruses. *Virology* 167: 653-656.

Saito T, Yamanaka K, Okada Y (1990). Long-distance movement and viral assembly of tobacco mosaic virus mutants. *Virology* 176: 329-336.

Schmitz I, Rao ALN (1996). Molecular studies on bromovirus capsid protein. 1. Characterization of cell-to-cell movement-defective RNA3 variants of brome mosaic virus. *Virology* 226: 281-293.

Schmitz I, Rao ALN (1998). Deletions in the conserved amino-terminal basic arm of cucumber mosaic virus coat protein disrupt virion assembly but do not abolish infectivity and cell-to-cell movement. *Virology* 248: 323-331.

Scholthof HB, Scholthof K-BG, Kikkert M, Jackson AO (1995). Tomato bushy stunt virus spread is regulated by two nested genes that function in cell-to-cell movement and host-dependent systemic invasion. *Virology* 213: 425-438.

Schoumacher F, Erny C, Berna A, Godefroy-Colburn T, Stussi-Garaud C (1994). Nucleic acid binding properties of the alfalfa mosaic virus movement protein produced in yeast. *Virology* 188: 896-899.

Sgro J, Jacrot B, Chroboczek J (1986). Identification of regions of brome mosaic virus coat protein chemically cross-linked in situ to viral RNA. *European Journal of Biochemistry* 154: 69-96.

Solovyev AG, Zelenina DA, Savenkov EI, Grdzelishvili V Z, Morozov S, Maiss, E. Casper R, Atabekov JG (1997). Host-controlled cell-to-cell movement of a hybrid barley stripe mosaic virus expressing a dianthovirus movement protein. *Intervirology* 40: 1-6.

Speir JA, Munshi S, Wang G, Baker TS, Johnson JE (1995). Structure of the native and swollen forms of cowpea chlorotic mottle virus determined by X-ray crystallography and cryo-electron microscopy. *Structure* 3: 63-78.

Stussi-Garaud C, Garaud J-C, Berna A, Godfroy-Colburn T (1987). In situ location of an alfalfa mosaic virus non-structural protein in plant cell walls: Correlation with virus transport. *Journal of General Virology* 68: 1779-1784.

Suzuki M, Kuwata S, Kataoka J, Masuta, C, Nitta N, Takanami Y (1991). Functional analysis of deletion mutants of cucumber mosaic virus RNA3 using an in vitro transcription system. *Virology* 183: 106-113.

Taliansky ME, Atabekova TI, Kaplan IB, Morozov SY, Malyshenko SI, Atakov JG (1982). A study of TMV TS mutant NI2519. 1. Complementation experiments. *Virology* 76: 701-708.

Taliansky ME, Malyshenko SI, Pshennikova ES, Kaplan IB, Ulanova EF, Atakov JG (1982). Plant virus-specific transport function. II. A factor controlling virus host range. *Virology* 122: 327-331.

Taliansky ME, Garcia-Arenal F (1995). Role of cucumovirus capsid protein in long distance movement within the infected plant. *Journal of Virology* 69: 916-922.

Tomenius K, Clapham D, Meshi T (1987). Localization by immunogold cytochemistry of the virus-encoded 30K protein in plasmodesmata of leaves infected with tobacco mosaic virus. *Virology* 160: 363-371.

Traynor P, Young BM, Ahlquist P (1991). Deletion analysis of brome mosaic virus 2a protein: Effects on RNA replication and systemic spread. *Journal of Virology* 65: 2807-2815.

Van der Kuyl AC, Neeleman L, Bol JF (1991). Role of alfalfa mosaic virus coat protein in regulation of the balance between plus and minus strand RNA synthesis. *Virology* 176: 346-354.

Van der Vossen EAG, Neeleman L, Bol JF (1994). Early and late functions of alfalfa mosaic virus coat protein can be mutated separately. *Virology* 202: 891-903.

Van Lent J, Storms MMH, Van der Meer F, Wellink J, Goldbach RW (1991). Tubular structures involved in movement of cowpea mosaic virus are also formed in infected cowpea protoplasts. *Journal of General Virology* 72: 2615-2623.

Van Lent J, Wellink J, Goldbach RW (1990). Evidence of the involvement of the 58K and 48K proteins in the intercellular movement of cowpea mosaic virus. *Journal of General Virology* 71: 219-223.

Vaquero C, Turner AP, Demangeat G, Sanz A, Serra MT, Roberts K, Garcia-Luque I (1994). The 3a protein from cucumber mosaic virus increases the gating capacity of plasmodesmata in transgenic tobacco plants. *Journal of General Virology* 38: 3193-3197.

Von Arnim A, Frischmuth T, Stanley J (1993). Detection and possible functions of African cassava mosaic virus DNA B gene products. *Virology* 192: 264-272.

Waigmann E, Lucas W, Citovsky V, Zambryski P (1994). Direct functional essay for tobacco mosaic virus cell-to-cell movement protein and identification of a domain involved in increasing plasmodesmal permeability. *Proceedings of the National Academy of Sciences, USA* 91: 1433-1437.

Weiland JJ, Edwards MC (1994). A single nucleotide substitution in the alpha-a gene confers oat pathogenicity to barley stripe mosaic virus strain ND18. *Molecular Plant-Microbe Interactions* 9: 62-67.

Wellink J, van Kammen A (1989). Cell-to-cell transport of cowpea mosaic virus requires both the 54/48K proteins and the capsid proteins. *Journal of General Virology* 51: 317-325.

Wieczorek A, Sanfacon H (1993). Characterization and subcellular location of tomato ringspot nepovirus putative movement protein. *Virology* 194: 734-743.

Wolf S, Deom CM, Beachy RN, Lucas WJ (1989). Movement protein of tobacco mosaic virus modifies plasmodesmatal size exclusion limit. *Science* 246: 377-379.

Ziegler-Graff V, Guilford PJ, Baulcombe DC (1991). Tobacco rattle virus RNA-1 29K gene product potentiates viral movement and also affects symptom induction in tobacco. *Virology* 182: 145-155.

SECTION III:
MOLECULAR BIOLOGY
OF RNA VIRUSES

Chapter 8

Mechanism of RNA Synthesis by a Viral RNA-Dependent RNA Polymerase

Kailayapillai Sivakumaran
Jin-Hua Sun
C. Cheng Kao

INTRODUCTION

Although RNA replication is of fundamental importance for pathogenesis in many RNA viruses, the mechanism remains poorly understood. The RNA-dependent RNA polymerase (RdRp) complex plays a central role in the replication process, but few RdRps that direct this process have been characterized at the biochemical level. Thus, structural and functional studies of RdRps have lagged significantly behind those of other DNA and RNA polymerases. The objective of this chapter is to review some of the current progress made in our laboratory in elucidating the mechanism of RNA replication.

Brome mosaic virus (BMV, genus *Bromovirus*), a minor pathogen of cereal crops of family *Bromoviridae*, has been studied as a model system for RNA replication. The genome of BMV is divided into three positive (+)-strand RNAs: RNA1 (3.2 kb), RNA2 (2.8 kb), and RNA3 (2.1 kb). RNA4 is transcribed through internal initiation from the negative (−)-strand copy of RNA3 (Ahlquist, 1992). RNA1 and 2 are monoistronic and encode the helicase-like 1a and polymerase-like 2a proteins, respectively. RNA3 is dicistronic and encodes the BMV moveent and capsid proteins (Ahlquist, 1992). The 1a and 2a proteins and as yet unidentified host proteins are complexed within the plant endoplasmic reticulum to form the BMV-RdRp complex (Restrepo-Hartwig and Ahlquist, 1996). BMV-RdRp complex solubilized from membranes of infected barley is able to accurately synthesize (−)-strand and subgenomic (+)-strand RNA from exogenously added specific templates (Miller et al., 1985; Quadt and Jaspars, 1990; Adkins et al., 1997). An overview of BMV RNA replication using RNA3 is shown in Figure 8.1A. The template specificity provided by the BMV-RdRp complex enables us to analyze the different phases of RNA synthesis, such as initiation, transition to elongation, and termination, in mechanistic detail. Genomic (−)-strand and subgenomic RNA synthesis by BMV-RdRp have been demonstrated to be similar overall. This review will focus on the mechanism of (−)-strand RNA synthesis. Readers are referred to Adkins et al. (1998) for additional information on the mechanism of subgenomic RNA synthesis.

INITIATION OF RNA SYNTHESIS

Initiation of RNA synthesis is perhaps the most distinct aspect of viral RNA replication. In contrast, the mechanism of nucleotidyl transfer that characterizes the elongation phase of RNA synthesis is very similar in all polymerases (Joyce and Steitz, 1995). The distinctiveness of the initiation process thus makes it an attractive target for antiviral treatment. Furthermore, the specificity of the initiation process prevents the recognition of unrelated templates, including those of other viral and cellular RNAs.

Many of our results were established with variations of an in vitro RNA synthesis assay using BMV-RdRp prepared from infected barley as described previously (Sun and Kao, 1997b). Standard assays consisted of 25 nM of template RNA with 10 μl of RdRp in a 40 μl reaction mixture containing 20 mM sodium glutamate (pH 8.2), 4 mM $MgCl_2$, 12.5 mM dithiothreitol, 0.5 percent (v/v) Triton X-100, 1 nM $MnCl_2$, 200 μM ATP and UTP, 500 μM GTP, and 250 nM $[\alpha\text{-}^{32}P]CTP$ (Amersham). Reaction mixtures were incubated at 30°C for 90 minutes and stopped by phenol/chloroform extraction, followed by ethanol precipitation in the presence of 5 μg of glycogen and 0.4 M ammonium acetate. Products were separated by electrophoresis and quantified by phosphorimage analysis.

The 3'-end of BMV genomic RNAs folds into a transfer (t) RNA-like structure. This structure has been shown to be aminoacylated with tyrosine in vitro, although aminoacylation does not appear to be required for BMV RNA replication. Initiation of BMV (−)-strand RNA synthesis takes place from the penultimate cytidylate present at the 3'-end of (+)-strand genomic RNA. Mutation of the initiation cytidylate will decrease the ability of RdRp to direct (−)-strand RNA synthesis. However, RNA synthesis will take place from the neighboring cytidylate at a reduced level in vitro and in vivo (see Figure 8.1B). Mutation of the nucleotide (nt) on either side of the initia-

FIGURE 8.1. Schematic Diagram Illustrating *Brome mosaic virus* (BMV) RNA Replication

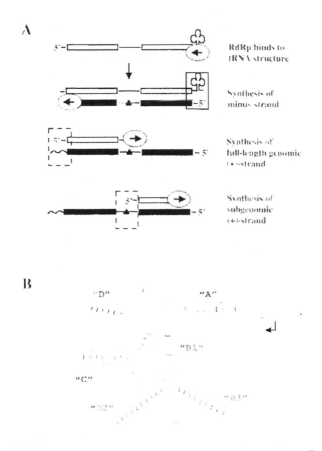

Note: (A) BMV-RdRp utilizes three RNA promoters during RNA replication. The viral RNA replication process is illustrated with BMV RNA3. The replicase complex and the direction of RNA synthesis are denoted by an oval and the enclosed arrow. Replication begins with recognition of the (−)-strand promoter (solid box) located at the 3'-tRNA-like (cloverleaf) end of the (+)-strand genomic RNA. The newly synthesized, complementary (−)-strand RNAs are depicted in black. (−)-strand RNA is then used as a template for both full-length, genomic (+)-strand synthesis and subgenomic (+)-strand synthesis. The approximate location of the genomic (+)-strand promoter is indicated with a dashed box. The subgenomic promoter is located internal to RNA3 and is responsible for RNA4 synthesis. The four domains of the BMV subgenomic promoter characterized by Marsh et al. (1988) are illustrated as an expansion within the (−)-strand RNA. (B) Diagram illustrating the sequence and structure of the 3'-end of (+)-strand genomic RNA. The folded tRNA-like structure contains stem-loop regions A to D, as marked. The initiation site for (−)-strand RNA synthesis is indicated by an arrow.

tion sequence has differential effects on RNA synthesis. Substitutions of the ultimate nt by any of the other three nts had relatively little effect on (–)-strand RNA synthesis; hence, the identity of the ultimate nt appears to be unimportant for efficient RNA synthesis. When the ultimate nt was deleted, initiation of (–)-strand RNA synthesis could still take place, perhaps from the +2 C that is now the penultimate nt. A change of nt at the +2 position from an adenylate to a guanylate reduced (–)-strand RNA synthesis, whereas the nt at the +3 position could be replaced by other nts without significant effect on (–)-strand RNA synthesis.

The penultimate nt, in addition to being recognized by the RdRp, is also required to direct the incorporation of the first nt, GTP. GTP could be replaced by guanine nt analogs during in vitro initiation. These analogs are useful for studying the initiation step as they are able to replace the function of GTP. However, these analogs are not useful for studying the elongation phase as they can only prime RNA synthesis but cannot be hydrolyzed to form a phosphodiester bond. The dimer GpG, which is complementary to the template RNA, can initiate (–)-strand synthesis. Radiolabeled GpG, but not GpA, can be incorporated into full-length BMV-RdRp products, demonstrating that the primer is not acting as an allosteric effector of RNA synthesis but is used as a substrate. In addition, GDP and GMP could replace GTP as the initiation nt. GDP, GMP, and the dinucleotide GpG could be used only in initiation and not in subsequent steps, as they do not contain phosphates that could be hydrolyzed during nucleotidyl transfer reaction.

The effect of oligonucleotide length and sequence on RdRp activity was also examined. Trimer 5'-GGU-3, and pentamer 5'-GGUCU-3', both of which are complementary to the template RNA, were tested. Addition of the trimer 5'-GGU-3' increased (–)-strand RNA synthesis eight to ten times and was similar to that observed with the dimer GpG. The pentamer 5'-GGUCU-3' had a limited stimulatory effect, increasing RdRp activity by 2.8 times.

In determining the K_M for the different nucleotides, we found that GTP was required at 50 µM, a concentration approximately 15 times higher than for CTP and UTP (see Table 8.1). In the presence of GpG, the K_M for GTP dropped to 3 µM. The addition of ApG and other dinucleotides had no effect on (–)-strand RNA synthesis or the K_M of GTP. GTP is required for both initiation and elongation, whereas the other nucleotides are only required for the elongation phase. The higher K_M value reflects the GTP requirement for initiation, while the lower K_M reflects its requirement during the elongation phase. These results could be interpreted as follows: the viral replicase has two nucleotide-binding pockets, one specific for the initiation nucleotide, GTP (which can be partially replaced with GpG or GDP), and a second nucleotide-binding pocket that recognizes and binds all four nucleotides for the elongation phase.

TABLE 8.1. Effect of Diribonucleotides on the K_M Values of RdRp

	K_M (µM) of RdRp*		
NTP	Control (n = 4)	With 500 µM GpG (n)	With 500 µM ApG (n)
GTP	50.27 ± 15.61	3.24 ± 1.44 (2)	51.37 ± 14.35 (2)
ATP	13.53 ± 2.65	12.78 ± 6.07 (2)	ND
UTP	3.27 ± 0.93	1.96 ± 0.68 (2)	ND
CTP	3.43 ± 3.24	2.70 ± 1.64 (4)	ND

* ^{32}P-CMP incorporation was used in measurements of K_Ms for GTP, ATP, and UTP. ^{32}P-UMP was used for CTP. Ranges indicate maximum and minimum variations from the mean.
ND = not determined.

Speculations on the Biological Significance of Primer Use

A unique feature of RdRp is that they are thought to initiate without primers. However, this difference is really one of semantics, as all polymerases use an initiation nucleotide as a primer; the difference is only one of length. Hence, all polymerases use the initiation nucleotide as a 1-nt primer to provide the 3'-hydroxyl group. The triphosphates are not hydrolyzed during their incorporation into the nascent strand during RNA replication (they may be hydrolyzed later during the capping reaction). The significance of primer use by viral replicases with respect to *in planta* infection is yet to be established, as such experiments are technically difficult. However, it is conceivable that dimers, trimers, and oligonucleotides could exist in the cell during viral infection due to abortive initiation by cellular and/or viral polymerases, or as a result of RNA degradation, and would be available to act as primers during RNA synthesis.

ABORTIVE SYNTHESIS

Following the initiation step, DNA-dependent RNA polymerases enter a phase in which dissociation of the enzyme-DNA-RNA ternary complex competes with the elongation step, resulting in the production of oligoribonucleotides in a process referred to as abortive cycling (Carpousis and

Gralla, 1980). Elongated transcripts are formed only when abortive cycling is overcome by a change in the RNA polymerase, resulting in processive RNA synthesis. Abortive cycling appears to be an innate property of all DNA-dependent RNA polymerases, including those from the T7 bacteriophage (Martin et al., 1988), *Escherichia coli* (Carpousis and Gralla, 1980), and mammalian cells (Ackerman et al., 1983). Abortive cycling has not been characterized with RdRps. To determine whether abortive RNA synthesis takes place with viral replicases, products made by BMV-RdRp were analyzed.

Characterization of an Initiation Product

Abortive initiation products would be of low molecular weight. To observe these, we electrophoresed the RdRp products on a denaturing 20 percent polyacrylamide gel. The newly synthesized full-length (−)-strand RNAs and other large but less than full-length RNAs migrated as heterogeneous products near the top of the gel and were collectively called the elongated (E) RNAs. The abortive products should be shorter than 20 nt for (−)-strand BMV RNA synthesis. A major low molecular weight product was 8 nt in length, but other smaller products were also observed. Characterization of the putative abortive initiation product was restricted to the 8 nt product, as accurate quantification of the shorter RNAs was difficult. Template specificity was determined by analyzing the products of RdRp reactions containing total tRNA from yeast, 60S ribosomal (r) RNA from wheat germ, transcript from a control template (T7C) provided by an in vitro transcription kit (Epicentre Technologies), poly(A-U)-RNA, and yeast lys- and trp-tRNAs. Only in the presence of BMV virion RNAs was the 8 nt product generated (see Figure 8.2, some data not shown). These results demonstrate that the BMV-RdRp is responsible for synthesis of the 8 nt product in a manner dependent on and specific to BMV virion RNA.

To determine whether the 8 nt product initiated from the penultimate cytidylate normally used during initiation of (−)-strand RNA synthesis, we characterized the nucleotide content of the 8 nt product by labeling different radionucleotides. Correct initiation of (−)-strand RNA synthesis has been shown to start at the penultimate cytidylate of (+)-strand RNA (see Figure 8.1B) with the end sequence of 5′ . . . AAGAGACCA$_{OH}$-3′ (Miller et al., 1986; Kao and Sun, 1996). An 8 nt product resulting from the correct initiation site should have the sequence 5′-GGUCUCUU-3′. We characterized the nucleotide content of the 8 nt product by carrying out RdRp reactions using different radionucleotides. The use of ^{32}P-CTP and ^{32}P-UTP is expected to result in the incorporation of four uridylates and two cytidylates. Consistent with this prediction, quantitation of the 8 nt product labeled with ^{32}P-UTP and ^{32}P-CTP showed a ratio of 2:1, whereas reactions performed with

FIGURE 8.2. *Brome mosaic virus* (BMV) RNA-Dependent-Specific 8-mer Generation by RdRp

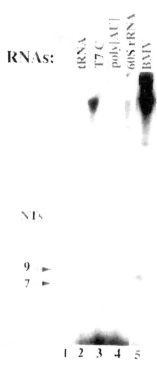

Note: Autoradiogram of 20 percent denaturing polyacrylamide gel of products from RdRp reactions performed using the RNA templates (1µg) indicated above the autoradiogram. BMV denotes RNA purified from BMV virions. The positions of 7 and 9 nucleotide (NTs) markers are indicated on the left of the autoradiogram.

^{32}P-ATP showed the presence of the elongated products (E RNAs), but not the 8 nt product, as predicted (see Figure 8.3). Last, labeling reactions performed with ^{32}P-ATP resulted in very few radiolabeled products (see Figure 8.3). A likely explanation is that RdRp needs c. 50 µM of GTP to half saturate the initiating nucleotide-binding site (Kao and Sun, 1996); hence, the reaction containing only 1 µM ^{32}P-GTP was unlikely to initiate synthesis. These results unambiguously show that the 8 nt product is initiated at the authentic cytidylate during (–)-strand RNA synthesis and is a product of the initiation mechanism.

FIGURE 8.3. Characterization of the 8 nt Product Sequence

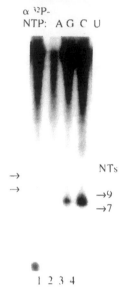

Note: Reactions were conducted using the standard protocol except that NTP concentrations were adjusted to 15 μM ATP, 50 μM GTP, 3 μM CTP, and 3 μM UTP. The radiolabel used in each reaction is indicated at the top of the autoradiogram. For reactions with a particular radiolabeled nucleotide, the corresponding unlabeled nucleotide was omitted. The positions of 7 and 9 nucleotide (NTs) markers are indicated on the right of the autoradiogram.

The molar ratios of the 8 nt product and E RNAs were also quantified. A higher amount (sixteenfold) of the 8 nt product was observed following five minutes of incubation, but the ratio dropped to eight to tenfold for all later time points. In addition, increasing amounts of unlabeled CTP were added to the RdRp reactions containing radiolabeled α^{32}P-CTP. As expected, increasing the amount of unlabeled CTP concentration decreased the radiolabeled E RNAs and the 8 nt product by equal rates. These results show that abortive initiation likely is not due to limiting amounts of CTP in the reaction but is part of the normal initiation process.

The 8 nt Product Is Aborted from the Replicase Complex

To be named aborted initiation products, the release of the 8 nt product by RdRp needs to be demonstrated. These experiments were performed by pausing the ternary complex during RNA synthesis. RNA synthesis by

RdRp was performed in the presence of three nucleotides: GTP, UTP, and radiolabeled CTP. Based upon the template sequence, the maximum length of the product would be 10 nt, after which the RdRp ternary complex could either remain on the template or dissociate. Following incubation the RdRp reaction products were fractionated through a gel filtration spin column and 15 fractions were collected. The different fractions were analyzed for the presence of nascent and template RNA. RdRp was found to elute in fractions 2 and 3 (see Figure 8.4B). The majority of the template, presumably unused by RdRp, eluted in fractions 5 through 11 (see Figure 8.4B). The

FIGURE 8.4. Products Synthesized by the RdRp of *Brome mosaic virus* and Separated by Gel Filtration Spin Columns

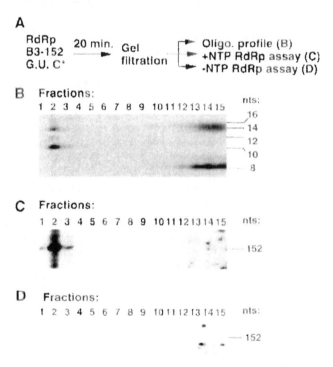

Note: (A) A brief protocol for the experiment. (B) Aliquots of the eluants from a CL-6B gel filtration column were extracted with phenol-chloroform and precipated with ethanol and then electrophoresed in a 20 percent denaturing polyacrylamide-urea gel. The remainder of the eluants were divided into two sets, one of which received NTPs (final concentration of 500 μM for each NTP (C), and the other of which received an equivalent volume of water (D).

early elution of the RdRp was consistent with the large size previously reported for the RdRp complex (Kao et al., 1992). Fractions 2 and 3 also had detectable amounts of the template RNA, indicating that some of the template RNA was associated with the RdRp complex. In addition, newly synthesized RNAs of 10, 12, and 14 nt were also observed in fractions 2 and 3 (see Figure 8.4B). The 12 and 14 nt RNAs were unexpected, since no exogenous ATP was added to the reactions and may have formed due to trace amounts of ATP contamination. Products longer than 14 nt were not observed. The 10, 12, and 14 nt products could be chased into full-length E RNAs by the addition of ATP to the reaction, indicating that they were still associated with the RdRp in a ternary complex (see Figure 8.4C). RNA synthesis was not observed in the absence of added NTPs (see Figure 8.4D). Last, the 8 nt products were observed in fractions 12 through 15 (see Figure 8.4B). The observation that the 8 nt product was present in fractions that did not contain enzymatically active RdRp shows that it is released by the RdRp. RdRp reactions performed with the full complement of NTPs also produced 8 nt RNA products and were separate from fractions containing RdRp activity (data not shown), indicating that the 8 nt RNA product is released by RdRp under normal reaction conditions.

Factors Affecting 8 nt and E RNA Synthesis

The effects of several DNA-dependent RNA polymerase inhibitors on RdRp synthesis were analyzed. These inhibitors included actinomycin-D, rifampicin, novobiocin, tagetitoxin, and heparin (Bautz, 1976; Johnson and McClure, 1976; Losick and Chamberlin, 1976; Webb and Jacob, 1988; Mathews and Durbin, 1994). While novobiocin, tagetitoxin, and heparin decreased the ability of the BMV-RdRp to synthesize full-length (−)-strand RNA, actinomycin-D and rifampicin did not affect RdRp activity (Hardy et al., 1979; see Figure 8.5). Inclusion of the polysulfonate heparin in RdRp reactions resulted in decreased E RNAs and 8 nt products (see Figure 8.6A). Quantifying the resulting products indicated that the synthesis of the 8 nt product was more sensitive to heparin than that of E RNAs. Increasing the concentration of heparin decreased the 8-mer to E RNA molar ratio, suggesting that polysulfonate heparin inhibits (−)-strand RNA synthesis at the initiation stage. Heparin likely competes with the template for polymerase recognition and thus prevents the initiation of transcription (Pfeffer et al., 1977; Carpousis and Gralla, 1985). Polymerase complexes that passed the initiation stage were less sensitive to heparin. Thus, heparin can be exploited to separate the initiation phase from the elongation phase during

FIGURE 8.5. Effects of RNA Polymerase Inhibitors on Synthesis of Full-Length and 8-mer RNAs

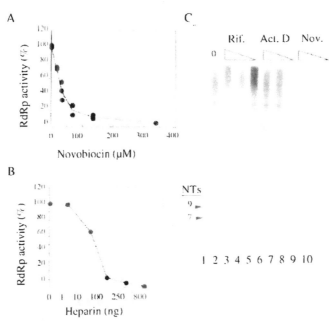

Note: Reactions were conducted by using the standard protocol with the inclusion of inhibitors. Full-length RNA products were resolved by 1 percent agarose gel electrophoresis and were quantified using a phosphorimager. (A) Effect of novobiocin (0, 18, 35, 68, 135, and 338 uM) on the synthesis of full-length (−)-strand RNA. (B) Effect of heparin (0, 1, 10, 100, 250, and 800 ng) on the synthesis of full-length (−)-strand RNAs. (C) Autoradiogram of results from a 20 percent denaturing polyacrylamide gel from RdRp reactions testing the effects of adding rifampicin, actinomycin-D, and novobiocin. Reactions were conducted using the standard protocol except that $0.3 \mu M$ [α-^{32}P]UTP was used, which resulted in the 8-mer and an additional oligonucleotide of 9 nts. Results were quantified by liquid scintillation counting of razor blade-excised bands. Inhibitors were added at the following concentrations: rifampicin (Rif.), 488, 244, 122 μM; actinomycin-D (Act. D), 320, 160, and 80 μM; novobiocin (Nov.), 200, 100, and 50 μM.

RNA synthesis. We will examine the transition from initiation to elongation in more detail later in this chapter.

We observed that Mn^{2+} in the form of either $MnCl_2$ or $MnSO_4$ had differential effects on the initiation and elongation processes. Lower amounts of Mn^{2+} (1 μM) caused an increase in the synthesis of both the E RNA and the 8 nt RNA products (see Figure 8.6B), whereas higher concentrations of Mn^{2+} resulted in marked inhibition of E RNA synthesis while increasing the synthesis of the 8 nt and other shorter oligoribonucleotide products. In reactions lacking

FIGURE 8.6. Effect of Heparin and Mn^{2+} on the Accumulation of 8-mer and E RNAs

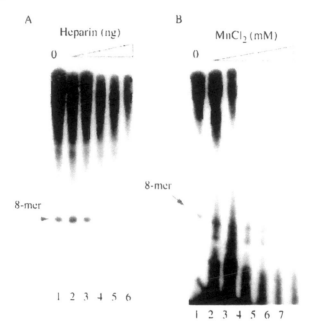

Note: Reactions were conducted using the standard protocol and analyzed by denaturing 20 percent polyacrylamide gels. (A) Autoradiogram demonstrating the effect of heparin added to the reactions at 0, 10, 20, 30, 50, and 100 ng (lanes 1 to 6, respectively). (B) Autoradiogram demonstrating the effect of Mn^{2+} added to the reactions at 0, 1, 3, 5, 7, 10, and 20 mM (lanes 1 to 7, respectively).

Mg^{2+}, only 10 percent of the E RNAs were produced at 3 to 10 mM Mn^{2+}, whereas higher amounts of the 8 nt product, along with additional oligoribonucleotides of smaller length, were generated at this concentration (see Figure 8.6B). At 10 mM MnCl$_2$, synthesis of the elongated product was completely abolished but significant amounts of the 8 nt and other shorter oligoribonucleotide products were still present. These results indicate that lower amounts of Mn^{2+} (1 to 5 mM) can replace Mg^{2+} during the initiation phase of (−)-strand RNA synthesis but cannot replace Mg^{2+} during the elongation phase.

Presence of the 8 nt product in tenfold molar excess of E RNAs suggests that dissociation of the initiation complex is a more common phenomenon than progression into the stable elongation complex. Hence, an additional event following the incorporation of the first 8 nt may be required to form a more stable ternary complex. Studies with bacteriophage T7 RNA polymerase (Martin et al., 1988) indicate that the extent of the dissociation depends on the sequence context near the initiation region. This is especially interesting with RdRp, as there is only one nucleotide 3' of the initiation nucleotide for (–)-strand RNA synthesis. When we deleted the terminal adenylate and assayed for abortive and elongative synthesis, we found reduced levels of both initiation and elongation products, but no change in the molar ratio of the products (see Table 8.2). This result suggests that the terminal adenylate can affect the interaction between RdRp and RNA, and its removal hampers the process of RNA synthesis. Substitution of the terminal adenylate with a cytidylate had a more profound effect on abortive initiation and a lesser effect on elongation, as evidenced by a decrease in the molar ratio (see Table 8.2). This suggests that the terminal adenylate influences the initiation process even though it is not used to direct the incorporation of the initiating nucleotide. Deletion of the 3'-most nucleotides, CA, or its substitution by CC, CU, AC, or UU resulted in a marked decrease in (–)-strand RNA synthesis (Dreher et al., 1984; Miller et al., 1986). Substitutions at the third and fourth positions had a more significant effect on E RNA synthesis than on initiation (see Table 8.2). Extensions of more than three nucleotides at the 3'-end of the template RNA favored the synthesis of E RNAs at the expense of the 8 nt products. Initiation of RNA synthesis from the end of an RNA molecule could be a physically difficult task for RdRp due to its proximity to the free 3'-end. Extending the 3'-end could thus provide a scaffold for a more stable binding of the RdRp and promote progression into the elongation mode. This may also explain why subgenomic RNA synthesis is more efficient as compared to (–)-strand RNA synthesis.

TRANSITION FROM INITIATION TO ELONGATION

The observation that synthesis of abortive products is more sensitive to heparin than the synthesis of elongation products suggests that RdRp makes a transition from an initiation phase to an elongation phase during RNA synthesis. To study this transition, we require RdRp complexes to be paused during specific stages in the process of RNA synthesis. In studies with *E. coli* and bacteriophage T7 DNA-dependent RNA polymerases (Levin et al., 1987), one or more NTPs were withheld in the reaction in order to pause the polymerase at specific positions along the DNA template. A similar strategy was adapted to characterize the transition of BMV-RdRp during (–)-strand RNA synthesis. To identify the ternary RdRp complexes capable

TABLE 8.2. Effect of Different Template RNA 3'-Ends on Initiation and Elongation (E)

RNA tested[a]	8-mer (percent)[b]	E RNA (percent)[c]	Molar ratio of 8-mer/E RNA
Wt	100 ± 21 (n = 12)[b]	100 ± +2	10.0 ± 4.1
Δ1	29 ± 7 (n = 4)	49 ± 8	8.6 ± 2.0
+ 1G	52 ± 8 (n = 4)	57 ± 13	11.6 ± 2.5
+ GAU	21 ± 15 (n = 4)	101 ± 29	2.7 ± 1.7
+ 13	20 ± 17 (n = 17)	139 ± 27	1.8 ± 1.6
+ 17	53 ± 13 (n = 4)	176 ± 18	3.3 ± 0.7
mu-1	14 ± 7 (n = 4)	46 ± 8	3.9 ± 2.1
mu-3	35 ± 8 (n = 4)	31 ± 9	12.3 ± 6.0
mu-4	40 ± 13 (n = 4)	21 ± 5	14.7 ± 5.5

[a] RNAs tested: Wt, wild-type; Δ1, deletion of 3' adenylate; +1G, addition of a guanylate to the 3'-end; +GAU, addition of these three nucleotides to the 3'-end; +13 and +17 indicates the addition of either 13 or 17 nt to the 3'-end; mu-1 changes the 3'-most adenylate to a guanylate; mu-3 changes the +2 cytidylate to an uridylate; mu-4 changes the +3 adenylate to a guanylate.

[b] Quantitation of the 8-mer, the E RNA, and the molar ratio. The numbers were converted to percentage of Wt.

[c] n = number of independent trials used to determine the amount of products produced by RdRp for each template. Values after "±" denote the standard deviation.

of elongation, reactions were incubated in the presence of GTP, CTP, and limiting concentrations of both ATP (0.01 µM) and ^{32}P-UTP (0.3 µM). The limiting ATP would pause RdRp at different positions during synthesis. After 20 minutes, one-half of the reaction was aliquoted into new tubes and UTP and ATP were added. The other half of the reaction received an equal volume of water. After incubation at 25°C for one hour, the reactions were treated with calf intestinal phosphatase before electrophoresis on a denaturing 20 percent polyacrylamide gel. A number of products less than full length and longer than 8 nt were apparent in reactions with limiting UTP and ATP (see Figure 8.7, lanes 1 and 2). Following the addition of ATP and UTP, the abundance of products longer than 8 nt decreased concomitant with an increase in full-length products that migrated at the top of the gel (see Figure 8.7, lanes 3 and 4). The disappearance of RNAs longer than 10 nt was likely due to continued elongation of the nascent strand to full-length products by the paused RdRp. The 8 nt RNA product did not extend into the full-length product, suggesting that it was not stably associated with the ternary RdRp complex.

FIGURE 8.7. Nascent RNAs Longer Than 10 nt Can Be Elongated into Full-Length Products

Note: BMV-RdRp was incubated with B3-152, GTP, CTP, [α-^{32}P]UTP, and ATP (final concentrations of 500, 150, 0.3, 0.01, and 0.01 μM respectively), for 20 minutes. The lower amount of ATP used in the reaction is denoted by the letter A in a smaller font. The reaction was subsequently divided into two equal aliquots and UTP and ATP (final concentrations of 1 mM each) were added to one aliquot while the other received an equal volume of water. All samples were incubated for one hour followed by treatment with two units of calf intestinal phosphatase (CIP) for ten minutes prior to electrophoresis in a 20 percent denaturing gel. The sizes of the products were determined by comparison to the migration of the dyes and molecular weight markers.

Progression Toward a Heparin-Resistant Ternary Complex

We wanted to determine whether RdRp is resistant to heparin inhibition following the synthesis of nascent RNA. The protocol used is shown in Figure 8.8A. Briefly, RdRp was preincubated for ten minutes with or without

FIGURE 8.8. Level of Resistance to Heparin Is Influenced by the Stage of the Initiation

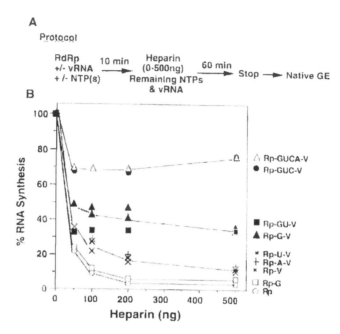

Note: (A) A brief protocol for the experiment. RdRp was incubated with nine different sets of conditions as follows: no additional components (Rp); GTP (Rp-G); vRNA (virion RNA; Rp-V); ATP and vRNA (Rp-A-V); UTP and vRNA (Rp-U-V); GTP and vRNA (Rp-G-V); GTP, UTP, and vRNA (Rp-GU-V); GTP, UTP, CTP, and vRNA (Rp-GUC-V); and all four NTPs and vRNA (Rp-GUCA-V). After a ten-minute preincubation, each set was divided into five aliquots and heparin from 0, 50, 100, 200, and 500 ng were added along with all the remaining components needed for elongation. All the reactions were further incubated for 60 minutes to allow the completion of RNA elongation before processing and electrophoresis in a native agarose gel (GE). (B) Plot of RNA synthesis versus the heparin concentration. The RNA products on the agarose gel were quantified with a phosphorimager and then normalized to the percentage of RNA produced in the absence of the heparin. The combination of components used in each set of reactions is noted on the right of the figure.

template RNA and with different combinations of NTPs. Following pre-incubation, components needed to complete one round of RNA synthesis were added, along with increasing amounts of heparin. After an additional hour of incubation, the reactions were terminated and processed for electro-phoresis. The amount of RNA synthesized by RdRp in 60 minutes in the absence of heparin was designated 100 percent, and the effect of heparin on each set of reactions was normalized to this value.

The results suggest several levels of resistance to heparin. RdRp preincubated without virion RNA (vRNA) or NTPs ("Rp" in Figure 8.8B) was greatly inhibited in the presence of 50 ng heparin. Increasing heparin to 200 ng further reduced RNA synthesis to near background levels. A similar inhibition profile was obtained with RdRp preincubated with 500 mM GTP in a reaction lacking vRNA ("Rp-G" in Figure 8.8B). The addition of vRNA to the preincubation mixture (Rp-V) was expected to result in a binary complex. Rp-V retained 15 percent of RNA synthesis in the presence of 200 ng heparin. RNA synthesis in the presence of heparin was improved to 40 percent when 100 μM GTP was added ("Rp-G-V" in Figure 8.8B). Guanylate serves as the initiation and second nucleotide during RNA synthesis and will allow the formation of the first phosphodiester bond. The level of RNA synthesis was not affected by the inclusion of 500 mM UTP or ATP, two nucleotides that cannot serve as initiation nucleotides. Addition of both UTP and GTP in the reaction ("Rp-GU-V" in Figure 8.8B) did not significantly alter the amount of synthesis even though the second phosphodiester bond might have formed. However, preincubation with template and nucleotides GTP, UTP, and CTP ("Rp-GUC-V" in Figure 8.8B) resulted in a ternary complex that was very similar to that formed with template and all four nucleotides ("Rp-GUCA-V" in Figure 8.8B) in their susceptibility to heparin. The limited decrease is likely due to heparin limiting RNA to one round of synthesis, even though more than one round could take place during the 60-minute incubation period. These results suggest three levels of stability in the transitions from template binding to a ternary complex with a nascent RNA of 10 to 14 nt.

Since preincubation of RdRp with template and 100 μM GTP alone improved RNA synthesis in the presence of heparin, we further analyzed the role of GTP in the formation of heparin-resistant RdRp complexes. GTP can function in RNA synthesis by acting as a primer as well as a substrate, resulting in the formation of a nascent RNA of two nucleotides. The effect of preincubation with GTP was compared to preincubation with GpG and GMP. GpG and GMP can serve as primers (Kao and Sun, 1996) but cannot be incorporated internally within an RNA chain. Heparin was added at 500 ng per reaction in this experiment (see Figure 8.9A). Consistent with the amount of synthesis in the presence of 500 ng of heparin, we found that preincubation of RdRp with RNA and GTP resulted in 50 percent of maximal RNA synthesis (see Figure 8.9B). The inclusion of GMP at 100 or 500 μM did not improve the stability of the complex above the 30 percent level. GpG at 200 μM or a combination of GpG and UTP failed to increase the stability of the complex. Therefore, the presence of a mono- or dinucleotide

FIGURE 8.9. Effect of GTP and Primers on the Level of Resistance to Heparin

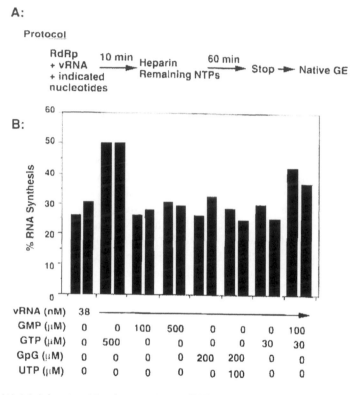

A:

Protocol

RdRp
+ vRNA
+ indicated
nucleotides → (10 min) → Heparin
Remaining NTPs → (60 min) → Stop → Native GE

B:

% RNA Synthesis

vRNA (nM)	38							
GMP (μM)	0	0	100	500	0	0	0	100
GTP (μM)	0	500	0	0	0	0	30	30
GpG (μM)	0	0	0	0	200	200	0	0
UTP (μM)	0	0	0	0	0	100	0	0

Note: (A) A brief protocol for the experiment. RdRp, vRNA, and different nucleotides or oligonucleotide primers were preincubated for ten minutes prior to the addition of 500 ng of heparin and the remaining NTPs for elongation. The reactions were then incubated for 60 minutes before processing for native gel electrophoresis. (B) The RNA products were quantified with the phosphorimager and then normalized to the percentage of products made in a reaction, after preincubation with all components needed for RNA synthesis, and subsequent incubation with the remaining NTPs for elongation in the presence of heparin. The relevant components added to the preincubation reaction and their final concentrations are noted below the bar graph. Each bar represents the result of one assay.

primer is not sufficient to improve the stability of the RdRp complex binding to the template RNA.

Next, we examined whether the GTP-induced improvement in stability is due to the formation of the first phosphodiester bond. Since GMP can serve as a primer for initiation of RNA synthesis, we compared the effects of preincubation of RdRp and RNA with (1) 100 μM GMP, (2) 30 μM GTP, or (3) 100 μM GMP and 30 μM GTP. GMP at 100 μM is in excess of the K_M value for GTP (Kao and Sun, 1996) and can serve as a substitute for the first nucleotide. GTP at 30 μM concentration would be insufficient to contribute significantly to the synthesis of the first phosphodiester bond in the absence of GMP. This was confirmed in reactions preincubated with vRNA and 30 μM GTP, but no GMP (see Figure 8.9B). Complexes formed by pre-incubation with GMP and GTP increased the stability of the complex to approximately 40 percent, an improvement over reactions preincubated with only GMP or other primers. However, the synthesis was not restored to the 50 percent level observed with GTP, probably because GMP is not the initiation nucleotide preferred by RdRp (Kao and Sun, 1996). Nonetheless, the improvement in synthesis in the presence of GMP and GTP suggests that the formation of the first phosphodiester bond increased the commitment of RdRp to the template RNA.

The formation of the first phosphodiester bond may trigger a change in the conformation of the RdRp complex, directly or indirectly, through the formation of an RNA-RNA hybrid. The latter hypothesis would suggest that an increase in the length of the nascent RNA product would result in a linear increase in the stability of the ternary complex. We have only limited data to address this issue. Preincubation of RdRp with the dinucleotide primer GpG and UTP (capable of forming three base pairs with the template RNA) did not result in increased levels of heparin resistance as compared to GTP alone (allowing formation of two base pairs; see Figure 8.9B). Furthermore, preincubation with GTP and UTP did not result in a more stable ternary complex either (allowing formation of three base pairs with template RNA). Thus, we favor the hypothesis that the change in stability is due to a conformational change in the RdRp complex brought about by the progressive formation of phosphodiester bonds.

Another transition in RdRp-RNA interaction occurs between the synthesis of the third and the thirteenth phosphodiester bonds. In the presence of GTP, UTP, and CTP, nascent RNA synthesis should result in a 10 nt product. RNA of 10 nt was observed as the predominant product, but lesser amounts of 12 and 14 nt products were also observed. Thus, it is difficult to pinpoint where further changes in stability occur between the third and the thirteenth phosphodiester bonds. It is interesting to note that the second transition resulted in the same level of stability (resistance to heparin), as the complex that was allowed to synthesize for ten minutes in the presence of all four NTPs. A model for the transitions of RNA synthesis from initiation to elongation by RdRp is presented in Figure 8.10. The *E. coli* RNA

FIGURE 8.10. A Model for the Transitions of RNA Synthesis from Initiation to Elongation by RdRp

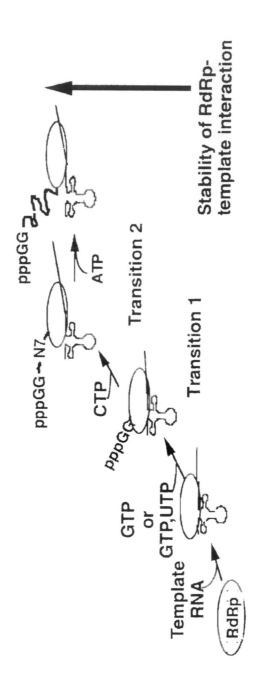

polymerase ternary complex was reported to be stable after the formation of approximately the tenth phosphodiester bond with several promoters of different strengths (Carpousis and Gralla, 1985; Levin et al., 1987; Strancy and Crothers, 1987). Martin et al. (1988) observed that the incorporation of approximately eight nucleotides allowed the T7 bacteriophage RNA polymerase to proceed from abortive to productive RNA synthesis. The sequence of the template may also influence the timing of the transition in T7 bacteriophage RNA synthesis, as nascent RNAs shorter than 10 nt have been reported to form stable ternary complexes (Ikeda and Richardson, 1986).

TERMINATION OF RNA SYNTHESIS

We have yet to study the termination of RNA synthesis by RdRp in detail. However, two scenarios at the end of the template can be envisioned: (1) RdRp pauses and remains on the template; (2) RdRp terminates synthesis by releasing itself from the template. In experiments carried out, when an additional sequence was added at the 3'-end of the template, we found that synthesis continues to the end of the extended template, suggesting that the termination of RNA synthesis is not directed by the sequence alone. Furthermore, the enzyme was able to initiate RNA synthesis from a second template, indicating that the RdRp can recycle to another template *in trans*. However, it is not known whether a free 3'-end would cause the release of the RdRp complex.

COMPARISON OF RNA-DEPENDENT
AND DNA-DEPENDENT RNA SYNTHESIS

Recent elucidation of the structures of several classes of polymerases reveals that the overall structure of polymerases is more similar than their primary sequences (Joyce and Steitz, 1995). This striking observation suggests that all polymerases may function in a mechanistically similar manner. In fact, it now appears that the mechanism of phosphoryl transfer is conserved in all polymerases (Steitz, 1998). In light of these observations, it is of interest to determine possible parallels between the polymerases. Our characterization of the steps involved in RNA synthesis by BMV-RdRp now permits a comparison with RNA synthesis by DNA-dependent RNA polymerase DdRp(s).

Initiation of RNA synthesis by RdRp and DdRp have several features in common (see Table 8.3). A purine nucleotide is strongly preferred and is used at a higher concentration for initiation (Blumenthal, 1980; McClure,

TABLE 8.3. Comparison of the Mechanism of RNA Synthesis by RdRp and DdRp

Characteristic	DdRp	RdRp
Initiation	Purine[a] Primer-inducible[e] Sequence-specific[i] High K_M for initiation nucleotide[k]	Purine[bcd] Primer-inducible[fgh] Sequence-specific [i] High K_M for initiation nucleotide[gl]
Abortive initiation	DdRp remains on template[m]	RdRp dissociates from template[n]
Template commitment	8-12 nt[o]	c. 10 nt[hp]
cis-acting	Enhancers[q]	Activators[c]
Regulatory sequences	Repressors[q]	Repressors[r]
Termination	Sequence-dependent[s]	Sequence-independent[h]

[a]McClure, 1985
[b]Blumenthal, 1980;
Miller et al., 1985;
Miller et al., 1986
[c]Akins et al., 1997
[d]Siegal et al., 1997
[e]Terao et al., 1972; Cazenave
and Uhlenbeck, 1994
[f]Blumenthal and Carmichael,
1979; Honda et al., 1986
[g]Kao and sun, 1996
[h]Sun et al., 1996
[i]Martin and Coleman, 1987;

Milligan et al., 1987;
Maslak et al., 1993; Schick
and Martin, 1993
[j]Siegel et al., 1997
[k]Patra et al., 1992
[l]Mitsunari and Hori, 1973
[m]Carpousis and Gralla, 1980
[n]Sun et al., 1996; Sun and Kao, 1997b
[o]Levin et al., 1987
[p]Sun and Kao, 1997a
[s]Tjian, 1996
[r]Bujarski et al., 1985
[q]Richardson, 1996

1985; Kao and Sun, 1996). BMV-RdRp prefers GTP as the initiation nucleotide but is able to use ATP with lower efficiency (Siegel et al., 1997). The K_M for GTP is approximately ten times higher for BMV-RdRp during initiation than during elongation of (−)-strand RNA synthesis (Kao and Sun, 1996). The Qβ RdRp also initiates with a guanylate with a K_M of 200 μM GTP (Mitsunari and Hori, 1973). In the case of bacteriophage T7 DdRp, the K_M for the initiation nt GTP was 234 μM, whereas other nucleotides used for the elongation step were required at a two to four times lower concentration (Patra et al., 1992). The initiation nucleotide can be considered a primer since it is not hydrolyzed during the formation of the first phosphodiester bond. Primers of a few nucleotides can be substituted for the initiation nucleotide in BMV-RdRp (Blumenthal and Carmichael, 1979; Honda et al., 1986; Kao and Sun, 1996), and also in T7 bacteriophage and *E. coli* DdRps (Terao et al., 1972). In fact, template specificity during RNA synthesis can be overcome by the presence of primers in Qβ RdRp and bacteriophage T7 DdRp (Blumenthal and Carmichael, 1979; Cazenave and Uhlenbeck, 1994).

The process of abortive initiation is shared by DdRps and RdRps. For bacteriophage T7 RNA polymerase, Muller et al. (1988) and Sousa et al. (1992) have determined that the N-terminal c. 175 residues of the polymerase are required for binding to the nascent RNA. Mutant polymerase proteins with single amino acid changes or deletions of the N-terminus retained the ability to initiate RNA synthesis but not enter the elongation phase. These observations led Sousa et al. (1992) to hypothesize that synthesis of a nascent RNA of a specific length and its subsequent binding to the N-terminus of the polymerase alter the conformation of the polymerase to a structure competent for the elongation step. This model could also explain the termination of RNA synthesis. The formation of a stable hairpin in the nascent RNA following synthesis of a termination sequence may separate the nascent RNA from the polymerase, thus causing the polymerase to revert to its initiation conformation that is not as stably associated with the template RNA.

The length of the abortive initiation products formed by both DdRp and RdRp are strikingly similar. Although the amount and length of the abortive initiation products may depend on the promoter and template used, the maximum lengths tend to be between 9 and 12 nt for bacteriophage T7 (Martin et al., 1988) as well as *E. coli* DdRps (Levin et al., 1987). This is in agreement with the size of abortive products (8 and 9 nt) synthesized by BMV (Sun et al., 1996; Sun and Kao, 1997a) and reovirus RdRps (Furuichi, 1981; Yamakawa et al., 1981). In addition, abortive synthesis from the BMV subgenomic promoter demonstrated that the formation of abortive products by RdRp is not due to initiation near the ends of RNAs. Previously we observed that the ratio of elongated to abortive products during (–)-strand RNA synthesis decreased when additional 3'-nucleotides were provided. This observation suggested that the unstable interaction between RdRp and the 3'-end of the template RNA may contribute to abortive initiation. Martin et al. (1988) have previously observed that abortive products synthesized by the bacteriophage T7 DdRp preferentially terminated following the incorporation of UMP into the nascent RNA. Hence, abortive synthesis appears to be an integral part of RNA synthesis by RdRp.

Despite the overall similarities in RNA synthesis by DdRps and RdRps, several differences between them should also be noted. First, RdRps usually initiate RNA synthesis from the ends of RNA templates, whereas DdRps initiate synthesis from an internal promoter using DNA as the template (Miller et al., 1986; Ishihama and Nagata, 1988; Kao and Sun, 1996). Second, RdRps appear to dissociate from the template during abortive synthesis (Sun and Kao, 1997a), whereas bacteriophage T7 RNA polymerase remains stably bound to supercoiled DNA (Diaz et al., 1996). Third, stability of the DdRp ternary complex is maintained primarily by RNA-protein and DNA-protein interactions, and not by RNA-DNA interactions (Altmann et al., 1994). However, some RdRp systems show evidence of an intermediate of (–)-strand RNA synthesis, namely, a double-stranded hybrid composed of the nascent and template RNAs (Baltimore, 1968; Takeda et al., 1986; Bienz et al., 1992; De Graaff et al., 1995); thus, the duplex may contribute to the stability of the RdRp ternary complex. Although we did observe a correlation between ternary complex stability and nascent RNA length, additional studies are required to assess its role.

FINAL COMMENTS

Synthesis of RNA from a DNA template includes at least six discrete steps: (1) specific binding of enzyme to the promoter; (2) melting of DNA to form an open complex; (3) initiation of RNA synthesis; (4) transition from the initially transcribing complex to an elongating complex; (5) elongation; and (6) termination of RNA synthesis (Carpousis and Gralla, 1985; McClure, 1985; Krummel and Chamberlin, 1992a,b). Our work establishes that viral RdRps follow essentially the same ordered set of steps.

It is becoming increasingly apparent that RdRps and DdRps share many biochemical activities and follow a highly parallel series of steps, including initiation, abortive initiation, transition to elongation, and processive elongation. Documenting the steps in RNA synthesis by RdRp is necessary to compare the mechanism of action of different polymerases and to better understand viral RNA replication. The great store of knowledge and techniques amassed by those studying transcription by DdRps and our present understanding of RdRps should enable us to make rapid advances in elucidating viral RNA replication.

REFERENCES

Ackerman S. Bunick D. Zandomeni R. Weinmann R (1983). RNA polymerase II ternary transcription complexes generated in vitro. *Nucleic Acids Research* 11: 6041-6064.

Adkins S, Siegel RW, Sun JH, Kao CC (1997). Minimal templates directing accurate initiation of subgenomic RNA synthesis in vitro by the brome mosaic virus RNA-dependent RNA polymerase. *RNA* 3: 634-647.

Adkins S, Stawicki S, Faurote G, Siegel R, Kao CC (1998). Mechanistic analysis of RNA synthesis RNA-dependent RNA polymerase from two promoters reveals similarities to DNA-dependent RNA polymerase. *RNA* 4: 455-470.

Ahlquist P (1992). Bromovirus RNA replication and transcription. *Current Opinion on Genetics Development* 2: 71-276.

Altmann CR, Solow-Cordero DE, Chamberlin MJ (1994). RNA cleavage and chain elongation by *Escherichia coli* DNA-dependent RNA polymerase in a binary enzyme RNA complex. *Proceedings of the National Academy of Sciences, USA* 91: 3784-3788.

Baltimore D (1968). Structure of the poliovirus replicative intermediate RNA. *Journal of Molecular Biology* 32: 359-368.

Bautz EKF (1976). Bacteriophage-induced DNA-dependent RNA polymerase. In Losick R, Chamberlin M (eds.), *RNA Polymerase*, Cold Spring Harbor Laboratory, Cold Spring Harbor, NY, pp. 273-284.

Bienz K, Egger D, Pfister T, Troxler M (1992). Structural and functional characterization of the poliovirus replication complex. *Journal of Virology* 66: 2740-2747.

Blumenthal T (1980). Qβ replicase template specificity: Different templates require different GTP concentrations for initiation. *Proceedings of the National Academy of Sciences, USA* 77: 2601-2605.

Blumenthal T, Carmichael GG (1979). RNA replication: Function and structure of Qβ replicase. *Annual Review of Biochemistry* 48: 525-548.

Bujarski JJ, Dreher TW, Hall TC (1985). Deletions in the 3'-terminal tRNA-like structure of brome mosaic virus RNA differentially affect aminoacylation and replication in vitro. *Proceedings of the National Academy of Sciences, USA* 82: 5636-5640.

Carpousis AJ, Gralla J (1980). Cycling of ribonucleic acid polymerase to produce oligonucleotides during initiation in vitro at the lacUV5 promoter. *Biochemistry* 19: 3245-3253.

Carpousis AJ, Gralla J (1985). Interaction of RNA polymerase with lacUV5 promoter DNA during mRNA initiation and elongation: Footprinting, methylation and rifampicin-sensitivity changes accompanying transcription initiation. *Journal of Molecular Biology* 183: 165-177.

Cazenave C, Uhlenbeck OC (1994). RNA template-directed RNA synthesis by T7 RNA polymerase. *Proceedings of National Academy of Sciences, USA* 91: 6972-6976.

De Graaff MC, Houwing J, Lukas N, Jaspars EMJ (1995). RNA duplex unwinding activity of alfalfa mosaic virus RNA-dependent RNA polymerase. *FEBS Letter* 371: 219-222.

Diaz GA, Rong M, McAllister WT, Durbin RK (1996). The stability of abortive cycling T7 RNA polymerase complexes depends upon template conformation. *Biochemistry* 35: 10837-10843.

Dreher TW, Bujarski JJ, Hall TC (1984). Mutant viral RNAs synthesized in vitro show altered aminoacylation and replicase template activities. *Nature (London)* 311: 171-175.

Furuichi Y (1981). Allosteric stimulatory effects of S-adenosylmethionine on the RNA polymerase in cytoplasmic polyhedrosis virus. *Journal of Biological Chemistry* 256: 483-493.

Hardy SF, German TL, Loesch-Fries LS, Hall TC (1979). Highly active template-specific RNA-dependent RNA polymerase from barley leaves infected with brome mosaic virus. *Proceedings of the National Academy of Sciences, USA* 76: 4956-4960.

Honda A, Mizumoto K, Ishihama A (1986). RNA polymerase of influenza virus: Dinucleotide-primed initiation of transcription at specific positions on viral RNA. *Journal of Biological Chemistry* 261: 5987-5991.

Ikeda RA, Richardson CC (1986). Interaction of the RNA polymerase of bacteriophage T7 with its promoter during binding and initiation of transcription. *Proceedings of the National Academy of Sciences, USA* 83: 3614-3618.

Ishihama A, Nagata K (1988). Viral RNA polymerases. *CRC Critical Review in Biochemistry* 23: 27-76.

Johnson D, McClure W (1976). Abortive initiation of in vitro RNA synthesis on bacteriophageλ DNA. In Losick R, Chamberlin MJ (eds.). *RNA Polymerase*. Cold Spring Harbor Laboratory, Cold Spring Harbor, NY, pp. 413-428.

Joyce CM, Steitz TA (1995). Polymerase structures and function: Variations on a theme? *Journal of Bacteriology* 177: 6321-6329.

Kao CC, Quadt R, Hershberger R, Ahlquist P (1992). Brome mosaic virus RNA replication proteins 1a and 2a form a complex in vitro. *Journal of Virology* 66: 6322-6329.

Kao CC, Sun JH (1996). Initiation of minus-strand RNA synthesis by the brome mosaic virus RNA-dependent RNA polymerase: Use of oligoribonucleotide primers. *Journal of Virology* 177: 6826-6930.

Krummel B, Chamberlin MJ (1992a). Structural analysis of ternary complexes of *Escherichia coli* RNA polymerase. Deoxyribonuclease I footprinting of defined complexes. *Journal of Molecular Biology* 225: 230-250.

Krummel B, Chamberlin MJ (1992b). Structural analysis of ternary complexes of *Escherichia coli* RNA polymerase. Individual complexes halted along different transcription units have distinct and unexpected biochemical properties. *Journal of Molecular Biology* 225: 221-237.

Levin JR, Krummel B, Chamberlin MJ (1987). Isolation and properties of transcribing ternary complexes of *Escherichia coli* RNA olymerase positioned at a single template base. *Journal of Molecular Biology* 196: 85-100.

Losick R, Chamberlin M (1976). RNA Polymerase. Cold Spring Harbor Laboratory, Cold Spring Harbor, NY.

Marsh LE, Dreher TW, Hall TC (1988). Mutational analysis of the core and modulator sequences of the BMV RNA3 subgenomic promoter. *Nucleic Acids Research* 16: 981-995.

Martin CT, Coleman JE (1987). Kinetic analysis of T7 RNA polymerase-promoter interactions with small synthetic promoters. *Biochemistry* 26: 2690-2696.

Martin CT, Muller DK, Coleman JE (1988). Processivity in early stages of transcription by T7 RNA polymerase. *Biochemistry* 27: 3966-3974.

Maslak M, Jaworski MD, Martin CT (1993). Tests of a model for promoter recognition by T7 RNA polymerase: Thymine methyl group contacts. *Biochemistry* 32: 4270-4274.

Mathews D, Durbin RD (1994). Mechanistic aspects of tagetitoxin inhibition of RNA polymerase from *Escherichia coli*. *Biochemistry* 33: 11987-11992.

McClure WR (1985). Mechanism and control of transcription initiation in prokaryotes. *Annual Review of Biochemistry* 54: 171-204.

Miller WA, Bujarski JJ, Dreher TW, Hall TC (1986). Minus-strand initiation by brome mosaic virus replicase within the 3' tRNA-like structure of native and modified RNA templates. *Journal of Molecular Biology* 187: 537-546.

Miller WA, Dreher TW, Hall TC (1985). Synthesis of brome mosaic virus subgenomic RNA in vitro by internal initiation on (−)-sense genomic RNA. *Nature (London)* 313: 68-70.

Milligan JF, Groebe DR, Witherell GW, Uhlenbeck OC (1987). Oligoribonucleotide synthesis using T7 RNA polymerase and synthetic DNA templates. *Nucleic Acids Research* 15: 8783-8798.

Mitsunari Y, Hori K (1973). Qβ replicase associated polycytidylic acid dependent polyguanylic acid polymerase. *Journal of Biochemistry* 74: 263-271.

Muller DK, Martin CT, Coleman JE (1988). Processivity of proteolytically modified forms of T7 RNA polymerase. *Biochemistry* 27: 5763-5771.

Patra D, Lafer EM, Sousa R (1992). Isolation and characterization of mutant bacteriophage T7 RNA polymerases. *Journal of Molecular Biology* 224: 307-318.

Pfeffer SR, Stahl SJ, Chamberlin MJ (1977). Binding of *Escherichia coli* RNA polymerase to T7 DNA. *Journal of Biological Chemistry* 252: 5403-5407.

Quadt R, Jaspars EMJ (1990). Purification and characterization of brome mosaic virus RNA-dependent RNA polymerase. *Virology* 178: 189-194.

Restrepo-Hartwig M, Ahlquist P (1996). Brome mosaic virus helicase- and polymerase-like proteins colocalize on the endoplasmic reticulum at sites of viral RNA synthesis. *Journal of Virology* 70: 8908-8916.

Richardson JP (1996). Structural organization of transcription termination factor Rho. *Journal of Biological Chemistry* 271: 1251-1254.

Schick C, Martin CT (1993). Identification of specific contacts in T3 RNA polymerase-promoter interactions: Kinetic analysis using small synthetic promoters. *Biochemistry* 32: 4275-4280.

Siegel RW, Adkins S, Kao CC (1997). Sequence-specific recognition of a subgenomic promoter by a viral RNA polymerase. *Proceedings of the National Academy of Sciences, USA* 94: 11238-11243.

Sousa R, Patra D, Lafer EM (1992). Model for the mechanism of bacteriophage T7 RNAP transcription initiation and termination. *Journal of Molecular Biology* 224: 319-334.

Steitz TA (1998). A mechanism for all polymerases. *Nature* 391: 231-232.

Straney DC, Crothers DM (1987). Comparison of the open complexes formed by RNA polymerase at the *Escherichia coli lac* UV5 promoter. *Journal of Molecular Biology* 195: 279-292.

Sun JH, Adkins S, Faurote G, Kao CC (1996). Initiation of (−)-strand RNA synthesis catalyzed by the BMV RNA-dependent RNA polymerase: Synthesis of oligonucleotides. *Virology* 226: 1-12.

Sun JH, Kao CC (1997a). Characterization of RNA products associated with or aborted by a viral RNA-dependent RNA polymerase. *Virology* 236: 348-353.

Sun JH, Kao CC (1997b). RNA synthesis by the brome mosaic virus RNA-dependent RNA polymerase: Transition from initiation to elongation. *Virology* 233: 63-73.

Takeda N, Kuhn RJ, Yang CK, Takegami T, Wimmer E (1986). Initiation of poliovirus plus-strand RNA synthesis in a membrane complex of infected HeLa cells. *Journal of Virology* 60: 43-53.

Terao T, Dahlberg JE, Khorana HG (1972). Studies on polynucleotides. *Journal of Biological Chemistry* 247: 6157-6166.

Tjian R (1996). The biochemistry of transcription in eukaryotes: A paradigm for multisubunit regulatory complexes. *Philosphical Transactions of the Royal Society of London Series B* 351: 491-499.

Webb ML, Jacob S (1988). Inhibition of RNA polymerase I-directed transcription by novobiocin—Potential use of novobiocin as a general inhibitor of eukaryotic transcription initiation. *Journal of Biological Chemistry* 263: 4745-4748.

Yamakawa M, Furuichi Y, Nakashima K, LaFiandra AJ, Shatkin AJ (1981). Excess synthesis of viral mRNA 5'-terminal oligonucleotides by reovirus transcriptase. *Journal of Biological Chemistry* 256: 6507-6514.

Chapter 9

Gene Expression Strategies
of RNA Viruses

Uli Commandeur
Wolfgang Rohde
Rainer Fischer
Dirk Prüfer

INTRODUCTION

Plant viral genomes encode a limited set of proteins that are required for viral replication, particle formation, movement of the virus or its RNA from cell to cell, polyprotein maturation, and interaction of the virus with its vector or host.

Because of their limited genomic size, plant RNA viruses exploit strategies for gene expression that are not commonly used in the plant kingdom. This includes transcriptional strategies such as the production of subgenomic RNAs (sgRNAs). SgRNAs are messenger RNAs (mRNAs) coding for the 3'-proximal genes of a polycistronic viral RNA and therefore are identical copies of the corresponding region on the viral genome (genomic RNA [gRNA]). When several genes are present at the 3'-end of the gRNA, a family of 3'-colinear sgRNAs are produced such that each expressed gene is located at the 5'-end of one sgRNA, a situation that is required for proper translation in eukaryotes.

At the level of translation initiation, leaky scanning of the initiation complex allows expression of more than one protein from the same RNA. In this mechanism, a portion of the 40S ribosomal subunit bypasses the 5'-proximal AUG and initiates at a downstream AUG (Kozak, 1989c), if this is

The authors would like to thank Dr. Neil Emans for critical reading of the manuscript.

placed in a better context for initiation of translation. One alternative to this distal entry of part of the translational machinery and its linear migration is the internal entry of ribosomes, guided by "internal ribosome entry sites" (IRES; Herman, 1989). During protein elongation, ribosomal frameshifting allows the production of two proteins (the frame protein and the transframe protein) that are identical at the N-terminus but differ downstream from the frameshifting point. The frame proteins are more abundant than transframe proteins since ribosomal frameshifting occurs at only very low frequencies. At the termination step, suppression of a stop codon involving specific suppressor transfer RNAs (tRNAs) produces a carboxy-terminal (C-terminal) extended protein. In addition, viruses use expression of a polyprotein precursor that is cleaved into different proteins using the protease activity of one or more of the viral products (Matthews, 1991; Maia et al., 1996).

For most of these expression strategies the underlying mechanism is well understood and reviewed elsewhere (Rohde et al., 1994; Maia et al., 1996; Drugeon et al., 1999). In this chapter, we focus on novel aspects of plant virus gene expression, and, therefore, strategies such as translation initiation mediated by "internal ribosomal entry sites" are highlighted.

SUBGENOMIC RNAs AND GENOME SEGMENTATION

A eukaryotic mRNA is monocistronic; that is, only the first open reading frame (ORF) is translated. Genomes of the positive-strand RNA viruses have mRNA polarity, and they are capable of directing protein synthesis both in vivo and in vitro. RNA viral genomes usually carry a set of genes necessary for the viral infection cycle. RNA viruses can express more than one gene from a given genome by using different expression mechanisms:

- *Genome segmentation:* The entire genome is distributed over several genomic RNAs, each of which carries only one or a few genes so that most of the genome segments become monocistronic (see Figure 9.1A).
- *Ambisense transcription strategy:* Viruses whose genes are encoded by both viral strand RNA and complementary strand RNA can be considered a special case of genome segmentation (see Figure 9.1B).
- *Subgenomic RNA (sgRNA) synthesis:* Some viral genes located internally in the viral RNA are transcribed from the gRNA into sgRNAs. These genes will then become the first ORF on the sgRNA and are accessible to ribosomes. Subgenomic promoter sequences are present on the minus strand of the gRNA, and the RNA-dependent RNA polymerase (RdRp) utilizes this strand as a template (see Figure 9.1C).

FIGURE 9.1. Gene Expression Strategies of Plant RNA Viruses

A. Genome segmentation

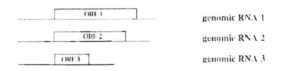

B. Ambisense strategy

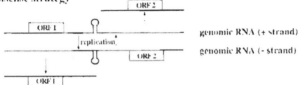

C. Subgenomic RNA synthesis

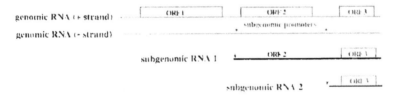

D. Premature termination

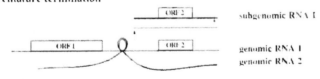

E. Proteolytic processing

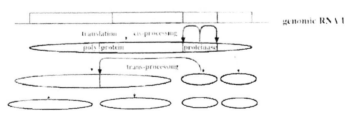

Note: ORF = open reading frame.

Genome Segmentation

Division of the genome into more than one segment satisfies the monocistronic requirement of the eukaryotic translation system. The genomes of some RNA viruses are split into several genomic components; each piece of the genome ideally contains one ORF and thus resembles a monocistronic mRNA in behavior.

Viruses falling into this category are members of the *Bromoviridae* family, for example, bromoviruses, cucumoviruses, and ilarviruses. All of them have a tripartite genome; each RNA in the viral genome is a one-gene mRNA, with the exception of RNA3, which contains a 3'-capsid protein (CP) gene. The CP gene is expressed from a subgenomic messenger (RNA4) and becomes encapsidated, and the CP itself is needed for the initiation of infection by *Alfalfa mosaic virus* (AMV, genus *Bromovirus*) and by ilarviruses (Van Vloten-Doting, 1975; Bol, 1999).

Segmented genomes can also potentially facilitate intervirus recombination. Reassortment of different genomic segments provides a simple mechanism for the exchange of genetic information between viruses (Fraile et al., 1997; see Chapter 10 of this book).

Ambisense Transcription Strategy

In RNA viruses using the ambisense (ambiguous sense) transcription strategy, genes are encoded on both the positive and negative strands.

Tomato spotted wilt virus (TSWV) was the first plant virus identified that uses such a transcription mechanism (Kormelink et al., 1992). TSWV is a member of the family *Bunyaviridae* and has been classified as the type species of the *Tospovirus* genus. These viruses are principally negative-strand RNA viruses, and the viral polymerase (L protein) that is necessary for the synthesis of the positive strand is present in the particle (Van Poelwijk et al., 1996). The L protein is also responsible for the transcription of viral mRNAs, and the termination of transcription is accomplished by the presence of a central hairpin loop. All mRNAs are capped at the 5'-ends and contain nonviral sequences captured from host mRNAs, which serve as primers for transcription. This "cap-snatching" mechanism has also been seen in other members of the *Bunyaviridae* as well as in members of the *Arenaviridae* and *Orthomyxoviridae*, two other families that include negative-strand viruses with segmented genomes (Bishop, 1986). Cap structures are cleaved from host mRNAs by a virus-encoded endonuclease and are subsequently used to prime transcription. In mixed infections of plants with TSWV and AMV, even AMV RNAs can function as cap donors for TSWV, paving the way to study the sequence specificity of the viral endonuclease in vivo (Duijsings et al., 1999).

From comparison of processes involved in the infection cycle of these viruses (Banerjee and Barik, 1992; Baudin et al., 1994), the cytoplasmic concentration of the N protein, encoded on the viral complementary RNA of the S segment, controls the switch from replication to transcription. In the beginning of infection, the concentration of N protein is low because of the low amount of its mRNA. Upon transcription, the replicase will produce more mRNA that will in turn be translated and lead to the accumulation of the N protein and other viral proteins. With the increase of N protein concentrations, the polymerase switches from transcription to replication and multiplies the viral gRNAs.

Subgenomic RNA Synthesis

Transcription of sgRNA from viral gRNA is one of the ways RNA viruses to express internally located genes. Similar to mRNA transcription from a DNA template, this process requires a promoter sequence located on the genomic minus strand. Such a subgenomic promoter could potentially regulate the timing and the level of viral gene expression. The 5'-ends of subgenomic promoters have been determined for different viruses by various means, such as direct RNA sequencing, primer extension analyses, RNase protection experiments, deletion analyses, and site-directed mutagenesis. A number of sequences have been determined and compiled in a review by Maia et al. (1996). Recently, it was shown for *Barley yellow dwarf virus* (BYDV), a luteovirus, that primary and secondary structural elements are required for the synthesis of sgRNA1 (Koev et al., 1999). This subgenomic promoter partially overlaps the coding region of the viral replicase; thus, its genetic information simultaneously serves as a regulatory RNA element and encodes part of a protein. From the data, the authors suggested that subgenomic promoters evolved independently at appropriate genomic locations while allowing overlapping ORFs to maintain their functions.

Long-distance RNA-RNA interaction seems to be involved in the synthesis of sgRNA2 of *Tomato bushy stunt virus* (TBSV), a tombusvirus (Zhang et al., 1999). Deletion analysis revealed a 12-nucleotide (nt)-long RNA sequence, located ~1 kb upstream of the initiation site of sgRNA2 synthesis, necessary for accumulation of sgRNA2. This sequence has the potential for base pairing with a region immediately upstream of the sgRNA2 synthesis initiation site. Introduced mutations, which either disrupted or maintained this interaction, resulted in a positive correlation between the predicted stability of the base pairing interaction and the efficiency of sgRNA2 accumulation.

Furthermore, sequences involved in this interaction seem to be evolutionarily conserved in the genomes of other tombusviruses.

This mechanism, which involves an RNA-RNA interaction *in trans*, was suggested for *Red clover necrotic mosaic virus* (RCNMV), a dianthovirus (Sit et al., 1998). RCNMV exhibits RNA-mediated regulation of transcription, which is unusual among RNA viruses that typically rely on protein regulators. A 34 nt sequence in RNA2 was shown to be required for transcription of the sgRNA. Mutations that prevented basepairing between the RNA1 subgenomic promoter and the 34 nt transactivator abolished expression of a reporter gene. The authors presented a model in which direct binding of RNA2 to RNA1 transactivates sgRNA synthesis. In this model, synthesis of the negative strand by the RdRp is prematurely terminated at the combined RNA1/RNA2 secondary structure; the resulting truncated minus strand then serves as the template for sgRNA synthesis (see Figure 9.1D).

RNA and DNA transcription, recently reviewed by Lai (1998), differ in that there is no termination signal to stop the transcription from a viral RNA template. Consequently, viral sgRNAs are coterminal with the viral RNA at the 3'-ends. The 3'-termini of the gRNAs of many positive-strand RNA viruses can fold into a tRNA-like structure (TLS) that is recognized by tRNA-specific enzymes (for review, see Haenni and Chapeville, 1997). For TLS, the formation of so-called pseudoknots is required, and in these pseudoknots, loop regions base pair with regions outside the loop (Pleij, 1995; Koenig et al., 1998). The inherently low stability of this structure results in the formation of alternative structures, and this allows pseudoknots to function as "molecular switches." Olsthoorn et al. (1999) provided evidence for the existence of two mutually exclusive conformations for the 3'-termini of alfamovirus and ilarvirus RNAs that enable the viruses to switch from translation to replication and vice versa. The pseudoknot conformation is required for minus-strand synthesis, but the linear conformation has been implicated in CP binding. CP binding presumably induces a transition from the pseudoknot conformer to the CP-binding conformer, which is not recognized by the RdRp. To test this hypothesis, the authors created a mutation that interfered with CP binding but had little effect on the template activity in vitro (Reusken et al., 1997). Whereas addition of CP to wild-type (wt) RNA strongly inhibited template recognition by the RdRp, the mutant RNA was not affected. These data support the hypothesis that binding of CP requires the formation of a stem loop that in turn disrupts the pseudoknot necessary for template recognition of the RdRp.

Strategies for the expression of sgRNAs are the same as those described for gRNAs, and these are outlined in this chapter. Some genes, such as those coding for CPs, are expressed from sgRNAs at high levels. Temporal regulation of expression of sgRNA-encoded genes has been described for *Tobacco mosaic virus* (TMV), a tobamovirus (Dawson and Lehto, 1990), and recently for BYDV (Wang et al., 1999) and *Beet yellows virus* (BYV), a closterovirus (Hagiwara et al., 1999). In the latter case, the authors used reporter gene fusions with ß-glucuronidase (GUS) and a proteinase domain to tag individual genes of BYV. When inserted at the very N-terminus of different ORFs, these self-processing reporter constructs permitted the expression of the reporter and the original viral product from a single expression unit. The strategy allowed preservation of all genes and control regions, thus minimizing alteration of the viral genetic makeup. The data indicated that the temporal regulation of BYV gene expression includes early and late phases.

INITIATION OF TRANSLATION

Cap-Independent Translation

The only mRNAs produced in plants known to naturally lack a cap structure (m^7GpppN) are of viral origin. The gRNA of *Tobacco etch virus* (TEV), member of the genus *Potyvirus* (family *Potyviridae*), and sometimes unofficially classified in the superfamily *Picornaviridae*, is a polyadenylated mRNA that is naturally uncapped and yet a highly competitive mRNA during translation. By deletion analysis, Niepel and Gallie (1999) have identified the elements in the TEV 5'-leader that are required for cap-independent translation. Through the introduction of stem-loop structures, the authors identified two centrally located cap-independent regulatory elements (CIREs); the 5'-proximal CIRE was less 5'-end dependent than the 5'-distal element, suggesting that they are functionally distinct. Furthermore, these elements may not be redundant control regions because they exhibit a combinatorial effect and both appear to be required for full cap-independent translation. Moreover, the introduction of a stable stem loop upstream of the TEV leader sequence or upstream of either CIRE in dicistronic constructs markedly enhanced their regulatory function. These data suggest that the TEV 5'-leader contains two elements that together promote internal initiation but that the function of one element, in particular, is facilitated by proximity to the 5'-end.

In BYDV, highly efficient cap-independent translation initiation at the 5'-proximal AUG is facilitated by the 3'-translation enhancer sequence (3'-TE) located near the 3'-end of the gRNA. The 3'-TE is required for translation and thus replication of the gRNA that lacks a 5'-cap (Allen et al., 1999). Wang et al. (1999) showed that the 3'-TE also mediated translation of uncapped viral subgenomic mRNAs (sgRNA1 and sgRNA2). A 109-nt-long viral sequence is sufficient for 3'-TE activity in vitro, but an additional viral sequence is necessary for cap-independent translation in vivo. The 5'-extremity of the sequence required in the 3'-untranslated region (UTR) for cap-independent translation in vivo coincides with the 5'-end of sgRNA2. Thus, sgRNA2 has the 3'-TE in its 5'-UTR. Competition studies using physiological ratios of viral RNAs showed that, *in trans*, the 109 nt 3'-TE alone, or in the context of 869 nt sgRNA2, inhibited translation of gRNA much more than it inhibited translation of sgRNA1. The divergent 5'-UTRs of gRNA and sgRNA1 contribute to this differential susceptibility to inhibition. The authors propose that sgRNA2 serves as a novel regulatory RNA to carry out the switch from early to late gene expression. Thus, this new mechanism for temporal translation control involves a sequence that stimulates translation *in cis* and acts to selectively inhibit translation of viral mRNA *in trans*.

Leaky Scanning

In the leaky-scanning mechanism, the 40S ribosome subunits partially bypass the 5'-proximal AUG and proceed to the next AUG with a more favorable initiation codon context for initiation of translation (Kozak, 1986a,b). In addition to the initiation codon context, other factors, such as the length of the untranslated leader sequence, the distance between the AUGs, and the presence of RNA secondary structures, are involved (Kozak, 1991, 1994).

A leaky-scanning mechanism has been suggested for the expression of overlapping genes in different RNA viruses. In luteoviruses, a second gene is entirely encompassed by the CP ORF on sgRNA1 (Tacke et al., 1990; Dinesh-Kumar and Miller, 1993). So-called "triple gene blocks" (TGBs), which contain partially overlapping genes, are found in a number of viruses. In *Potato virus X* (PVX), a potexvirus (Verchot et al., 1998), and *Barley stripe mosaic virus* (BSMV), a hordeivirus (Zhou and Jackson, 1996), the start codon is in a less favorable context for initiation than is that of the third ORF. In *Cucumber necrosis virus* (CNV), a tombusvirus, two factors. leader length and codon context, influence the efficiency of expression of overlapping genes (Johnston and Rochon, 1996). Context-dependent leaky scanning has recently been described for TBSV for the expression of nested genes (Scholthof et al., 1999).

Evidence for leaky scanning was also provided by Sivakumaran and Hacker (1998) with the cowpea strain of *Southern bean mosaic virus*, a sobemovirus, and by Simón-Buela et al. (1997) with *Plum pox virus* (PPV). a potyvirus. RNAs of potyviruses, and also of all members of the *Picornaviridae* family, lack a 5'-cap; instead, a genome-linked viral protein (VPg) is present at the 5'-end. The genome of PPV contains a single ORF that is translated into a large polyprotein. Although the ORF starts at nucleotide 36 (AUG_{36}), it is translated from the second, AUG_{147}, which is in a more favorable position for translation initiation. The authors have carried out in vitro translation and transient expression analysis in protoplasts of a nested set of substitution and deletion mutants, and these results show that no internal structure in the 5'-UTR of PPV is necessary for efficient translation initiation. On the other hand, when the cryptic AUG_{36} was placed in a favorable context, it turned into an efficient initiation codon in vitro. Furthermore, AUGs that were placed in a favorable context, initiating short intraleader ORFs, repressed translation initiation from the AUG_{147} in vitro and in vivo. These results point to leaky scanning as the mechanism of translation initiation of PPV RNA. This is in contrast to the internal initiation of translation described for picornaviruses, which usually have stable secondary structures in long untranslated leader sequences.

Internal Ribosome Entry

Posttranscriptional mechanisms that control plant gene expression (Filipowicz and Hohn, 1996) have been reviewed in detail (Browning, 1996; Fütterer and Hohn, 1996; Gallie, 1996). In addition to translation initiation strategies such as leaky scanning, internal reinitiation, or ribosome shunting that escape the "first AUG" rule of 40S complexes scanning eukaryotic mRNA, direct ribosome entry at the site of translation initiation represents a particular example of cap-independent initiation. This internal initiation is mediated by so-called ribosome landing pads, or internal ribosomal entry sites (IRES), and provides an alternative initiation strategy to the generally accepted scanning mechanism (Kozak, 1989c, 1999). Although the existence of scanning-independent strategies of translation initiation has been questioned on the basis of the location of IRES elements at the 5'-region of natural mRNAs (genomic and subgenomic mRNAs in the case of viruses; Kozak, 1999), the activity of IRES at internal positions in artificial dicistronic mRNAs containing hairpins with the capacity to block ribosome scanning (see Mountford and Smith, 1995) or on circular, covalently closed RNA (Chen and Sarnow, 1995) provides ample evidence for cap-independent internal ribosome entry.

Translation initiation via IRES was first discovered and studied in detail with picornaviruses (Jang et al., 1988; Pelletier and Sonenberg, 1988; Jackson et al., 1990; see, for recent reviews, Belsham and Sonenberg, 1996; Ehrenfeld, 1996). Further examples for IRES-mediated protein initiation are *Hepatitis C virus* (HCV; Tsukiyama-Kohara et al., 1992; Wang et al., 1993; Reynolds et al., 1995; Rijnbrand et al., 1995), RNA tumor viruses (Berlioz and Darlix, 1995; Berlioz et al., 1995; Vagner, Waysbort, et al., 1995), or the tricistronic sgRNA of infectious bronchitis virus (Liu and Inglis, 1992; Le et al., 1994). Functional IRES elements have been identified in capped nonviral RNAs as well (Jackson, 1991; Macejak and Sarnow, 1991; Oh et al., 1992; Vittorioso et al., 1994; Vagner, Gensac, et al., 1995; Gan and Rhoads, 1996; Nanbru et al., 1997; Akiri et al., 1998; Miller et al., 1998; Stein et al., 1998; Stoneley et al., 1998). A further interesting example of an IRES element in cellular mRNA is that of eIF4G, the eukaryotic initiation factor that is cleaved by picornaviral proteases during the switch from cap-dependent to IRES-mediated initiation (discussed in the following pages; Gan and Rhoads, 1996).

Few examples have been found of the existence and function of IRES elements in plant viruses (for a review, see Fütterer and Hohn, 1996). These include the potyviruses, e.g., TEV (Carrington and Freed, 1990; Niepel and Gallie, 1999), *Turnip mosaic virus* (TuMV; Basso et al., 1994), and *Potato virus Y* (Levis and Astier-Manifacier, 1993), and the comoviruses, such as *Cowpea mosaic virus* (Thomas et al., 1991). More recently, IRES activity was observed in TMV, both in a crucifer-infecting TMV (crTMV; Ivanov et al., 1997) and in the TMV U1 strain (Skulachev et al., 1999).

The functional characterization of plant-specific IRES domains has just begun. The 5'-UTR of TEV gRNA is 143 nt in length and promotes cap-independent translation (Carrington and Freed, 1990). Niepel and Gallie (1999) have dissected this leader sequence by deletion analysis and identified two centrally located CIREs that are present in regions of 39 nt (CIRE-1) and 53 nt (CIRE-2), respectively; both are required for optimal translation efficiency. A further requirement is the presence of the poly(A)-tail as in natural TEV RNA, since the activity of CIRE-1 and CIRE-2 was downregulated in poly(A)-RNA. Such a functional characterization of domains is not yet available for the recently described IRES elements in crTMV (Ivanov et al., 1997) or the TMV U1 strain (Skulachev et al., 1999). The originally characterized IRES of 148 nt (IRES$_{CP148}^{CR}$) for the internal initiation of the CP gene of crTMV, but not of TMV U1 (Ivanov et al., 1997), was complemented by identifying IRES activity in the 228-nucleotide sequence preceding the movement protein (MP) gene on the I$_2$ sgRNA of crTMV. Surprisingly, and in contrast to the CP-specific IRES, both TMV U1 and crTMV contain IRES that mediate MP translation. Although IRES activity was well documented, questions concerning the biological relevance of TMV IRES-mediated translation initiation remain, since concurrent MP and CP synthesis from the corresponding sgRNAs by a scanning mechanism is not excluded.

IRES elements of vertebrate picornaviruses, such as poliovirus, are also part of 5'-leader sequences that assume complex secondary and tertiary structures. In general, active IRES sequences are located upstream of the internal AUG from which they direct initiation. A second feature is a pyrimidine-rich tract that binds host factors (polypyrimidine tract-binding protein |PTB|). PTB is a component of the initiation complexes, and it is necessary for complex formation and translational efficiency, as demonstrated for foot-and-mouth disease virus (FMDV) IRES (Niepmann et al., 1997). The specificity of this PTB/IRES complex formation has been investigated with poliovirus by RNA mobility shift experiments. The poly(rC)-binding protein 2 (PCBP2) from HeLa cells binds specifically to stem loop IV, a central domain of the poliovirus IRES (Blyn et al., 1996). PTBs are of nonviral origin and may thus play specific roles in the host cell. One example is the eukaryotic initiation factor 4G (eIF4G) mRNA: A polypyrimidine tract is required for eIF4G mRNA IRES activity that resides on a 101 nt segment of the 5'-UTR (Gan et al., 1998). This tract is located at the 3'-end of the eIF4G 5'-UTR, and its presence resembles one of the functional requirements for initiation complex formation with picornaviral IRES elements.

Although by primary sequence, picornaviral IRES do not show any significant homology, RNA-folding analyses have revealed a structural core that is conserved in the 5'-UTR of all picornaviruses (Le et al., 1996; Le and Maizel, 1998). For HCV, a pseudoknot structure upstream of the AUG initiation codon has been identified as a structural requirement for IRES activity by mutation analysis (Wang et al., 1995). The importance of secondary or tertiary RNA structures for IRES activity was further revealed by mutation analysis. With aphthovirus RNA, secondary structure-destabilizing mutations had a detrimental effect on IRES activity, which could be rescued by restoring the RNA structure (Martinez Salas et al., 1996). A 3 nt insertion into stem loop IV of poliovirus IRES prevented PCBP2 binding and eliminated the translational activity and virus viability (Blyn et al., 1996). In contrast, a single point mutation in the c-myc IRES greatly stimulated the binding of cellular factors (Paulin et al., 1998).

Various mechanisms operate to regulate IRES activity. Poliovirus infection severely impedes 5'-cap-dependent translation of cellular mRNAs in infected cells by proteolytic inactivation of eIF4G, thereby favoring polioviral mRNA translation by IRES-mediated translation initiation. This has been studied in some detail with FMDV. FMDV encodes cysteine proteinases that cleave the eIF4G into N- and C-terminal fragments containing the domains for binding to the initiation factors 4A, F3, and 4E, respectively. These eIF4G cleavage products enhance translation of uncapped and IRES-containing mRNAs (Borman et al., 1997). eIF4G cleavage is inhibited by the host protein PHAS-1 that binds to eIF4E and, by interaction of this complex with eIF4G, presumably renders eIF4G inaccessible to proteolysis (Ohlman et al., 1997). Whereas in most picornaviruses eIF4G proteolysis is a prerequisite for IRES-mediated picornaviral mRNA translation, the *Hepatitis A virus* (HAV) IRES require intact eIF4G for activity (Borman and Kean, 1997). This may explain why HAV and *Encephalomyocarditis virus* (EMCV), both members of the genus *Cardiovirus* (family *Picornaviridae*), do not interfere with the translation of cellular mRNAs by 5'-cap-mediated ribosome binding.

Other IRES elements are subject to developmental regulation: the homeotic *Drosophila* genes *Ultrabithorax* (*Ubx*) and *Antennapedia* (*Antp*) both contain IRES activities in their respective 5'-UTRs, which are functionally not equivalent and subject to temporal and spatial regulation, as evident from IRES-mediated reporter gene expression in transgenic *Drosophila* (Ye et al., 1997).

Normally, IRES sequences direct translation initiation at authentic AUG codons (Reynolds et al., 1996), but they can also mediate gene regulation in unexpected ways. One example is shown by the IRES of *Plautia stali intestine virus* (PSIV), a picorna-like insect virus. The PSIV IRES direct non-canonical translation initiation at a CUU codon, which is not recognized by the scanning complex after deletion of the 5′-IRES sequence (Sasaki and Nakashima, 1999). The IRES present in human fibroblast growth factor 2 (FGF-2) mRNA mediate initiation both at an AUG and at three CUG codons (Vagner, Gensac, et al., 1995). The four FGF-2 isoforms may have different cellular functions, since the CUG-initiated FGF-2s are targeted to the nucleus, whereas the AUG initiation product remains predominantly in the cytoplasm. Constitutive expression of the four FGF-2 forms leads to different cell phenotypes and reveals a distinct role of the FGF-2 IRES in affecting the control of cell proliferation and differentiation.

Both *cis*- and *trans*-acting regulators of IRES activity have been identified. In the case of HCV RNA, the 3'-UTR increases translation efficiency from HCV IRES. The positive-sense HCV RNA 3'-UTR consists of 98 nt which assume a stem loop structure and bind the PTB. Mutations in the 3'-UTR that abolished the PTB binding capacity reduced IRES activity, suggesting that the 3'-UTR positively regulates IRES-directed translation, with both HCV and the unrelated EMCV IRES (Ito et al., 1998). Negative regulation of IRES-mediated translation has also been described. The absence of translation of poliovirus mRNA in *Saccharomyces cerevisiae* has been traced to a small inhibitory yeast RNA (IRNA) of 60 nt that also inhibits HCV IRES-mediated initiation. A chimeric poliovirus in which the polioviral IRES had been replaced by HCV IRES could not replicate in a hepatoma cell line that constitutively expressed I RNA (Das et al., 1996; Das, Ott, Yamane, Tsai, et al., 1998; Das, OH, Yamane, Venkatesan, et al., 1998).

These few examples may serve to demonstrate how IRES can affect translation initiation and the mechanisms by which IRES activity is regulated. The various control mechanisms established for the animal and yeast systems provide an outlook on what we can expect for plant viral IRES or IRES of plant mRNAs once they have been identified. In addition, internal ribosome entry may not necessarily result in protein translation. Two functions have been attributed to the poliovirus IRES: internal translation initiation capacity and the regulation of RNA replication (Borman et al., 1994). Thus, the internal entry of ribosomes at IRES elements may activate control mechanisms at the levels of both translation and transcription. As a consequence of observations from the animal virus field, point mutations in plant viral IRES sequences may up- or downregulate translational efficiency for viral proteins and thus severely interfere with plant virus viability or pathogenicity.

For the animal system, the importance of IRES elements in the construction of di- or tricistronic mRNAs for both basic and applied research has been discussed (Mountford and Smith, 1995). In transgenic plants, the coexpression of two or more genes from the identical mRNA would represent a valuable tool for molecular breeding through the pyramiding of traits under the provision that transcriptional control of the transgenes by the identical promoter is desirable. Further applications, such as gene trapping or gene targeting, are equally important in plants, since these could allow the identification of developmentally regulated genes.

Non-AUG initiation

According to the scanning model, the 40S ribosomal subunit in association with Met-tRNA, initiation factors bind to the 5'-end of mRNA and migrate to the first AUG embedded in an optimal context for translation initiation. On the other hand, in vitro as well as in vivo studies indicated that non-AUG codons can also facilitate translation initiation (Kozak, 1989b; Peabody, 1989; Dasso and Jackson, 1989; Kozak, 1991), although non-AUG initiation was found to be less efficient when compared to canonical translation efficiencies (Kozak, 1989b; Mehdi et al., 1990; Gordon et al., 1992). Initiation at non-AUG codons seems to be rare and appears to be controlled by flanking sequences and/or RNA secondary structures (Kozak, 1989a; Hann et al., 1992; Boyd and Thummel, 1993; Boeck and Kolakofsky, 1994; Kozak, 1997).

The use of non-AUG codons in gene expression has been shown for pro- and eukaryotes (Kozak, 1989b). For the prokaryotic system, translation initiation at a UUG codon could be observed for the *carbamoylphosphate synthetase* gene of *Escherichia coli* (Weyens et al., 1985). For animal viruses, translation initiation of viral proteins takes place at an AUC codon for the adeno-associated virus CP (Becera et al., 1985), at an ACG codon for the *Sendai virus* C protein (Curran and Kolakofsky, 1988), and at a CUG codon for translation of the cell surface antigen of *Murine leukemia virus* (Prats et al., 1989). For the plant kingdom, the first indirect evidence for non-AUG initiation was provided by Hohn and co-workers (Schultze et al., 1990) and confirmed in plant protoplasts by transient expression of appropriate reporter gene fusions (Gordon et al., 1992). The first example of non-AUG initiation in plant RNA viruses was found for *Soil-borne wheat mosaic virus*, the type species of the genus *Furovirus*. Here, translation initiates in addition to the CP AUG at an upstream-located CUG triplet (Shirako and Wilson, 1993). Recently, Schmitz et al. (1996) demonstrated that an AUU codon located within the Ω sequence of TMV is capable of translation initiation when this leader was used as a translational enhancer (see also Tacke et al., 1996). In yeast, non-AUG codons are recognized inefficiently (Zitomer et al., 1984; Olsen, 1987; Clements et al., 1988; Donahue and Cigan, 1988).

ELONGATION OF TRANSLATION

−1/+1 Ribosomal Frameshifting

Expression of a single protein from two or more overlapping ORFs by ribosomal frameshifting is a translational mechanism that has been studied in detail using viruses as model systems (for review, see Farabaugh, 1996). The most common frameshift in the −1 direction was first discovered in retro- and coronaviruses (Jacks and Varmus, 1985; Brierley et al., 1987) but has subsequently been identified in many plant viruses (Brault and Miller, 1992; Prüfer et al., 1992; Garcia et al., 1993; Kujawa et al., 1993; Kim and Lommel, 1994; Makinen et al., 1996; Kim and Lommel, 1998). Mutational analysis of corresponding genomic areas, known to be responsible for the frameshift event, revealed the presence of two characteristic elements: the frameshift site (also known as slippery sequence or heptanucleotide signal; for a detailed list of characterized plant virus frameshift sites see Maia et al., 1996) and a stem-loop or pseudoknot structure. For most plant viruses, the structural requirement (stem-loop or pseudoknot structure) has been identified. Studies on the frameshift region in a German isolate of *Potato leafroll virus* (PLRV) indicated the presence of a stable stem-loop structure (Prüfer et al., 1992); however, a pseudoknot structure was found for the Polish PLRV isolate (Kujawa et al., 1993). Further extensive mutational analysis of the frameshift region in the German PLRV isolate demonstrated that both structures are possible and may suggest that −1 ribosomal frameshifting in PLRV underlies a coherent system that guarantees protein expression if one of the structures is disrupted by mutation (Prüfer et al., in preparation).

Interestingly, plant virus frameshift efficiencies range from ~ 1 (Brault and Miller, 1992; Prüfer et al., 1992) to 7 percent (Kim and Lommel, 1998) and are low compared to animal viruses (e.g., 25 to 30 percent for the infectious bronchitis coronavirus). For PLRV, the obtained low efficiency of approximately 1 percent can be explained by the weak RNA structure (stem loop or pseudoknot) downstream of the heptanucleotide signal. Mutational analysis studies in which this structure was replaced by a more stable stem loop showed a very strong increase in frameshift efficiency (up to 15 percent; Prüfer et al., in preparation). Very little is known about +1 ribosomal frameshifting in plant viruses, and only closteroviruses might make use of this strategy during protein expression (Agranovsky, 1996; Jelkmann et al., 1997). Recently, Rennecke and Jelkmann (1998) reported that the ORF 1b of little cherry closterovirus is probably expressed by a +1 frameshift. To investigate the in vivo +1 frameshift efficiency, the corresponding region was subcloned into full-length PVX complementary DNA (cDNA) clones either expressing the CP or the GUS gene as reporters. With both systems, they could demonstrate that +1 frameshifting takes place in the expression of ORF1b with an efficiency of less than 1 percent.

TERMINATION OF TRANSLATION

Stop Codon Suppression

Readthrough of leaky termination codon is a well-known noncanonical translation mechanism in plant RNA viruses. Suppression of an amber stop codon in TMV RNA leads to the expression of an 183 kDa protein (Bruening et al., 1976), and Skuzeski et al. (1991) could detect an approximately 5 percent readthrough efficiency in plant protoplasts using the GUS reporter system. Mutational analysis indicated a strong influence on leakiness of the 3'-sequences flanking the TMV UAG. This result was supported by experiments in which the TMV 3'-flanking sequences were replaced by their counterparts from PLRV, TMV, and *Tobacco rattle virus*, a tobravirus, and consequently tested for their ability to mediate readthrough into the GUS gene in vivo. None of these constructs gave a comparable readthrough efficiency (Prüfer, unpublished results).

Suppression of the amber stop codon separating the PLRV CP gene (ORF3) and the in-frame ORF5 was low in vivo (Tacke et al., 1990) when compared to TMV, suggesting that more information is necessary for the efficient readthrough event. Recently, Miller and co-workers (Brown et al., 1996) analyzed *cis*-acting sequences for the BYDV genome and two regions downstream of the corresponding stop codon were identified as being important for readthrough in vitro and in vivo. Readthrough was significantly reduced when a section containing the first 5 of 16 CCNNNN nt repeats was deleted. In addition, a second element located some 697 to 758 nts downstream of the stop codon was required, and they demonstrated that this element also functioned in vivo in oat protoplasts, when placed more than 2 kb from the CP gene stop in the untranslated region following the GUS gene. However, in vitro translation studies of full-length or truncated versions of PLRV sgRNA1 in wheat germ extracts and reticulocyte lysates did not show readthrough of ribosomes into ORF5 coding sequences. In addition, also with naturally occurring tyrosine-specific suppressor tRNAs (Beier et al., 1984) as well as purified tRNA fractions from PLRV-infected and healthy potato plants, no suppression of the PLRV amber stop codon was detected in vitro (Prüfer and Beier, unpublished observations).This observation might be explained by the fact that both viruses do not belong to the same subgroup and therefore have different requirements for expressing their genes.

POLYPROTEIN PROCESSING AND HOST FACTORS

Virus-encoded proteinases, including those of plant viruses, and polyprotein processing have been discussed in detail elsewhere (Dougherty and Semler, 1993; Ryan and Flint, 1997; Spall et al., 1997). Plant RNA viruses that express polyproteins include viruses from diverse genera, such as como-, nepo-, poty-, carla-, clostero-, and tymoviruses (Spall et al., 1997). By activity analyses and sequence comparison with cellular proteinases the viral encoded proteinases are grouped into chymotrypsin- and papain-like proteinases. Polyproteins can be processed either post- or cotranslational and *in trans* or *in cis* (see Figure 9.1E). All proteinases are multifunctional proteins that play additional important roles in the viral infection cycle. In potyviruses, proteinases also exhibit RNA-binding activity (Merits et al., 1998, and references therein), indicating their functions in the RNA replication process. P1 acts as a transactive accessory factor during genome amplification (Verchot and Carrrington, 1995). The helper component proteinase (HC-Pro) is needed for virus transmission (Atreya et al., 1992) and systemic movement in plants (Klein et al., 1994). Moreover HC-Pro plays an important role in suppression of a natural antiviral defense mechanism in plants, posttranscriptional gene silencing (Anandalakshmi et al., 1998; Brigneti et al., 1998; Kasschau and Carrington, 1998).

Efficient translation of most eukaryotic mRNAs requires that a 5'-cap structure and the 3'-poly(A)-track act in concert (Gallie, 1998). Moreover, the 5''-cap binds translation factors prior to the recruitment of ribosomal subunits (Merrick, 1992). For TuMV that lacks a 5'-cap, it has been shown by means of a yeast two-hybrid system that the VPg can interact with the eukaryotic initiation factor eIF(iso)4E from *Arabidopsis thaliana* (Wittmann et al., 1997). Since eIF(iso)4E plays an essential role in the initiation of the translation of capped mRNAs, its association with VPg points to a similar role of the viral protein in the translation of the viral genome.

OUTLOOK

In 1996, Maia and co-workers awarded the prize for the "translational expert" in the plant RNA virus field to the Luteovirus Group. They finished their laudatio with the words "we have not given up the hope that data will soon emerge demonstrating that they [the luteoviruses] also use posttranslational cleavage." Indeed, one year later Van der Wilk et al. (1997) demonstrated that the VPg of PLRV is located at an internal position of the ORF1 protein and probably released by posttranslational cleavage. More recently, another truncated protein, P1-C25, representing the C-terminus of ORF1 could be detected in PLRV-infected plants (Prüfer et al., 1999) and it might be a product of protease activity.

Review articles concerning luteovirus gene expression were published earlier (Rohde et al., 1994; Miller et al., 1995; Mayo and Ziegler-Graff, 1996), and a summary of all used strategies is given in Figure 9.2. As is

FIGURE 9.2. Strategies Used in Luteoviral Gene Expression

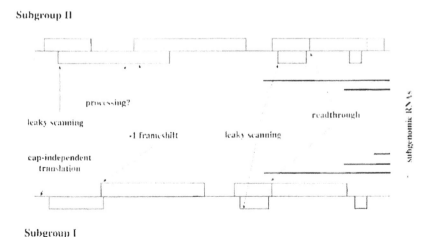

Subgroup II

processing?

leaky scanning

readthrough

-1 frameshift

leaky scanning

subgenomic RNAs

cap-independent translation

Subgroup I

clear, a large number of canonical and noncanonical strategies are involved in the expression of luteoviral genes. We believe that luteoviruses are still a "translational gold mine" where new insights in plant viral gene expression will soon be unearthed.

REFERENCES

Agranovsky AA (1996). Principles of molecular organization, expression, and evolution of closteroviruses: Over the barriers. *Advances in Virus Research* 47: 119-158.

Akiri G, Nahari D, Finkelstein Y, Le SY, Elroy Stein O, Levy BZ (1998). Regulation of vascular-endothelial growth factor (VEGF) expression is mediated by internal initiation of translation and alternative initiation of transcription. *Oncogene* 17: 227-236.

Allen E, Wang S, Miller WA (1999). Barley yellow dwarf virus RNA requires a cap-independent translation sequence because it lacks a 5' cap. *Virology* 253: 139-144.

Anandalakshmi R, Pruss GJ, Ge X, Marathe R, Mallory AC, Smith TH, Vance VB (1998). A viral suppressor of gene silencing in plants. *Proceedings of the National Academy of Sciences, USA* 95: 13079-13084.

Atreya CD, Atreya PL, Thornbury DW, Pirone TP (1992). Site-directed mutations in the potyvirus HC-Pro gene affect helper component activity, virus accumulation, and symptom expression in infected tobacco plants. *Virology* 191: 106-111.

Banerjee AK, Barik S (1992). Gene expression of vesicular stomatitis virus genome RNA. *Virology* 188: 417-428.

Basso J, Dallaire P, Charest PJ, Daventier Y, Laliberte JF (1994). Evidence for an internal ribosome entry site within the 5'-nontranslated region of turnip mosaic potyvirus RNA. *Journal of General Virology* 75: 3157-3165.

Baudin F, Bach C, Cusack S, Ruigrok RW (1994). Structure of influenza virus RNP. I. Influenza virus nucleoprotein melts secondary structure in panhandle RNA and exposes the bases to the solvent. *EMBO Journal* 13: 3158-3165.

Becera SP, Rosa JA, Hardy M, Baroudy BM, Anderson CW (1985). Direct mapping of adeno-associated virus capsid protein B and C: A possible AUC initiation codon. *Proceedings of the National Academy of Sciences, USA* 76: 7919-7923.

Beier H, Barciszewska M, Krupp G, Mitnacht R, Gross HJ (1984). UAG readthrough during TMV translation: Isolation and sequence of two tRNAs[Tyr] with suppressor activity from tobacco plants. *EMBO Journal* 3: 351-356.

Belsham GJ, Sonenberg N (1996). RNA-protein interactions in regulation of picornavirus RNA translation. *Microbiological Reviews* 60: 499-511.

Berlioz C, Darlix JL (1995). An internal ribosomal entry mechanism promotes translation of murine leukemia virus *gag* polyprotein precursors. *Journal of Virology* 69: 2214-2222.

Berlioz C, Torrent C, Darlix JL (1995). An internal ribosomal entry signal in the rat VL30 region of the Harwey murine sarcoma virus leader and its use in dicistronic retroviral vectors. *Journal of Virology* 69: 6400-6407.

Bishop DH (1986). Ambisense RNA viruses: Positive and negative polarities combined in RNA virus genomes. *Microbiological Sciences* 3: 183-187.

Blyn LB, Swiderek KM, Richards O, Stahl DC, Semler BL, Ehrenfeld E (1996). Poly(rC) binding protein 2 binds to stem-loop IV of the poliovirus RNA 5' noncoding region: Identification by automated liquid chromatography-tandem mass spectrometry. *Proceedings of the National Academy of Sciences, USA* 93: 11115-111120.

Boeck R, Kolakofsky D (1994). Positions +5 and +6 can be major determinants of the efficiency of non-AUG initiation codons for protein synthesis. *EMBO Journal* 13: 3608-3617.

Bol JF (1999). Alfalfa mosaic virus and ilarviruses: Involvement of coat protein in multiple steps of the replication cycle. *Journal of General Virology* 80: 1089-1102.

Borman AM, Deliat FG, Kean KM (1994). Sequences within the poliovirus internal ribosome entry segment control viral RNA synthesis. *EMBO Journal* 13: 3149-3157.

Borman AM, Kean KM (1997). Intact eukaryotic initiation factor 4G is required for hepatitis A virus internal initiation of translation. *Virology* 237: 129-136.

Borman AM, Kirchweger R, Ziegler E, Rhoads RE, Skern T, Kean KM (1997). EIF4G and its proteolytic cleavage products: Effects on initiation of protein synthesis from capped, uncapped, and IRES-containing mRNAs. *RNA* 3: 186-196.

Boyd L, Thummel CS (1993). Selection of CUG and AUG initiator codons for Drosophila E74A translation depends on downstream sequences. *Proceedings of the National Academy of Sciences, USA* 90: 9164-9167.

Brault V, Miller WA (1992). Translational frameshifting mediated by a viral sequence in plant cells. *Proceedings of the National Academy of Sciences, USA* 89: 2262-2266.

Brierley I, Boursnell ME, Binns MM, Bilimoria B, Blok VC, Brown TD, Inglis SC (1987). An efficient ribosomal frame-shifting signal in the polymerase-encoding region of the coronavirus IBV. *EMBO Journal* 6: 3779-3785.

Brigneti G, Voinnet O, Li WX, Ji LH, Ding SW, Baulcombe DC (1998). Viral pathogenicity determinants are suppressors of transgene silencing in *Nicotiana benthamiana*. *EMBO Journal* 16: 6739-6746.

Brown CM, Dinesh-Kumar SP, Miller WA (1996). Local and distant sequences are required for efficient readthrough of the barley yellow dwarf virus PAV coat protein gene stop codon. *Journal of Virology* 70: 5884-5892.

Browning KS (1996). The plant translational apparatus. *Plant Molecular Biology* 32: 107-144.

Bruening G, Beachy TN, Scalla R, Zaitlin M (1976). In vitro and in vivo translation of the ribonucleic acids of a cowpea strain of tobacco mosaic virus. *Virology* 71: 498-517.

Carrington JC, Freed DD (1990). Cap-independent enhancement of translation by a plant potyvirus 5'-untranslated region. *Journal of Virology* 64: 1590-1597.

Chen C-Y, Sarnow P (1995). Initiation of protein synthesis by the eukaryotic translational apparatus on circular RNAs. *Science* 268: 415-417.

Clements JM, Laz TM, Sherman F (1988). Efficiency of translation by non-AUG codons in *Saccharomyces cerevisiae. Molecular and Cellular Biology* 8: 4533-4536.

Curran J, Kolakofsky D (1988). Scanning independent ribosomal initiation of the Sendai virus X protein. *EMBO Journal* 7: 245-251.

Das S, Kenan DJ, Bocskai D, Keene JD, Dasgupta A (1996). Sequences within a small yeast RNA required for inhibition of internal initiation of translation: Interaction with La and other cellular proteins influences its inhibitory activity. *Journal of Virology* 70: 1624-1632.

Das S, Ott M, Yamane A, Tsai W, Gromeier M, Lahser F, Gupta S, Dasgupta A (1998). A small yeast RNA blocks hepatitis C virus internal ribosome entry site (HCV IRES)-mediated translation and inhibits replication of a chimeric poliovirus under translational control of the HCV IRES element. *Journal of Virology* 72: 5638-5647.

Das S, Ott M, Yamane A, Venkatesan A, Gupta S, Dasgupta A (1998). Inhibition of internal entry site (IRES)-mediated translation by a small yeast RNA: A novel strategy to block hepatitis C virus protein synthesis. *Frontiers in Bioscience* 3: D250-D268.

Dasso MC, Jackson RJ (1989). Efficient initiation of mammalian mRNA translation at a CUG codon. *Nucleic Acids Research* 17: 6485-6497.

Dawson WO, Lehto KM (1990). Regulation of tobamovirus gene expression. *Advances in Virus Research* 38: 307-342.

Dinesh-Kumar SP, Miller WA (1993). Control of start codon choice on a plant viral RNA encoding overlapping genes. *Plant Cell* 5: 679-692.

Donahue TF, Cigan AM (1988). Genetic selection for mutations that reduce or abolish ribosomal recognition of the HIS4 translational initiator region. *Molecular and Cellular Biology* 8: 2955-2963.

Dougherty WG, Semler BL (1993). Expression of virus-encoded proteinases: Functional and structural similarities with cellular enzymes. *Microbiological Reviews* 57: 781-822.

Drugeon G, Urcuqui-Inchima S, Milner M, Kadaré G, Valle RPC, Voyatzakis A, Haenni A-L, Schirawski J (1999). The strategies of plant virus gene expression: Models of economy. *Plant Science* 148: 77-88.

Duijsings D, Kormelink R, Goldbach R (1999). Alfalfa mosaic virus RNAs serve as cap donors for tomato spotted wilt virus transcription during coinfection of *Nicotiana benthamiana. Journal of Virology* 73: 5172-5175.

Ehrenfeld E (1996). Initiation of translation by picornavirus RNAs. In Hershey JWB, Matthews, MB, Sonenberg N (eds.), *Translational Control*, Cold Spring Harbor Laboratories Press. Cold Spring Harbor, New York, pp. 549-573.

Farabaugh PJ (1996). Programmed translational frameshifting. *Annual Reviews of Genetics* 30: 507-528.

Filipowicz W, Hohn T, eds. (1996). Post-transcriptional control of gene expression. *Plant Molecular Biology* 32: 1-405.

Fraile A, Alonso-Prados JL, Aranda MA, Bernal JJ, Malpica JM, Garcia-Arenal F (1997). Genetic exchange by recombination or reassortment is infrequent in natural populations of a tripartite RNA plant virus. *Journal of Virology* 71: 934-940.

Fütterer J. Hohn T (1996). Translation in plants: Rules and exceptions. *Plant Molecular Biology* 32: 159-189.

Gallie DR (1996). Translational control of cellular and viral mRNAs. *Plant Molecular Biology* 32: 145-158.

Gallie DR (1998). Controlling gene expression in transgenics. *Current Opinion in Plant Biology* 1: 166-172.

Gan W, La Celle M, Rhoads RE (1998). Functional characterization of the internal ribosome entry site of eIF4G mRNA. *Journal of Biological Chemistry* 273: 5006-5012.

Gan W, Rhoads RE (1996). Internal initiation of translation directed by the 5'-untranslated region of the mRNA for eIF4G, a factor involved in the picornavirus-induced switch from cap-dependent to internal initiation. *Journal of Biological Chemistry* 271: 623-626.

Garcia A, Van Duin J, Pleij CW (1993). Differential response to frameshift signals in eukaryotic and prokaryotic translational systems. *Nucleic Acids Research* 21: 401-406.

Gordon K, Fütterer J, Hohn T (1992). Efficient initiation of translation at non-AUG triplets in plant cells. *Plant Journal* 25: 809-813.

Haenni AL, Chapeville F (1997). An enigma: The role of viral RNA aminoacylation. *Acta Biochimica Polonica* 44: 827-837.

Hagiwara Y, Peremyslov VV, Dolja VV (1999). Regulation of closterovirus gene expression examined by insertion of a self-processing reporter and by northern hybridization. *Journal of Virology* 73: 7988-7993.

Hann SR, Sloan-Brown K, Spotts GD (1992). Translational activation of the non-AUG-initiated c-myc 1 protein at high cell densities due to methionine deprivation. *Genes and Development* 6: 1229-1240.

Herman RC (1989). Alternatives for the initiation of translation. *Trends in Biochemical Sciences* 14: 219-222.

Ito T, Tahara SM, Lai MMC (1998). The 3'-untranslated region of hepatitis C virus RNA enhances translation from an internal ribosome entry site. *Journal of Virology* 72: 8789-8796.

Ivanov PA, Karpova OV, Skulachev MV, Tomashevskaya OL, Rodionova NP, Dorokhov YuL, Atabekov JG (1997). A tobamovirus genome that contains an internal ribosome entry site functional in vitro. *Virology* 232: 32-43.

Jacks T, Varmus HE (1985). Expression of the Rous sarcoma virus pol gene by ribosomal frameshifting. *Science* 230: 1237-1242.

Jackson RJ (1991). Initiation without an end. *Nature* 353: 14-15.

Jackson RJ, Howell MT, Kaminski A (1990). The novel mechanism of initiation of picornavirus RNA translation. *Trends in Biochemistry* 15: 477-483.

Jang SK, Krausslich HG, Nicklin MJH, Duke GM, Palmenberg AC, Wimmer E (1988). A segment of the 5' untranslated region of encephalomyelitis virus RNA directs internal entry of ribosomes during in vitro translation. *Journal of Virology* 62: 2636-2643.

Jelkmann W, Fechtner B, Agranovsky AA (1997). Complete genome structure and phylogenetic analysis of little cherry virus, a mealybug-transmissible closterovirus. *Journal of General Virology* 78: 2067-2071.

Johnston JC, Rochon DM (1996). Both codon context and leader length contribute to efficient expression of two overlapping ORFs of a cucumber necrosis virus bifunctional subgenomic mRNA. *Virology* 221: 232-239.

Kasschau KD, Carrington JC (1998). A counterdefensive strategy of plant viruses: Suppression of posttranscriptional gene silencing. *Cell* 95: 461-470.

Kim KH, Lommel SA (1994). Identification and analysis of the site of –1 ribosomal frameshifting in red clover necrotic mosaic virus. *Virology* 200: 574-582.

Kim KH, Lommel SA (1998). Sequence element required for efficient –1 ribosomal frameshifting in red clover necrotic mosaic dianthovirus. *Virology* 250: 50-59.

Klein PG, Klein RR, Rodriguez-Cerezo E, Hunt AG, Shaw JG (1994). Mutational analysis of the tobacco vein mottling virus genome. *Virology* 204: 759-769.

Koenig R, Pleij CW, Beier C, Commandeur U (1998). Genome properties of beet virus Q, a new furo-like virus from sugarbeet, determined from unpurified virus. *Journal of General Virology* 79: 2027-2036.

Koev G, Mohan BR, Miller WA (1999). Primary and secondary structural elements required for synthesis of barley yellow dwarf virus subgenomic RNA1. *Journal of Virology* 73: 2876-2885.

Kormelink R, De Haan P, Meurs C, Peters D, Goldbach R (1992). The nucleotide sequence of the M RNA segment of tomato spotted wilt virus, a bunyavirus with two ambisense RNA segments. *Journal of General Virology* 73: 2795-2804. Published erratum appeared in *Journal of General Virology* (1993) 74: 790.

Kozak M (1986a). Influences of mRNA secondary structure on initiation by eukaryotic ribosomes. *Proceedings of the National Academy of Sciences, USA* 83: 2850-2854.

Kozak M (1986b). Point mutations define a sequence flanking the AUG initiator codon that modulates translation by eukaryotic ribosomes. *Cell* 44: 283-292.

Kozak M (1989a). Circumstances and mechanisms of inhibition of translation by secondary structure in eukaryotic mRNAs. *Molecular and Cellular Biology* 9: 5134-5142.

Kozak M (1989b). Context effects and inefficient initiation at non-AUG codons in eukaryotic cell free translation systems. *Molecular and Cellular Biology* 9: 5073-5080.

Kozak M (1989c). The scanning model for translation: An update. *Journal of Cell Biology* 108: 229-241.

Kozak M (1991). Structural features in eukaryotic mRNAs that modulate the initiation of translation. *Journal of Biological Chemistry* 266: 19867-19870.

Kozak M (1994). Determinants of translational fidelity and efficiency in vertebrate mRNAs. *Biochemie* 76: 815-821.

Kozak M (1997). Recognition of AUG and alternative initiator codons is augmented by G in position +4 but is not generally affected by the nucleotides in positions +5 and +6. *EMBO Journal* 16: 2482-2492.

Kozak M (1999). Initiation of translation in prokaryotes and eukaryotes. *Gene* 234:178-208.

Kujawa AB, Drugeon G, Hulanicka D, Haenni AL (1993). Structural requirements for efficient translational frameshifting in the synthesis of the putative viral RNA-dependent RNA polymerase of potato leafroll virus. *Nucleic Acids Research* 21: 2165-2171.

Lai MM (1998). Cellular factors in the transcription and replication of viral RNA genomes: A parallel to DNA-dependent RNA transcription. *Virology* 244: 1-12.

Le SY, Maizel JV (1998). Evolution of a common structural core in the internal ribosome entry sites of picornaviruses. *Virus Genes* 16: 25-38.

Le SY, Siddiqui A, Maizel JV (1996). A common structural core in the internal ribosome entry sites of picornavirus, hepatitis C virus, and pestivirus. *Virus Genes* 12: 135-147.

Le SY, Sonenberg N, Maizel JV (1994). Distinct structural elements and internal entry of ribosomes in mRNA 3 encoded by infectious bronchitis virus. *Virology* 198: 405-411.

Levis C, Astier-Manifacier S (1993). The 5' untranslated region of PVY RNA, even located in an internal position, enables initiation of translation. *Virus Genes* 7: 367-379.

Liu DX, Inglis SC (1992). Internal entry of ribosomes on a tricistronic mRNA encoded by infectious bronchitis virus. *Journal of Virology* 66: 6143-6154.

Macejak D, Sarnow P (1991). Internal entry of translation mediated by the 5'-leader of a cellular mRNA. *Nature* 353: 90-94.

Maia IG, Séron K, Haenni A-L, Bernardi F (1996). Gene expression from viral RNA genomes. *Plant Molecular Biology* 32: 367-319.

Makinen K, Tamm T, Naess V, Truve E, Puurand U, Munthe T, Saarma M (1996). Characterization of cocksfoot mottle sobemovirus genomic RNA and sequence comparison with related viruses. *Journal of General Virology* 76: 2817-2825.

Martinez Salas E, Regalado MP, Domingo E (1996). Identification of an essential region for internal initiation of translation in the aphthovirus internal ribosome entry site and implications for viral evolution. *Journal of Virology* 70: 992-998.

Matthews REF (1991). *Plant Virology* (Third Edition). Academic Press, San Diego, London, 835 pp.

Mayo MA, Ziegler-Graff V (1996). Molecular biology of luteoviruses. *Advances in Virus Research* 46: 413-460.

Mehdi H, Ono E, Gupta KC (1990). Initiation of translation at CUG, GUG and ACG codons in mammalian cells. *Gene* 91: 173-178.

Merits A, Guo D, Saarma M (1998). VPg, coat protein and five non-structural proteins of potato A potyvirus bind RNA in a sequence-unspecific manner. *Journal of General Virology* 79: 3123-3127.

Merrick WC (1992). Mechanism and regulation of eukaryotic protein synthesis. *Microbiological Reviews* 56: 291-315.

Miller DL, Dibbens JA, Damert A, Risau W, Vadas MA, Goodall GJ (1998). The vascular endothelial growth factor mRNA contains an internal ribosome entry site. *FEBS Letters* 434: 417-420.

Miller WA, Dinesh-Kumar SP, Paul CP (1995). Luteovirus gene expression. *Critical Reviews in Plant Sciences* 14: 179-211.

Mountford PS, Smith AG (1995). Internal ribosome entry sites and dicistronic mRNAs in mammalian transgenesis. *Trends in Genetics* 11: 179-184.

Nanbru C, Lafon I, Audigier S, Gensac MC, Vagner S, Huez G, Prats AC (1997). Alternative translation of the proto-oncogene c-myc by an internal ribosome entry site. *Journal of Biological Chemistry* 272: 32061-32066.

Niepel M, Gallie DR (1999). Identification and characterization of the functional elements within the tobacco etch virus 5' leader required for cap-independent translation. *Journal of Virology* 73: 9080-9088.

Niepmann M, Petersen A, Meyer K, Beck E (1997). Functional involvement of polypyrimidine tract-binding protein in translation initiation complexes with the internal ribosome entry site of foot-and-mouth disease virus. *Journal of Virology* 71: 8330-8339.

Oh SK, Scott MP, Sarnow P (1992). Homeotic gene *Antennapedia* mRNA contains 5'-noncoding sequences that confer translational initiation by internal ribosome binding. *Genes and Development* 6: 1643-1653.

Ohlman T, Pain VM, Wood W, Rau M, Morley SJ (1997). The proteolytic cleavage of eukaryotic initiation factor (eIF)4G is prevented by eIF4E binding protein (PHAS-1; eIF4E-BP1) in the reticulocyte lysate. *EMBO Journal* 16: 844-855.

Olsen Q (1987). Yeast cells may use AUC or AAG as initiation codon for protein synthesis. *Carlsberg Research Communications* 52: 83-90.

Olsthoorn RC, Mertens S, Brederode FT, Bol JF (1999). A conformational switch at the 3'-end of a plant virus RNA regulates viral replication. *EMBO Journal* 18: 4856-4864.

Paulin FEM, Chappell SA, Willis AE (1998). A single nucleotide change in the c-myc internal ribosome entry segment leads to enhanced binding of a group of protein factors. *Nucleic Acids Research* 26: 3097-3103.

Peabody DS (1989). Translation initiation at non-AUG triplets in mammalian cells. *Journal of Biological Chemistry* 264: 5031-5035.

Pelletier J, Sonenberg N (1988). Internal initiation of translation of eukaryotic mRNA directed by a sequence derived from poliovirus RNA. *Nature* 334: 320-325.

Pleij CW (1995). Structure and function of RNA pseudoknots. *Genetic Engineering* (New York) 17: 67-80.

Prats AC, DeBilly G, Wang P, Darlix JL (1989). CUG initiation codon used for the synthesis of a cell surface antigen coded by the murine leukemia virus. *Journal of Molecular Biology* 205: 363-372.

Prüfer D, Kawchuk L, Monecke M, Nowok S, Fischer R, Rohde W (1999). Immunological analysis of potato leafroll luteovirus (PLRV) P1 expression identifies a 25 kDa RNA-binding protein derived via P1 processing. *Nucleic Acids Research* 27: 421-425.

Prüfer D, Tacke E, Schmitz J, Kull B, Kaufmann A, Rohde W (1992). Ribosomal frameshifting in plants: A novel signal directs the −1 frameshift in the synthesis of the putative replicase of potato leafroll luteovirus. *EMBO Journal* 11: 1111-1117.

Rennecke B, Jelkmann W (1998). Expression of little cherry closterovirus RNA-polymerase by a +1 ribosomal frameshift. Abstract on "Arbeitskreis Virologie and Nederlandse Kring voor Plantevirologie," Wageningen, The Netherlands.

Reusken CB, Neeleman L, Brederode FT, Bol JF (1997). Mutations in coat protein binding sites of alfalfa mosaic virus RNA 3 affect subgenomic RNA 4 accumulation and encapsidation of viral RNAs. *Journal of Virology* 71: 8385-8391.

Reynolds JE, Kaminski A, Carroll AR, Clarke BE, Rowlands DJ, Jackson RJ (1996). Internal initiation of translation of hepatitis C virus RNA: The ribosome entry site is at the authentic initiation codon. *RNA* 2: 867-878.

Reynolds JE, Kaminski A, Kettinen HJ, Grace K, Clarke BE, Carroll AR, Rowlands DJ, Jackson RJ (1995). Unique features of internal initiation of hepatitis C virus RNA translation. *EMBO Journal* 14: 6010-6020.

Rijnbrand R, Bredenbeek P, Van der Straaten T, Whetter L, Inchauspe G, Lemon S, Spaan W (1995). Almost the entire 5'-nontranslated region of hepatitis C virus is required for cap-independent translation. *FEBS Letters* 365: 115-119.

Rohde W, Gramstat A, Schmitz J, Tacke E, Prüfer D (1994). Plant viruses as a model system for the study of non-canonical translation mechanisms in higher plants. *Journal of General Virology* 75: 2141-2149.

Ryan MD, Flint M (1997). Virus-encoded proteinases of the picornavirus super-group. *Journal of General Virology* 78: 699-723.

Sasaki J, Nakashima N (1999). Translation initiation at the CUU codon is mediated by the internal ribosome entry site of an insect picorna-like virus in vitro. *Journal of Virology* 73: 1219-1226.

Schmitz J, Prüfer D, Rohde W, Tacke E (1996). Non-canonical translation mechanisms in plants: Efficient in vitro and in planta initiation at AUU codons of the tobacco mosaic virus enhancer sequence. *Nucleic Acids Research* 24: 257-263.

Scholthof HB, Desvoyes B, Kuecker J, Whitehead E (1999). Biological activity of two tombusvirus proteins translated from nested genes is influenced by dosage control via context-dependent leaky scanning. *Molecular Plant-Microbe Interactions* 12: 670-679.

Schultze M, Hohn T, Jiricny J (1990). The reverse transcriptase gene of cauliflower mosaic virus is translated separately from the capsid gene. *EMBO Journal* 9: 1177-1185.

Shirako Y, Wilson TMA (1993). Complete nucleotide sequence and organization of the bipartite RNA genome of soil-borne wheat mosaic virus. *Virology* 195: 16-32.

Simón-Buela L, Guo HS, Garcia JA (1997). Cap-independent leaky scanning as the mechanism of translation initiation of a plant viral genomic RNA. *Journal of General Virology* 78: 2691-2699.

Sit TL, Vaewhongs AA, Lommel SA (1998). RNA-mediated trans-activation of transcription from a viral RNA. *Science* 281: 829-832.

Sivakumaran K, Hacker DL (1998). The 105-kDa polyprotein of southern bean mosaic virus is translated by scanning ribosomes. *Virology* 246: 34-44.

Skulachev MV, Ivanov PA, Karpova OV, Korpela T, Rodionova NP, Dorokhov YuL, Atabekov JG (1999). Internal initiation of translation by the 5'-untranslated region of the tobamovirus subgenomic RNA I$_2$. *Virology* 263: 139-154.

Skuzeski JM, Nichols LM, Gesteland RF, Atkins JF (1991). The signal for a leaky UAG stop codon in several plant viruses includes the two downstream codons. *Journal of Molecular Biology* 218: 365-373.

Spall VE, Shanks M, Lomonossoff GP (1997). Polyprotein processing as a strategy for gene expression in RNA viruses. *Seminars in Virology* 8: 15-23.

Stein I, Itin A, Einat P, Skaliter R, Grossman Z, Keshet E (1998). Translation of vascular endothelial growth factor mRNA by internal ribosome entry: Implications for translation under hypoxia. *Molecular and Cellular Biology* 18: 3112-3119.

Stoneley M, Paulin FEM, Le Quesne JPC, Chappell SA, Willis AE (1998). C-Myc 5' untranslated region contains an internal ribosome entry segment. *Oncogene* 16: 423-428.

Tacke E, Prüfer D, Salamini F, Rohde W (1990). Characterization of potato leafroll luteovirus subgenomic RNA: Differential expression by internal translation initiation and UAG suppression. *Journal of General Virology* 71: 2265-2272.

Tacke E, Salamini F, Rohde W (1996). Genetic engineering of potato for broad-spectrum protection against virus infection. *Nature Biotechnology* 14: 1597-1601.

Thomas AM, Ter Haar E, Wellink K, Voorma HO (1991). Cowpea mosaic virus middle component RNA contains sequence that allows internal binding of ribosomes and that requires eukaryotic initiation factor 4F for optimal translation. *Journal of Virology* 65: 2953-2959.

Tsukiyama-Kohara K, Iizuka N, Kohara M, Nomoto A (1992). Internal ribosome entry site within hepatitis C virus RNA. *Journal of Virology* 66: 1476-1483.

Vagner S, Gensac, M-C, Maret A, Bayard F, Amalric F, Prats H, Prats A-C (1995). Alternative translation of human fibroblast growth factor 2 mRNA occurs by internal entry of ribosomes. *Molecular and Cellular Biology* 15: 35-44.

Vagner S, Waysbort A, Marenda M, Gensac M-C, Amalric F, Prats A-C (1995). Alternative translation initiation of the Moloney murine leukemia virus mRNA controlled by internal ribosome entry involving the p57/PTB splicing factor. *Journal of Biological Chemistry* 270: 20376-20383.

Van der Wilk F, Verbeek M, Dullemans AM, Van den Heuvel JFJM (1997). The genome-linked protein of potato leafroll virus is located downstream of the putative protease domain of the ORF1 product. *Virology* 234: 300-303.

Van Poelwijk F, Kolkman J, Goldbach R (1996). Sequence analysis of the 5'-ends of tomato spotted wilt virus N mRNAs. *Archives of Virology* 141: 177-184.

Van Vloten-Doting L (1975). Coat protein is required for infectivity of tobacco streak virus: Biological equivalence of the coat proteins of tobacco streak and alfalfa mosaic viruses. *Virology* 65: 215-225.

Verchot J, Angell SM, Baulcombe DC (1998). In vivo translation of the triple gene block of potato virus X requires two subgenomic mRNAs. *Journal of Virology* 72: 8316-8320.

Verchot J, Carrington JC (1995). Evidence that the potyvirus P1 proteinase functions *in trans* as an accessory factor for genome amplification. *Journal of Virology* 69: 3668-3674.

Vittorioso P, Carattoli A, Londei P, Macino C (1994). Internal translation initation in the mRNA from the *Neurospora crassa* albino-3 gene. *Journal of Biological Chemistry* 269: 26650-26654.

Wang C, Le SY, Siddiqui A (1995). An RNA pseudoknot is an essential structural element of the internal ribosome entry site located within the hepatitis C virus 5' noncoding region. *RNA* 1: 526-537.

Wang C, Sarnow P, Siddiqui A (1993). Translation of human hepatitis C virus RNA is mediated by an internal ribosome binding mechanism. *Journal of Virology* 67: 3338-3344.

Wang S, Guo L, Allen E, Miller WA (1999). A potential mechanism for selective control of cap-independent translation by a viral RNA sequence *in cis* and *in trans*. *RNA* 5: 728-738.

Weyens G, Rose K, Falmagne P, Glansdorff N, Piérard A (1985). Synthesis of *Echerichia coli* carbamoylphosphate synthetase initiates at a UUG codon. *European Journal of Biochemistry* 150: 111-115.

Wittmann S, Chatel H, Fortin MG, Laliberte JF (1997). Interaction of the viral protein genome linked of turnip mosaic potyvirus with the translational eukaryotic initiation factor (iso)4E of *Arabidopsis thaliana* using the yeast two-hybrid system. *Virology* 234: 84-92.

Ye X, Fong P, Iizuka N, Choate D, Cavener DR (1997). *Ultrabithorax* and *Antennapedia* 5' untranslated regions promote developmentally regulated internal translation initiation. *Molecular and Cellular Biology* 17: 1714-1721.

Zhang G, Slowinski V, White KA (1999). Subgenomic mRNA regulation by a distal RNA element in a (+)-strand RNA virus. *RNA* 5: 550-561.

Zhou H, Jackson AO (1996). Expression of the barley stripe mosaic virus RNA beta "triple gene block." *Virology* 216: 367-379.

Zitomer RS, Walthall DA, Rymond BC, Hollenberg CP (1984). *Saccharomyces cerevisiae* ribosomes recognize non-AUG initiation codons. *Molecular and Cellular Biology* 4: 1191-1197.

Chapter 10

Recombination in Plant RNA Viruses

Chikara Masuta

INTRODUCTION

The great majority of plant viruses are positive-sense RNA viruses depending on the virus-encoded RNA-dependent RNA polymerase (replicase) for RNA replication. RNA viruses thus undergo rapid genetic change because of the high error rate of viral replication, resulting in heterogeneous populations of related RNA sequences (Simon and Bujarski, 1994; Rubio et al., 1999). Furthermore, RNA viruses containing segmented genomes have the advantage of reassortment of the RNA segments.

The first report of RNA recombination in animal viruses was an exchange of genetic markers between two polioviruses (Ledinko, 1963). The most probable mechanism for RNA recombination (template switching, copy choice) was first proposed for poliovirus (Cooper et al., 1974). The role of RNA recombination in virus evolution has been considered since that time. The first direct evidence for RNA recombination was demonstrated by the biochemical analysis of RNA sequences of foot-and-mouth disease virus (King et al., 1982). Since then, sequence comparisons have shown that the past recombinant events occurred intergenically or intragenically in many RNA viruses. We have learned that recombinant viruses could be created in experimental conditions under certain selection pressures. For plant RNA viruses, the occurrence of de novo RNA recombination was first observed in *Brome mosaic virus* (BMV, genus *Bromovirus;* Bujarski and Kaesberg, 1986). We now have a large number of reports describing RNA recombination in many plant RNA viruses.

OVERVIEW OF RESEARCH HISTORY
ON THE RECOMBINATION OF PLANT RNA VIRUSES

Classification of RNA Recombination

RNA recombination can be generally classified into three types (Lai, 1992). *Homologous recombination* occurs between two similar or closely related RNAs with extensive sequence homology (see Figure 10.1A and B). Crossovers are found at sites precisely matched, and the resultin recombinant RNAs must retain the exact sequence from the parental RNAs. Although homology of the entire sequence is not necessarily required between two molecules, sequence homology must be present around the crossover sites. In contrast to homologous recombination, crossovers in *aberrant homologous recombination* occur not only at the homologous sites but also at unrelated sites (see Figure 10.1C). Sequence duplication, deletion, and insertion of nucleotides of unknown origin are often observed in the recombinant progeny. *Nonhomologous recombination* occurs on RNAs that do not share any sequence homology (see Figure 10.1D). Crossover sites sometimes share similar secondary structures. The formation of a partial heteroduplex between two RNA segments of a plant virus, BMV, as one of the mechanisms of nonhomologous recombination has been proposed (Bujarski and Dzianott, 1991).

However, a new classification grouping types of RNA recombination has been recently proposed, as the term homologous cannot be clearly defined for RNA-based recombination (Nagy and Simon, 1997). They pointed out that even very short sequence similarity, such as 5 to 15 nucleotides (nt), between two dissimilar RNAs could facilitate RNA recombination (Nagy and Bujarski, 1993, 1995). In other words, we do not have any concrete criteria to distinguish between "homologous" and "nonhomologous" recombination. In addition, according to the old classification, recombination driven by 'complementarities' between two RNAs must be regarded as nonhomologous recombination; however, the two RNAs actually make a duplex that triggers the template switching mechanism. This leads to some confusion, since the term nonhomologous implies that the two molecules are not totally related. Nagy and Simon (1997) have offered a more understandable classification with the following three categories: In *similarity-essential recombination,* substantial sequence similarity between two RNAs is required, and recombination occurs either precisely or imprecisely; thus, both homologous and aberrant homologous recombination are included in this class. *Similarity-nonessential recombination* does not apparently require either sequence similarity or base pairing between two RNAs. Rather than sequence similarity, this class of recombination is determined by the existence of an internal promoter sequence, a secondary structure, a heteroduplex formation between parental RNAs, and others. *Similarity-assisted recombination,* which is an intermediate of the previous two, requires additional RNA determinants, such as a secondary structure and a replicase-binding site, as well as sequence similarity.

In this chapter, RNA recombination in plant viruses is briefly reviewed, whereafter the RNA-RNA recombination between two cucumoviruses, *To-*

FIGURE 10.1. Classification of RNA Recombination by Replicase-Mediated Template Switching

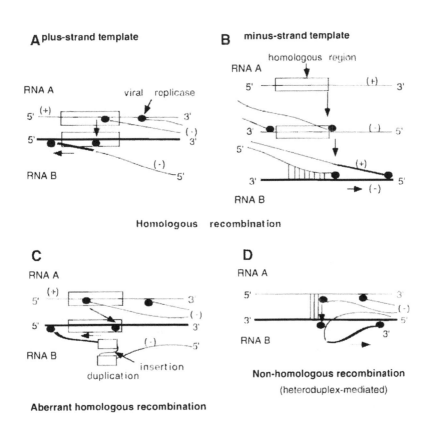

A plus-strand template

B minus-strand template

Homologous recombination

C

Aberrant homologous recombination

D

Non-homologous recombination
(heteroduplex-mediated)

Note: The homologous recombination may take place at the level of either minus- (A) or plus-strand (B) synthesis. For aberrant homologous recombination, only plus-strand synthesis is shown (C). As an example of nonhomologous recombination, the heteroduplex-mediated recombination is shown (D). The vertical lines between the two lines represent base pairing. The open box on each strand is a homologous region between the two RNAs. The template switching is shown by an arrow. The black circle is the replicase complex.

mato aspermy virus (TAV) and *Cucumber mosaic virus* (CMV), is discussed in terms of the template-switching mechanism. Template switching is the most probable mechanism for RNA recombination in plant viruses. Since single-stranded positive-sense RNA plant viruses have relatively simple genome structures, their RNA recombination events will be a very good model system for understanding RNA virus evolution.

Natural Recombination in the Evolution of Plant Viruses

Comparative studies on nucleotide sequences of numerous plant viruses indicate evidence that RNA recombination events have occurred in the past. Some examples are shown in Figure 10.2. Interspecific recombination was suggested between members of family *Luteoviridae, Cucurbit aphid-borne yellows virus* and *Pea enation mosaic virus-1* RNA1. Two possible recombination sites were identified in the readthrough protein following the coat protein (CP). In tobraviruses, *Tobacco rattle virus* (TRV) RNA2 varies in size and sequence, showing the chimeric nature with the 3'-end of RNA1. An anomalous isolate, TRV TCM strain RNA2, was generated by double RNA recombination between two tobraviruses, *Pea early-browning virus* (PEBV) RNA2 and TRV RNA1 (RNA2).

Defective interfering RNAs (DI RNAs) are found to be associated with many plant RNA viruses. They were thought to originate from the helper virus by a series of truncations and are unable to replicate autonomously. In addition, they are synthesized by errors made during viral replication. The DI RNAs of *Tomato bushy stunt virus* (TBSV) a tombusvirus, could participate in replicase-mediated RNA recombination under certain selection pressures, suggesting that natural DI RNAs have potential to serve as sequence sources for viral evolution via recombination (White and Morris, 1994).

Recombination in Experimental Conditions

Many de novo RNA recombination events between two RNAs have been created in experimental conditions. In particular, the detection of wild-type recombined progeny was considerably enhanced when two defective parental RNAs complementing each other were used.

Bromoviruses

RNA recombination under selection pressure was first demonstrated on BMV (Bujarski and Kaesberg, 1986). This is the type species of the genus *Bromovirus*, and it contains three divided positive-strand RNA genomes. RNA1 and RNA2 participate in viral replication and transcription, and RNA3

FIGURE 10.2. Evidence for Natural Recombination Events in Several Plant RNA Viruses

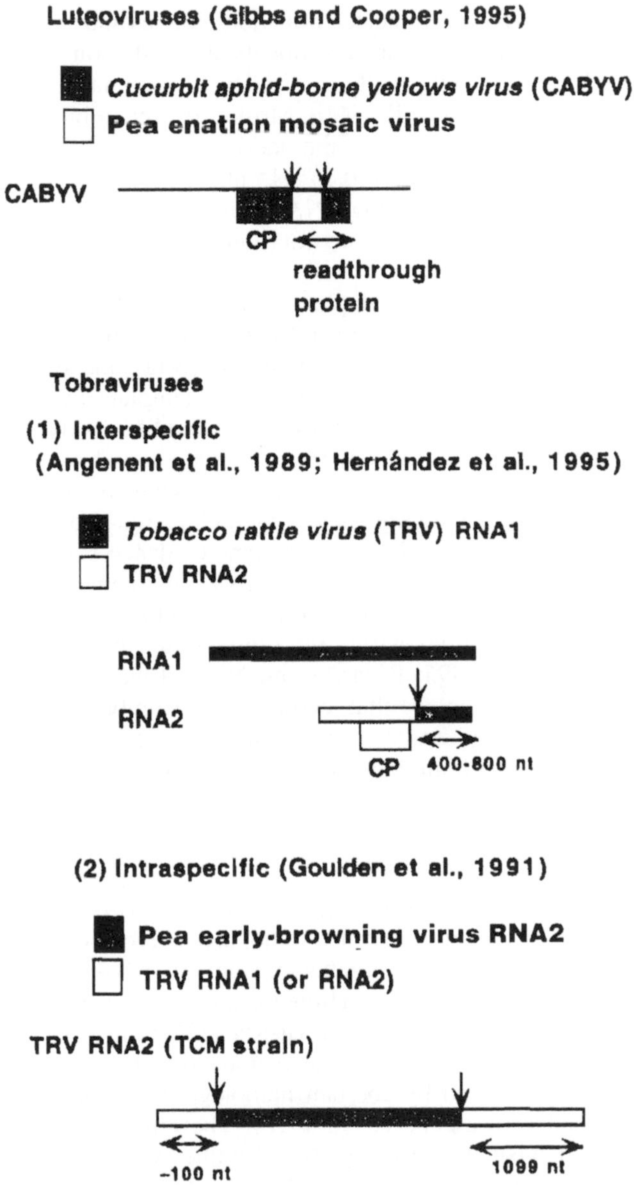

Luteoviruses (Gibbs and Cooper, 1995)

☐ *Cucurbit aphid-borne yellows virus* (CABYV)
☐ **Pea enation mosaic virus**

CABYV

CP ⟷ readthrough protein

Tobraviruses

(1) Interspecific
(Angenent et al., 1989; Hernández et al., 1995)

☐ *Tobacco rattle virus* (TRV) RNA1
☐ **TRV RNA2**

RNA1

RNA2

CP 400-800 nt

(2) Intraspecific (Goulden et al., 1991)

☐ **Pea early-browning virus RNA2**
☐ **TRV RNA1 (or RNA2)**

TRV RNA2 (TCM strain)

-100 nt   1099 nt

is necessary for cell-to-cell movement and the production of coat protein. A deletion at the 3'-end of RNA3 was repaired by the corresponding domains of RNA1 or RNA2 during systemic infection. The sequence analysis around the crossover sites indicated both homologous and aberrant homologous recombination. In similar experiments, it was demonstrated that a defective RNA2 could be efficiently repaired by recombination with the other RNA segments (Rao and Hall, 1990). Most of the recombinants turned out to result from homologous recombination. Nagy and Bujarski (1995) showed that a very short (as short as a 15 nt sequence) identity resulted in homologous recombination using RNA3 mutants, and that secondary structures around the crossover sites are not required for homologous recombination.

The formation of a heteroduplex between two RNAs could induce nonhomologous recombination when the RNA3 mutant with 3'-domain of RNA1 was employed in antisense orientation (Nagy and Bujarski, 1993). This recombination mechanism actually favors template switching by viral replicase because a heteroduplex formation would be likely to induce the replicase to jump from one to another. As template switching must be mediated by a viral replicase complex, it can be assumed that mutations in replicase would affect the frequency of recombination and in the recombination sites. Recent studies have given some evidence supporting the hypothesis. It was demonstrated that mutations within the helicase resulted in an increase in the frequency of recombination and the distribution of crossover sites, while mutations within the replicase decreased recombination (Nagy and Bujarski, 1995; Bujarski and Nagy, 1996; Figlerowitz et al., 1997). These results suggest that, certainly, a recombination type is supported by a secondary structure (or even sequence complementarity) at crossover sites besides homologous recombination based on sequence similarity.

Carmoviruses

Another well-studied virus is a monopartite virus, *Turnip crinkle virus*, a carmovirus, which contains a single-stranded, positive-sense RNA. TCV is frequently associated with a number of subviral RNAs, satellite RNAs (sat-RNAs), and DI RNAs. The chimeric nature between TCV genomic RNA and sat-RNA D was observed on sat-RNA C (Zhang et al., 1991), apparently deriving from nonhomologous recombination. On the other hand, new sat-RNAs were generated by aberrant homologous recombination between sat-RNA C and D (Cascone et al., 1990). The replicase-driven template-switching mechanism was also suggested for this TCV system. Concerning the recombination between TCV sat-RNAs, when the replicase reaches the natural endpoint of the negative strand of sat-RNA D, it binds to a stem-loop structure (motif I-hairpin) on the negative strand of sat-RNA C and then reinitiates synthesis from the recognition site (Carpenter et al., 1995; Nagy and Simon, 1997). Sequence similarity at the crossover site, although not critical for recombination, is thought to participate in template switching by helping to bring the sat-RNA with a motif into proximity. Nagy and Simon (1997) thus gave this recombination the name of "similarity-assisted recombination."

Recombination Between Transgenes and RNA Viruses

Recently, several experiments have demonstrated that virus genes in transgenic plants would be available for recombination with the viral genomes under infection (Greene and Allison, 1994; Borja et al., 1999). The first report with RNA viruses came from a bioassay using a deletion mutant of *Cowpea chlorotic mottle virus*, a bromovirus, that lacked the 3'-one-third portion of the coat protein (CP) gene and transgenic plants containing the 3'-two-thirds portion of the CP gene. A systemic viral infection indicated that a viable virus was recovered by RNA recombination with the transgene. Another clear demonstration was the recovery of a wild-type virus via a double recombination event between the CP-defective TBSV and the CP transgene (Borja et al., 1999).

RECOMBINATION IN CUCUMOVIRUSES

Cucumoviruses

Members of the genus *Cucumovirus*, which belongs to the family Bromoviridae, essentially have a genome organization similar to that of members of the genus *Bromovirus*. Currently, the genus *Cucumovirus* includes three virus species, CMV, TAV, and *Peanut stunt virus* (PSV). The relationships between the three viruses can be estimated by phylogeny analysis. Figure 10.3 shows a phylogenic tree of the 2a (or 1a) proteins constructed by the neighbor-joining (N-J) method, indicating that V-TAV branched at the earliest time among the three viruses.

Between CMV-Y and V-TAV, nucleotide sequence similarities over the whole RNAs 1 and 2 are 66 and 61 percent, respectively. At the level of the amino acid sequence, the methyl transferase and the helicase of the 1a protein and the polymerase central core region of the 2a protein exhibit 83, 70, and 72 percent homology, respectively, between CMV-Y and V-TAV (Masuta et al., 1998).

FIGURE 10.3. Phylogenic Tree of the 2a Proteins of Cucumoviruses

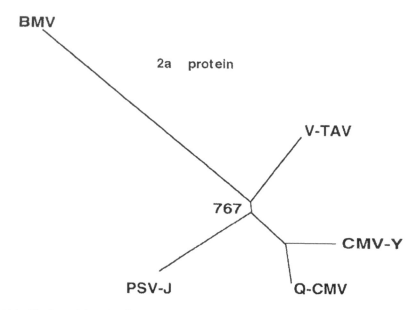

Note: The branch length reflects nucleotide substitution values per site. The dendrogram was constructed by neighbor-joining on the program provided by DNA Research Center, National Institute of Genetics, Mishima, Japan. *Brome mosaic virus* (BMV) was included as an outgroup to assess the exact root placement. CMV-Y and Q-CMV are strains of *Cucumber mosaic virus;* PSV-J is a strain of *Peanut stunt virus;* V-TAV is a strain of *Tomato aspermy virus.* The number indicates the percentage of bootstrap replicates (1,000 times).

The genome structure of CMV-Y is illustrated in Figure 10.4. As described for BMV, these viruses contain three genomic RNAs, namely RNAs 1, 2, and 3. RNA3 encodes two proteins, 3a and CP; CP is expressed via RNA4, a subgenomic (sg) RNA from RNA3. It is noteworthy that cucumoviruses produce an additional protein, 2b, that is absent in bromoviruses. RNA4A, an sgRNA for the 2b protein, is generated from the 3'-end of RNA2.

Cucumovirus Pseudorecombinants

In many experiments, artificial pseudorecombinants were created between CMV and TAV (see Table 10.1). Due to recent developments in molecular biology, the availability of infectious transcripts (or infectious complementary

FIGURE 10.4. Genome Structure of *Cucumber mosaic virus* Strain Y

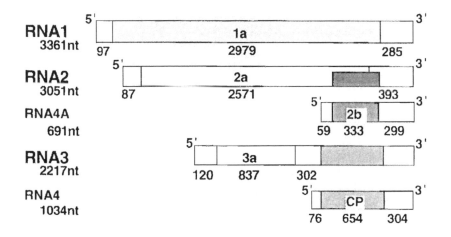

Note: The open reading frames are labeled. The boxes represent the genomic RNAs and subgenomic RNAs.

(c) DNAs) has greatly facilitated the production of pseudorecombinants and given credibility to their analyses. It was reported that RNAs 1 and 2 were not exchangeable, whereas RNA3 may be exchanged giving pseudorecombinants (Rao and Francki, 1981). This incompatibility of the heterogeneous combination of RNAs 1 and 2 makes cucumoviruses separate species.

White et al. (1995) demonstrated by phylogeny estimation that reassortment events among three cucumoviruses had given rise to present forms. They also showed a natural cucumovirus isolate that possesses genomic segments from CMV and PSV. These results suggest that pseudorecombination may be a mechanism of speciation in cucumoviruses.

NEWLY EVOLVED RECOMBINANT VIRUSES BETWEEN CMV AND TAV UNDER SELECTION PRESSURE

As shown in Table 10.1, pseudorecombinants of RNA3 can be supported by both TAV and CMV RNAs 1 and 2, but viable combinations have not been recovered from any reciprocal exchanges of RNAs 1 and 2. For convenience, genomic RNAs are hereafter designated T1 to T3 for V-TAV and C1 to C3 for CMV-Y.

TABLE 10.1. Characteristics of the Reported Pseudorecombinants Among Cucumoviruses

| Cucumoviruses | | | Pseudorecombinants | Biological properties of pseudorecombinants | | |
CMV	TAV	PSV		RNA or Protein	Phenotype	Reference
R-CMV	P-TAV		T1T2C3, C1C2T3	RNAs 1 and/or 2	Cell-to-cell movement in cucumber	Salánki et al. (1997)
			T1T2(C3-T3), C1C2(T3-C3)	Combination of CP and MP	Cell-to-cell movement in tobacco. N. glutinosa and N. benthamiana. and in cucumber	
				RNA3 3'-part	Symptom severity in N. glutinosa	
Q-CMV	V-TAV		C1C2T3	RNA3B, RNA5	RNA3B and RNA5 generation	Shi et al. (1997)
Kin-CMV	1-TAV		1-TAV + C3	RNA3, CP	Long-distance movement in cucumber	Taliansky and Garcia-Arenal (1995)
CMV		PSV	C1C2P3		Naturally occurring reassortant	White et al. (1995)
Trk7-CMV	P-TAV		T1T2C3, C1C2T3	RNAs 1+2	Accumulation of CMV sat-RNAs	Moriones el al. (1994)
Trk7-CMV	P-TAV		C1C2T3, C1C2(T3-C2)	RNAs 2 and 3	RNA-RNA recombination in RNA3	Fernández-Cuartero et al. (1994)
Q-CMV	V-TAV		T1T2C3, C1C2T3	RNAs 1+2	Not interchangeable	Rao and Francki (1981)
CMV-L	CMMV		T1T2T3, C1C2C3	RNAs 1+2	Symptoms on cucumber and cowpea	Hanada and Tochihara (1980)
Q-CMV	V-TAV		T1T2C3,C1C2T3	RNAs 1+2	T1+T2: unable to infect cucumber	Habili and Francki (1974)

Note: CMV = *Cucumber mosaic virus*, PSV = *Peanut stunt virus*; TAV = *Tomato aspermy virus*; C1C2C3 = genomic RNAs of CMV; P3 = genomic RNA of PSV; T1T2T3 = genomic RNAs of TAV.

Infectivity of Pseudorecombinants and Analysis of the Progeny

When we inoculated combinations of C1T1T2C3 and T1T2C2C3 onto *Nicotiana benthamiana*, systemic infections with different symptoms were observed (see Figure 10.5; Masuta et al., 1998). To identify the origins of each segment from the progeny viruses, RT-PCR (reverse transcription–polymerase chain reaction) and Northern blot hybridization were performed. The results revealed that all the inoculated segments could survive at least in primarily infected tissues (see Table 10.2). Both pseudorecombinants could infect *N. benthamiana* but not *N. tabacum* systemically. After single-lesion isolation on the inoculated leaves of *N. tabacum* followed by three passages through *N. benthamiana*, we found that T2 and C1 apparently had been eliminated from T1T2C2C3 and C1T1T2C3, respectively (see Table 10.2 and Figure 10.6).

Evolution of a Hybrid Virus

The progeny virus derived from T1T2C2C3 inoculation was purified, and the cDNA clones for each RNA segment were synthesized. The sequence analyses revealed that the virus apparently consisted of T1, C2, and

FIGURE 10.5. Symptoms Induced on *Nicotiana benthamiana* Inoculated with Pseudorecombinants

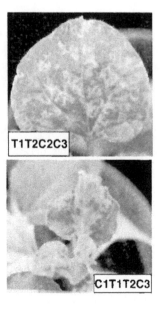

T1T2C2C3

C1T1T2C3

TABLE 10.2. Detection of Each Viral Segment in the Pseudorecombinants

Inoculum	Treatment		RT-PCR or Northern			
Pseudorecombinants	Single-lesion isolation	No. of passages	T1	T2	C1	C2
T1T2C2C3	–	1	+	+	–	+
	:	3	+	–	–	+
C1T1T2C3	–	1	+	+	+	–
	+	3	+	+	–	–

Note: C1C2C3 and T1T2T3 = genomic RNA segments of *Cucumber mosaic virus* and *Tomato aspermy virus,* respectively.

FIGURE 10.6. Northern Blot Analysis of Encapsidated Viral RNAs from Pseudorecombinants After Single-Lesion Transfer Followed by Three Passages Through *Nicotiana benthamiana.*

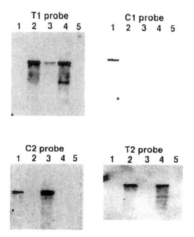

Note: The probes are indicated above each blot. Lanes 1 to 4 contain RNAs isolated from tissues infected with CMV-Y, a strain of *Cucumber mosaic virus;* V-TAV, a strain of *Tomato aspermy virus;* and pseudorecombinants T1T2C2C3 and C1T1T2C3, respectively. Lane 5 is a mock inoculation; C1C2C3 and T1T2 are RNA segments of CMV and TAV, respectively.

C3. In addition, an additional C2 molecule was found whose 3'-terminal 142 nt had been replaced by the corresponding region of T2 (322 nt), suggesting that an RNA recombination event had occurred (see Figure 10.7).

The junction at the crossover site is shown in Figure 10.8. Intact wild-type C2 was also surviving together with the recombinant C2 containing the 3'-end of T2 (designated C2-T2). This hybrid virus therefore multi-

FIGURE 10.7. Genomic Organization of T1C2(C2-T2)C3 Pseudorecombinant (Chimera)

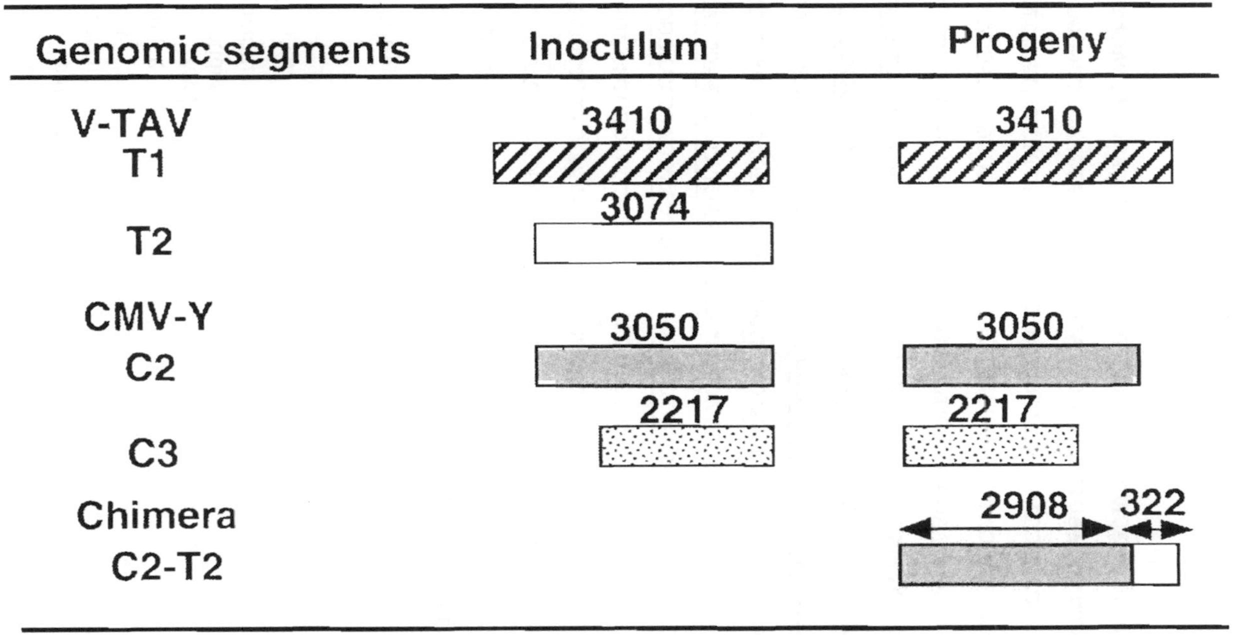

Notes: CMV-Y is a strain of *Cucumber mosaic virus*. V-TAV is a strain of *Tomato aspery virus*. C2C3 and T1T2 are RNA segments of CMV-Y and V-TAV, respectively. Number of nucleotides is shown over RNA.

FIGURE 10.8. Nucleotide Sequence Around the Crossover Site in Recombinant C2-T2

2555

AATCTCAGACTGTTCCGCTTCCTACCGTTCTATCAAGTAGATG

GTTCGGAACTGACAGGGTCATGCCGCCATGCGAACGTGGCGGA

GTCACCCGAGCCTGAGGCCTCTCGTTTAGAGTTATCGGCGGAA

GACCATGATTTTGACGATACAGATTGGTTCGCCGGTAACGAAT

GGGCGGAAGGTGCTTTCTGAAACCTCCCCTTCCGCATCTCCCT

CCGGTTTTCTGTGGCGGGAGCTGAGTTGGCAGTATTGCTATAA

ACTGTCTGAGGTCACTAAACACATTGTGGTGAACGGGTTGTCC

ATCCAGCTTACGGCTAAAATGGTCAGTCGTAGAGAAATCTACG

CCAGCTAGTCCGAAGACGTTAAACTACGTTCGAACCGTGTTCG

C2 ◄—┘ └—►T2

AATGTCTGAGTTGGTAGTATTGCTCTAAACTATCTGAAGTCAC

TAAACGCTTGTGCGGTGAACGGGTTGTCCATCCAGCTAACGGC

TAAAATGGTCAGTCATGTCGGAAGACATGCCGTCGGTCTTTGA

TCGATGAGGTGCCTTTGAACCCTTTATCCCGGGGTTCTTCGGA

AGGTGAGACTTGAATTCCATGTAGAGTCTCGCCGTGCACGGTA

TCACACTGATGATACCTTCAGAGTGCAGGCATCGCTACGGTTT

TCCGTAGGTTCCCCCTAGGGGTCCCA 3230

Note: The homologous region is underlined. A 20 nt sequence (see the text) is boxed.

plies via an interspecific hybrid replicase complex with the TAV helicase subunit and the CMV polymerase subunit. Interestingly, the chimeric replicase acquired new properties; for example, the hybrid could support replication of a satellite RNA of CMV that is not replicated by V-TAV.

We speculate on the steps in the evolution of the hybrid virus, T1C2(C2-T2)C3, as follows. After inoculation of T1T2C2C3, a replicase complex initially supplied from T1 and T2 amplified small amounts of C2 as well as T1, T2, and C3. Then, a primary recombinant that generated a precursor to C2-T2 appeared. This element then evolved to adapt well to the T1a-T2a replicase. As C2-T2 began to supply increasing amounts of C2a, the probability for productive interactions between C2a and T1a increased.

Initially, C2a may have been involved in replication through relatively weak transient associations between C2a and T1a. Subsequently, T1 mutated to produce a 1a protein with the ability to bind more specifically and with higher affinity to C2a than to T2a. Then, the heterogeneous T1a-C2a replicase became dominant in the viral replication system, and T2 was eliminated after several host passages, whereas C2 survived. These events or permutations led to stable establishment of the quadripartite virus, T1C2(C2-T2)C3.

Mechanism of RNA Recombination

Relatively long (95 nt) sequence identity (93 percent homology) between T2 and C2 exists in the 3'-terminal sequence (positions 2795-2889 in numbering C2). The recombinant C2-T2 contains a duplication of this region that is flanked by the crossover site (see Figure 10.8). Furthermore, the 3'-proximal region at the crossover site starts with a 20 nt sequence (GUCCGAAGACGUUAAACUAC) that also appears at crossover sites on the recombinant RNA3 molecules of C1C2T(C)3 and C1C2T(C)3(-4) (Fernández-Cuartero et al., 1994). Those pseudorecombinants contain TAV RNA3 with the 3'-end of CMV RNA2, and the hybrid RNA3s could outcompete authentic RNA3 from both CMV and TAV, suggesting that a recombined RNA can show an increase in its relative fitness. Interestingly, the 20 nt sequence at the crossover site is also found at the 5'-extreme end of an sgRNA, RNA5, derived from RNA 3 or 2 of CMV in subgroup II (Blanchard et al., 1996), and also at the 5'-end of sgRNAs, RNA3B and RNA5, derived from V-TAV RNA3 (Shi et al., 1997). These findings indicate that the sequence region including the 20 nt may be an internal promoter for the generation of an sgRNA from the negative strand, and, thus, the replicase complex is likely to bind in the vicinity of the region. Surprisingly, an sgRNA from RNA3 of *Beet necrotic yellow vein virus* (BNYVV), a benyvirus, contains the same 20 nt sequence at the 5'-end (Bouzoubaa et al., 1991). The promoter region mapped for sgRNA synthesis extends to no more than position −16 in the 5'-direction and to between +100 and +208 in the 3'-direction, relative to the sgRNA transcription initiation site (Balmori et al., 1993). Thus, it certainly includes the 20 nt sequence. The promoter had no obvious sequence homologies to any of the known sgRNA promoter sequences (Balmori et al., 1993). The 5'-end sequences of the sgRNAs of Q-CMV, V-TAV, and BNYVV are shown in Figure 10.9. Since BNYVV seems to have evolved differently from cucumoviruses, although they have a common ancestor that is thought to be derived from an alphavirus, at this point, we cannot reasonably explain why BNYVV and some cucumoviruses share the same 20 nt sequence at the 5'-ends of their sgRNAs.

FIGURE 10.9. Comparison of the Nucleotide Sequence of the 5'-Ends of the Subgenomic RNAs from Three Viruses

-10

BNYVV RNA3sub AUUUAUUCUUUUGUGU AAUCUCCGAAGACGUUAACUAC ACAUGAUUUCACGGUGUUCGG

V-TAV RNA5 GUGAGAUAUGCGUCGGUCU AGUUGCGAAGACGUUAAACUAG GCUUGAACGGUGUUCGAGUGU

Q-CMV RNA2sub GUUUGAUUUCCGACCUUCGUC GUCGGAAGACGUUAAACUAG GCUCUCUUUAUUGCGAGUGC

Note: BNYVV = Beet necrotic yellow vein virus; V-TAV = a strain of Tomato aspermy virus; Q-CMV = a strain of Cucumber mosaic virus.

In brief, we would like to propose a model for RNA recombination found on C2-T2 (see Figure 10.10). This RNA recombination adequately fits the criteria for Class 3 recombination (similarity-assisted recombination) proposed by Nagy and Simon (1997). During the positive-strand synthesis from the template (CMV RNA2 negative strand in Figure 10.10B), the nascent strand pauses within the homologous region due to an internal secondary structure (see Figure 10.10A). If the polymerase is at the recombination mode due to modification in the RNAs or the replicase components, it slides from the template strand to the acceptor RNA (TAV RNA2 negative strand in Figure 10.10B). The sequence complementarity between the nascent strand and the acceptor must help the polymerase to bring the acceptor into proximity. In this event, the 20 nt sequence (described earlier) may be essential for the replicase-driven template switching because it may serve as a polymerase-binding site. The recombinant RNA contains duplication of the homologous region with an insertion of two unknown nucleotides (see Figure 10.8); in this manner, this recombination can be also classified as aberrant homologous recombination.

SUMMARY AND CONCLUSION

We have shown that RNA recombination events may occur naturally in plants infected with two or more viruses, and that accumulating variability in the viral genomes is important for viral evolution. We have also seen that viral genetic sources can be exchanged interspecifically or intraspecifically, the most probable mechanism for such RNA recombination being replicase-driven template switching.

However, we should also realize that the high levels of de novo recombination observed in experimental conditions do not necessarily reflect the degree of natural occurrence of recombinant viruses: these events are actually quite rare in the lapse of a few years. RNA recombination is certainly a mechanism for plant viruses to evolve and adapt to their environment, but the high error rate of viral polymerase must have deleterious effects on the integrity of viral genomes. In nature, RNA recombination may play a role mainly in repairing errors in RNA synthesis, and viruses containing segmented genomes can overcome these effects by reassortment of RNA segments. Viruses seem to keep RNA recombination at the minimum levels required for survival. A new virus species generated by RNA recombination may only survive if it is better adapted to the given environment than parental viruses.

FIGURE 10.10. Hypothetical Model for RNA Recombination Observed in C2-T2 (B)

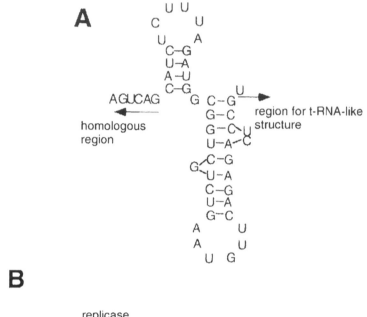

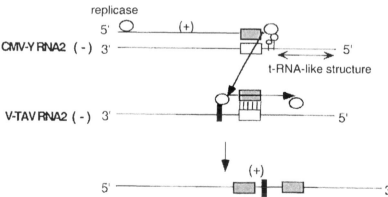

Note: A probable secondary structure (A), which may suspend the RNA synthesis of the replicase, is found just downstream of the complementary region (open box) in the negative strand of strain Y of *Cucumber mosaic virus,* (CMV-Y) RNA 2. This secondary structure appears only on CMV and not on strain V of *Tomato aspermy virus* (V-TAV). The black box represents the 20 nt sequence, which may be part of the polymerase-binding site.

REFERENCES

Angenent GC, Posthumus E, Brederode FT, Bol JF (1989). Genome structure of tobacco rattle virus strain PLBV: Evidence on the occurrence of RNA recombination among tobraviruses. *Virology* 171: 271-274.

Balmori E, Gilmer D, Richards K, Guilley H, Jonard G (1993). Mapping the promoter for subgenomic RNA synthesis on beet necrotic yellow vein virus RNA3. *Biochimie* 75: 517-521.

Blanchard CL, Boyce PM, Anderson BJ (1996). Cucumber mosaic virus RNA5 is a mixed population derived from the conserved 3'-terminal regions of genomic RNAs 2 and 3. *Virology* 217: 598-601.

Borja M, Rubio T, Scholthof HB, Jackson AO (1999). Restoration of wild-type virus by double recombination of tombusvirus mutants with a host transgene. *Molecular Plant-Microbe Interactions* 12: 153-162.

Bouzoubaa S, Niesbach-Klösgen U, Guilley IJH, Richards K, Jonard G (1991). Shortened forms of beet necrotic yellow vein virus RNA-3 and -4: Internal deletions and a subgenomic RNA. *Journal of General Virology* 72: 259-266.

Bujarski JJ, Dzianott AM (1991). Generation and analysis of nonhomologous RNA-RNA recombinants in brome mosaic virus: Sequence complementarities at crossover sites. *Journal of Virology* 65: 4153-4159.

Bujarski JJ, Kaesberg P (1986). Genetic recombination between RNA components of a multipartite plant virus. *Nature* 321: 4153-4159.

Bujarski JJ, Nagy PD (1996). Different mechanism of homologous and nonhomologous recombination in brome mosaic virus: Role of RNA sequences and replicase proteins. *Seminars in Virology* 7: 363-372.

Carpenter CD, Oh JW, Zhang CX, Simon AE (1995). Involvement of a stem-loop structure in the location of junction sites in viral RNA recombination. *Journal of Molecular Biology* 245: 608-622.

Cascone PJ, Carpenter CD, Li Z-H, Simon AE (1990). Recombination between satellite RNAs of turnip crinkle virus. *EMBO Journal* 9: 1706-1715.

Cooper PD, Steiner-Pryor S, Scotti PD, Delong D (1974). On the nature of poliovirus genetic recombinations. *Journal of General Virology* 42: 153-164.

Fernández-Cuartero B, Burgyán J, Aranda MA, Salánki K, Moriones E, García-Arenal F (1994). Increase in the relative fitness of a plant virus RNA associated with its recombinant nature. *Virology* 203: 373-377.

Figlerowitz M, Nagy PD, Bujarski JJ (1997). A mutation in the putative RNA polymerase gene inhibits nonhomologous, but not homologous, genetic recombination in an RNA virus. *Proceedings of the National Academy of Sciences, USA* 94: 2073-2078.

Gibbs MJ, Cooper JI (1995). A recombinational event in the history of luteoviruses probably induced by base-pairing between the genomes of two distinct viruses. *Virology* 206: 1129-1132.

Goulden MG, Lomonossoff GP, Wood KR, Davies JW (1991). A model for the generation of tobacco rattle virus (TRV) anomalous isolates: Pea early browning virus RNA-2 acquires TRV sequences from both RNA-1 and RNA-2. *Journal of General Virology* 72: 1751-1754.

Greene AE, Allison RF (1994). Recombination between viral RNA and transgenic plant transcripts. *Science* 263: 1423-1425.

Habili N, Francki RIB (1974). Comparative studies on tomato aspermy and cucumber mosaic viruses. III. Further studies on relationship and construction of a virus from parts of the two viral genomes. *Virology* 61: 443-449.

Hanada K, Tochihara H (1980). Genetic analysis of cucumber mosaic, peanut stunt and chrysanthemum mild mottle viruses. *Annals of Phytopathological Society of Japan* 46: 159-168.

Hernández C, Mathis A, Brown DJF, Bol JF (1995). Sequence of RNA 2 of a nematode-transmissible isolate of tobacco rattle virus. *Journal of General Virology* 76: 2847-2851.

King AMQ, McCahon D, Slade WR, Newman JWI (1982). Recombination in RNA. *Cell* 29: 921-928.

Lai MMC (1992). RNA recombination in animal and plant viruses. *Microbiological Reviews* 56: 61-79.

Ledinko N (1963). Genetic recombination with poliovirus type 1: Studies of crosses between a normal horse serum-resistant mutant and several guanidine-resistant mutants of the same strain. *Virology* 20: 107-119.

Masuta C, Ueda S, Suzuki M, Uyeda I (1998). Evolution of a quadripartite hybrid virus by interspecific exchange and recombination between replicase components of two related tripartite RNA viruses. *Proceedings of the National Academy of Sciences, USA* 95: 10487-10492.

Moriones E, Díaz I, Fernández-Cuartero B, Fraile A, Burgyán J, García-Arenal F (1994). Mapping helper virus functions for cucumber mosaic virus satellite RNA with pseudorecombinants derived from cucumber mosaic and tomato aspermy viruses. *Virology* 205: 574-577.

Nagy PD, Bujarski JJ (1993). Targeting the site of RNA-RNA recombination in brome mosaic virus with antisense sequences. *Proceedings of the National Academy of Sciences, USA* 90: 6390-6394.

Nagy PD, Bujarski JJ (1995). Efficient system of homologous RNA recombination in brome mosaic virus: Sequence and structure requirements and accuracy of crossovers. *Journal of Virology* 69: 131-140.

Nagy PD, Simon AE (1997). New insights into the mechanisms of RNA recombination. *Virology* 235: 1-9.

Rao ALN, Francki RIB (1981). Comparative studies on tomato aspermy and cucumber mosaic viruses. VI. Partial compatibility of genome segments from the two viruses. *Virology* 114: 573-575.

Rao ALN, Hall TC (1990). Requirement for a viral trans-acting factor encoded by brome mosaic virus RNA-2 provides strong selection in vivo for functional recombinants. *Journal of Virology* 64: 2437-2441.

Rubio T, Borja M, Scholthof HB, Jackson AO (1999). Recombination with host transgenes and effects on virus evolution: An overview and opinion. *Molecular Plant-Microbe Interactions* 12: 87-92.

Salánki K, Carrère I, Jacquemond M, Balázs E, Tepfer M (1997). Biological properties of pseudorecombinant and recombinant strains created with cucumber mosaic virus and tomato aspermy virus. *Journal of Virology* 71: 3597-3602.

Shi B-J, Ding S-W, Symons RH (1997). Two novel subgenomic RNAs derived from RNA 3 of tomato aspermy cucumovirus. *Journal of General Virology* 78: 505-510.

Simon AE, Bujarski JJ (1994). RNA-RNA recombination and evolution in virus-infected plants. *Annual Review of Phytopathology* 32: 337-362.

Taliansky ME, García-Arenal F (1995). Role of cucumovirus capsid protein in long-distance movement within the infected plant. *Journal of Virology* 69: 916-922.

White KA, Morris TJ (1994). Recombination between defective tombusvirus RNAs generates functional hybrid genomes. *Proceedings of the National Academy of Sciences, USA* 91: 3642-3646.

White PS, Morales F, Roossinck MJ (1995). Interspecific reassortment of genomic segments in the evolution of cucumoviruses. *Virology* 207: 334-337.

Zhang CX, Cascone PJ, Simon AE (1991). Recombination between satellite and genomic RNAs of turnip crinkle virus. *Virology* 184: 791-794.

Chapter 11

Variability and Evolution of *Potato Virus Y,* the Type Species of the *Potyvirus* Genus

Laurent Glais
Camille Kerlan
Christophe Robaglia

INTRODUCTION

This chapter describes the main characteristics of the *Potyvirus* genus, its relationships with other groups of positive-strand RNA viruses, and the relationships between potyvirus species. In the second part, the diversity of *Potato virus Y* (PVY) isolates at the biological and molecular levels is mentioned. Last, the possible role of recombination events in the emergence and evolution of new PVY strains is discussed in detail.

THE FAMILY POTYVIRIDAE

Potato Virus Y is the type species of the *Potyvirus* genus, which, together with other genera, namely, *Rymovirus, Bymovirus,* and *Ipomovirus* (not formerly officially recognized by the International Committee on Taxonomy of Viruses [ICTV]), once formed the family *Potyviridae* (Brunt, 1992). These genera were originally distinguished on the basis of their vector transmission: potyviruses are transmitted by aphids, rymoviruses by mites, bymoviruses by root-infecting fungi, and ipomoviruses by whiteflies. In 1998, the genus *Ipomovirus* was recognized by the ICTV and new genera

The authors wish to express their gratitude to Suzanne Astier-Manifacier, to whom this chapter is dedicated. Although in retirement, Suzanne remains a specialist in plant virology, especially interested in molecular biology of potyviruses.

were added to this family, namely. *Macluravirus* (Badge et al., 1997) and *Tritimovirus* (Salm et al., 1996; Pringle, 1999), which are transmitted by aphids and mites, respectively (see Figure 11.1).

From the fossil history of hosts and vectors of viruses in the family *Potyviridae*, bymoviruses infecting grasses and transmitted by fungal vectors are thought to have appeared first (Ward et al., 1995). Rymoviruses also infect grasses but are transmitted by the more recent eriophyid mite vectors presumably derived from them. Aphid-transmitted potyviruses then appeared simultaneously with the diversification of herbaceous angiosperms and polyphagous aphids, in the Early Cretaceous, between about 130 and 90 million years ago (Crane et al., 1995). However, recent sequence-based phylogenetic analysis suggests that no correlation exists between vector type and potyviral lineages; thus, a revised classification of the family *Potyviridae* was proposed (Hall et al., 1998).

FIGURE 11.1. Phylogenetic Tree of the Full-Length Derived Amino Acid Sequences Representative of Each Genus in the *Potyviridae* Family

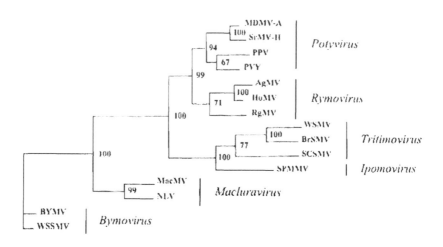

Source: Modified from Hall et al., (1998: 329).

Note: The values at the forks indicate the number of times out of 100 trees that this grouping occurred after bootstrapping the data. AgMV = *Agropyron mosaic virus;* BYMV = *Barley yellow mosaic virus;* BrSMV = *Brome streak mosaic virus;* HoMV = *Hordeum mosaic virus;* MacMV = *Maclura mosaic virus;* MDMV-A = *Maize dwarf mosaic virus-strain A;* NLV = *Narcissus latent virus;* PPV = *Plum pox virus;* PVY = *Potato virus Y;* RgMV = *Ryegrass mosaic virus;* SCSMV = Sugarcane streak mosaic virus; SPMMV = *Sweet potato mild mottle virus;* SrMV-H = *Sorghum mosaic virus-strain H;* WSMV = *Wheat streak mosaic virus;* WSSMV = *Wheat spindle streak mosaic virus.*

The *Potyvirus* genus includes probably more than 200 members or possible members, making *Potyviridae* the largest plant virus family, accounting for nearly 25 percent of the known plant viruses.

Virions of potyviruses are filamentous, nonenveloped flexuous rods, 680 to 900 nm long, 12 to 15 nm in diameter, with helical symmetry. Each virion is composed of about 2,000 copies of coat protein (28 to 34 kDa), constituting 95 percent of the mass of the viral particle (Hollings and Brunt, 1981). Viruses of the family *Potyviridae* induce the formation of nuclear and cytoplasmic inclusion bodies in infected cells (Edwardson, 1974).

Potyviruses have monopartite genomes composed of single-stranded positive-sense RNA, 9,500 to 10,000 nucleotides long. A viral protein (VPg) is covalently linked at the 5'-end of the genome, which is polyadenylated at the 3'-end (Hollings and Brunt, 1981). RNA of potyviruses is translated into a large precursor polyprotein that is cleaved co- and post-translationally into mature proteins (Dougherty and Carrington, 1988; Riechmann et al., 1992) (see Figure 11.2).

FIGURE 11.2. Diagram of Polyprotein Processing

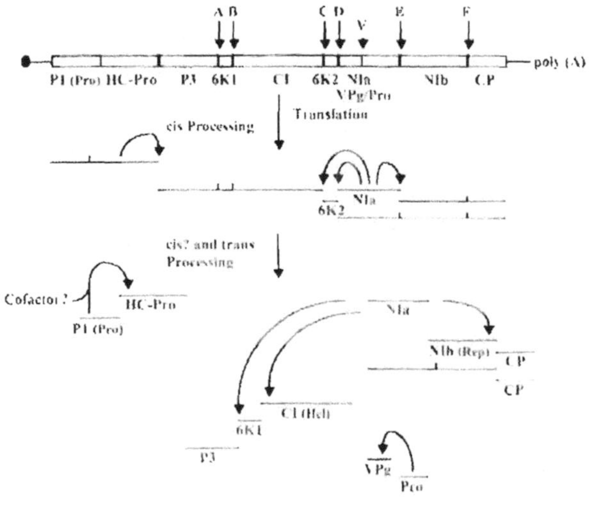

Source: Reproduced from Riechmann et al. (1992: 7).

Note: Open reading frames are represented as open bars and VPg as a black circle. Arrows A to F and V define NIa cleavage sites of the polyprotein. The primary events are probably cotranslational and autocatalytic, yielding precursors and mature products. P1 = P1 protein; HC-Pro = aphid transmission helper component; P3 = P3 protein; 6K1/6K2 = 6 kDA proteins; CI = cytoplasmic inclusion; NIa/NIb = nuclear inclusions a or b; CP = capsid protein; Hel = helicase; Pro = protease; Rep = replicase.

The gene products of the potyviral genome are from 5'- to 3': The P1 protein contains a protease domain at its C-terminal region, cleaving itself from the adjacent helper component protease (HC-Pro) protein; its other functions are presently unknown. The HC-Pro protein is multifunctional; it is involved in aphid transmission of the viral particle and in movement of the virus within the plant, and it is a suppressor of the RNA-dependent gene silencing that recently emerged as a defense mechanism against viruses (Brigneti et al., 1998; Kasschau and Carrington, 1998; Revers et al., 1999). The C-terminal part of HC-Pro is also a protease, cleaving itself from the precursor polyprotein, the P3 protein of unknown function, which is a "replication complex block" composed of the cytoplasmic inclusion (CI) protein, VPg protease, and polymerase. The CI protein forms typical pinwheels in the cytoplasm of infected cells; it has an associated helicase activity and is postulated to be involved in viral cell-to-cell movement. CI is bordered by two small proteins, 6K1 and 6K2, of unknown functions. The NIa protein forms nuclear inclusions; it is a two-domain protein containing the VPg, presumably involved in replication and translation, and a protease domain that cleaves all proteins of the C-terminal half of the precursor. The NIb protein is the putative RNA-dependent RNA polymerase that also forms nuclear inclusions. The coat protein (CP) is involved in cell-to-cell movement and vector transmission, in association with HC-Pro.

At the molecular level, the genome of family members of *Potyviridae* (genera *Potyvirus* and *Bymovirus*) share many similarities with animal members of the families *Picornaviridae* (genus *Enterovirus*) and *Caliciviridae* (genus *Calicivirus*), and with plant viruses in the family *Comoviridae* (genera *Comovirus* and *Nepovirus*), leading to the creation of a supergroup of picorna-like viruses (Goldbach, 1987; Gorbalenya, 1995). Also, fungal viruses in the family *Hypoviridae* show phylogenetic relatedness to members of the genus *Potyvirus* (Barnett et al., 1995).

Conserved structural features of potyvirus RNA genomes are the presence of a protein covalently linked at the 5'-end of the genome and of a polyadenylated tail at the 3'-end. Genes for proteins putatively involved in the replication complex (helicase, VPg, protease, polymerase) are similarly organized as a polyprotein precursor, and their encoded amino acid sequences bear conserved motifs. These features are possibly linked to a particular mode of replication (see Figure 11.3).

Further sequence comparisons of viral RNA-dependent RNA polymerase domains lead to the assignment of potyviruses within an extended supergroup comprising picorna-, noda-, como-, nepo-, sobemo-, hypo-, calici-, corona-, toro-, arteri-, and some luteoviruses (Koonin and Dolja, 1993). However, when coat protein sequences are considered, *Potyviridae* viruses show evolutionary

FIGURE 11.3. Schematic Representation of the Conserved Motifs Within the Picorna-like Supergroup

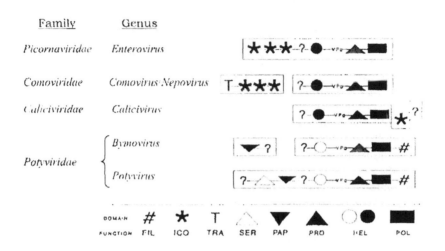

Source: Modified from Gorbalenya, 1995: 52.

Note: The domains derived from the same polyprotein are enclosed in a common box, and those encoded in the different RNA segments are spaced. POL= RNA-dependent RNA polymerase; HEL = helicase; PRO = chymotrypsin-like protease. Other domains drawn are FIL and ICO = filamentous and icosahedral capsid proteins, respectively; TRA = domain ensuring virus spread throughout plant; SER and PAP = unclassified and papain-like proteases, respectively; ? = domain(s) with an unassigned function.

relationships with a distinct group of plant filamentous viruses comprising potex-, carla-, and closteroviruses (Dolja et al., 1991).

Moreover, the analysis of the helicase protein sequences suggests an evolutionary relationship between the families *Potyviridae* and *Flaviridae* (Koonin and Dolja, 1993). Last, in addition to protein-coding regions presenting homology with other RNA virus families, a highly conserved motif (UCAACACAACAU) in the 5'-nontranslated region (5'-NTR), named "potybox," was found to be specific to potyviruses (Marie-Jeanne-Tordo et al., 1995) and bymoviruses (Atreya et al., 1992), suggesting its specificity to the family *Potyviridae*.

These data strongly support the hypothesis that the *Potyviridae* genome may have evolved from a combination of genes and gene fragments from different evolutionary origins.

RELATIONSHIPS BETWEEN POTYVIRUSES SPECIES

The taxonomic relationship between potyvirus species has always been a matter of debate.

An optimal amino acid sequence alignment of four distinct virus polyproteins revealed that P1, P3, and CP N-terminal domain are the most variable regions (Domier et al., 1989; Shukla et al., 1991; Ward et al., 1992). Indeed, between distinct potyviruses, only 20 percent sequence identity is scored in the P1 region (Domier et al., 1989). Despite this high variability, a pentapeptide in the P1 C-terminal region (FIVRG), possibly involved in P1 protease activity, appears conserved in all potyviruses (Shukla et al., 1991; Marie-Jeanne-Tordo et al., 1995). The second most variable gene product is the P3 protein, which reveals about 30 percent homology between distinct potyviruses (Shukla et al., 1991).

An interesting feature of the potyviral CPs is that they have an extremely variable N-terminal sequence linked to a well-conserved core domain. The C-terminal region is also variable. Coat protein is by far the major product of the virion, representing 95 percent of the particle weight; its N-terminal sequence, which is found at the surface of the particle (Shukla et al., 1988), is the first to be exposed to a host's recognition and defence mechanisms. Potyvirus species and strains can be classified on the basis of the coat protein size resulting primarily from differences in the length of the N-terminal regions (Shukla et al., 1988; Shukla and Ward, 1989a). Between distinct potyviruses, the CP size ranges from 263 to 330 amino acids, whereas within a particular species, such as PVY, all isolates share a similar CP length of, in PVY, 267 amino acids (Shukla and Ward, 1989a). Comparative nucleotide sequence analysis of coat protein from 17 strains of 8 distinct potyviruses has revealed a bimodal distribution (Shukla and Ward, 1988). Isolates belonging to distinct potyvirus species share sequence homologies ranging from 38 to 71 percent, whereas strains of the same potyvirus species present sequence homologies ranging from 90 to 99 percent (see Figure 11.4). On the basis of this bimodal distribution, the coat protein gene is a marker enabling the distinction between species and strains.

On the basis of the taxonomic relationships defined by coat protein sequences and host specificities, Ward et al. (1995) defined three major clusters for aphid-transmitted potyviruses. Two clusters contain viruses infecting rosid and caryophyllid plant hosts, while the third infects asterids and contains PVY together with *Lettuce mosaic virus* (LMV). Coat protein sequence analyses also show that PVY is closely related to *Pepper mottle virus* (PepMoV), *Pepper severe mosaic virus*, and *Potato virus V* (Berger et al.,

FIGURE 11.4. Frequency Distribution of Amino Acid Sequence Homologies for the Coat Proteins from 17 Strains of 8 Distinct Potyviruses

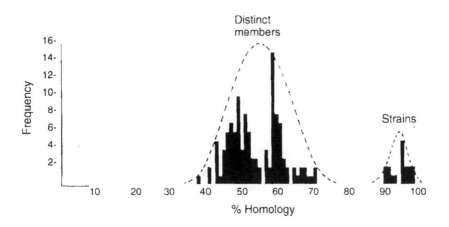

Source: Reproduced from Shukla and Ward, 1988: 2706.

Note: The homologies between distinct viruses display a mean value of 54.1 percent, whereas the homologies between strains of individual viruses show a mean of 95.4 percent.

1997). The taxonomic proximity with PepMoV is also found when sequences of the 3'- nontranslated region (3'-NTR) (Frenkel et al., 1989; Van der Vlugt et al., 1993) and of the CI protein (Lee et al., 1997), which contains the helicase domain, are considered. However, these potential evolutionary relationships between PVY and other potyviruses are not always conserved when other genes or gene segments are considered. For example, the close proximity with LMV, defined by using the coat protein sequence as indicator, is not found when the adjacent 3' noncoding sequences are analyzed (Berger et al., 1997). This suggests that recombination events have occurred during evolution of potyviruses, as was shown for the evolution of viral strains (Revers et al., 1996; Glais et al., 1998).

BIOLOGICAL DIVERSITY OF PVY

Since its identification, PVY has been seen as a complex of different isolates (Smith, 1931). Potato virus C (PVC), recognized as a PVY isolate (Bawden,

1943) when identified in the 1930s (Salaman, 1930; Bawden, 1936; Dykstra, 1936), was the first of a strain group later named PVY^C. Another strain group, now designated PVY^N (De Bokx, 1961), was found first in 1935 in a tobacco plant growing close to experimental potato plants (Smith and Dennis, 1940), then in Peruvian and Bolivian potato cultivars in 1941-1942 (Nobrega and Silberschmidt, 1944; Silberschmidt, 1960). This strain group was associated with severe epidemics in potato and tobacco crops in Europe in the 1950s (Klinkowski and Schmelzer, 1960; Horvath, 1967a; Weidemann, 1988). PVY was recognized in the 1950s as damaging to tomato crops in Australia (Sturgess, 1956) and in South and North America (Silberschmidt, 1956; Silberschmidt, 1957; Simons, 1959), and to pepper crops in Florida (Simons, 1959) and Israel (Nitzany and Tanne, 1962). Due to the emergence of necrotic isolates in the 1970s and 1980s, PVY has become a major threat to both of these crops (Tomlinson, 1987; Legnani, 1995; Marchoux et al., 1995). Last, PVY has emerged as a severe problem in trailing petunias (Boonham et al., 1999).

In connection with its large host range, which includes not only widely cultivated solanaceous species but also many solanaceous and nonsolanaceous weeds, PVY displays high variability. In many studies, the original host species was the first trait used to characterize PVY isolates as different strains. Thus, potato isolates are the oldest and most-studied PVY isolates. This has caused potato and tobacco isolates to have their own classifications. Tomato and pepper isolates, often considered a single group, are now more and more studied separately.

Potato isolates historically have been divided into three main strain groups; PVY^O, PVY^N, and PVY^C, according to symptoms induced in *Nicotiana tabacum* cv. Samsun and *Solanum tuberosum* ssp. *tuberosum* (Delgado-Sanchez and Grogan, 1970). Unlike PVY^O and PVY^N isolates, some (but not all) PVY^C isolates are non-aphid transmissible (Watson, 1956; De Bokx and Piron, 1978; Blanco-Urgoiti, Sanchez, et al., 1998).

PVY tobacco isolates are broadly defined as "severe" and "mild," based on whether they induce necrosis on any tobacco genotype (Gooding, 1985). However, historically, three strain groups, MsMr, MsNr, and NsNr, have been identified according to their reaction in tobacco cultivars resistant or susceptible to the root-knot nematode (RKN) *(Meloidogyne incognita)*. A resistance-breaking group of isolates designated VAM-B was then detected in tobacco crops (Latore and Flores, 1985; Gooding, 1985; Reddick and Miller, 1991; Blancard et al., 1994) (Horvath, unpublished data). Latore and Flores (1985) proposed that the tobacco genotype VAM may serve as an additional host for typing PVY tobacco isolates. Recently, Blancard (1998) proposed a new nomenclature differentiating six pathotypes according to their behavior on four genotypes, including VAM and another genotype (NG TG52) resistant to the VAM-B pathotype. In this classification, the Gooding and Tolin's MsNr and MsMr strain groups are ranged in the "mosaic group," whereas the "necrotic group" is subdivided into four pathotypes, designated 0, 1, 2, and 1-2, respectively; the pathotype 0 corresponds to the NsNr strain group.

Tomato isolates of PVY also show great biological variability (Walter, 1967). Gebre Selassie et al. (1985) and Marchoux et al. (1995) distinguished isolates causing only mosaic from those inducing mosaic and necrosis. Recently, Legnani (1995) defined seven PVY tomato pathotypes according to their virulence in a range of *Lycopersicon* genotypes, the most widespread pathotype including both necrotic and nonnecrotic isolates.

Pepper isolates of PVY infecting pepper have been classified into three pathotypes, designated PVY-0, PVY-1 and PVY-1-2, in accordance with their ability to overcome resistance genes *(pvr1, pvr2)* present in several pepper cultivars (Gebre Selassie et al., 1985). Within these three groups, pepper isolates have been further defined as "common" or "necrotic" (D'Aquino et al., 1995).

However, some isolates belong to none of the PVY^O, PVY^N, and PVY^C strain groups. Indeed, in the past, a fourth group, called PVY^{An} (Horvath, 1967b), was described that included particular potato and tomato isolates, among that was an isolate sharing both PVY^O and PVY^N properties. In potato, many variants that were more or less characterized (De Bokx et al., 1975; Thompson et al., 1987; Chrzanowska, 1994) were reported, among which the non-aphid-transmissible PVC can be considered the first (Cockerham, 1943; Bawden and Kassanis, 1947). Two isolates, previously typed as PVY^C but serologically distant (De Bokx et al., 1975; Calvert et al., 1980), were detached from PVY and a new distinct virus, named *Potato virus V*, was generated (Fribourg and Nakashima, 1984). During the past two decades, three variants have emerged: (1) PVY^{NTN} (Le Romancer et al., 1994), first detected in Hungary in 1978 (Beczner et al., 1984), is now widespread in many potato-growing areas and is characterized by its necrotic properties in potato tubers. (2) PVY^N-Wi, found in Poland in 1984, was described as differing in virulence and aggressiveness from the earlier PVY^N isolates (Chrzanowska, 1991) and then was shown to be serologically related to PVY^O isolates (Chrzanowska, 1994); similar isolates were also reported in Canada (McDonald and Singh, 1996), Spain (Blanco-Urgoiti, Tribodet, et al., 1998), and France (Kerlan et al., 1999). (3) Last, a less-characterized pathotype, PVY^Z, found in Britain in 1984, differed from PVY^O and PVY^C by its ability to overcome the hypersensitive genes *Nytbr* and *Nc* and, furthermore, to target a hypothetical *Nz* gene (Jones, 1990).

The frequent and almost simultaneous emergence of new necrotic strains in potato, tobacco, tomato, or pepper in the same growing areas, for instance, in the 1990s in Canada (McDonald and Kristjansson, 1993; Stobbs et al., 1994) or in Mediterranean countries such as Slovenia (Pepelnjak, 1993), has led to questions regarding whether these newly emerged strains were able to migrate from one crop to another and to spread in each of them.

Host specificity for PVY is highly variable. Some studies demonstrated a high level of host specificity (Gebre Selassie et al., 1985). McDonald and Kristjansson (1993) described a pepper isolate infecting tomato and tobacco, but not six potato cultivars. D'Aquino et al. (1995), studying a pepper necrotic isolate, underlined the specificity of pepper isolates and introduced the notion of a separate pepper strain group.

In contrast, it was demonstrated that various PVY isolates are able each to infect several cultivated species, at least upon mechanical inoculation. Thus, several isolates from potato and tobacco were proved to infect several cultivars of *Capsicum annuum* (Horvath, 1966; Horvath, 1967c; Marte et al., 1991; McDonald and Kristjansson, 1993; Le Romancer et al., 1994). Many tomato cultivars were found susceptible to a nontomato PVY^N isolate (Stobbs et al., 1994) and to a large range of PVY isolates from various host species (Legnani, 1995). Similarly, a necrotic isolate from tomato (Kerlan and Tribodet, 1996) and a color-breaking isolate from petunia (Boonham et al., 1999) were shown able to induce tuber necrosis in potato. It was further suggested that tomato (Legnani, 1995) and tobacco (Blancard, 1998) can be infected by most, if not all, PVY isolates, including those which originate from potato and pepper.

The current tendency in regard to host specificity of PVY strains has impact on epidemiology and control. Thus, it is now accepted in Canada that tomato crops could provide a seasonal reservoir host for PVY^N that could infect nearby potato or tobacco crops (Stobbs et al., 1994). Similarly, in France, the risks generated by tobacco and potato plantings in the same region have been recently highlighted (Blancard, 1998). Such changes can undoubtedly be seen as reflecting PVY evolution.

Comparisons of results from investigations using different strains are not easy because the original host plant is not always mentioned and sometimes even the identity of the strain is uncertain (Stobbs et al., 1994). The basic problem is the correlation between potato, tobacco, tomato, and pepper strain groups. For example, the possible relationship between the group of necrotic isolates in tobacco and the potato PVY^N strain group, as described by De Bokx and Huttinga (1981), has not yet been clarified, even though PVY^N was identified almost simultaneously in both plants (under the name of tobacco veinal necrosis virus [TVNV]). More recently, McDonald and

Kristjansson (1993) underlined many discrepancies in the typing of several PVY isolates, especially from tobacco and pepper, and demonstrated that some isolates inducing necrosis in tobacco (NsNr and MsNr isolates) (Gooding and Tolin, 1973) cannot be classified within the PVY^N strain group since they fail to induce systemic infection in several potato cultivars.

Pathotyping is also not sufficient to accurately define an isolate. For example, PVY^N-Wi isolates are still typed as PVY^N, though they share properties with both PVY^N and PVY^O (McDonald and Singh, 1996). Otherwise, PVY^{NTN} was defined as a PVY^N subgroup inducing superficial necrosis in potato tubers; however, Kerlan and Tribodet (1996) demonstrated that, at least in artificial conditions, most, if not all, PVY^N isolates are able to induce such tuber necrosis.

Although biological indexing has been applied in the past to large-scale testing, it is time-consuming. As an alternative to avoid drawbacks and the sometimes uncertain results of bioassays, molecular characterization of the genome is preferred and recommended for a more reliable classification of PVY. Simple tests linking genome polymorphism to potential for disease have already been developed (Glais et al., 1996; Weidemann and Maiss, 1996).

GENETIC VARIABILITY OF PVY

Complete nucleotide sequences are available for several PVY isolates: PVY^N-Fr ([Robaglia et al., 1989]; accession number D00441), PVY^{NTN}-H ([Thole et al., 1993]; accession number M95491), PVY^N-605 (accession number X97895), and PVY^O-139 ([(Singh and Singh, 1996]; accession number U09509). This provides the necessary information to estimate genetic variability within PVY strains. This analysis reveals that the variability is not uniform along the genome. The more conserved regions are the nuclear inclusion b (NIb) gene coding for the putative RNA-dependent RNA polymerase and the gene for the cytoplasmic inclusion (CI) coding for the helicase.

The 5'-Nontranslated Region (5'-NTR) and P1 Protein

The region presenting the higher interstrain variability codes for the P1 protein (Marie-Jeanne-Tordo et al., 1995). This region displays 2 to 3 percent variability within the N or O groups of the potato strain (Chachulska, 1998), as well as within the pepper strain (Marie-Jeanne-Tordo et al., 1995). In contrast, 28 percent variability exists between the N and O groups of the potato strain, with a high level of variability also between any isolates of the potato, pepper, and tobacco strains (Marie-Jeanne-Tordo et al., 1995). The P1 gene region is also the most variable between potyvirus species (Ward et al., 1995). P1 amino acid and 5'-NTR sequence analyses lead to the identification of three groups. Potato PVY^O isolates and nonpotato PVY isolates form two related groups (Group II and Group III) and another distinct one (Group I) with PVY^N isolates (Marie-Jeanne-Tordo et al., 1995) (see Figure 11.5). However, one PVY^N isolate each from potato (PVY^N-Fr), tobacco (PVY-NN), and tomato (PVY-LYE84) are grouped with potato PVY^O isolates (Group II).

FIGURE 11.5. Phylogenetic Trees Derived from the 5'-Nontranslated Sequences (a) and the P1 Coding Region Amino Acid Sequences (b) of *Potato virus Y* (PVY) Isolates

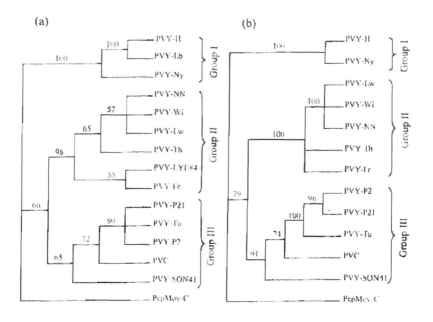

Source: Reproduced from Marie-Jeanne-Tordo et al., 1995: 943.

Note: Trees were rooted by including strain C of *Pepper mottle virus* (PepMoV-C) as an outgroup. Group I = potato PVY^N isolates; Group II = potato PVY^O isolates; Group III = nonpotato isolates. The values at the forks indicate the number of times out of 100 trees that this grouping occurred after bootstrapping the data.

The Coat Protein (CP)

Except for the conserved motif (DAG) involved in aphid transmission, the N-terminal sequences are variable (Van der Vlugt et al., 1993). This provides the basis for the immunospecificity of strain groups (Shukla and Ward, 1988, 1989b). Monoclonal antibodies targeting the coat protein can specifically recognize PVY^N (Gugerli and Fries, 1983) and PVY^O isolates (Hataya et al., 1994). Pepper and tobacco strains react negatively with monoclonal antibodies recognizing PVY^N isolates, but most of them react positively with mono- and polyclonal antibodies recognizing PVY^O isolates (McDonald and Kristjansson, 1993; Soto et al., 1994), suggesting that these two strains, at the CP N-terminal region, are more related to the PVY^O strain group (Sudarsono et al., 1993; Soto et al., 1994). This is also in accordance with the fact that the coat protein amino acid sequences of pepper and tobacco viruses (Shukla and Ward, 1989a; Fakhfakh et al., 1995) begin with an alanine residue that is also correlated with membership in the O group (Van der Vlugt et al., 1993).

Phylogenetic analysis based on overall CP amino acid sequences leads to the classification of PVY isolates into two major groups (Van der Vlugt et al., 1993), generally following the classification according to the pathotypes: PVY^O isolates from potato are gathered into Group I and PVY^N isolates from potato correspond to Group II. However, there is one exception, namely, PVY^N-Fr, which induces veinal necrosis in tobacco but is clustered in Group I. Most pepper and tobacco isolates form an unreliable third group (Group III) closely related to Group II (Blanco-Urgoiti et al., 1996) (see Figure 11.6).

The CP sequence is therefore correlated to pathogenicity, at least for isolates of the potato strain (PVY^O, PVY^N) (Shukla and Ward, 1988, 1989b; Van der Vlugt et al., 1993).

The 3'-Nontranslated Region (3'-NTR)

As with the CP, the size of the 3'-nontranslated region is found to be a reliable indicator of the existence of potyvirus species (Frenkel et al., 1989; Atreya, 1992). Indeed, the 3'-NTR region follows a similar bimodal distribution between viruses and strains and is useful in potyvirus taxonomy (Frenkel et al., 1989; Van der Vlugt et al., 1993). All PVY isolates display a 3'-NTR close to 335 nucleotides in length (Van der Vlugt et al., 1993), whereas distinct potyvirus species reveal a difference in length (189 to 337 nucleotides, for TEV and PVY, for instance) (Frenkel et al., 1989). In addition to size differences, the 3'-NTR of distinct potyviruses presents a lower degree of homology (39 to 53 percent), whereas homology ranges from 83 to 99 percent within the same species (Frenkel et al., 1989). In potato PVY strains, a dis-

FIGURE 11.6. Phylogenetic Tree Derived from the *Potato virus Y* (PVY) Coat Protein Gene Sequences

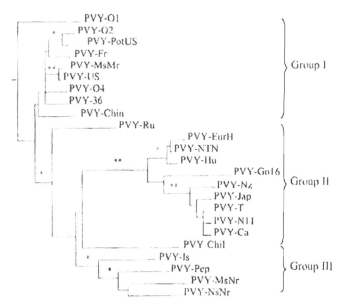

Source: Reproduced from Blanco-Urgoiti et al., 1996: 2432.

Note: Group I = PVY^O isolates; Group II = potato PVY^N isolates; Group III = nonpotato isolates; ' = statistically significant branch (>95 percent); '' = highly statistically significant branch (>99 percent). See text for details.

tinct clustering of PVY^N (Group I) and PVY^O isolates (Group II) is found (Van der Vlugt et al., 1993; Chachulska, 1998) (see Figure 11.7). However, some exceptions are observed. Indeed, three PVY^N isolates (PVY^N-Wi, PVY-H, PVY^N-Ru) are clustered into Group II with PVY^O isolates. In the pepper PVY strain, for which the 3'-NTR sequence homology is 99 percent (Fakhfakh et al., 1995), a greater nucleotide identity occurred with PVY^O isolates in Group II (about 95 percent homology) than with PVY^N isolates in Group I (about 80 percent homology) (Chachulska, 1998). Secondary RNA structure analysis of the 3'-NTR sequence shows that all PVY potato isolates display four major stem-loop RNA structures at the same location. Nucleotide sequence differences in the RNA loops allow PVY^O and PVY^N isolates to be placed into two different clusters. This clustering corresponds with that based on CP and 3'-NTR sequences (Van der Vlugt et al., 1993). Isolates of PVY-P2 and PVY-P21 of the PVY pepper strain display five and three stem loops, respectively, in the secondary structure of the 3'-NTR se-

FIGURE 11.7. Phylogenetic Tree Based on the 3'-Nontranslated Region of *Potato virus Y* (PVY)

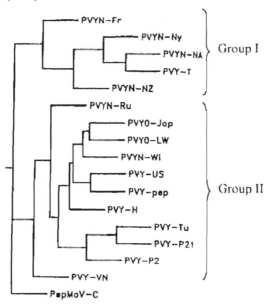

Source: Modified from Chachulska, 1998.

Note: Group I = potato PVY[N] isolates; Group II = potato (PVY[O]) pepper, and tobacco isolates. Trees were rooted by including strain C of *Pepper mottle virus* (PepMoV-C) as an outgroup.

quence. One stem loop, present in both pepper isolates, is characteristic of the PVY[O] group (Van der Vlugt et al., 1993; Fakhfakh et al., 1995).

All these findings show a rather high level of correlation between biological and molecular classification of PVY potato isolates, though some exceptions have yet to be explained. Further studies at the genomic level are also needed to precisely outline and define the nonpotato group and, last, to better understand the relationship between both of these groups.

MECHANISMS LEADING TO RNA GENOME POLYMORPHISMS

Mutations

Unlike DNA-dependent polymerases, RNA-dependent polymerases (reverse transcriptases and RNA-dependent RNA polymerases) promote a high mutation rate due to their low fidelity and lack of 3'-5'-exonuclease activity (Ramirez et al., 1995). For RNA viruses, the error rate was found to be in the order of 10^{-4} to 10^{-5} (one error per 10 kb to 100 kb) (Domingo et al., 1995). In consequence, a replicating virus does not correspond to one genome, but to a population of variable sequences, leading to the concept of quasi-species (Smith et al., 1997). It should be noted that the term *quasi-species* applies to a population of different chemical species, submitted to biological selection, not to the existence of different biological species (Domingo et al., 1995). The consequence of this population structure of viral genomes is that, although a virus can maintain population equilibrium for a long time when the selection pressure remains constant, rapid changes can occur following modifications of environmental selective conditions, such as host changes (Domingo and Holland, 1997).

Recombinations

Recombination events can be classified as homologous and nonhomologous or illegitimate.

Homologous recombination is the result of a crossover between two closely related RNA molecules. The crossover sites can occur between sites that match precisely on the two molecules, leading to a recombinant RNA with similar sequences and structural organization compared to the parental RNAs, or between homologous RNAs at noncorresponding sites, leading to sequence insertions or deletions in the recombinant progeny. This latter case is named *aberrant homologous recombination* (Simon and Bujarski, 1994). This type of recombination is thought to facilitate the exchange of genes or gene portions, contributing, on the one hand, to the stability of viral genomes by allowing the correction of errors caused by mutations between quasi-species (Chao, 1997) and, on the other hand, to the evolution of viral genomes by allowing the reassortment of sequences between related viruses.

Nonhomologous or *illegitimate recombination* occurs between two unrelated RNA molecules with some sequence complementarities at the crossover sites. This type of recombination is thought to be involved in the acquisition of new genes from the hosts or from other viruses (Mayo and Jolly, 1991; Meyers et al., 1991).

Recombination occurs by a copy-choice process involving template switching by RNA-dependent RNA polymerases. Specific RNA sequences and structures at or close to the recombination junctions were found to be necessary (Nagy and Bujarski, 1997).

RNA recombination events have been particularly observed in experimental situations in which evidence of recombination is generally obtained from the comparison of nucleotide sequences for many plant viruses and subviral molecules, such as satellite RNAs and viroids (Lai, 1992; Miller and Koev, 1998; Aaziz and Tepfer, 1999). In the case of sequence analysis of distinct genomic regions for isolates of the same species, different clusterings may appear for the regions under study. Practically, the comparison of the phenetic tree from each different genome region can reveal that one isolate switches from one branch to another. Trees can be drawn from sequences or from restriction fragment length polymorphisms (Blanco-Urgoiti et al., 1996).

Plum pox virus was the first potyvirus for which a recombination event between two strains was demonstrated (Cervera et al., 1993). In an extensive study of the sequences of 109 different potyvirus species, it was found that as many as 17 percent of them had undergone a recombination event (Revers et al., 1996). Recently, evidence of simple and multiple crossover recombination events were reported in a natural population of another potyvirus, *Yam mosaic virus* (Bousalem et al., 2000).

Multiple alignment of CP and 5'-NTR sequences suggests that PVY^{NTN} isolates form a homogeneous group related to PVY^N isolates (Revers et al., 1996; Blanco-Urgoiti et al., 1996; Chachulska et al., 1997; Blanco-Urgoiti, Tribodet, et al., 1998) (see Figures 11.5 and 11.6). Clustering analysis of the CP gene and of the 3'-NTR of a large number of PVY isolates shows that PVY^{NTN} isolates switch from one cluster to another, depending on the considered region (Revers et al., 1996). In the CP N-terminal (ter) and core regions, PVY^{NTN} isolates cluster within the PVY^N group, in accordance with their pathological and serological features. In contrast, they cluster with PVY^O isolates when the CP C-ter gene region and the 3'-NTR region are considered. RFLP (restriction fragment length polymorphism) analyses further show that two other recombination events occur in PVY^{NTN} genomes: (1) between the helper component (HC-Pro) and the P3 genes and (2) between the 6K2 and the NIa genes. From the 5'-NTR region up to the HC-Pro gene, the PVY^{NTN} genome is closely related to the PVY^N sequence; from the P3 up to the 6K2 genes, it is related to the PVY^O sequence; from NIa gene to the middle of the CP gene, it is related to the PVY^N sequence; and up to the 3'-extremity, it is related to PVY^O sequence (Glais, unpublished data).

The analysis of the CP sequence of PVY^N-W isolates reveals a high degree of homogeneity within this pathotype (97.8 percent homology) (McDonald et al., 1997) and suggests that these isolates are more closely related to PVY^O (96 percent homology) than to PVY^N (92 percent homology), as in the 3'-NTR region (Chachulska et al., 1997) (see Figure 11.7). However, when looking at the 5'-NTR, PVY^N-W isolates from Poland (PVY^N-Wi) (Chrzanowska, 1991) and from Canada (I-136 and I-L56) (McDonald et al.,

1997) are distantly related. These isolates display an inverse relationship with PVY^N and PVY^O isolates. The relationship of the PVY^N isolate with the Canadian isolates I-136 and I-L56 is closer than that observed between PVY^O isolate and PVY^N-W isolate from Poland (Marie-Jeanne-Tordo et al., 1995). A conserved sequence motif (UUUCA) located at position 124 to 128 of the 5'-NTR, correlated with belonging to the PVY^{NTN} subgroup (Marie-Jeanne, 1993), is also present in both Canadian PVY^N-W isolates (McDonald et al., 1997). A wide genetic variation within PVY^N-W isolates is also observed in the 5'-NTR.

Extensive RFLP analysis, using ten restriction enzymes, of the 5'-NTR from other PVY^N-W isolates confirms the division of this group into two subgroups (Glais et al., 1999). Moreover, other RFLP studies (Glais et al., 1998) have shown that in the first 4,063 nucleotides, PVY^N-W isolate genomes do not belong to the PVY^O, whereas in the last 5,670 nucleotides, they are related to PVY^O, suggesting the presence of a recombination phenomenon (see Figure 11.8).

PVY^N-W isolates thus appear to constitute a heterogeneous group. Their genomes may result from one or two recombination events. In all examined PVY^N-W isolates, a potential recombination breakpoint lies between the HC-Pro and the P3 genes, at the same location as in PVY^{NTN} isolates. Another potential crossing-over site lies in the P1 coding sequence in half of the isolates that were analyzed. Therefore, in these genomes, the region from the 5'-NTR region to the middle of the P1 gene can be related either to PVY^O or to PVY^N sequences; then from this point up to the HC-Pro gene, the sequence appears related to the PVY^N sequence, and then the remainder of the genome is related to the PVY^O sequence (Glais, unpublished data).

In the CP gene, a recombination event can be discerned in a particular isolate belonging to potato PVY^N strain group, the PVY^N-Fr isolate. In this region, the CP N-ter and C-ter regions are related to PVY^O, whereas the 3'-NTR displays a higher relationship with PVY^N sequences (Revers et al., 1996).

The genome of PVY^Z has not yet been thoroughly analyzed; however, an initial RFLP analysis of the PVY^Z CP gene shows that these isolates are clustered with PVY^O isolates (Blanco-Urgoiti, Tribodet, et al., 1998). RFLP analysis of the 5'-NTR of two PVY^Z isolates shows that they are related to PVY^{NTN} (Blanco-Urgoiti, Tribodet, et al., 1998). Complementary studies with PVY^Z isolates using a larger set of restriction enzymes are in progress (Kerlan, unpublished data).

Homologous recombination also has been discovered in the pepper strain of PVY (Fakhfakh et al., 1995). Phenetic analysis based on the sequence alignment of different PVY isolates shows that PVY-P21 clusters with

FIGURE 11.8. Phenetic Tree Based on the Restriction Fragment Length Polymorphism Analyses of the First 4,063 Nucleotides (a) and the Last 5,670 Nucleotides (b) of the *Potato virus Y* (PVY) Genome

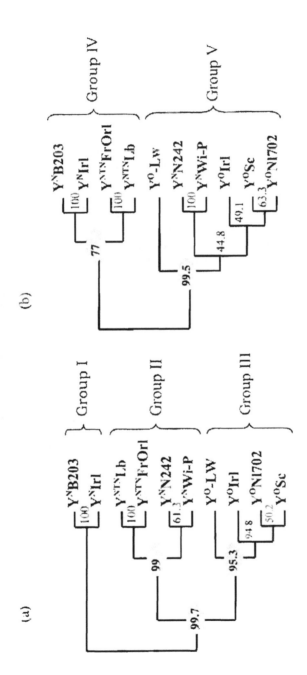

Source: Reproduced from Glais et al., 1998: 2086-2087.

Note: Group I = PVY^N isolates; Group II = PVY^{NTN} and PVY^N-W isolates; Group III = PVY^O isolates; Group IV = PVY^N and PVY^{NTN} isolates; Group V = PVY^N-W and PVY^O isolates. The values at the forks indicate the number of times out of 100 trees that this grouping occurred after bootstrapping the data.

potato and tobacco isolates and not with pepper isolates such as in 5'-NTR, P1 (Marie-Jeanne-Tordo et al., 1995), and 3'-NTR (Fakhfakh et al., 1995) regions.

CONCLUSIONS

Natural PVY hosts include economically important crops, which explains why PVY infections attracted the attention of pathologists as soon as they were recognized as crop-damaging diseases. From this continuous control of different plant hosts, it can be reasonably deduced that new PVY isolates were identified quite early after their emergence, allowing PVY to be a particularly interesting virus to watch for evolutionary developments. PVY evolution was first noticed in the 1950s because new isolates displayed particular biological features and, more recently, thanks to genetic studies that have revealed the generation of new genomes by recombination events. For example, recombination between PVY^O and PVY^N isolates is apparently under way in potato, leading to the emergence of PVY^{NTN} and PVY^N-W isolates. Historically, PVY^{NTN} isolates were considered to have appeared first; however, genetic studies showing the presence of three recombination events in PVY^{NTN} isolates and only one (or two) in PVY^N-W isolates suggest an opposite order of appearance.

Stating that a good correlation exists between genotype and phenotype for the potato PVY^N and PVY^O strain groups, genome analysis now suggests the existence of a third, and possibly fourth, group of strains (Blanco-Urgoiti, Sanchez, et al., 1998). Further studies on a larger range of PVY isolates from these groups are necessary to investigate their relationships with PVY^N and PVY^O and to evaluate their homogeneity.

Although PVY strains generally have a large host range, one interesting observation is that several isolates from pepper cannot infect potato. This may indicate that two different PVY lineages are evolving independently in these hosts.

Concerning the origin of PVY^N isolates, several hypotheses can be expressed: (1) PVY^O has long been known to infect potato and is the most common PVY strain group in this plant, whereas PVY^N may be a more recent potato virus, suggesting that potato PVY^O may also progressively invade other plants and give rise to a new PVY isolate called PVY^N. (2) On the other hand, PVY^N was firstly discovered in a potato plant in 1941 and was found similar to a tobacco virus (TVNV) isolated in 1935, suggesting that PVY^N may spread from tobacco to potato. (3) Another possibility is that PVY^N isolates might have infected potato long before being discovered because they only induced mild symptoms in this host plant. The problem is discovering why this virus waited until 1935 to infect tobacco. (4) Otherwise, we can speculate that PVY^O and PVY^N isolates diverged from a common viral ancestor that has followed two different evolutionary paths.

Concerning the geographical origin of this virus, most, if not all, of its natural hosts originate from South and Central America, suggesting that PVY is also native to these regions. The importation of major solanaceous crops to Europe and other continents has possibly played an important role in its recent evolution. Clearly, combined genetic and epidemiological studies of many more extant viruses are needed. In addition, it would certainly be useful to have more knowledge on potyviruses related to PVY infecting pepper, tobacco, and tomato as well as wild and cultivated *Solanum* species in South America and possibly in East Africa. In these regions, PVY infections were reported in pepper that has been traditionally cultivated and for which natural barriers impede long-distance migration of aphid vectors. Inspections in these areas, together with the use of potato tuber collections or herbariums, from the beginning of century, might give some indication about the existence of PVY.

There is increasing evidence of the modification of virus epidemiology due to changes in the environment from either natural or human-linked causes (Murphy, 1999). In the case of plant viruses, these can be the proximity of previously separated crop species; the increased intercontinental exchange of plants, agricultural products, and viral vectors; and the modification of vector populations by increased use of pesticides and climatic changes. Also, the effect of elevated concentrations of carbon dioxide in the atmosphere may modify the response of plants to virus infection (Malmström and Field, 1997). All of these environmental causes modify selection pressures and can lead to the rapid evolution of virus genomes, especially in the case of RNA viruses (Holland and Domingo, 1998).

Recent changes in PVY virus populations may be representative of these evolutionary tendencies. Increased monitoring of plant crops as well as an increased knowledge of the forces driving virus evolution would certainly be profitable for sustainable agriculture.

REFERENCES

Aaziz R, Tepfer M (1999). Recombination in RNA viruses and in virus-resistant transgenic plants. *Journal of General Virology* 80: 1339-1346.
Atreya CD (1992). Application of genome sequence information in potyvirus taxonomy: An overview. *Archives of Virology Supplementum* 5: 17-23.

Atreya CD, Atreya PL, Thornbury DW, Pirone TP (1992). Site-directed mutations in the potyvirus HC-Pro gene affect helper component activity, virus accumulation, and symptom expression in infected tobacco plants. *Virology* 191: 106-111.

Badge J, Robinson DJ, Brunt AA, Foster GD (1997). 3'-Terminal sequences of the RNA genomes of narcissus latent and maclura mosaic viruses suggest that they represent a new genus of the Potyviridae. *Journal of General Virology* 78: 253-257.

Barnett OW, Adam G, Brunt AA, Dijkstra J, Dougherty WG, Edwardson JR, Goldbach R, Hammond J, Hill JH, Jordan RL, et al. (1995). Family *Potyviridae*. In Murphy FA, Fauquet CM, Bishop DHL, Ghabrial SA, Jarvis AW, Martelli GP, Mayo MA, Summers MD (eds.), *Virus Taxonomy*. Springer-Verlag, Wien, New York, pp. 348-351.

Bawden FC (1936). The viruses causing top necrosis (acronecrosis) of the potato. *Annals of Applied Biology* 23: 487-497.

Bawden FC (1943). Some properties of the potato viruses. *Annals of Applied Biology* 30: 82-83.

Bawden FC, Kassanis B (1947). The behaviour of some naturally occurring strains of potato virus Y. *Annals of Applied Biology* 34: 503-516.

Beczner L, Horvath J, Romhanyi I, Forster H (1984). Studies on the etiology of tuber necrotic ringspot disease in potato. *Potato Research* 27: 339-352.

Berger PH, Wyatt SD, Shiel PJ, Silbernagel MJ, Druffel K, Mink GI (1997). Phylogenetic analysis of the Potyviridae with emphasis on legume-infecting potyviruses. *Archives of Virology* 142: 1979-1999.

Blancard D (1998). *Maladies du tabac. Observer, identifier, lutter.* INRA-Paris, 375 pp.

Blancard D, Ano G, Cailleteau B (1994). Principaux virus affectant le tabac en France. *Annales du Tabac, Seita Section* 2: 39-50.

Blanco-Urgoiti B, Sanchez F, Dopazo J, Ponz F (1996). A strain-type clustering of potato virus Y based on the genetic distance between isolates calculated by RFLP analysis of the amplified coat protein gene. *Archives of Virology* 141: 2425-2442.

Blanco-Urgoiti B, Sanchez F, Perez de San Roman C, Dopazo J, Ponz F (1998). Potato virus Y group C isolates are a homogeneous pathotype but two different genetic strains. *Journal of General Virology* 79: 2037-2042.

Blanco-Urgoiti B, Tribodet M, Leclere S, Ponz F, Perez de San Roman C, Legorburu FJ, Kerlan C (1998). Characterization of potato Potyvirus Y (PVY) isolates from seed potato batches. Situation of the NTN, Wilga and Z isolates. *European Journal of Plant Pathology* 104: 811-819.

Boonham N, Hims M, Barker I, Spence N (1999). Potato virus Y from petunia can cause symptoms of potato tuber necrotic ringspot disease (PTNRD). *European Journal of Plant Pathology* 105: 617-621.

Bousalem M, Douzery EJP, Fargette D (2000). High genetic diversity, distant phylogenetic relationships and intraspecies recombination events among natural populations of *Yam mosaic virus*: A contribution to understanding potyvirus evolution. *Journal of General Virology* 81: 243-255.

Brigneti G, Voinnet O, Li WX, Ji LH, Ding SW. Baulcombe DC (1998). Viral pathogenicity determinants are suppressors of transgene silencing in *Nicotiana benthamiana*. *EMBO Journal* 17: 6739-6746.

Brunt AA (1992). The general properties of Potyviruses. *Archives of Virology Supplementum* 5: 3-16.

Calvert EL, Cooper P, Mc Clure J (1980). An aphid transmitted strain of PVY[C] recorded in potatoes in Northern Ireland. *Record of Agricultural Research* 28: 63-74.

Cervera MT, Riechmann JL, Martin MT, Garcia JA (1993). 3'-Terminal sequence of the Plum pox virus PS and Ö6 isolates: Evidence for RNA recombination within the potyvirus group. *Journal of General Virology* 74: 329-334.

Chachulska AM (1998). Potato virus Y phylogeny and engineered virus resistance. PhD thesis, University of Warsaw, Poland, 106 pp.

Chachulska AM, Chrzanowska M, Robaglia C, Zagorski W (1997). Tobacco veinal necrosis determinants are unlikely to be located within the 5' and 3' terminal sequences of the potato virus Y genome. *Archives of Virology* 142: 765-779.

Chao L (1997). Evolution of sex and the molecular clock in RNA viruses. *Gene* 205: 301-308.

Chrzanowska M (1991). New isolates of the necrotic strain of Potato virus Y (PVY[N]) found recently in Poland. *Potato Research* 34: 179-182.

Chrzanowska M (1994). Differentiation of potato virus Y (PVY) isolates. *Phytopathologica Polonica* 8: 15-20.

Cockerham G (1943). The reaction of potato varieties to viruses X, A, B and C. *Annals of Applied Biology* 30: 338-344.

Crane PR, Friis EM, Pedersen KR (1995). The origin and early diversification of angiosperms. *Nature* 374: 27-33.

D'Aquino L, Dalmay T, Burgyan J (1995). Host range and sequence analysis of an isolate of potato virus Y inducing veinal necrosis in pepper. *Plant Disease* 79: 1046-1050.

De Bokx JA (1961). Waardplanten van het aardappel-Y[N]-virus. *Tijdschrift over Plantenziekten* 67: 273-277.

De Bokx JA, Huttinga H (1981). Potato virus Y. *CMI/AAB Descriptions of plant viruses* No. 242. CMI/AAB, Slough, England. 6pp.

De Bokx JA, Kratchanova B, Maat ZD (1975). Some properties of a deviating strain of potato virus Y. *Potato Research* 18: 38-51.

De Bokx JA, Piron PGM (1978). Transmission of potato virus Y[C] by aphids. In *Abstracts of Conference Papers of the 7th Triennial Conference of the European Association for Potato Research*, Ziemnaka Institute, Warsaw, Poland, pp. 244-245.

Delgado-Sanchez S, Grogan RG (1970). Potato virus Y. *CMI/AAB Descriptions of Plant Viruses* No. 37. CMI/AAB, Kew, Surrey, England, 4 pp.

Dolja VV, Boyko VP, Agranovsky AA, Koonin EV (1991). Phylogeny of capsid proteins of rod-shaped and filamentous RNA plant viruses: Two families with distinct patterns of sequence and probably structure conservation. *Virology* 184: 79-86.

Domier LL, Franklin KM, Hunt AG, Rhoads RE, Shaw JG (1989). Infectious in vitro transcripts from cloned cDNA of a Potyvirus, tobacco vein mottling virus. *Proceedings of the National Academy of Sciences, USA* 86: 3509-3513.

Domingo E, Holland JJ (1997). RNA virus mutations and fitness for survival. *Annual Review of Microbiology* 51: 151-178.

Domingo E, Holland JJ, Briebricher C, Eigen M (1995). Quasi-species: The concept and the word. In Gibbs AJ, Calisher CH, Garcia-Arenal F (eds.), *Molecular Basis of Virus Evolution*, Cambridge University Press, Cambridge, pp. 181-191.

Dougherty WG, Carrington JC (1988). Expression and function of potyviral gene products. *Annual Review of Phytopathology* 26: 123-143.

Dykstra TP (1936). Comparative studies of some European and American potato viruses. *Phytopathology* 26: 597-606.

Edwardson JR (1974). Some properties of the potato virus Y group. *Florida Agricultural Experiment Stations Monograph Series* 4: 398.

Fakhfakh H, Makni M, Robaglia C, Elgaaied A, Marrakchi M (1995). Polymorphisme des régions capside et 3' NTR de 3 isolats tunisiens du virus Y de la pomme de terre (PVY). *Agronomie* 15: 569-579.

Frenkel MJ, Ward CW, Shukla DD (1989). The use of 3' non-coding nucleotide sequences in the taxonomy of potyviruses: Application to watermelon mosaic virus 2 and soybean mosaic virus-N. *Journal of General Virology* 70: 2775-2783.

Fribourg CE, Nakashima J (1984). Characterization of a new potyvirus from potato. *Phytopathology* 74: 1363-1369.

Gebre Selassie K, Marchoux G, Delecolle B, Pochard E (1985). Variabilité naturelle des souches du virus Y de la pomme de terre dans les cultures de piment du sud-est de la France. Caractérisation et classification en pathotypes. *Agronomie* 5: 621-630.

Glais L, Kerlan C, Tribodet M, Marie-Jeanne-Tordo V, Robaglia C, Astier-Manifacier S (1996). Molecular characterization of potato virus Y^N isolates by PCR-RFLP. *European Journal of Plant Pathology* 102: 655-662.

Glais L, Tribodet M, Gauthier JP, Astier-Manifacier S, Robaglia C, Kerlan C (1998). RFLP mapping of the whole genome of ten viral isolates representative of different biological groups of potato virus Y. *Archives of Virology* 143: 2077-2091.

Glais L, Tribodet M, Kerlan C (1999). Le phénomène de recombinaison observé chez le virus Y de la pomme de terre (PVY) est—il impliqué dans l'apparition des nécroses sur tabac et sur les tubercules de pomme de terre? In *Rencontres de Virologie Végétales*, Aussois, France, p. 81.

Goldbach R (1987). Genome similarities between plant and animal RNA viruses. *Microbiological Sciences* 4: 197-202.

Gooding GV, Jr. (1985). Relationship between strains of potato virus Y and breeding for resistance, cross-protection and interference. *Tobacco Science* 29: 99-104.

Gooding GV, Jr., Tolin SA (1973). Strains of potato virus Y affecting flue-cured tobacco in the southeastern United States. *Plant Disease Report* 57: 200-204.

Gorbalenya AE (1995). Origin of RNA viral genomes: Approaching the problem by comparative sequence analysis. In Gibbs AJ, Calisher CH, Garcia-Arenal F (eds.). *Annual Review of Phytopathology.* Cambridge University Press. Cambridge, pp. 49-66.

Gugerli P, Fries P (1983). Characterization of monoclonal antibodies to potato virus Y and their use for virus detection. *Journal of General Virology* 64: 2471-2477.

Hall JS, Adams B, Parsons TJ, French R, Lane LC, Jensen SG (1998). Molecular cloning, sequencing, and phylogenetic relationships of a new Potyvirus, sugarcane streak mosaic virus, and a reevaluation of the classification of the Potyviridae. *Molecular Phylogenetics and Evolution* 10: 323-332.

Hataya T, Inoue AK, Ohshima K, Shikata E (1994). Characterization and strain identification of a potato virus Y isolate non-reactive with monoclonal antibodies specific to the ordinary and necrotic strains. *Intervirology* 37: 12-19.

Holland J, Domingo E (1998). Origin and evolution of viruses. *Virus Genes* 16: 13-21.

Hollings M, Brunt AA (1981). Potyvirus group. *CMI/AAB Descriptions of plant viruses* No. 245, CMI/AAB, Kew, Surrey, England, 8pp.

Horvath J (1966). Studies on strains of potato virus Y. 2. Normal strain. *Acta Phytopathologica Academiae Scientiarum Hungaricae* 1: 334-352.

Horvath J (1967a). Studies on strains of potato virus Y. 3. Strain causing browning of midribs in tobacco. *Acta Phytopathologica Academiae Scientiarum Hungaricae* 2: 95-108.

Horvath J (1967b). Studies on strains of Potato virus Y. 4. Anomalous strain. *Acta Phytopathologica Academiae Scientiarum Hungaricae* 2: 195-210.

Horvath J (1967c). Virulenz differenzen verschiedener Stämme und Isolate des Kartoffel-Y-virus an Capsicum-Arten und Varietäten. *Acta Phytopathologica Academiae Scientiarum Hungaricae* 2: 17-37.

Jones RAC (1990). Strain group specific and virus specific hypersensitive reactions to infection with potyviruses in potato cultivars. *Annals of Applied Biology* 117: 93-105.

Kasschau KD, Carrington JC (1998). A counterdefensive strategy of plant viruses: Suppression of posttranscriptional gene silencing. *Cell* 95: 461-470.

Kerlan C, Tribodet M (1996). Are all PVY[N] isolates able to induce potato tuber necrosis ringspot disease? In *Proceedings of the 13th Triennial Conference of the European Association of Potato Research.* European Association of Potato Research. Veldhoven, The Netherlands, pp. 65-66.

Kerlan C, Tribodet M, Glais L, Guillet M (1999). Variability of potato virus Y in potato crops in France. *Journal of Phytopathology* 147: 643-651.

Klinkowski M, Schmelzer K (1960). A necrotic type of potato virus Y. *American Potato Journal* 37: 221-227.

Koonin EV, Dolja VV (1993). Evolution and taxonomy of positive-strand RNA viruses: Implications of comparative analysis of amino acid sequences. *Critical Reviews in Biochemistry and Molecular Biology* 28: 375-430.

Lai MM (1992). RNA recombination in animal and plant viruses. *Microbiological Reviews* 56: 61-79.

Latore BA, Flores V (1985). Strain identification and cross-protection of potato virus Y affecting tobacco in Chile. *Plant Disease* 69: 930-932.

Lee KC, Mahtani PH, Chng CG, Wong SM (1997). Sequence and phylogenetic analysis of the cytoplasmic inclusion protein gene of zucchini yellow mosaic Potyvirus: Its role in classification of the Potyviridae. *Virus Genes* 14: 41-53.

Legnani R (1995). Evaluation and inheritance of the *Lycopersicon hirsutum* resistance against Potato virus Y. *Euphytica* 86: 219-226.

Le Romancer M, Kerlan C, Nedellec M (1994). Biological characterization of various geographical isolates of potato virus Y inducing superficial necrosis on potato tubers. *Plant Pathology* 43: 138-144.

Malmström CM, Field CB (1997). Virus-induced differences in the response of oat plants to elevated carbon dioxide. *Plant Cell and Environment* 20: 178-188.

Marchoux G, Palloix P, Gebre Selassie K, Caranta C, Legnani R, Dogimont C (1995). Variabilité du virus Y de la pomme de terre et des Potyvirus voisins. Diversité des sources de résistance chez le piment (*Capsicum* sp.). *Annales du Tabac, Seita Section* 2: 25-34.

Marie-Jeanne V (1993). Région 5'du génome du virus Y de la pomme de terre: Etude par séquençage de son polymorphisme et des motifs conservés. Contribution à l'étude de la protéine P1. PhD thesis, University of Paris XI-Orsay, p. 118.

Marie-Jeanne-Tordo V, Chachulska AM, Fakhfakh H, Le Romancer M, Robaglia C, Astier-Manifacier S (1995). Sequence polymorphism in the 5' NTR and in the P1 coding region of potato virus Y genomic RNA. *Journal of General Virology* 76: 939-949.

Marte M, Belleza G, Polverari A (1991). Infective behaviour and aphid-transmissibility of Italian isolates of potato virus Y in tobacco and peppers. *Annals of Applied Biology* 118: 309-317.

Mayo MA, Jolly CA (1991). The 5'-terminal sequence of potato leafroll virus RNA: Evidence for recombination between virus and host RNA. *Journal of General Virology* 72: 2591-2595.

McDonald JG, Kristjansson GT (1993). Properties of strains of *Potato virus* Y^N in North America. *Plant Disease* 77: 87-89.

McDonald JG, Singh RP (1996). Host range, symptomology, and serology of isolates of *Potato virus* Y (PVY) that share properties with both the PVY^N and PVY^O strain groups. *American Potato Journal* 73: 309-315.

McDonald JG, Wong E, Henning D, Tao T (1997). Coat protein and 5' non-translated region of a variant of *Potato virus* Y. *Canadian Journal of Plant Pathology* 19: 138-144.

Meyers G, Tautz N, Duhovi EJ, Thiel HJ (1991). Viral cytopathogenicity correlated with integration of ubiquitin-coding sequences. *Virology* 180: 602-616.

Miller WA, Koev G (1998). Getting a handle on RNA virus recombination. *Trends in Microbiology* 6: 421-423.

Murphy FA (1999). The evolution of viruses, the emergence of viral diseases: A synthesis that Martinus Beijerinck might enjoy. *Archives of Virology Supplementum* 15: 73-85.

Nagy PD, Bujarski JJ (1997). Engineering of homologous recombination hotspots with AU-rich sequences in brome mosaic virus. *Journal of Virology* 71: 3799-3810.

Nitzany FE, Tanne E (1962). Virus diseases of peppers in Israël. *Phytopathologia Mediterranea* 4: 180-182.

Nobrega NR, Silberschmidt K (1944). Sobre una provavel variante do virus "Y" da batatinha (Solanum virus 2, Orton) que tem a peculiaridade de provocar necroses em plantas de fumo. *Arquivos do Instituto Biológico* 15: 307-330.

Pepelnjak M (1993). Potato virus Y^{NTN} on tomato. In *Proceedings of the 12th Triennial Conference of the European Association of Potato Research*. European Association of Potato Research, Paris, France, p. 350.

Pringle CR (1999). Virus taxonomy. The Universal System of Virus Taxonomy, updated to include the new proposals ratified by the International Committee on Taxonomy of Viruses during 1998 [news]. *Archives of Virology* 144: 421-429.

Ramirez BC, Barbier P, Seron K, Haenni AL, Bernardi F (1995). Molecular mechanisms of point mutations in RNA viruses. In Gibbs AJ, Calisher CH, Garcia-Arenal F (eds.), *Molecular Basis of Virus Evolution*, Cambridge University Press, Cambridge, pp. 105-118.

Reddick BB, Miller RD (1991). Identification of virus population from TN 86 vs. non-TN 86 Burley tobacco field in East Tennessee. *Tobacco Abstracts* 35: 650.

Revers F, Le Gall O, Candresse T, Le Romancer M, Dunez J (1996). Frequent occurrence of recombinant potyvirus isolates. *Journal of General Virology* 77: 1953-1965.

Revers F, Le Gall O, Candresse T, Maule AJ (1999). New advances in understanding the molecular biology of plant/potyvirus interactions. *Molecular Plant-Microbe Interactions* 12: 367-376.

Riechmann JL, Lain S, Garcia JA (1992). Highlights and prospects of potyvirus molecular biology. *Journal of General Virology* 73: 1-16.

Robaglia C, Durand-Tardif M, Tronchet M, Boudazin G, Astier-Manifacier S, Casse-Delbart F (1989). Nucleotide sequence of potato virus Y (N Strain) genomic RNA. *Journal of General Virology* 70: 935-947.

Salaman RN (1930). Virus diseases of potato: Streak. *Nature* 126: 241.

Salm SN, Rey ME, Rybicki EP (1996). Phylogenetic justification for splitting the Rymovirus genus of the taxonomic family Potyviridae. *Archives of Virology* 141: 2237-2242.

Shukla DD, Frenkel MJ, Ward CW (1991). Structure and function of the potyvirus genome with special reference to the coat protein coding region. *Canadian Journal of Plant Pathology* 13: 178-191.

Shukla DD, Strike PM, Tracy SL, Gough KH, Ward CW (1988). The N and C termini of the coat proteins of potyviruses are surface-located and the N-terminus contains the major virus-specific epitopes. *Journal of General Virology* 69: 1497-1508.

Shukla DD, Ward CW (1988). Amino acid sequence homology of coat proteins as a basis for identification and classification of the potyvirus group. *Journal of General Virology* 69: 2703-2710.

Shukla DD, Ward CW (1989a). Identification and classification of potyviruses on the basis of coat protein sequence data and serology: Brief review. *Archives of Virology* 106: 171-200.

Shukla DD, Ward CW (1989b). Structure of potyvirus coat proteins and its application in the taxonomy of the potyvirus group. *Advances in Virus Research* 36: 273-314.

Silberschmidt K (1956). Una doenca do tomateiro em piedade causada pelo virus Y da batatinha. *Arquivos do Instituto Biológico* 23: 125-150.

Silberschmidt K (1957). Cross-protection ("premunity") tests with two strains of potato virus Y in tomatoes. *Turrialba* 7: 34-43.

Silberschmidt KM (1960). Types of potato virus Y necrotic to tobacco: History and recent observation. *American Potato Journal* 37: 151-159.

Simon AE, Bujarski JJ (1994). RNA-RNA recombination and evolution in virus-infected plants. *Annual Review of Phytopathology* 35: 337-362.

Simons JN (1959). Potato virus Y appears in additional areas of peppers and tomato production in South Florida. *Plant Disease Reporter* 43: 710-711.

Singh M, Singh RP (1996). Nucleotide sequence and genome organization of a Canadian isolate of the common strain of potato virus Y (PVY^O). *Canadian Journal of Plant Pathology* 18: 209-224.

Smith DB, McAllister J, Casino C, Simmonds P (1997). Virus "quasispecies": Making a mountain out of a molehill? *Journal of General Virology* 78: 1511-1519.

Smith KM (1931). On the composite nature of certain potato virus diseases of the mosaic group as revealed by the use of plant indicators. *Proceedings of the Royal Society, London B* 109: 251-267.

Smith KM, Dennis RWG (1940). Some notes on a suspected variant of *Solanum* virus 2 (potato virus Y). *Annals of Applied Biology* 27: 65-70.

Soto MJ, Arteaga ML, Fereres A, Ponz F (1994). Limited degree of serological variability in pepper strains of potato virus Y as revealed by analysis with monoclonal antibodies. *Annals of Applied Biology* 124: 37-43.

Stobbs LW, Poysa V, Van Schagen JG (1994). Susceptibility of cultivars of tomato and pepper to a necrotic strain of potato virus Y. *Canadian Journal of Plant Pathology* 16: 43-48.

Sturgess OW (1956). Leaf shrivelling virus diseases of the tomato. *Queensland Journal of Agricultural Science* 13: 175-220.

Sudarsono, Woloshuk SL, Xiong Z, Hellmann GM, Wernsman EA, Weissinger AK, Lommel SA (1993). Nucleotide sequence of the capsid protein cistrons from six potato virus Y (PVY) isolates infecting tobacco. *Archives of Virology* 132: 161-170.

Thole V, Dalmay T, Burgyan J, Balazs E (1993). Cloning and sequencing of potato virus Y (Hungarian isolate) genomic RNA. *Gene* 123: 149-156.

Thompson GJ, Hoffman DCA, Prins PJ (1987). A deviant strain of potato virus Y infecting potatoes in South Africa. *Potato Research* 30: 219-228.

Tomlinson JA (1987). Epidemiology and control of virus diseases of vegetables. *Annals of Applied Biology* 110: 661-681.

Van der Vlugt RA, Leunissen J, Goldbach R (1993). Taxonomic relationships between distinct potato virus Y isolates based on detailed comparisons of the viral coat proteins and 3' non-translated regions. *Archives of Virology* 131: 361-375.

Walter JM (1967). Hereditary resistance to disease in tomato. *Annual Review of Phytopathology* 5: 131-144.

Ward CW. McKern NM, Frenkel MJ, Shukla DD (1992). Sequence data as the major criterion for potyvirus classification. *Archives of Virology Supplementum* 5: 283-297.

Ward CW, Weiller GF, Shukla DD, Gibbs AJ (1995). Molecular systematics of the *Potyviridae*, the largest plant virus family. In Gibbs AJ, Calisher CH, Garcia-Arenal F (eds.), *Molecular Basis of Virus Evolution*, Cambridge University Press, Cambridge, pp. 477-500.

Watson MA (1956). The effect of different host plants of potato virus C in determining its transmission by aphids. *Annals of Applied Biology* 44: 599-607.

Weidemann HL (1988). Importance and control of potato virus Y^N (PVY^N) in seed potato production. *Potato Research* 31: 85-94.

Weidemann HL, Maiss E (1996). Detection of the potato tuber necrotic ringspot strain of potato virus Y (PVY^{NTN}) by reverse transcription and immunocapture polymerase chain reaction. *Journal of Plant Diseases and Protection* 103: 337-345.

SECTION IV:
MOLECULAR BIOLOGY
OF DNA VIRUSES

Chapter 12

Geminivirus Replication and Gene Expression

Zulma I. Monsalve-Fonnegra
Gerardo R. Argüello-Astorga
Rafael F. Rivera-Bustamante

INTRODUCTION

Higher plants can be infected by a large number of viruses, most of which have RNA genomes. The family *Geminiviridae* is one of the only two plant virus families that have DNA genomes and replicate through DNA intermediates. These pathogens are a large and diverse group of angiosperm viruses that are characterized by twin icosahedral capsids and circular single-stranded DNA (ssDNA) genomes. Each geminate particle encapsidates one ssDNA molecule (2.5 to 3.0 kb) that replicates in the nuclei of the infected cells via a double-stranded DNA (dsDNA) intermediate (Lazarowitz, 1992).

Geminiviruses are responsible for devastating diseases in a variety of cereal, fiber, and vegetable crops worldwide, including maize, wheat, sugarcane, tomato, pepper, tobacco, bean, cotton, squash, beet, and cassava. Besides their importance as plant pathogens, geminiviruses are ideal model systems for the study of plant DNA replication and transcription because of their simple genome structure and their extensive reliance on the host molecular machinery to carry out these processes.

In this chapter, we describe some aspects of the molecular biology of geminiviruses, with particular emphasis on those related to the regulation of viral gene expression. The regulatory elements for both gene expression and virus replication are located in the same region, and several *cis*-acting elements and viral proteins participate in both processes. Therefore, these processes are inextricably linked in geminiviruses and cannot be properly understood if they are separately described. Accordingly, we also include discussion of the functional organization of the geminivirus replication origin.

GEMINIVIRUS CLASSIFICATION

The members of the family *Geminiviridae* are classified into three genera on the basis of their insect vector, host range, and genome organization (Van Regenmortel et al., 1997). Table 12.1 summarizes this classification. Members of the genus *Mastrevirus*, previously known as Subgroup I, have not been reported in America. The genus *Curtovirus* (Subgroup II), the smallest of the three genera, arose from a recombination event between a mastrevirus and a begomovirus (compare the genomic organization of the three genera in Figure 12.1, and see Chapter 15 of this book). The third genus, *Begomovirus* (Subgroup III), is the most diverse and widely distributed *Geminiviridae* lineage. Phylogenetic analyses of the begomoviruses have demonstrated distinct geographic lineages, and two major clusters have emerged: the New World cluster for American viruses and the Old World cluster for viruses from Europe, Africa, and Asia. A third smaller, but growing cluster includes the American squash leaf curl virus (SgLCV) and Texas pepper viruses (SqLCV cluster) (Argüello-Astorga, Guevara-González, et al., 1994).

GENOME STRUCTURE

The genomic structures of the three genera are shown in Figure 12.1. The viral proteins required for the different processes of the infection cycle are

TABLE 12.1. Classification of the Family *Geminiviridae*

	Mastrevirus	*Curtovirus*	*Begomovirus*
Virus type species	MSV	BCTV	BGMV
Genomic organization	Monopartite	Monopartite	Mono- or bipartite
Host Range	Monocotyledonous	Dicotyledonous	Dicotyledonous
Insect vector	Leafhoppers	Leafhoppers, Treehoppers	Whiteflies
Other members	WDV, SSV	HCTV, TPCTV	ACMV, PHV, SqLCV, TGMV, TLCV, TYLCV

Note: MSV = *Maize streak virus;* BCTV = *Beet curly top virus;* BGMV = *Bean golden mosaic virus;* WDV = *Wheat dwarf virus;* SSV = *Sugarcane streak virus;* HCTV = *Horseradish curly top virus;* TPCTV = *Tomato pseudo-curly top virus;* ACMV = *African cassava mosaic virus;* PHV = *Pepper huasteco virus;* SqLCV = *Squash leaf curl virus;* TGMV = *Tomato golden mosaic virus;* TLCV = *Tomato leaf curl virus;* TYLCV = *Tomato yellow leaf curl virus.*

FIGURE 12.1. Genomic Structure of Geminiviruses

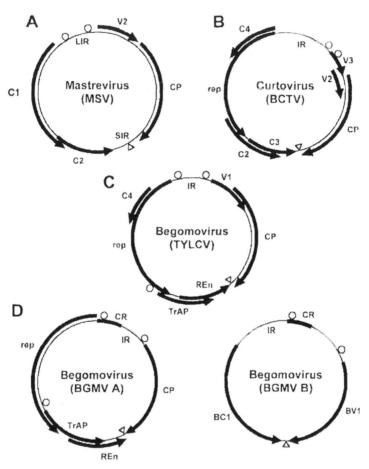

Note: The genomic structures of the three genera of the family *Geminiviridae* are shown. All viruses present circular genomes with the genes arranged into two divergent clusters and an intergenic region (IR, LIR) that contains the origin of replication and the promoters. Mastreviruses (A) also have an additional, smaller IR (SIR). Curtoviruses (B) present up to three genes in the virion-sense orientation. Begomoviruses can be either monoparite (C) or bipartite (D). In the case of bipartite begomoviruses, both components share a highly identical segment called a common region (CR, filled box). Open circles represent the location of identified TATA boxes, whereas triangles show the location of bidirectional polyadenylation signals. *BC1* and *C1-C4* = genes encoded in the complementary sense; *BV1* and *V1-V3* = virion-sense genes; BCTV = *Beet curly top virus;* BGMV = *Bean golden mosaic virus;* CP = coat protein; REn = replication enhancer protein; rep = replication initiation protein; TrAP = transactivator protein; MSV = *Maize streak virus;* TYLCV = *Tomato yellow leaf curl virus.*

presented in Table 12.2. Monopartite geminiviruses (i.e., mastreviruses, curtoviruses, and some Old World begomoviruses) contain four to six overlapping genes encoding all the viral proteins necessary for replication, transcription, virus movement, and encapsidation. These genes are arranged in two divergent clusters separated by a large intergenic region (LIR) that contains the promoters for both transcription units. The mastreviruses also present a second smaller intergenic region (SIR) located at the end of the transcription units (Figure 12.1). The genes on the right side (clockwise orientation) are called virion sense because they are transcribed from a DNA strand with the same polarity as the encapsidated viral DNA (also calledplus strand). Similarly, genes on the left side (counterclockwise orientation) are called complementary sense because they are transcribed from a DNA

TABLE 12.2. Geminivirus Proteins and Their Functions

Virus	Process	Proteins*
Mastrevirus	Replication	Rep (C1:2), RepA (C1')
	Transcription (activator)	Rep (C1:2)
	Encapsidation	CP (V1)
	Movement	CP, V2
	Host activation	RepA (C1')
Curtovirus	Replication	Rep (C1), C3
	Transcription activator	TrAP (C2)
	Encapsidation	CP (V2)
	Movement	CP (V2), V1
	Host activation	C4
Begomovirus	Replication	Rep (C1, AC1, AL1),
		REn (C3, AC3, AL3)
	Transcription activator	TrAP C2, AC2, AL2)
	Transcription repressor	Rep (C1, AC12, AL12)
	Encapsidation	CP (V2, AR1, AV1)
	Movement	CP (V2, AR1, AV1)
		BC1 (BL1), BV1 (BR1)
	Host activation	Rep (C1, AC1, AL1)

*A standard nomenclature for geminivirus genes and proteins was recently suggested (Van Regenmortel et al., 1997). However, it has not been completely adopted yet. In this chapter, we primarily use the suggested nomenclature, although we also include in parentheses alternative nomenclatures.

strand that is complementary to the encapsidated DNA (minus strand) and present only in the dsDNA intermediate. In general, genes encoded in the complementary sense are involved in virus replication and transcription (*C1* and *C2* for mastreviruses and *rep(C1)*, *TrAP(C2)*, and *REn(C3)* for curtoviruses and monopartite begomoviruses), whereas genes encoded in the virion sense have functions in encapsidation (*CP*) and in virus movement (*V2* and *V3*) (Lazarowitz, 1992; Hanley-Bowdoin et al., 1999).

Bipartite begomoviruses have a genome composed of two ssDNAs (designated as components A and B), both of which are required for successful infection. They also display two divergent sets of genes separated by an intergenic region (IR) that includes a segment of about 180 to 200 nucleotides (nt), called the common region (CR), which is the only region highly conserved between both components. All of the *cis* elements required for viral replication reside within the CR, which varies from virus to virus, with the exception of a highly conserved 30 nt element with the potential to form a stem-loop (hairpin) structure. A typical bipartite geminivirus is shown in Figure 12.1D. Component A contains four or five genes. The coat protein gene *(CP)* is the only gene found in the virion-sense orientation, whereas genes such as *rep (replication initiation protein)*, *TrAP (transactivator protein)*, and *REn (replication enhancer protein)* are found in the complementary sense. A fourth gene *(AC4)* has been described in some geminiviruses. However, its function is still not well understood. Component B, on the other hand, contains two genes, one in the complementary sense *(BC1)* and another in the virion sense *(BV1)*. Both genes encode proteins involved in the movement of viral DNA and are necessary for systemic movement and symptom production. The BC1 protein seems to be involved in cell-to-cell transport of the virus, probably by changing the plasmodesmata size exclusion limit. The BV1 protein binds both ss- and dsDNA and is able to move viral DNA out of the nucleus. The evidence suggests that BV1 is also involved in long-distance movement of the virus, allowing the spread of the virus through the vascular system (Lazarowitz, 1992; Hanley-Bowdoin et al., 1999).

GEMINIVIRUS INFECTION CYCLE: A BRIEF OVERVIEW

The main aspects of the geminivirus infection cycle are summarized in Figure 12.2. The first stage in the infection cycle involves the injection of viral ssDNA into a plant cell by its insect vector. After its uncoating, the virus ssDNA genome is transported into the nucleus by an unknown process that probably involves the CP. Once in the nucleus, the viral ssDNA is converted into dsDNA, which is the form that acts as a template for both trans-

FIGURE 12.2. Geminivirus Infection Cycle

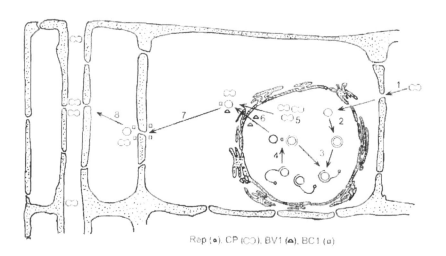

Rep (●), CP (◯), BV1 (▲), BC1 (□)

Note: Geminivirus cycle starts when the insect vector introduces the viral particle into the plant. The viral ssDNA is released from the particle and directed to the nucleus (1), where it is converted into dsDNA by host machinery (2). Viral dsDNA is expressed to produce viral proteins (e.g., Rep) required to initiate rolling-circle replication. Rep nicks the viral DNA and binds to the free 5'-end (3). Host DNA polymerase uses the free 3'-end to start DNA synthesis, and the newly synthesized strand displaces the parental strand (bound to Rep). After one round of replication, Rep performs a second nick and ligates the displaced parental strand to form a circular ssDNA molecule. Rep is also released and the dsDNA molecule can start the cycle again (4). Late expression produces enough coat protein (CP) to encapsidate viral ssDNA (5). Viral particles or ssDNA are taken out of the nucleus by a process involving BV1 protein (6). Viral complexes (ssDNA, BV1, and/or CP) are recognized by the viral protein BC1 and directed to the cell membrane (plasmodesmata) (7) for cell-to-cell movement. Finally, viral particles reach the phloem, where they are acquired by the insect vector for transmission (8).

cription and replication. This viral dsDNA is associated with histones and packaged into so-called minichromosomes (Abouzid et al., 1988).

Similar to other viral systems, the expression of geminiviral genes seems to follow a finely tuned temporal sequence. In this case, it is believed that the genes involved in replication and transcription (e.g., *rep, TrAP,* and *REn*) are expressed earlier than the virion-sense genes (e.g., *CP* and *BV1*). After the expression of early viral genes (left side or complementary sense), the multiplication of the virus genome by a rolling-circle (RC) mechanism generates new viral ssDNA molecules from the dsDNA intermediate. An ssDNA molecule produced in this process has two fates, depending upon the stage of infection. In an early stage, viral ssDNA can be converted, by host machinery, to dsDNA forms. These molecules will then be directed to the transcription and/or replication processes, amplifying the virus genome within the cells—a necessary condition to initiate a systemic infection (Timmermans et al., 1994). The second alternative occurs in a later stage when the late gene products CP and BV1 are present and can bind viral ssDNA either to encapsidate it or to transport it out of the cell nucleus. Once in the cytoplasm, the virus moves to neighboring cells through the plasmodesmata, and finally to the phloem for long-distance transport. It has been shown that some viruses require the CP for systemic spread in some hosts, whereas in other hosts the CP is dispensable. For example, *Pepper huasteco virus* (PHV) mutants that cannot produce CP can systemically infect pepper and *Nicotiana benthamiana* plants. However, when the same mutants are inoculated into *N. tabacum* plants, the virus replicates only in the inoculated cells; it does not spread throughout the plant (Guevara-González et al., 1999). This suggests two types of host-dependent virus movement. The last stage of the cycle corresponds to the uptake of the virions by the insect vector. In this case, it has been shown that the CP and, probably, virus particles are indispensable for insect transmission.

GEMINIVIRUS REPLICATION

Geminivirus replication can be divided into two steps: (1) the conversion of ssDNA (virion DNA) into a dsDNA form and (2) the production of virion-sense ssDNA from the dsDNA intermediate. Little is known about the first process in begomoviruses and curtoviruses. However, in the case of the mastreviruses, a small oligoribonucleotide complementary to sequences located in the SIR (Figure 12.1) is found associated with the virions. It is accepted that this oligoribonucleotide may prime the synthesis of the complementary or minus strand (Donson et al., 1987). Analogous primers have not been found in the other two genera, although it is presumed that they could have equivalent mechanisms.

The process of generation of new viral ssDNA molecules from dsDNA intermediates is better known. Geminiviruses use an RC replication mechanism in which the viral protein Rep plays a leading role. This key protein is encoded by the *rep* gene in begomoviruses and curtoviruses, and by *C1* and *C2* genes in mastreviruses. Although geminivirus Rep proteins are multifunctional polypeptides, they do not display DNA polymerase activity.

They are considered replication initiation proteins and are related to the proteins found in the other RC-replicating ssDNA systems, such as ssDNA viruses (e.g., ϕX-174 bacteria, parvoviruses, and circoviruses) and some plasmids (e.g., pT181) (Gros et al., 1987; Khan et al., 1988; Lee and Kornberg, 1992). In most cases, the proteins show sequence-specific DNA-binding properties as well as DNA nicking-closing and topoisomerase-like activities (Fontes et al., 1992; Desbiez et al., 1995; Laufs et al., 1995; Orozco et al., 1997).

To initiate RC replication, Rep cleaves the viral plus-strand DNA between positions 7 and 8 of the nonameric motif (5'-TAATATT$^\downarrow$AC-3') universally present in all geminiviruses. This sequence is invariably found in the loop of the conserved 30 bp hairpin element located in the virus IR. After the nicking of the origin, the protein becomes covalently linked to the 5'-end (-PO4$^-$) of the nicked strand, via a phosphotyrosine linkage. The 3'-end (-OH$^-$) of the nicked DNA primes the plus-strand synthesis by host DNA polymerases, which use the minus strand as a template, and displaces the parental plus strand linked to the Rep protein. After the origin of replication has been reconstituted and a full-length, linear ssDNA (plus strand) has been displaced from the dsDNA intermediate, Rep performs both a second cleavage in the new nonanucleotide sequence and a ligation of the two ends of the linear ssDNA. This ligation of the Rep-linked 5'-PO4 and the 3'-OH releases Rep and generates a circular ssDNA molecule (Heyraud-Nitschke et al., 1995).

Composition of Geminivirus Replication Origin

The minimal origin of replication has been mapped in two bipartite begomoviruses, *Tomato golden mosaic virus* (TGMV) and SqLCV. Its analysis has revealed an unexpected structural and functional complexity. Earlier, the minimal replication origin of SqLCV was mapped to a 90 bp fragment that includes the conserved hairpin element where Rep cleaves the viral plus strand and about 60 bp of the region proximal to the *rep* gene (Lazarowitz et al., 1992). More recently, a similar result was obtained for TGMV, for which the minimal replication origin was mapped to an 89 bp region (Fontes et al., 1994). This fragment was able to support direct replication of a bacterial plasmid when Rep and REn were supplied *in trans*. This minimal origin is located on the left side of the TGMV CR and includes, besides the hairpin element, five additional *cis*-acting elements that contribute to replication origin function (Figure 12.3). The key elements in the origin are the specific Rep-binding sites, whose nt sequence greatly varies among geminiviruses. In the case of TGMV, Rep recognizes a 13 bp directly

FIGURE 12.3. Geminivirus Origin of Replication

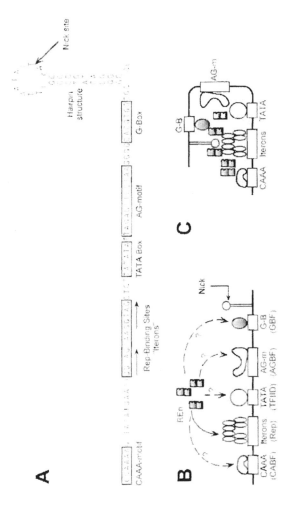

Note: The origin of replication of *Tomato golden mosaic virus* (A) includes the replication initiation protein (Rep)-binding sites (iterons), the hairpin structure, and the nicking site. Several binding sites for transcription factors, including the TATA box for the *rep* gene, are adjacent to the replication elements. The putative factors that bind those elements are shown in parentheses (B). A model that tries to explain how Rep can make the nick in the hairpin structure after binding the iterons suggests a folding of the DNA with the possible involvement of host transcription factors and/or viral replication enhancer protein (REn), which is known to interact with Rep (C). Rep and REn proteins probably act as oligomers. AGBF = AG-motif binding factor; CABF = CAAA-motif binding factor; GBF = G-box binding factor.

repeated motif, 5'-GGTAGtaaGGTAG-3' closely associated to the *rep* TATA box element (Fontes et al., 1994). Directly repeated motifs with different nucleotide sequences are positioned similarly in all the known begomoviruses and curtoviruses (Argüello-Astorga, Guevara-González, et al., 1994). The 5 bp core sequence of those iterated motifs displays the consensus GGNRN and is the critical element for Rep recognition. Several copies (three to six) of those iterated core sequences, or iterons, are present in the IR of all geminiviruses and display a lineage-specific arrangement. Evidence suggests that the iterons are the major *cis*-acting replication specificity determinants in geminiviruses (Eagle and Hanley-Bowdoin, 1997). For example, pseudorecombinant viruses formed by mixing components A and B from different viruses are infectious only if the sequences of the iterons of both viruses are similar (Argüello-Astorga, Guevara-González, et al., 1994).

Two binding sites for host transcription factors are components of TGMV origin, a TATA box element adjacent to the Rep-binding site and a G-box motif next to the hairpin. Functional analysis of TGMV origin, in which those elements were mutated, revealed that they are not essential for viral replication but significantly contribute to efficient origin utilization (Eagle and Hanley-Bowdoin, 1997). Additional *cis*-acting sequences in the origin region do not have detectable roles in transcription but greatly contribute to replication. For example, the AG-1 motif is esential for origin function, whereas a second element, the CA motif, located next to the Rep-binding site, apparently functions as an efficiency element (Orozco et al., 1998).

The close arrangement of the six elements identified in the TGMV replication origin suggests that the protein factors that bind to them during the initiation of replication probably interact with one another (see Figure 12.3B). The observation that the Rep-binding sites are separated from the nonanucleotide cleavage-joining site by up to 80 to 90 nt led to suggesting a model for the initiation of geminivirus replication. In this model (see Figure 12.3C), plant transcription factors interact among themselves and induce a loop in the DNA that brings the Rep complex bound at the iterons in contact with the cleavage site. It has also been suggested that the REn protein could be involved in initiation of replication. REn can interact with itself or with Rep to form homo- or hetero-oligomers. It is possible that this REn-Rep multimeric complex might facilitate the interaction of Rep with the cleavage site after its binding to the iterons (Argüello-Astorga, Herrera-Estrella, and Rivera-Bustamante, 1994; Hanley-Bowdoin et al., 1999).

Although the *cis*-acting elements identified in the TGMV origin are well conserved in most American begomoviruses, some of them are not found in the IRs of other geminiviruses (Hanley-Bowdoin et al., 1999). This suggests that viruses from different lineages might have different genetic elements and protein factors contributing to the functionality of the origin.

CONTROL OF VIRAL GENE EXPRESSION

Geminiviruses, as do other plant and animal viruses, utilize cellular RNA polymerases to transcribe their genes, and they rely on plant transcription factors and coactivators to regulate gene expression. However, geminiviruses also encode their own regulatory proteins that interact with and modulate the activities of host factors (see Table 12.2). The concerted action of viral and cellular regulators makes possible the tight control of expression of viral gene products that are involved in different stages of the infection cycle. For geminiviruses, the primary level of gene expression seems to be at initiation of transcription, although regulatory processes at both posttranscriptional and posttranslational levels probably also occur and are functionally relevant. However, these regulatory processes are, as yet, uninvestigated in these pathogens (Palmer and Rybicki, 1998).

Transcription of Viral Genes

Geminivirus genomes are bidirectionally transcribed from the IR, leading to the synthesis of mRNAs that correspond to both virion- and complementary-sense genes. Diverse studies have shown that geminivirus transcription is complex, and multiple overlapping RNA species are usually produced. For example, in TGMV, six mRNAs for component A and four mRNAs for component B have been identified. The virion-sense genes (*CP* and *BV1*) are transcribed to give single mRNAs, whereas the pattern of transcription is more complicated for the complementary-sense genes. For example, at least three overlapping transcripts have been detected for the left side transcription of component A. All of these transcripts are polycistronic and share a common polyadenylation site (Accotto et al., 1989; Mullineaux et al., 1990; Frischmuth et al., 1991; see Figure 12.1).

The members of the three genera use different expression strategies, especially regarding RNA processing. For an in-depth discussion on this important subject, the reader is encouraged to consult the excellent reviews by Palmer and Rybicki (1998) and Hanley-Bowdoin et al. (1999).

Promoters of Complementary-Sense Genes

As mentioned earlier, the viral proteins required at an early stage in the viral cycle are Rep and REn. The transcription of these genes must be carried out in the absence of any viral factor, an assumption that has been experimentally verified in several cases (Ruiz-Medrano, 1996; Hanley-Bowdoin et al., 1999). TGMV *rep* promoter supports high levels of transcription in protoplasts, and deletion studies have shown that most of its activity can be attributed to the 60 bp immediately upstream of the *rep* transcription start site. This region contains a canonical TATA box element and a conserved G-box motif located next to the conserved hairpin sequence (see Figure 12.3A). Mutations in both *cis*-acting elements drastically affected leftward transcription (Eagle and Hanley-Bowdoin, 1997). Moreover, in the case of G-box mutants, the residual promoter activity was very low, indicating that this DNA motif is the primary transcriptional activating element of TGMV *rep* promoter. Interestingly, the G-box element is positionally conserved in most of the New World begomoviruses, but it is not found in *rep* promoters from Old World begomoviruses and members of the so-called SqLCV cluster. It is not known whether these viruses have another genetic element playing a similar role (Argüello-Astorga, Guevara-González, et al., 1994a).

Rep downregulates its own expression in begomoviruses, and probably in other geminiviruses as well (Eagle et al., 1994). It has been shown that the iterons are involved in the self-regulation of TGMV *rep* expression. The location of these elements between the *rep* TATA box and the transcription start site suggests a steric interference of Rep with the binding of the RNA polymerase or its progression (Eagle et al., 1994) (see Figure 12.3A). Similar mechanisms have been determined in other viral systems (e.g., SV40). This hypothesis, however, is not in complete agreement with several features of the Rep-mediated repression observed in TGMV. For example, repression is dependent on both the position and orientation of the Rep-binding site, as well as on the presence of an intact G-box element. These features suggest a possible active interference with the transcription apparatus (Eagle and Hanley-Bowdoin, 1997). More research is needed to determine the exact mechanism for Rep repression.

Transgenic plants expressing the reporter gene *uidA* (β-glucuronidase, GUS) under the control of PHV *rep* promoter (240 nt) have been used to investigate the tissue-specific expression of *rep*. These plants showed a low level of GUS expression in meristems and vascular tissues (Ruiz-Medrano and Rivera-Bustamante, unpublished results). Similar results have been observed recently with the *rep* promoter from *Taino tomato mottle virus* (TToMoV) (Ramos and Rivera-Bustamante, unpublished results). The expression of viral genes in vascular tissue correlates well with the phloem-limited nature of most geminiviruses. On the other hand, no evidence supports the presence of geminiviruses in apical meristems. Perhaps, the meristematic expression observed for PHV and TToMoV *rep* promoters represents only the metabolic stage of this cell type. Geminiviruses have evolved to use host DNA synthesis machinery, so all factors for virus replication and gene expression are present in meristematic tissues, or in any type of cell that is actively dividing and has a high DNA synthesis rate. However, in an infected plant, the virus never reaches the apical meristem cells, probably because of the lack of efficient virus transport in that tissue (Lucy et al., 1996).

Promoters of Virion-Sense (Late) Genes

The tissue specificity of PHV *CP* promoter also has been studied using transgenic plants that express the GUS gene under the direction of several versions of the promoter. Three versions have been extensively tested. Figure 12.4 shows a diagrammatic representation of the promoters: $-693CP$, $-235CP$, and $-115CP$. The number indicates the number of bases upstream of the initiation codon included as promoter. Plants with the $115CP$-GUS construction showed a weak expression in meristematic tissues, whereas plants with the other two promoter versions showed a strong expression in vascular and meristematic tissues (Ruiz-Medrano et al., 1999). This suggested that the elements directing the expression in meristematic tissues are located in the first 115 nt, whereas the elements needed for expression in the vascular tissue are probably located between the nucleotides -235 and -115. Similar results have been observed for TGMV, in which the elements required for expression in the phloem were located in the $-163/-103$ region. In addition, the region $-235/-115$ should contain elements that increase the strength of the promoter, an equivalent to the G-box found in the *rep* promoter (Argüello-Astorga, Guevara-González, et al., 1994). It is not known whether the elements that direct the expression in vascular tissues of both *rep* and *CP* promoters are the same. A more detailed analysis of the *rep* promoter with versions that eliminate the region from -235 to -115 is required (Ruiz-Medrano, 1996).

The proper temporal regulation of viral gene expression is a critical aspect of virus infection cycles. A transcriptional strategy used by many viruses is the control of so-called *late genes* (i.e., genes whose products are required only in the late stages of the infection) by viral proteins that are produced in the early stages of the cycle. Early viral proteins that are able to induce or transactivate the expression of late viral genes have been extensively studied in animal systems (e.g., VP16 in herpes simplex virus and E1A in adenoviruses) (Martin and Green, 1992).

The members of distinct geminivirus lineages execute their transcription programs in different ways, but all of them employ the aforementioned gen-

FIGURE 12.4. Regulatory Elements in *Pepper huasteco virus* Coat Protein (PHV CP) Promoter

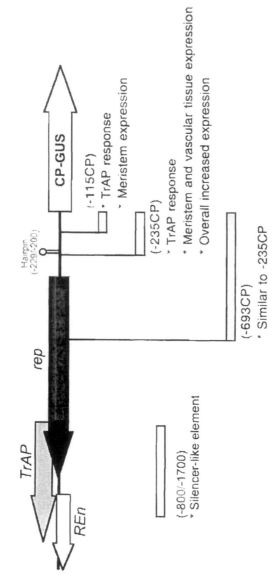

Note: A chimeric gene (GUS reporter fused to the 5'-end of PHV CP) was used to transform *Nicotiana tabacum* plants. Three different size promoters were used: –115*CP* (up to 115 nucleotides upstream from the initiation codon), –235*CP* and –693*CP*. The expression pattern of the promoters was obtained (GUS assay). The smallest version (–115*CP*) showed only a rather weak expression in meristematic tissues. The other two versions showed a strong expression in both vascular and meristematic tissues. The high similarity in the expression patterns of both types of plants, –235*CP*-GUS and –693*CP*-GUS, suggests that most regulatory elements for the *CP* gene are located in the first 235 nucleotides. On the other hand, all three versions were responsive by transactivator protein (TrAP). Transgenic plants with larger promoters (up to –2600 nucleotides) were used to map a silencer-like element similar to the one described for *Tomato golden mosaic virus* (TGMV). *CP* = coat protein; *REn* = replication enhancer protein; *rep* = replication initiation protein; *TrAP* = transactivator protein.

eral strategy. Early studies on TGMV gene expression showed that the product of the gene previously known as *AL2* was required for the expression of the *CP* and *BV1* genes (Sunter and Bisaro, 1991). It was also shown that this effect was at the level of transcription. Accordingly, the *AL2* ORF was renamed *TrAP* for *transcription activator protein*.

TrAP activation of the *CP* and *BV1* promoters is dependent on specific promoter sequences and is independent of replication. Transient expression assays showed that the IR of TGMV and *African cassava mosaic virus* (ACMV) containing the putative *CP* TATA box and the transcription start site are sufficient to support transcription of a reporter gene when TrAP is supplied *in trans* (Haley et al., 1992). In addition, a *CP* promoter from a specific virus can be transactivated by a TrAP obtained from other viruses. This showed that *TrAP* genes are functionally interchangeable among distantly related begomoviruses (Sunter et al., 1994).

The observation that TrAP functions in a virus-nonspecific way suggested that the DNA elements present in the *CP* and *BV1* promoters and involved in the transactivation process could also be conserved. This prediction was found to be true after a comparative analysis of the intergenic sequences of dicot-infecting geminiviruses that revealed the presence of a conserved DNA motif (CLE, for conserved late element; consensus: GTGGTCCC) in most of the *CP* and *BV1* promoters (Argüello-Astorga et al., 1994). Experimental evidence later showed that, at least in the case of PHV, CLEs were responsible for TrAP-mediated transactivation. The evidence included PHV *CP* promoter analysis to delimit the TrAP responsiveness to a 115 nt fragment that includes three CLE motifs (see Figure 12.4). Site-directed mutagenesis of those CLE motifs eliminated the ability to respond to TrAP. Finally, in gain-of-function experiments, oligonucleotides containing the CLE motifs and fused to heterologous promoters (e.g., minimal and truncated 35S promoters) were able to provide TrAP responsiveness (Ruiz-Medrano et al., 1999).

A Silencer Controls Late Promoters

As mentioned earlier, *CP* promoter versions, both from TGMV and PHV, that included up to 700 nt showed strong expression in vascular and meristematic tissues. However, when longer fragments were used as promoters, it was noticed that the expression was turned off (Ruiz-Medrano, 1996). This suggested the presence of a silencer-like element in the region beyond nucleotide −700. Interestingly, in the case of TGMV, such an element was further characterized and mapped to a fragment that overlaps the *TrAP* gene but does not seem to affect the transcription of any complementary-sense genes (*rep, TrAP,* and *REn*). The precise genomic location of the PHV silencer has not been determined yet, although preliminary data point to the same general location reported for TGMV (Diaz-Plaza and Rivera-Bustamante, unpublished results).

Mechanisms of TrAP Action

Taken as a whole, the studies on the begomovirus *CP* promoters have shown that sequences located in the IR are able to direct the transcription of late genes in some tissues, independently of any viral protein. Nonetheless, those potential promoter activities are normally repressed, in the full virus context, in the absence of TrAP, by the action of a promoter-specific silencer. Therefore, it has been suggested that the observed activation of viral late promoters by TrAP could occur by at least two different mechanisms: (1) derepression in phloem and (2) transactivation in other tissues (e.g., mesophyll), where the truncated *CP* promoters do not direct reporter gene expression; that is, TrAP functions as a tissue-specific derepressor/activator (see Figure 12.5; Sunter and Bisaro, 1997).

An unresolved question that is critical for our understanding of TrAP action is whether TrAP is recruited to viral late promoters by directly binding at specific DNA motifs, as with many cell transcriptional activators, or through protein-protein interactions with DNA-binding host factors, in a way analogous to other well-known viral transactivators, such as adenovirus E1A and herpesvirus VP16 proteins (Martin and Green, 1992). The former possibility was initially favored because TrAP displays features typical of transcription factors: it contains an N-terminal basic domain closely associated with a putative zinc finger domain, characteristic of DNA-binding domains, and an acidic C-terminus, characteristic of activation domains. However, several experimental observations strongly suggest that TrAP probably functions in a way analogous to VP16 and E1A viral transactivators, i.e., through interactions with DNA-binding plant proteins. First, recombinant TrAP proteins from diverse begomoviruses do not bind DNA in a sequence-specific way; second, the only reported *cis*-acting element that mediates TrAP action, CLE, is similar to DNA-binding sites for known transcription factors found in several plant genes (e.g., *PCNA* gene from rice; Kosugi et al., 1991); third, synthetic oligonucleotides containing either one or two copies of the CLE conferred strong transcriptional activity to a truncated (–90/+1) *Cauliflower mosaic virus* (CaMV) 35S promoter in transgenic plant systems (Ruiz-Medrano et al., 1999). This enhanced transcriptional activity is observed in absence of geminivirus proteins. Therefore, it can be concluded that the CLE is a binding site for plant transcription factors. It has been noticed that some promoters of plant genes involved in cell cycle

FIGURE 12.5. Regulation of Replication Initiation Protein *(rep)* and Coat Protein *(CP)* Promoters

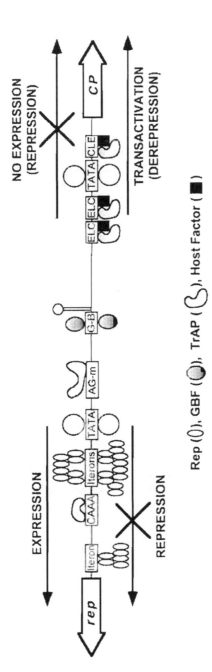

Rep (◊), GBF (◯), TrAP (◠), Host Factor (■)

Note: Leftward transcription: *rep* promoter is functional in the early stages of the infection cycle. Rep protein binds to the iterons near the TATA box to initiate replication. At some point, perhaps when the levels of Rep are high, Rep downregulates its own expression in a not-well-understood manner that probably involves the binding to a third iteron found near the rep initiation codon, or by forming larger oligomers. Rightward transcription: *CP* gene is not expressed in the early stages of the cycle. This promoter is maintained silently by the silencer-like element found further upstream (not shown). In a later stage, when the transcription activator protein *(TrAP)* gene has been expressed, TrAP interacts with the conserved late element (CLE) motif near the TATA box to activate the expression of *CP*. This interaction probably involves a still unknown host factor. The CLE copies in the opposite orientation (ELC) are apparently not essential but contribute to the transactivation process. The involvement of other host factors in the regulation is still not clear in any case. For other abbreviations, see note to Figure 12.3.

273

regulation present CLE-like elements (e.g., *PCNA* gene of rice; Kosugi et al., 1991). It will be interesting to verify the effect of TrAP in the regulation of these genes.

Regulation of Virion-Sense Gene Expression in Mastreviruses

Monocot-infecting geminiviruses do not have a gene homologous to *TrAP;* therefore, other viral proteins must regulate the activity of the *CP* gene promoter. In these viruses, two overlapping genes *(C1* and *C2)* encode the replication/transactivator protein. The *C1-C2* genes are transcribed as a single RNA, which is processed by the host machinery through differential splicing, and may be translated in either Rep (by spliced RNA) or RepA (unspliced RNA) proteins. Rep is homologous to the replication protein of curtoviruses and begomoviruses (see Table 12.2; Palmer and Rybicki, 1998).

Rep A, or a RepA/Rep complex, could be the viral factor involved in transactivation of virion-sense gene expression, and probably also responsible for the downregulation of *C1-C2* transcription. Little is known about the control of late *(V1/CP)* promoters in mastreviruses. In the case of *Maize streak virus,* the type species of these viruses, an element called rpe1 (or UAS) and consisting of two direct CG-box repeats was identified. This element is bound by maize nuclear factors, and it is involved in host-dependent transcriptional activity of the *V1/CP* promoter (Fenoll et al., 1990). It has not been determined whether rpe1 could have a role similar to the one shown by the CLE and be involved in transactivation by Rep/RepA.

Besides the rpe1 and TATA box elements, no other transcriptional elements in the IR of a mastrevirus have been experimentally characterized. Nevertheless, a recent sequence analysis of the IRs of all known mastreviruses has revealed the existence of an arrangement of conserved DNA motifs located at the 3'-part of the hairpin sequence. Interestingly, those conserved arrangements include sequences identical or similar to the begomovirus CLEs. Although no experimental evidence indicates a role of those conserved sequences in virus transcription, it is possible that CLEs could also be involved in mastrevirus late gene transactivation and be the functional targets for their transactivator proteins (Palmer and Rybicki, 1998). An interesting hypothesis implied from these observations is that the CLEs and the cell factors that bind to them would have remained functioning as targets for viral transactivators throughout the extended evolutionary history of geminiviruses, despite the dramatic changes in the nature and structural features of those viral proteins.

CONCLUDING REMARKS

Although the initial idea of using geminiviruses as vectors to express foreign genes in plants has proven difficult to achieve, the molecular characterization of several aspects of their infection cycle has opened other possibilities. One of these is the use of diverse geminivirus genetic elements in biotechnology processes. For example, the use of geminivirus genes (either wild-type or modified versions) in pathogen-derived resistance strategies is starting to be widely explored. Another important concept is the use of promoters with elements derived from geminiviruses. This concept will include tissue-specific promoters that will direct the expression of foreign genes in tissues normally invaded by these viruses. Finally, the most attractive idea would be the use of geminivirus-inducible promoters that are not expressed in normal conditions but are turned on when the virus infects a cell. Future research will elucidate the feasibility of these concepts.

REFERENCES

Abouzid AM, Frischmuth T, Jeske H (1988). A putative replicative form of the abutilon mosaic virus gemini group) in a chromatin-like structure. *Molecular and General Genetics* 212: 252-258.

Accotto GP, Donson J, Mullineaux PM (1989). Mapping of *Digitaria* streak virus transcripts reveals different RNA species from the same transcription unit. *The EMBO Journal* 8: 1033-1039.

Argüello-Astorga GR, Guevara-González RG, Herrera-Estrella LR, Rivera-Bustamante RF (1994). Geminivirus replication origins have a group-specific organization of iterative elements: A model for replication. *Virology* 203: 90-100.

Argüello-Astorga G, Herrera-Estrella L, Rivera-Bustamante R (1994). Experimental and theoretical definition of geminivirus origin of replication. *Plant Molecular Biology* 26: 553-556.

Desbiez C, David C, Mettouchi A, Laufs J, Gronenborn B (1995). Rep protein of tomato yellow leaf curl geminivirus has an ATPase activity required for viral DNA replication. *Proceedings of the National Academy of Sciences, USA* 92: 5640-5644.

Donson J, Morris-Krsinich BAM, Mullineaux PM, Boulton MI, Davies JW (1987). A putative primer for second-strand DNA synthesis of maize streak virus is virion associated. *The EMBO Journal* 3: 3069-3073.

Eagle PA, Hanley-Bowdoin L (1997). *cis* elements that contribute to geminivirus transcriptional regulation and the efficiency of DNA replication. *Journal of Virology* 71: 6947-6955.

Eagle PA, Orozco BM, Hanley-Bowdoin L (1994). A DNA sequence required for geminivirus replication also mediates transcriptional regulation. *The Plant Cell* 6: 1157-1170.

Fenoll C, Schwarz JJ, Black DM, Schneider M, Howell SH (1990). The intergenic region of maize streak virus contains a GC-rich element that activates rightward transcription and binds maize nuclear factors. *Plant Molecular Biology* 15: 865-877.

Fontes EPB, Gladfelter HJ, Schaffer RL, Petty ITD, Hanley-Bowdoin L (1994). Geminivirus replication origins have a modular organization. *The Plant Cell* 6: 405-416.

Fontes EPB, Luckow VA, Hanley-Bowdoin L (1992). A geminivirus replication protein is a sequence-specific DNA binding protein. *The Plant Cell* 4: 597-608.

Frischmuth S, Frischmuth T, Jeske H (1991). Transcript mapping of abutilon mosaic virus, a geminivirus. *Virology* 185: 595-604.

Gros MF, Te Riele H, Ehrlich SD (1987). Rolling circle replication of single-stranded DNA plasmid pC194. *The EMBO Journal* 6: 3863-3869.

Guevara-González RG, Ramos PL, Rivera-Bustamante RF (1999). Complementation of coat protein mutants of pepper huasteco geminivirus in transgenic tobacco plants. *Phytopathology* 89: 540-545.

Haley A, Zhan X, Richardson K, Head K, Morris B (1992). Regulation of the activities of African cassava mosaic virus promoters by the AC1, AC2, and AC3 gene products. *Virology* 188: 905-909.

Hanley-Bowdoin L, Settlage SB, Orozco BM, Nagar S, Robertson D (1999). Geminiviruses: Models for plant DNA replication, transcription, and cell cycle regulation. *Critical Reviews in Plant Sciences* 18: 71-106.

Heyraud-Nitschke F, Shumacher S, Laufs J, Schaefer S, Schell J, Gronenborn B (1995). Determination of the origin cleavage and joining domain of geminivirus Rep proteins. *Nucleic Acids Research* 23: 910-916.

Khan SA, Murray RW, Koepsel RR (1988). Mechanism of plasmid pT181 DNA replication. *Biochimica and Biophysica Acta* 20: 375-381.

Kosugi S, Susuka I, Ohashi Y, Murakami T, Arai Y (1991). Upstream sequences of rice proliferating cell nuclear antigen (PCNA) gene mediate expression of PCNA-GUS chimeric gene in meristems of transgenic tobacco plants. *Nucleic Acids Research* 19: 1571-1576.

Laufs J, Traut W, Heyraud F, Matzit V, Rogers SG, Schell J, Gronenborn B (1995). In vitro cleavage and joining at the viral origin of replication by the replication initiator protein of tomato yellow leaf curl virus. *Proceedings of the National Academy of Sciences, USA* 922: 3879-3883.

Lazarowitz SG (1992). Geminiviruses: Genome structure and gene function. *Critical Reviews in Plant Sciences* 11: 327-349.

Lazarowitz SG, Wu LC, Rogers SG, Elmer JS (1992). Sequence-specific interaction with the viral AL1 protein identifies a geminivirus DNA replication origin. *The Plant Cell* 4: 799-809.

Lee EH, Kornberg A (1992). Features of replication fork blockage by the *Escherichia coli* terminus-binding protein. *Journal of Biological Chemistry* 267: 8778-8784.

Lucy AP, Boulton MI, Davies JW, Maule AJ (1996). Tissue specificity of *Zea mays* infection by maize streak virus. *Molecular Plant-Microbe Interactions* 9: 22-31.

Martin KJ, Green MR (1992). Transcriptional activation by viral immediate early proteins: Variations on a common theme. In McKnight SL, Yamamoto KR (eds.), *Transcriptional Regulation*, Cold Spring Harbor Laboratory Press, Cold Spring Harbor, New York, pp. 695-725.

Mullineaux PM, Guerineau F, Accotto GP (1990). Processing of complementary sense RNAs of *Digitaria* streak virus in its host and in transgenic tobacco. *Nucleic Acids Research* 18: 7259-7265.

Orozco BM, Gladfelter HJ, Settlage SB, Eagle PA, Gentry RN, Hanley-Bowdoin L (1998). Multiple *cis*-elements contribute to geminivirus origin function. *Virology* 242: 346-356.

Orozco BM, Miller AB, Settlage SB, Hanley-Bowdoin L (1997). Functional domains of a geminivirus replication protein. *Journal of Biological Chemistry* 272: 9840-9846.

Palmer KE, Rybicki EP (1998). The molecular biology of mastreviruses. *Advances in Virus Research* 50: 183-234.

Ruiz-Medrano R (1996). Análisis de la expresión de los principales promotores del geminivirus huasteco del chile. PhD thesis, Irapuato, Guanajuato, Mexico, Centro de Investigación y de Estudios Avanzados del IPN.

Ruiz-Medrano R, Guevara-González RG, Argüello-Astorga GR, Monsalve-Fonnegra Z, Herrera-Estrella LR, Rivera-Bustamante RF (1999). Identification of a sequence element involved in AC2-mediated transactivation of the pepper huasteco virus coat protein gene. *Virology* 253: 162-169.

Sunter G, Bisaro DM (1991). Transactivation in a geminivirus: AL2 gene product is needed for coat protein expression. *Virology* 180: 416-419.

Sunter G, Bisaro DM (1997). Regulation of a geminivirus coat protein promoter by AL2 protein (TrAP): Evidence for activation and derepression mechanisms. *Virology* 232: 269-280.

Sunter G, Stenger DC, Bisaro DM (1994). Heterologous complementation by geminivirus AL2 and AL3 genes. *Virology* 203: 203-210.

Timmermans MCP, Prem Das O, Messing J (1994). Geminiviruses and their uses as extrachromosomal replicons. *Annual Review of Plant Physiology and Plant Molecular Biology* 45: 79-112.

Van Regenmortel MH, Bishop DH, Fauquet CM, Mayo MA, Maniloff J, Calisher CH (1997). Guidelines to the demarcation of virus species [news]. *Archives of Virology* 142: 1505-1518.

Chapter 13

The Molecular Epidemiology of Begomoviruses

Judith K. Brown

INTRODUCTION

The molecular epidemiology of whitefly-transmitted geminiviruses in genus *Begomovirus* (family *Geminiviridae*) has attracted much recent attention owing to the growing recognition of their importance. Geminiviruses have become important deterrents to the production of food and industrial crops during the past twenty years, as cropping practices have changed to meet the growing demand for agricultural commodities (Brown, 1990, 1994, 2001; Brown and Bird, 1992).

Molecular epidemiology attempts to explain those processes that lead to a successful state of disease in a host plant, primarily to enable identification of a weak link in the biotic and molecular interactions or processes that may be capitalized upon to achieve disease control. A successful viral pathogen must necessarily multiply, a stage of the disease process that requires a "replication-susceptible" host plant, while the second stage of the process requires a host permissive to invasion of cells adjacent to the primary infection court and, ultimately, to systemic spread.

This chapter presents an overview of the begomoviruses as timely viral pathogens in the context of their distribution, host range, molecular features, virus-vector interactions, and disease control, in light of molecular epidemiological information about begomovirus-incited diseases. Examples have been selected to illustrate factors that may be involved in the emergence or demise of a begomovirus as an important plant pathogens and, similarly, in upsurges or the complete displacement of "biological types" or variants of the whitefly vector.

BEGOMOVIRUSES: DISTRIBUTION
AND CHARACTERISTICS

Whitefly-transmitted geminiviruses, or begomoviruses, have emerged as serious pathogens of agronomic and horticultural crops in subtropical and tropical regions of the Americas (Bird and Sanchez, 1971; Costa, 1976; Bird and Maramorosch, 1978; Mound, 1983; Brown and Bird, 1992; Brown, 1990, 1994; Polston and Anderson, 1997) and in tropical and subtropical Africa and Asia (Varma, 1963; Muniyappa, 1980; Bock and Harrison, 1985; Traboulsi, 1994; Czosnek and Laterrot, 1997; Mansoor et al., 1999). Figure 13.1 illustrates the nearly global distribution of well-studied begomoviruses, denoted by specific viral acronyms. Table 13.1 lists, for each virus, the viral acronym and corresponding virus name, the geographic origin, the host plant on which the virus was described, and the Gene Bank accession number of the A component or genome sequence of the virus.

Begomoviruses have circular, single-stranded DNA (ssDNA) genomes that replicate through double-stranded DNA (dsDNA) intermediates by a rolling-circle mechanism. The genomes of begomoviruses are approximately 2.6 to 2.8 kb in size. Bipartite genomes share a common region (CR) of approximately 200 nucleotides (nt) that is highly conserved among cognate components of a viral species, while the analogous region in monopartite viruses is referred to as the large intergenic region (LIR). The CR and LIR contain modular *cis*-acting elements of the origin of replication *(ori)* and promoter elements (Lazarowitz, 1992; Bisaro, 1994; Hanley-Bowdoin et al., 1999). Four to six open reading frames (ORFs) capable of encoding proteins >10 kDa in size are present on the A component or monopartite genome. The viral capsid protein (CP) is encoded by the ORF *AV1*, which is the most highly conserved begomovirus gene (Harrison, 1985; Padidam et al., 1995). The CP of begomoviruses is required for encapsidation of ssDNA and for whitefly-mediated transmission. The CP is also necessary for systemic spread of all monopartite and some bipartite viruses, depending upon the degree of virus-host adaptation (Pooma et al., 1996; Hou et al., 1998). The *AC1* ORF encodes Rep, a replication initiation protein, and specificity is mediated through sequence-specific interactions with *cis*-acting elements of the *ori*. The *AC2* ORF encodes a transcription activator (TrAP) required for rightward gene expression, and the *AC3* ORF encodes a replication enhancer (REn) (Bisaro, 1994; Hanley-Bowdoin et al., 1999; Lazarowitz, 1992). Movement and systemic infection in bipartite viruses is accomplished by B component ORFs *BV1* and *BC1*, which are responsible

FIGURE 13.1. Distribution of Begomoviruses for Which the Full-Length Genome or the A Component Nucleotide Sequence Is Available

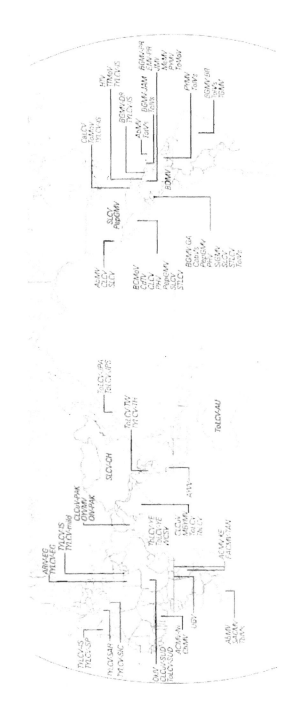

Note: Corresponding virus names and acronyms, geographic origin, host plant of origin, and Gene Bank accession numbers are provided in Table 13.1.

TABLE 13.1. Begomoviruses Within the Family *Geminiviridae* for Which a Full-Length A Component or a Monopartite Genome Sequence Were Included in Phylogenetic Analysis

Virus Name	Acronym on Cladogram	Geographic Origin	Host Plant on Which Described	Gene Bank Accession Number
Abutilon mosaic	AbMV-WI	Kenya	*Abutilon* spp.	X15983
African cassava mosaic	ACMV-KE	Kenya	Cassava	J02057
African cassava mosaic	ACMV-NG	Nigeria	Cassava	X68318
Ageratum yellow vein	AYVV	Singapore	*Ageratum* spp.	X74516
Althaea rosea-infecting	ARIV-EG	Egypt	*Althaea rosea*	AF014881
Bean calico mosaic	BCMoV	Sonora, Mexico	Common bean	AF110189
Bean dwarf mosaic	BDMV	Colombia	Common bean	M88179
Bean golden mosaic	BGMV-BR	Brazil	Common bean	M88686
Bean golden mosaic	BGMV-DR	Dominican Republic	Common bean	L01635
Bean golden mosaic	BGMV-GA	Guatemala	Common bean	M91604
Bean golden mosaic	BGMV-PR	Puerto Rico	Common bean	M10070
Cabbage leaf curl	CaLCV	Florida	Cabbage	U65529
Chayote mosaic	ChMV	Nigeria	Chayote	AJ223191
Chino del tomate	CdTV	Sinaloa, Mexico	Tomato	AF226665
Cotton leaf crumple	CLCV	Arizona	Cotton	Not released
Cotton leaf curl	CLCuV-PK	Pakistan	Cotton	AJ002458; AJ002459
East African cassava mosaic	EACMV-TZ	Tanzania	Cassava	Z83256
Havana tomato	HTV	Cuba	Tomato	Y14874
Indian cassava mosaic	ICMV	India	Cassava	Z48182
Mung bean yellow mosaic	MBYMV	Thailand	Mung bean	D14703
Okra leaf-curl	OLCV-PK	Pakistan	Okra	AJ002451
Okra infecting	OIV-SD	Sudan 235	Okra	Not released

Pepper golden mosaic (TPV-TAM)	PepGMV Complex	Tamaulipas, Mexico	Pepper	U57457
Pepper huasteco	PHV	Mexico	Pepper	X70418
Potato yellow mosaic	PYMV-VE	Venezuela	Potato	D00940
Sida golden mosaic	SiGMV-CR	Costa Rica	*Sida* spp.	X99550
Sida golden mosaic associated	SiGMV-HN	Honduras	*Sida* spp.	Y11097
Squash leaf curl E strain	SLCV-E	Southwest, United States	Squash	M38183
Tomato golden mosaic	TGMV	Brazil	Tomato	K02029
Tomato mottle	ToMoV	Florida	Tomato	L14460
Taino tomato mottle	TTMoV	Cuba	Tomato	AF012300
Tomato-infecting	ToIV-PAN	Panama	Tomato	Y15034
Tomato leaf curl	ToLCV-AU-D1, D2	Australia	Tomato	AF084006; AF084007
Tomato leaf curl	ToLCV-Ban 2,3,4	Bangalore, India	Tomato	Z48182; ZU38239; AF165098
Tomato leaf curl associated	ToLCV-Nde-1,2	New Delhi, India	Tomato	U15015; U15016
Tomato leaf curl associated	ToLCV-JP A	Japan	Tomato	AB014347
Tomato leaf curl associated	ToLCV-TW	Taiwan	Tomato	U88692
Tomato yellow leaf curl	TYLCV-Is	Israel	Toma	X15656
Tomato yellow leaf curl	TYLCV-Is (DR)	Dominican Republic	Tomato	AF024715
Tomato yellow leaf curl	TYLCV-Is mild	Israel	Tomato	X76319
Tomato yellow leaf curl	TYLCV-Is mild (ES)	Spain	Tomato	AF071228
Tomato yellow leaf curl	TYLCV-Sar	Sardinia, Italy	Tomato	X61153
Tomato yellow leaf curl	TYLCV-Sic	Sicily, Italy	Tomato	Z28390
Tomato yellow leaf curl	TYLCV-ES (ALM)	Almeria, Spain	Tomato	L27708
Tomato yellow leaf curl	TYLCV-TH	Thailand	Tomato	M59838
Uganda cassava associated	UgV	Uganda	Cassava	Z83252

for nuclear transport of ssDNA and cell-to-cell movement functions (Sanderfoot and Lazarowitz, 1996). For monopartite viruses, several ORFs are involved in movement functions, including the *V1 (CP)*, *V2*, and translation products of the *C4* ORF, which exert host-specific effects on symptom severity, virus accumulation, and movement (Krake et al., 1998).

DETECTION, IDENTIFICATION,
AND CLASSIFICATION OF BEGOMOVIRUSES

With the recognition of diversity within begomoviruses has come the development of standard approaches to identify new and resurgent viruses. For most plant viruses, serology has traditionally been the method of choice for identification of new and resurgent viruses. However, virus-specific antiserum for begomoviruses is not available because the CP is extremely conserved across all species. DNA-based diagnostics, including polymerase chain reaction (PCR) and nucleotide sequencing of viral amplicons, have supplanted serology for detection, identification, and classification of this group. Sequence alignments are conveniently presented as cladograms to illustrate predicted relationships, and distances can be calculated to estimate divergence (Rybicki, 1994; Padidam et al., 1995; see Figure 13.2).

Because substantial information on begomovirus genome organization and gene function is now available, as is knowledge of numerous genomic and *CP* gene sequences, it is possible to scan partial or entire viral sequences and develop hypotheses regarding the prospective utility of a particular ORF or nontranslated region for inferring phylogenetic relationships. Sequences associated with functional domains of viral polypeptides, nucleotide sequence motifs or regulatory elements that are conserved within the genus or groups of species or strains, and sequences that are highly variable have been explored for this purpose (Argüello-Astorga et al., 1994; Rybicki, 1994; Padidam et al., 1995; Brown, 2001).

The *CP* is the only begomovirus sequence formally approved by the International Committee on Taxonomy of Viruses for predicting begomovirus identity (Mayo and Pringle, 1998). The basis for the high degree of sequence conservation in this ORF is likely due to its multifunctional role. Its encoded protein exhibits hypervariability within the N-terminal ~55 to 60 amino acids (aa) (Padidam et al., 1995), while the central portion of the ORF contains both variable and conserved regions, and C-terminal amino acids are nearly identical for all viruses (Brown et al., 2001). An arbitrary value of 90 percent nt sequence identity has been suggested as a guideline for predicting the identity of a distinct begomoviral species (< 90 percent) or a viral strain (> 90 percent) (Rybicki, 1998), though, clearly, evidence suggests that alterations in one or several amino acids can lead to altered symptom or host range phenotypes (Lazarowitz, 1991; Petty et al., 1995; Brown et al., 2000). Comparative nt sequence analysis of the middle two-thirds of the *CP* (core *CP*) or of the complete *CP* predicts similar clustering of begomovviruses within one large group, and two major subclusters furtherdelineate Old and New World viruses (Brown et al., 2001; see Figure 13.2).

FIGURE 13.2. Single Most Parsimonious Tree Showing Begomovirus Relationships Based on the Full-Length Genome or A Component Nucleotide Sequence As Predicted by a Maximum Parsimony Method (PAUP)

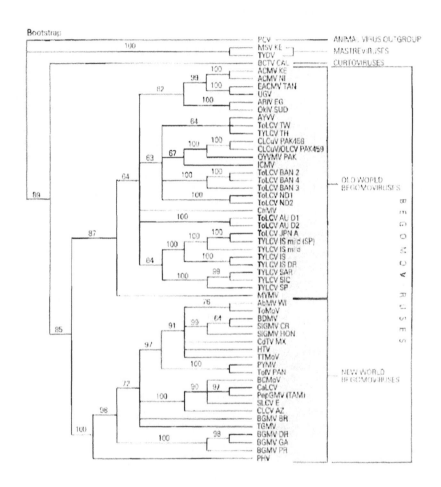

Note: Outgroups are the animal-infecting *Porcine circovirus* (PCV); the monocot-infecting *Maize streak virus* (MSV), type species of the genus *Mastrevirus;* the dicot-infecting mastrevirus *Tobacco yellow dwarf virus* (TYDV); and the dicot-infecting *Beet curly top virus* (BCTV)-California, type species of the genus *Curtovirus*. Shaded area contains the cluster of begomoviruses and two subclusters of Old and New World representatives, respectively. Begomovirus names associated with the respective acronyms shown on the tree are listed in Table 13.1.

A second potentially informative sequence is located within the *AL1/C1* ORF, which encodes the Rep-associated protein. The N-terminal 189 to 200 aa region of this polypeptide contains "virus-specific" motifs, while more conserved sequences are present in the midregion and C-terminal amino acids. Using computer-based predictions, two "virus-specific" sets of alpha helices separated by several aa residues were identified in the N-terminal 160 aa of the *AL1* ORF of *Tomato golden mosaic virus* (TGMV), while studies with *Bean golden mosaic virus* (BGMV) revealed an ATP-binding domain downstream from the N-terminal 220 aa that is conserved within the genus. The N-terminal 200 aa product of this multifunctional gene contains sequences involved in oligomerization of AL1 protein, DNA binding, and DNA cleavage and joining functions. Because sequences in this region are essential for recognizing virus-specific sequence elements in the CR/LIR to initiate replication (Bisaro, 1996; Hanley-Bowdoin et al., 1999), this sequence is potentially informative for establishing viral identification, predicting phylogenetic relationships, and possibly for discriminating between biological strains. Although the *AC1/C1* ORF itself is less conserved than the *CP* ORF, this ORF may throw new light on begomoviral genealogy, particularly when compared to *CP* trees. Incongruencies between *Rep* and cladograms reconstructed with other informative begomovirus sequences may assist in recognizing important recombination sites and events (Padidam et al., 1999), and they could provide evidence of genetic drift or modular protein evolution (Fuchs and Buta, 1997).

A third prospective informative region is a nontranslated CR/LIR sequence of begomoviruses. All geminiviruses contain a conserved consensus sequence of approximately 30 nt in length in this region. This sequence contains the T/AC cleavage site at which rolling-circle replication is initiated, and this conserved feature led to early speculation that recombination might occur frequently at this site. In addition, virus-specific, directly repeated sequences that bind Rep during replication initiation are also present in this region and are considered highly useful for determining virus identity, identifying virus relationships, and providing clues as to the likelihood that bipartite genomes may interact with others *in trans* to form a reassortant. The directly repeated element in bipartite and most monopartite genomes contains two 4 to 5 nt directly repeated sequences separated by 2 to 4 "spacer" nt that act in a virus-specific manner to bind Rep (Hanley-Bowdoin et al., 1999). As a result, these elements have been hypothesized to constitute informative indicators of viral genealogy useful in defining species and strains when traced through viral lineages (Argüello-Astorga et al., 1994; Paximadis et al., 1999; Brown, 2001). However, it is not yet clear how much plasticity in sequence or spacing can be tolerated at this site without interfering with replication.

Alignment of begomoviral A component and monopartite viral genome sequences or of several viral ORFs yields similar or congruent trees. Phylogenetic analysis of the complete genome (monopartite) and A component (bipartite) viral sequences or of the *CP* gene nucleotide sequences indicates strict clustering of begomoviruses by extant geographic origin, and not by monopartite or bipartite genome organization, crop of origin, or host range. Certain Old World viruses may be somewhat more divergent than the majority of New World taxa, perhaps because they have been more isolated by geographic barriers and/or patterns of human travel, having fewer opportunities for interactions between viruses in different regions but, at the same time, great potential for localized diversity (see Figure 13.1). All begomoviruses in the New World have bipartite genomes, and many are distributed as overlapping geographic clusters, suggesting a lesser role of geographic barriers, fewer genotypes that have diverged in isolation, and, consequently, a more confounded evolution (see Figures 13.1 and 13.2). Similar virus relationships are also predicted by an alignment of the N-terminal 200 aa of Rep, although a comparison of virus-specific, directly repeated sequences in the CR does not readily support taxa separation on the basis of recent geographic origin (Brown, 2001).

BEGOMOVIRUS SUSCEPTS AND HOSTS IN RELATION TO EPIDEMIOLOGY

Natural Host Range

Natural hosts of begomoviruses are plant species in which the virus can replicate, cause systemic infection, and encapsidate, and from which virions are ingested and transmitted to a susceptible host by the whitefly vector. Because the genomes of plant viruses are exceedingly small and limited in the number and function of coding and regulatory sequences, their replication depends nearly entirely on plant regulatory and metabolic processes (Dawson and Hilf, 1992). Typically, plant viruses have limited and very specific host ranges, though several possibilities exist for altering the host range, namely, through recombination and, in bipartite genomes, through reassortment.

The extent of genome variation among begomoviruses, around which virus classification has been established, does not clearly correlate with the documented host ranges of these viruses (Harrison and Robinson, 1999). At one end of the spectrum are two host-adapted bean-infecting begomoviruses, represented by Caribbean BGMV-PR and its regional variants and by BGMV-BR from Brazil. Although *Macroptilium lathyroides* was thought to be a natural host of BGMV-PR, recent evidence has revealed that this weed is infected by *Macroptilium golden mosaic virus* (MGMV), a newly discovered virus that causes golden mosaic symptoms in this host that were once attributed to BGMV (Idris et al., 1999). Year-round cultivation of bean in the Caribbean region and in Central America may have been responsible for the host adaptation of BGMV to bean. Apparently BGMV-PR and BGMV-BR have coevolved with bean to the extent that they do not naturally infect any other plant species and, in theory, would become extinct if bean was removed permanently from these agroecosystems, making cultivated bean a "dead-end host."

Cotton leaf crumple virus (CLCV), found primarily in Sonoran Desert agroecosystems in the southwestern United States and northwestern Mexico, has a restricted host range, and bean, cotton, and *Malva parviflora* are the only hosts found infected naturally. Although *M. parviflora* is a ubiquitous, perennial weed in cotton-growing areas of the Sonoran Desert in the United States and Mexico that is commonly infected with CLCV, there is no information regarding its importance as a primary or overwintering virus. Presumably, this weed must serve as a natural perennial reservoir of CLCV because, in this region, cotton is cultivated as an annual, making necessary the reinfection of cotton plantings each season. The CP sequences of more than 40 isolates collected over four seasons from cotton and *M. parviflora* in the Sonoran Desert share greater than 99 percent CP sequence identity (Brown, 1998), suggesting that CLCV may be highly adapted to two or several species. At the other end of the spectrum, begomoviruses from the Eastern Hemisphere, including cotton leaf curl associated virus and its variants in Pakistan and India, have broad host ranges (Nateshan et al., 1996), infecting members of Compositae, Cucurbitaceae, Malvaceae, and Solanaceae, among others, suggesting that a broad host range may be characteristic of the most divergent of Old World begomovirus species.

Much attention has been given to determining experimental host ranges of begomoviruses, but little is known about the natural reservoirs or the role they may play in the epidemiology of begomovirus-incited diseases. In a study of the flora in south Florida to identify prospective reservoirs of *Tomato mottle virus* (ToMoV) shortly after it emerged as an important pathogen, only one noncultivated species, tropical soda apple *(Solanum viarum)*, could be identified as a naturally infected virus host. Although ToMoV was experimentally transmissible from tomato to *S. viarum* by *Bemisia tabaci* type B, it could not be transmitted from the weed to tomato, suggesting that tomato is the primary source of infection in all hosts. Consequently, this weed species is probably a "dead-end host" of the virus and, further, is not preferred by the B-type vector (McGovern et al., 1994). In Israel, *Cynanchum acutum* is a natural weed reservoir of *Tomato yellow leaf curl virus* (TYLCV-IS) that plays an important role in the repeated introduction of virus to tomato in the Jordan Valley (Cohen et al., 1988). Another reservoir for TYLCV in southern Spain (TYLCV-SP) is *Solanum nigrum*, from which the virus is experimentally transmissible to and from tomato (Bedford et al., 1998). However, it has not been determined whether *S. nigrum* is an indigenous host of TYLCV-SP, or if virus has instead spread to this host from infected tomato. For the majority of begomoviruses, weed reservoirs remain unexplored or unidentified, making it difficult to draw important epidemiological and ecological conclusions in relation to disease spread. Many extant begomoviruses appear to have adapted rather quickly to the cultivated host plants in which they are presently found because few have been documented in indigenous weed hosts. The one exception is the large group of begomoviruses in Puerto Rico that were first described from their indigenous weed hosts (Bird, 1957; Bird and Sanchez, 1971). For most other begomoviruses known in crops today, it seems likely that they have been associated with that crop species for some time, and that cultivated, not wild, hosts are the primary sources of inocula for subsequent infections of crop and weed hosts. When an infected weed host is identified, it may be possible to test this hypothesis by examining the extent of variability within weed- and crop-associated populations.

Interactions Between Viral-Encoded and Plant Determinants of Host Range

Host-specific factors are required to initiate replication and to redirect host cells to synthesize begomoviral DNA. Consequently, such factors directly influence viral host range. Replication of begomoviruses requires host factors for expression of *Rep* and for formation of replication complexes in the *ori*. Expression of TGMV *Rep* induces the accumulation of a DNA polymerase accessory factor (PCNA) in infected plant cells that binds to a host-encoded retinoblastoma-like (Rb) protein (RRB1), but unlike mastreviruses, TGMV lacks the recognizable Rb-binding motif (Ach et al., 1997). Binding of this protein to Rep redirects the cell, which has already entered S phase, resulting in conditions suitable for viral replication. Abrogation of this regulatory mechanism is thought to permit the initiation of viral DNA synthesis in what were terminally differentiated cells. As such, these proteins are key regulators of the cell cycle and cell differentiation (Ach et al., 1997; Hanley-Bowdoin et al., 1999). Consequently, host range may in part be dictated by the ability of the begomoviral Rep to control the expression of a limited number of host genes. Indeed, evidence from complementable and noncomplementable host adaptation studies indicates that begomoviral host ranges are limited by more than one kind of incompatible interaction (Petty et al., 1995; Gillette et al., 1998).

Excluding species that are not natural hosts because of vector-imposed constraints, and assuming replication is unhindered, host range determination in begomoviruses also involves viral- and host-encoded proteins that mediate viral DNA movement to and from the nucleus and intercellularly in phloem and/or nonphloem cells (Sanderfoot and Lazarowitz, 1996). The classical cell-to-cell movement protein of bipartite viruses (*BC1* ORF) and the nuclear shuttle protein (*BV1* ORF) may do so by interacting with viral DNA based on size and topology, rather than sequence (Rojas et al., 1998), suggesting that constraints on genome size may be dictated by size limitations in systemic movement instead of encapsidation (Rojas et al., 1998). The BC1 protein functions by modifying the size exclusion limit of the plasmodesmatal pore in advance of viral transport through the channel and it chaperones viral ssDNA-BV1 complexes originating in the cell nucleus to the cytoplasm of an uninfected cell, a key requisite to symptom development. This is substantiated by evidence that expression of *BC1* in transgenic tobacco plants mimics disease symptoms and is localized to plant cell wall and plasma membrane fractions. This process requires the association of BC1 with host endoplasmic reticulum membranes and use of resultant tubules as conduits for cell-to-cell spread. In recipient cells, the BV1-ssDNA complexes are released from BC1, allowing BV1 to target the ssDNA to the nucleus to initiate a new round of infection (Pascal et al., 1994; Sanderfoot and Lazarowitz, 1996; Ward et al., 1997). The host range of *Squash leaf curl virus*, extended host range strain (SLCV-E), can be altered by introducing a single missense mutation (Cys 98 to Arg) in the *BV1* ORF, disallowing infection of the typically universally permissive host *Nicotiana benthamiana* (Ingham and Lazarowitz, 1993; Ingham et al., 1995).

The mechanisms underlying the ability of certain geminiviruses, but not others, to invade nonphloem cells of their hosts are not yet known. However, this feature has important implications for begomovirus-host plant interactions and, hence, for viral host range. Begomoviruses that do not exist outside phloem tissues may remain host restricted but might nonetheless "compete," by some unknown mechanism, by occupying the niche so tenaciously that others cannot invade. Further, interactions between begomoviruses and other unrelated plant viruses in phloem and nonphloem tissues may have interesting epidemiological implications, including those stemming from prospects of synergism, helper viruses, opportunities to participate in transmission by a foreign vector, exposure to "new" prospective hosts, and horizontal transmission of foreign viral sequences.

Exploring Host Range Through Artificial Reassortants

Begomovirus variability depends ultimately on mutation but can also be influenced by other routes of genome alteration, including recombination and reassortment. Either or both of these latter mechanisms may be involved when multiple viruses infect the same plant host (Bisaro, 1994; Harrison and Robinson, 1999).

Aspects of host range, defined primarily by an ability to replicate and move inter- and intracellularly, can be examined in vivo by constructing reassortants between distantly and closely related bipartite begomoviruses. In this way, it is possible to examine interactions between viral- and host-specific factors involved in replication and movement that are capable of operating *in trans* between noncognate A and B components in whole plants. It has been hypothesized that viral strains of the same species should be capable of transactivating replication and movement functions in a reciprocal manner. In some instances, closely related viral species may also be capable of *trans*-complementation in one or both directions. This is expected because replication and movement require both virus-specific and broadly conserved viral and host sequences.

To predict the outcome of reassortment experiments, it is necessary to consider overall virus relationships with respect to predicted lineages and the specific sequence elements and amino acid motifs involved in replication and movement. For example, phylogenetic analyses of *African cassava mosaic virus* (ACMV) and *Indian cassava mosaic virus* (ICMV) indicate that they cluster with Old World begomoviruses, whereas *Bean dwarf mosaic virus* (BDMV) in Colombia and ToMoV in Florida are New World begomoviruses from the Caribbean Basin/Central America and Mexico. This New World lineage has many other members, including *Abutilon mosaic virus* (AbMV), *Chino del tomaté virus* (CdTV), Havana tomato virus (HTV), *Sida golden mosaic virus* (SiGMV), and taino tomato mottle virus (TTMoV) (see Figure 13.2). On the other hand, *Bean calico mosaic virus* (BCMoV), cabbage leaf curl virus (CaLCV), and SLCV are members of a lineage from the southwestern United States and northwestern Mexico (see Figure 13.2). Finally, BGMV and TGMV are found in separate clusters within the New World group (Rybicki, 1994; Padidam et al., 1995; see Figure 13.2).

Reciprocal exchanges of the A and B components of AbMV, ACMV, ICMV, and TGMV using the permissive experimental host *N. benthamiana* revealed interesting results concerning host specificity of replication functions and, perhaps, clues about begomovirus evolution. In no combination was *trans*-complementation of replication or movement observed. However, when cognate A and B components were coinoculated with all possible heterologous DNAs, ACMV was the only virus capable of mediating systemic movement of the other three viruses. Only TGMV and AbMV complemented movement reciprocally (Frischmuth et al., 1993). Subsequent studies revealed an incompatibility between BGMV-PR and TGMV in reciprocal exchange experiments and concluded that host specificity was not solely attributable to DNA B-encoded ORFs, and that both host and viral factors were essential for systemic movement (Pooma et al., 1996; Qin et al., 1998).

The two New World viruses BDMV and ToMoV were capable of forming viable reassortants by reciprocal exchange in *N. benthamiana,* despite their placement in different lineages (Gilbertson et al., 1993). Compared to wild-type infections, symptoms were attenuated and DNA B levels were reduced and serially passaged BDMV A and ToMoV B components were not viable beyond three passages. In contrast, after three to five passages of ToMoV A and BDMV B reassortants through *N. benthamiana,* symptom severity was increased compared to early passages and DNA levels were wild type. Examination of progeny virus indicated recombination had occurred in which the BDMV *ori* was replaced with that of ToMoV A, ultimately permitting more effective replication of the chimeric B component (Hou and Gilbertson, 1996). This provided the first experimental evidence for reassortment and intermolecular recombination between two begomoviruses, clearly demonstrating the rapidity with which a "new" virus can be generated when interacting sequences are highly conserved between distinct separate species. Factors involved in host adaptation of these viruses were further investigated by constructing reassortants in bean (BDMV-host adapted) and tomato (ToMoV-host adapted). Reassortants infected the respective "nonadapted" host, but infectivity was reduced, plants were asymptomatic, and DNA accumulation was reduced (Hou et al., 1998). Here, the B components of the nonadapted viruses were the deciding factor in forming viable reassortants in respective nonadapted hosts (Hou et al., 1998).

SLCV naturally infects pumpkin, CaLCV infects cabbage and *Arabidopsis,* and both viruses experimentally infect *N. benthamiana.* Both viruses are placed in the Sonoran Desert lineage of New World begomoviruses. In reciprocal exchange experiments, CaLCV A × SLCV B yielded infectious reassortants in *N. benthamiana,* but not in *Arabidopsis* or in pumpkin. Here, CaLCV A *trans*-replicated SLCV B, but no viable reassortants were produced in the reciprocal exchange in any of the three test species. These results suggest that although virus-specific *ori* and Rep sequences essential for replication may be sufficiently similar and permit *trans*-replication between distinct species of the same lineage, the ability to replicate is not the sole criterion for establishing a viable infection, given constraints are also imposed by movement functions in relation to host factors (Hill et al., 1999). In a similar experiment with two viruses from the same phylogenetic cluster, AbMV and SiGMV, reassortment was possible in one direction, but not in both. Here, although AbMV A × SiGMV B produced a compatible reassortant, the reassortant was more limited in host range than either parental virus, and accumulation of B component DNA was less than in wild type. Although replication was demonstrated for SiGMV A and AbMV B pseudorecombinants, systemic infection was not observed for the six hosts examined, despite the ability of (cognate) parental components to infect all six species. These results suggest highly specific, but as yet unknown interactions between host factors and viral sequences (Höfer et al., 1997).

WHITEFLY VECTOR BIOLOGY
THAT INFLUENCES EPIDEMIOLOGY

Transmission Characteristics

Whitefly-mediated transmission of begomoviruses by *B. tabaci* is accomplished through a circulative process in which virions pass barriers within the vector, ultimately reaching the accessory salivary glands, the site from which particles are thought to be released or "transmitted" in saliva during whitefly feeding. This type of interaction is termed persistent, or one in which the virus persists for at least 100 hours following acquisition, but usually for the life of the vector. Successful transmission typically requires an acquisition access period on infected source plants of one to several hours, a latent period of 6 to 12 hours during which the virus is not yet transmissible, and an inoculation access period on test plants of 30 minutes or more. During the latent period, although virions are detectable in the whitefly, inoculation access feeding on a host plant does not result in virus transmission. Once virions enter the accessory salivary gland, whiteflies are capable of transmission for 10 to 15 days or for the life of the vector.

Viral Determinants of Specificity

B. tabaci is the only known vector of begomoviruses worldwide (Bird and Maramorosch, 1978; Muniyappa, 1980; Bedford et al., 1994), suggesting a highly conserved mechanism for the *B. tabaci*-mediated transmission of begomoviruses (Rosell et al., 1999). Three lines of evidence support the exclusive role of the CP as the viral determinant in vector-mediated transmission: (1) the extreme conservation in the amino acid sequence of the CP shared among all begomoviruses, compared to the notable divergence among CP sequences of leafhopper-transmitted geminiviruses that are transmitted by one to several vector species; (2) evidence that replacement of a begomovirus *CP* with that of a leafhopper-transmitted virus rendered the begomovirus leafhopper transmissible (Briddon et al., 1990); and (3) *CP* deletion mutants, or a chimeric virus in which *CP* has been replaced by that of a non-whitefly-transmissible virus, that are not whitefly transmissible, even though the modified viruses cause a systemic infection. There is no evidence for a viral-encoded helper component; thus, the CP most likely is the viral protein that binds putative whitefly receptors (Noris et al., 1998).

Whitefly Determinants of Transmission Specificity

The basis for begomovirus-vector specificity within the transmission pathway is thought to involve a cascade of recognition events between viral CP and vector-encoded proteins that ultimately contribute to a transmission event, beginning with ingestion of virions by whitefly stylets, from where they pass into the whitefly gut. Virions enter the hemocoel after crossing the gut membrane, possibly by receptor-mediated endocytosis, and are carried in hemolymph until they enter the membrane-bound accessory salivary glands, thereby completing the acquisition phase. Begomoviruseses accumulate in the filter chamber (Hunter et al., 1998) and can be detected in the hemolymph, saliva, and honeydew of *B. tabaci* (Rosell et al., 1999). An aphid/luteovirus model suggests that phloem-inhabiting luteoviruses employ receptor-mediated endocytosis to permit movement across cellular membranes (Gildow, 1982, 1993). If a similar mechanism is applicable to the whitefly/begomovirus system, the most likely regions in which virus uptake could occur are the filter chamber, the anterior portion of the descending midgut, or the internal ileum. However, the midgut and internal ileum epithelia have a brush border at the apical membrane that appears ideal for virus absorption (Cicero et al., 1995; Ghanim et al., 2000). After the vector locates host phloem tissues, virions in saliva are targeted to companion cells during feeding.

SLCV is readily detectable in hemolymph of the B-type vector, but not in hemolymph of a nonvector whitefly capable of ingesting virus from the same plant (Rosell et al., 1999). It is possible that SLCV virions are incapable of crossing the gut membrane of the nonvector, or that they are rapidly degraded once on the other side. These features are nonetheless suggestive of the requirement for specific interactions between virions and whitefly gut membranes, and, further, this interface may be one main line of specificity in the transmission pathway. In either case, apparently, virions are "protected" while associated with vector hemolymph, but the putative "protectant" is not sufficiently present in nonvector whiteflies (Hunter et al., 1996, 1998; Rosell et al., 1999). Recent evidence points to the involvement of a chaperonin protein, GroEL, in whitefly hemolymph that binds begomovirus CP (Morin et al., 1999), possibly suggesting that whitefly GroEL is responsible for "protection" of virions while in the hemolymph enroute to the salivary glands.

The environment in the accessory salivary glands may be ideal for absorption of virions from the hemolymph, compartmentalizing virions in vesicles and delivering them with the watery secretions associated with this gland. These hypotheses are consistent with findings in a well-studied, aphid/luteovirus system (Gildow 1982, 1993) in which the virus-salivary gland/ASG interface is a site of transmission specificity. An understanding of the sites of specificity and the processes involved in whitefly-mediated transmission may facilitate novel approaches for neutralizing begomovirus transmission, possibly in transgenic plants, to minimize vector-mediated virus spread and to reduce disease incidence.

THE INFLUENCE OF VECTOR BIOTYPES, RACES, AND VARIANTS ON VIRUS SPREAD

Biological "Races" or "Types" of B. tabaci, the Only Known Vector of Begomoviruses

Although begomoviruses are transmitted by a single species of whitefly, ample evidence supports biotic variation between isolated populations or variants, despite the absence of correspondingly useful morphological differences to permit their separation (Gill, 1990). Several early reports implicated biological races or strains of *B. tabaci* that limited or facilitated the transmission and, therefore, the host range of certain viruses (Brown and Bird, 1992; Bird and Maramorosch, 1978). Particular vector traits such as host preferences and host adaptiveness can feasibly influence the emergence or resurgence of begomoviruses (Bird, 1957). The recognition of "races" of *B. tabaci* that exhibited different host ranges or host preferences

(Bird, 1957; Bird and Sanchez, 1971; Bird and Maramorosch, 1978; Brown, Coats, et al., 1995) provided the first evidence of a prominent biotic feature that could drive selectivity of transmission, thereby influencing the dissemination and mixing of begomoviruses in common hosts. In Puerto Rico, the highly polyphagous "*Sida* race," which colonizes *Sida* and many other plant species in several families, was the only known vector of over 30 begomovirus diseases of weed and crop species (Bird and Sanchez, 1971; Bird and Maramorosch, 1978; Brown and Bird, 1992). In contrast, *Jatropha mosaic virus* (JMV) was restricted to *Jatropha gossypifolia* (and two close relatives), at least in part, because its vector, the monophagous "*Jatropha* race," was unable to colonize additional plant species on the island (Bird, 1957).

With widespread and frequent upsurges in *B. tabaci* populations in agroecoystems throughout the subtropical world (Mound and Halsey, 1978; Byrne et al., 1990; Brown and Bird, 1992; Cock, 1993; Brown, 1990, 1994; Traboulsi, 1994), much attention has been given to exploring biotic and genic differences between isolated *B. tabaci* populations (Costa and Brown, 1991; Burban et al., 1992; Wool et al., 1993; Brown, Coats, et al., 1995; Brown, Frohlich, and Rosell, 1995; Frohlich et al., 1999; De Barro and Driver, 1997; Guirao et al., 1997) toward understanding how such variability may influence begomovirus epidemiology. Biotic variation in *B. tabaci* is manifested primarily through host range or host specialization and through differences in fecundity, dispersal behavior, and virus transmission fficiency. In the case of the B biotype, the ability to induce phytotoxic dis - orders in certain hosts may also come into play because the physiology of the host is altered dramatically by the interaction (Bedford et al., 1994; Bird, 1957; Costa and Brown, 1991; Brown, Coats, et al., 1995; Brown, Frohlich, and Rosell, 1995). These phenotypes, and others, may directly or indirectly facilitate or undermine opportunities for virus dispersal, host range expansion or restriction, and the formation of "new viruses" through recombination and reassortment.

Transmission experiments with a well-characterized collection of *B. tabaci* variants and a number of Old and New World viruses revealed that restricted or negative transmission rarely occurred, while most virus-vector combinations examined were viable, the most important factor being compatibility (Bedford et al., 1994). This is remarkable, as the broad representation of begomoviruses and the host species belonging to multiple genera and plant families further suggests that most begomoviruses can be transmitted by this typically polyphagous vector in the majority of cases. Several host-adapted virus and/or vector populations are well-documented; among the best examples are JMV and the *Jatropha* race in Puerto Rico (Bird,

1957), the cassava-adapted *B. tabaci* race from Ivory Coast that transmits cassava-infecting begomoviruses (Burban et al., 1992), *Asystasia golden mosaic virus* and the Benin biotype from Benin (Bedford et al., 1994), and *B. tabaci* from sweet potato in Nigeria (NI), which was a poor vector of most viruses examimed. except for TYLCV from Yemen (TYLCV-YE), which it readily transmitted (Bedford et al., 1994).

That transmission of most begomoviruses is typically possible by most *B. tabaci* biotypes or populations, but does not rule out the differences in efficiency of transmission, also has been shown by Bedford et al. (1994). Certain vector biotypes or populations were clearly more efficient vectors of certain viruses than others (Bedford et al., 1994). This result was also reported for several Old World tomato-infecting begomoviruses and their vectors that were of the same host and geographic origin (McGrath and Harrison, 1995). In this study, transmission frequency to and from tomato was greatest with combinations from the same sites, and virus relationships were broadly correlated with viral CP epitope profiles for begomoviruses from similar geographic locations (McGrath and Harrison, 1995). Idris (1997) also showed that a New World *B. tabaci* (type A) was a more efficient vector of a New World tomato-infecting virus than the Old World B type, even though the B-type vector exhibits a stronger preference for tomato than the A type. In contrast, the B-type more efficiently transmitted an Old World tomato-infecting begomovirus, compared to the New World A-type vector. Despite evidence that biotypes of *B. tabaci* are capable of transmitting all begomoviruses to some degree to and from the host species upon which they are capable of feeding, these findings suggest that transmission may be most efficient between certain begomoviruses and their indigenous vector(s), owing to a close and constant association. These differences may be manifested through enhanced binding affinity or avidity of virions for whitefly receptors in the gut and/or accessory salivary glands, differential binding of virions by GroEL in vector hemolymph, and/or to other factors that influence the quantity of virions that is ingested or acquired.

A Cryptic Species: Biological Types and Topotypes

Exhaustive light microscopic (Bedford et al., 1994) and transmission electron microscopic (Rosell et al., 1997) examination of a collection of *B. tabaci* specimens from different geographic sites and hosts using key morphological characters revealed an absence of useful characters for distinguishing biological variants. This invariance in morphological characters, together with evidence of widespread biotic and genic polymorphism, has led to the hypothesis that *B. tabaci* is collectively a "cryptic species," or a species complex (Brown, Frohlich, and Rosell, 1995; Rosell et al., 1997; Frohlich et al., 1999). Owing to the cryptic nature of this vector, various molecular marker sequences have been explored for their potential to resolve relationships between variants of *B. tabaci* (Burban et al., 1992; Wool et al., 1993; Brown, Coats, et al., 1995; Brown, Frohlich, and Rosell, 1995; De Barro and Driver, 1997; Brown, 2001). the mitochondrial (mt) *COI* gene is the most informative molecular marker to date, having permitted a separation of several distinct groupings of *B. tabaci* with geographic and/or host of origin as the two primary overriding factors in their separation (Frohlich et al., 1999; Kirk et al., 2000; Rey et al., 2000; Brown, 2001). Five or six major groups that exhibit 10 percent or greater sequence divergence can be resolved by mt *COI* sequences (see Table 13.2 and Figure 13.3).

A notable example in which specific biotic features can be shown to greatly influence the importance of a vector population is the B type of *B. tabaci*, now the most important pest and vector of emerging begomoviruses in the Americas. This Old World variant was transported throughout the United States and Caribbean Basin on nursery plants, where it became established during the period 1986-1990 (Gill, 1992; Brown, Coates, et al., 1995; Brown, Frohlich, and Rosell, 1995). The B type is characterized by a wide host range and its ability to induce phytotoxic disorders and to transmit New and Old World begomoviruses. Certain New World begomoviruses have been identified only recently in plant species once considered nonhosts of indigenous begomoviruses (Bird and Sanchez, 1971; Brown, Frohlich, and Rosell, 1995), further emphasizing the importance of vectopreferences in the prospective host range of begomoviruses. In addition, the B type was difficult, if not impossible, to control with insecticides available at the time of its introduction (Coats et al., 1994), which led to the suggestion that the insecticide-resistant phenotype was a major factor in the original success of the B type in its Old World habitat, and in its subsequent establishment in the New World. Mt *COI* sequence comparisons now provide convincing evidence for affiliation of the B type with the Middle East/Mediterranean/Northern Africa cluster of *B. tabaci* (see Figure 13.3). Extremely low divergence between mt *COI* sequences of all B-type individuals examined from many sites worldwide indicates that the B type probably resulted from a single founder event sometime before 1985. The highly fecund, extremely polyphagous B type of *B. tabaci* appears to have arisen as a result of heavy pesticide use in arid, irrigated agroecosystems (Frohlich et al., 1999).

TABLE 13.2. Field and Reference Collections of *Bemisia tabaci*, Geographic Origin, and Plant Host Species for Which Mitochondrial *COI* Gene Sequences Were Compared by Maximum Parsimony Analysis

B. tabaci Collection	Geographic Orlgin	Host Species
ABA Benin	Benin[3]	*Asystasia* spp.
AZA Arizona	Arizona[2,6,11]	*Gossypium hirsutum*
AZB Arizona	Arizona[2,6,10]	*Euphorbia pulcherrima*
B Guatemala	Guatemala[2]	*Cucumis melo*
B Israel	Israel[2]	*Lycopersicon esculentum*
B Japan	Japan[2]	*E. pulcherrima*
Bolivia tomato	Bolivia[2]	*L. esculentum*
Cameroon cassava	Cameroon[13]	*Manihot esculenta*
CC (B) California	California[3]	*G. hirsutum*
CUL Mexico	Culiacan, Mexico[2,6]	*L. esculentum*
FC (B) Florida	Florida[3]	*Solanum nigrum*
HC China	China[7]	*G. hirsutum*
Israel Euphorbia	Israel[8]	*Euphorbia* spp.
Israel Lantana	Israel[8]	*Lantana camara*
Italy eggplant	Italy[8]	*S. melongena*
IW India	India[3]	*Citrullus vulgaris*
JAT Puerto Rico	Puerto Rico[5,6]	*Jatropha gossypifolia*
Malaysia Malva	Malaysia[1]	*Malva parviflora*
Malaysia Sonchus	Malaysia[1]	*S. oleraceus*
Morroco tomato	Morroco[2]	*L. esculentum*
Mozambique	Mozambique[13]	*M. esculenta*
NEW-Nepal	Nepal[4]	*C. vulgaris*
Pakistan eggplant	Pakistan[8]	*S. melongena*
Pakistan squash	Pakistan[8]	*Cucurbita maxima*
PC 91 Pakistan	Pakistan[3]	*G. hirsutum*
PC 92 Pakistan	Pakistan[3]	*G. hirsutum*
PC 95 Pakistan	Pakistan[3]	*G. hirsutum*
SC Sudan	Sudan[3]	*G. hirsutum*
SIDA Puerto Rico	Puerto Rico[5]	*Chamaesyces hypericifolia*
SP 95 Spain	Spain[4]	*Cucurbita melo*
SP 99 Spain	Spain[2]	*L. esculentum*
Spain Ipomoea	Spain[1]	*Ipomoea batatas*
TC Turkey	Turkey[3]	*G. hirsutum*
Thailand cowpea	Thailand[1]	*Vigna unguiculata*
Thailand cuke	Thailand[1]	*C. pepo*
Thailand Euphorbia	Thailand[1]	*E. pulchrima*

TABLE 13.2 *(continued)*

Uganda 1 cassava	Uganda[12]	*M. esculenta*
Uganda 2 cassava	Uganda[12]	*M. esculenta*
Uganda 3 cassava	Uganda[12]	*M. esculenta*
Uganda 4 Sw Pot	Uganda[12]	*I. batatas*

Note: Relationships predicted by *COI* sequences are shown in Figure 13.3.

[1] Kirk et al., 2000.

[2] From Arizona collection or culture; J. K. Brown et al.

[3] Bedford et al., 1994.

[4] Rosell et al., 1997.

[5] From culture of J. Bird, University of Puerto Rico, Rio Piedras.

[6] Costa et al., 1993.

[7] From culture of I. Bedford and P.G. Markham, John Innes Centre, Norwich, United Kingdom.

[8] Collectors A.Kirk and L. Lacey, USDA/ARS-Foreign Exploration.

[9] Brown, Coates, et al., 1995; Brown, Frohlich, and Rosell, 1995; Frohlich et al., 1999.

[10] Prototype "B biotype"; Costa and Brown, 1991.

[11] Prototype "A biotype"; Costa and Brown, 1991.

[12] Collectors J. Legg and W. Otim-Nape, IITA Cassava IPM Project, Kampala, Uganda.

[13] Collectors S. Berry and C. Rey; Rey et al., 2000.

NEW DISEASES, EPIDEMICS, AND PANDEMICS

Passiflora *Leaf Mottle Disease in Puerto Rico Caused by Local Begomovirus Mobilized by the Exotic B Vector*

Direct evidence of the influence of vector biotype alteration has been explored recently in the Americas, allowing insight into the molecular epidemiology of a new begomovirus-incited disease of *Passiflora edulis*. *Passiflora* leaf mottle disease was reported for the first time in Puerto Rico, shortly after the invasion of the B type (Brown et al., 1993). The begomovirus JMV that naturally infects two wild species of passion fruit and the weed *Jatropha gossypifolia* in Puerto Rico was transmissible only by the host-specific *Jatropha* biotype. However, JMV was suspected as a causal agent of the new disease of *Passiflora* spp. because symptoms resembled the unique, severe leaf distortion associated with JMV-infected *Jatropha* spp.

Experimental transmission of the *Passiflora*-associated virus with the *Jatropha* and B biotypes revealed that JMV was transmissible from *Jatropha* and the two wild *Passiflora* species to *Passiflora* by the *Jatropha* type, although transmission occurred at a very low frequency. Although the B type

FIGURE 13.3. Single Most Parsimonious Tree Showing Relationships Between Biotypes, Variants, and Topotypes of *Bemisia tabaci* Based on the Cytochrome Oxidase I Nucleotide Sequence

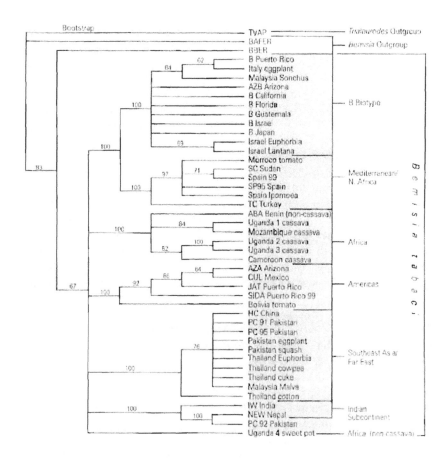

Note: Outgroups are the greenhouse whitefly, *Trialeurodes vaporariorum* (TVAP), *Bemisia afer* (BAFER), and *B. berbericola* (BBER). Shaded area highlights the single large cluster of *B. tabaci*, in which are delineated six main geographically based subgroups of representative *B. tabaci* and one noncassava specialist from sweet potato in Africa; all other African exemplars were collected from cassava. Acronyms collection host species, geographic origin, biotype reference designations, and collectors are shown in Table 13.2.

was unable to transmit JMV from *Jatropha* at all, it was a highly effective vector of the virus between *Passiflora* plants. Further, the *Jatropha* biotype was unable to transmit JMV between *Passiflora* plants, or from *Passiflora* spp. to either of its natural *Jatropha* spp. hosts. However, the B biotype could not transmit JMV from *Passiflora* to either wild *Jatropha* species. These studies led to the hypothesis that the *Jatropha* biotype probably transmitted JMV incidentally to *Plassiflora* spp. on rare occasions, but that JMV did not spread in plantations because the *Jatropha* biotype did not colonize *Passiflora* spp., even though *Passiflora* was a competent JMV host. Molecular analysis of virus nucleotide sequences from all hosts confirmed that JMV was associated with *Passiflora* spp. in plantations, two *Jatropha* species, and a wild *Plassifora* species adjacent to plantations (Brown et al., 1993; Brown and Bird, 1996).

These results suggested that the host-restricted *Jatropha* biotype had incidentally transmitted JMV to *Passiflora* on prior occasions, but that the virus was unable to spread within plantations until *Passiflora* was colonized by the exotic B biotype. Thus, the epidemic was initiated because the B type effectively colonized *Passiflora* and transmitted JMV to and from *Passiflora*. This is the first example of a begomovirus-incited disease resulting from interactions between two distinct vector biotypes from Old and New World sites, respectively, and a New World virus that was previously host restricted due to limitations imposed by its whitefly vector (Brown et al., 1993; Brown and Bird, 1996).

Mosaic Disease of Bean Caused by Macroptilium Golden Mosaic Virus *from Puerto Rico That Is Distinct from BGMV*

BGMV-PR has long been suspected to incite disease and cause the bright yellow mosaic symptoms in *M. lathyroides,* a common weed in Puerto Rico. Examination of *M. lathyroides,* a common weed plant exhibiting bright yellow mosaic symptoms collected recently in Puerto Rico, revealed the presence of a previously unidentified begomovirus pathogen of bean. Interestingly, inoculation of bean by viruliferous whiteflies exposed to symptomatic *M. lathyroides,* a weed native to Puerto Rico, yielded infected bean plants exhibiting a green-yellow mosaic and stunting, symptoms reminiscent of BGMV-PR infection. However, repeated inoculation of *M. lathyroides* with the classical BGMV-PR by whitefly transmission or biolistic means revealed that *M. lathyroides* is not a host of BGMV (Idris et al., 1999), indicating that previous assumptions regarding the host range of BGMV were incorrect. Examination of CP sequences of the *M. lathyroides* virus indicated 77.3 to

79.3 percent shared nucleotide sequence identity with BGMV from the Dominican Republic, Honduras, Guatemala, Jamaica, and Puerto Rico, leading to the provisional designation of MGMV-PR as a new species. Comparison of the *cis*-acting elements in the common region revealed that BGMV-PR and MGMV have distinct (theoretical) Rep-binding sites (Idris et al., 1999) and further suggests the existence of two distinct bean-infecting species.

The unexpected discovery of MGMV suggests that it may pose a serious new threat to bean production in the region if tolerant or resistant germplasm does not protect plants against infection by this new virus. Bean breeding programs and associated research in the Caribbean and Central America have traditionally been aimed at developing begomovirus resistance in bean to BGMV because it was the only bean-infecting begomovirus recognized in the region (Faria et al., 1994). Interestingly, BGMV is known to naturally infect only *Phaseolus vulgaris*, and no alternate cultivated or weed hosts have been identified to date (J. Bird, personal communication). The latter observation suggests that BGMV-PR coevolved with bean in the Caribbean/Central America for some time and is highly host adapted. One positive implication is that BGMV isolates throughout the Caribbean region exhibit little variation, as has been shown experimentally (Faria et al., 1994), and should theoretically be amenable to control by cultivation of resistant varieties, with minimal likelihood of selecting for isolates that overcome host resistance genes. On the contrary, these findings emphasize the importance of periodic surveillance of begomoviruses to permit early discovery of emergent viruses and to routinely ascertain the identity of virus isolates used in breeding programs, particularly in areas invaded by new vector types or following atypical outbreaks of local vector populations.

Evaluation of bean cultivar responses to BGMV in Florida has shown promise toward developing resistant varieties (Velez and Bassett, 1998); however, the begomovirus infecting beans in Florida constitutes a genotype that is still unique with respect to BGMV-PR/Caribbean and MGMV (Idris, Hiebert, and Brown, in preparation). Evaluation of identical bean lines in Florida and Caribbean sites would be highly useful, given that at least three bean-infecting begomoviruses are now recognized between the two locations. A second highly homogeneous bean-infecting begomovirus in Brazil, BGMV-BZ, that infects only bean (Gilbertson et al., 1993; J. Faria, personal communication) has been the target of a local bean breeding program in which resistant germplasm that confers oligogenic-based protection was identified based on additive gene action. Here, broad and narrow heritabilities were highly correlated with all response traits, suggesting that they were selected simultaneously for enhancement (Pessoni et al., 1997); however, it is likely that this resistance is virus specific.

Cassava-Infecting Diseases in Africa not Caused by ACMV

Mosaic diseases of cassava have been known in sub-Saharan Africa since the turn of the century. Until recently, ACMV and *East African cassava mosaic virus* (EACMV) were the only begomoviruses associated with the disease. ACMV occurs in Angola, the Ivory Coast, Kenya, and Nigeria, while EACMV predominates eastwardly in Kenya, Madagascar, Malawi, Mozambique, Tanzania, Zambia, and Zimbabwe (Berrie et al., 1998; Zhou, Robinson, and Harrison, 1998). However, in the late 1980s, a devastating epidemic caused by an unusually severe mosaic isolate was reported for the first time in north-central Uganda and, since that time, has spread to infect large portions of East and Central Africa (Harrison, Liu, Zhou, et al., 1997; Harrison, Zhou, et al., 1997). The epidemic spread southward at a rate of 20 to 30 km per year based on high disease incidence in northward localities and low incidence in the southern region of the study zone (Legg and Ogwal, 1998). Most cassava varieties are highly susceptible, and entire crops have been lost to the disease since its emergence. Analysis of symptomatic cassava revealed the presence of a novel, more virulent form of ACMV, provisionally referred to as the Uganda Variant (UgV). It is hypothesized that UgV arose by interspecific recombination between portions of the native ACMV and EACMV, based on evidence that UgV contains a large portion of *AV1* identical to ACMV, while the remainder of the genome is similar to that of EACMV (Zhou et al., 1997).

Studies to date implicate UgV as the causal agent of the disease, but it is not yet clear if symptom severity is due to coinfection by UgV and ACMV in West Africa or by UgV alone (Zhou, Robinson, and Harrison, 1998). In addition, the whitefly vector has been reported to interact synergistically with virus-diseased cassava host plants, reproducing more rapidly on diseased than on healthy plants (Colvin et al., 1999). Preliminary analysis of mt *COI* sequences for whitefly vector populations collected in transects across the spreading epidemic indicates that a previously unknown *B. tabaci* strain is associated with the epidemic, in that it appears to be distinct from the type occurring in adjacent nonepidemic areas (Fauquet et al., 1998; Legg, 1996; Legg and Brown, in preparation).

At least an additional two more cassava-infecting begomoviruses have been reported recently in Africa. South African cassava mosaic virus (SACMV) has a CP gene nucleotide sequence similar to members of the TYLCV-Is cluster, although it groups biologically with other cassava-infecting begomoviruses (Berrie et al., 1998). A second variant from Cameroon that may represent yet another putative recombinant begomovirus from cassava is currently under study (Fondong, 1999).

Collectively, evidence for an increased number of cassava-associated begomoviruses during the past decade further emphasizes the need for accurate virus identity and tracking of begomoviruses. The dynamics underlying the recent emergence of begomoviruses in cassava throughout Africa will likely be difficult to re-create, given the lack of information about the distribution and variation in begomovirus and whitefly vector populations at the outset.

Divergence in Cotton-Infecting Viruses from the Eastern and Western Hemispheres

Cotton-infecting begomoviruses have become economically important during the 1990s. One of the most dramatic examples was the outbreak of cotton leaf curl disease in the early 1990s in Pakistan when high-yielding varieties replaced lower-yielding, virus-tolerant cultivars (Mansoor et al., 1999). Leaf curl epidemics have also been reported in India, and although Koch's Postulates have not yet been completed, evidence suggests that more than one virus is associated with epidemics in India (Nateshan et al., 1996) and Pakistan (Harrison, Lui, Khalid, et al., 1997; Zhou, Lui, et al., 1998a).

Since multiple begomovirus genotypes are associated with symptomatic cotton (Liu et al., 1998; Zhou, Robinson, and Harrison, 1998) which exhibits diverse disease symptoms, such as upward and downward leaf curl, enations, thickening of veins, and various degrees of stunting, the exact nature of the causal agent(s) has not been elucidated (Briddon et al., 2000a). Further, the extent and distribution of viral variants are not clear, nor have relationships between distinct whitefly biotypes and transmission relationships with leaf curl variants been examined.

Recently, a novel circular ssDNA has been associated with symptomatic cotton from Pakistan. The ssDNA is half the size of a begomovirus genome but shares no sequence identity with the begomovirus genomes isolated from symptomatic cotton. Instead, the ssDNA contains an origin of replication, including the hallmark nonanucleotide (albeit, with a single base substitution) akin to that in begomoviruses, and encodes a protein that shares high sequence similarity with nanovirus-like Rep-associated proteins. The ssDNA autonomously replicates in tobacco and is encapsidated as a virion in begomovirus CP, suggesting that the foreign DNA is whitefly transmissible by association with a whitefly-transmissible begomovirus. Although an apparently viable, full-length begomoviral DNA was cloned from symptomatic cotton, inoculation of cotton or tobacco seedlins with the clone did not yield typical severe leaf curl symptoms (Mansoor et al., 1999; Briddon et al., 2000a). Quite recently, an ssDNA satellite molecule together with the cloned helper CLCuV genome were shown to be essential for reproducing wild-type disease symptoms (Briddon et al., 2000b).

In Egypt, a distinct begomovirus, hollyhock leaf crumple virus (HLCrV) has been cloned and its nucleotide sequence determined (Abdel-Salam, 1998), revealing a distinct new viral species affecting a malvaceous host. Also, a new begomoviral species was associated with cotton exhibiting leaf curl symptoms in Sudan. Based on CP, LIR, and Rep comparisons, nearly identical isolates of this same species were detected in *Sida* and okra from two different locations in Sudan. Comparison of full-length genomes of the leaf curl/crumple-associated viruses from malvaceous hosts in Pakistan, Egypt, and Sudan indicates that at least four separate species may be involved (Idris and Brown, 2000). These viruses have likely resurged owing to new whitefly vector pressures and to the recent cultivation of high-yielding, virus-susceptible cotton varieties throughout the world.

New World cotton-infecting begomoviruses in the Sonoran Desert (Arizona, California, and Sonora, Mexico) have been problematic since the 1950s. However, the establishment of the B-type vector in 1986 resulted in the resurgence of CLCV and the emergence of begomoviruses of cotton and kenaf *(Hibiscus cannabinus)* in Texas, Guatemala, and the Dominican Republic. CLCV from Arizona was previously investigated with respect to host range and vector transmission characteristics. Only recently have nucleotide sequence comparisons been carried out with these and other begomoviruses of malvaceous plants. Analysis of *CP, Rep,* and LIR sequences has revealed that begomoviruses of malvaceous hosts are dispersed across four or five clusters of New World begomoviruses, while cotton-infecting viruses from Arizona, Texas, and Guatemala constitute separate species and reside in the SLCV cluster with other begomoviruses from the Sonoran Desert and nearby agroecosystems.

Epidemiologically Based "Customizing" of Disease Resistance

Whether predictions can be made concerning the utility of highly conserved viral-encoded genes or partial sequences, based on phylogenetic predictions of virus relationships using those sequences, has not yet been explored. Careful selection of viral-derived sequences for strategies based on transgenic resistance is essential if useful and sustainable protection is to be achieved. For example, it is important to understand the distribution of particular begomoviruses in relation to varietal development efforts targeted for specific environments and markets.

Given the apparent high degree of resistance specificity revealed through recent plant breeding efforts and virus-derived resistance strategies explored for geminiviruses (Duan et al., 1997; Pessoni et al., 1997; Velez and Bassett, 1998), it may be important to consider alternate strategies when feasible. For example, conserved sequences in Rep or TrAP that encode functional motifs found within a single species and in strains of that species, or across entire viral lineages, may provide useful targets for transgenic resistance. Likewise, regions of the CP or functional motifs in movement genes that exhibit minimal divergence may present useful targets. In this light, closely related members of the BGMV-Caribbean virus cluster, variants of cotton-associated begomoviruses in Pakistan and India, and tomato-infecting viruses of the TYLCV-Is cluster (see Figure 13.2) may provide targets for exploring customized resistance by exploiting sequences shared across minimally divergent taxa. More challenging will be the AbMV or SLCV lineages that contain numerous species but that may share functionally important motifs sufficient to devise broadly based resistance in the same crop.

CONCLUSIONS

Much remains to be learned about the mechanisms underlying begomovirus pathogenesis, symptom induction, and host range, and how they influence disease epidemiology. The extent of diversity within and between begomoviral quasi-species is only now being realized. The rapidity with which begomovirus genomes are capable of changing through recombination, mutation, and reassortment has become apparent only in the past several years. Little is known about the intricacies of the begomovirus-whitefly transmission pathway that may influence disease spread and virus variability through variation in vector transmission efficiency. Preliminary observations suggest the likelihood that certain begomovirus-vector combinations coevolved, creating viruses capable of "competing" in mixtures for putative gut and salivary receptors and hemolymph-shuttled chaperonins. Only in the past ten years has definitive proof been provided for biological and genic variation among whitefly vector populations, but how these differences influence inoculum levels and pressures and dissemination, in light of emergent and resurgent begomoviruses, is not understood.

The recognition that begomoviruses are capable of rapidly diverging through multiple mechanisms underscores the need for accurate molecularly based methods that permit detection and tracking of biologically significant variants. Molecular approaches must combine knowledge of biology and ecology and the ability to monitor both conserved sequences and specific sites most likely to undergo alteration with phylogenetic predictions to facilitate accurate identification and tracking of begomovirus variants and to recognize new or resurgent viruses. Establishing databases of baseline sequences for extant viruses will permit future comparisons in establishing and interpreting disease patterns and associated trends for vector populations. Accurate molecular epidemiological information will assist plant breeding efforts in developing disease-resistant varieties by enabling the selection of timely and relevant viral species (and variants) for germplasm screening. In the long run, conserved and/or unique viral sequences will lend assistance in developing virus resistance through pathogen-derived resistance approaches.

REFERENCES

Abdel-Salam AM (1998). Biological, biochemical and serological studies on hollyhock leaf crumple virus (HLCrV): A newly discovered whitefly transmitted geminivirus. *Arab Journal of Biotechnology* 1: 41-58.

Ach RA, Durfee T, Miller AB, Taranto P, Hanley-Bowdoin L, Zambryski PC, Gruissem W (1997). RRB1 and RRB2 encode maize retinoblastoma-related proteins that interact with a plant D-type cyclin and geminivirus replication protein. *Molecular and Cellular Biology* 17: 5077-5086.

Argüello-Astorga GR, Guevara-Gonzalez RG, Herrera-Estrella LR, Rivera-Bustamante RF (1994). Geminivirus replication origins have a group-specific organization of iterative elements: A model for replication. *Virology* 203: 90-100.

Bedford ID, Briddon RW, Brown JK, Rosell RC, Markham PG (1994). Geminivirus transmission and biological characterisation of *Bemisia tabaci* (Genn-adius) biotypes from different geographic regions. *Annals of Applied Biology* 125: 311-325.

Bedford ID, Kelly A, Banks GK, Briddon RW, Cenis JL, Markham PG (1998). *Solanum nigrum:* An indigenous weed reservoir for a tomato yellow leaf curl geminivirus in southern Spain. *European Journal of Plant Pathology* 104: 221-222.

Berrie LC, Palmer KE, Rybicki EP, Rey MEC (1998). Molecular characterization of a distinct South African cassava infecting geminivirus. Archives of Virology 143: 2253-2260.

Bird J (1957). A whitefly-transmitted mosaic of *Jatropha gossypifolia*. Technical paper. Agricultural Experimental Station, University of Puerto Rico 22: 1-35.

Bird J, Maramorosch K (1978). Viruses and virus diseases associated with whiteflies. *Advances in Virus Research* 22: 55-110.

Bird J, Sanchez J (1971). Whitefly-transmitted viruses in Puerto Rico. Technical paper. Agricultural Experimental Station, University of Puerto Rico 55: 461-467.

Bisaro DM (1994). Recombination in geminiviruses: Mechanisms for maintaining genome size and generating genomic diversity. In J Paszkowski (ed.), *Homologous Recombination and Gene Silencing in Plants*. Kluwer, Dordrecht, The Netherlands, pp. 39-60.

Bisaro DM (1996). Geminivirus DNA replication. In Pamphilis De (ed.), *DNA Replication in Eukaryotic Cells*. Cold Spring Harbor Laboratory Press, Cold Spring Harbor, NY, pp. 833-854.

Bock KR, Harrison BD (1985). African cassava mosaic virus. *AAB Descriptions of Plant Viruses*. AAB, Wellesbourne, Warwick, UK, 6p.

Briddon RW, Mansoor S, Bedford ID, Pinner MS, Markham PG (2000a). Clones of cotton leaf curl geminivirus induce symptoms atypical of cotton leaf curl disease. *Viral Genes* 20: 19-26.

Briddon RW, Mansoor S, Bedford ID, Pinner MS, Markham PG (2000b). A novel component required for induction of cotton leaf curl disease symptoms. *Phytopathology* 90: S9 (abstr).

Briddon RW, Pinner MS, Stanley J, Markham PG (1990). Geminivirus coat protein gene replacement alters insect specificity. *Virology* 177: 85-94.

Brown JK (1990). An update on the whitefly-transmitted geminiviruses in the Americas and the Caribbean Basin. *FAO Plant Protection Bulletin* 39: 5-23.

Brown JK (1994). The status of *Bemisia tabaci* (Genn.) as a pest and vector in world agroecosystems. *FAO Plant Protection Bulletin* 42: 3-32.

Brown JK (1998). Global diversity and distribution of cotton-infecting geminiviruses: An essential requisite to developing sustainable resistance. In Duggar P, Richter D (eds.), *1998 Proceedings of Beltwide Cotton Conference*, San Diego, CA, January 5-9, 1998, National Cotton Council, Memphis, TN, pp. 155-160.

Brown JK (2001). Molecular markers for the identification and global tracking of whitefly vector-begomovirus complexes. *Virus Research* 71: 233-260.

Brown JK, Bird J (1992). Whitefly-transmitted geminiviruses in the Americas and the Caribbean Basin: Past and present. *Plant Disease* 76: 220-225.

Brown JK, Bird J (1996). Introduction of an exotic whitefly (*Bemisia*) vector facilitates secondary spread of *Jatropha* mosaic virus, a geminivirus previously vectored exclusively by the *Jatropha* biotype. In Gerling D, Mayer RT (eds.), *Bemisia '95: Taxonomy, Biology, Damage, Control and Management*. Intercept Publications, Wimborne, UK, pp. 351-353.

Brown JK, Coats SA, Bedford ID, Markham PG, Bird J, Frohlich DR (1995). Characterization and distribution of esterase electromorphs in the whitefly, *Bemisia tabaci* (Genn.) (Homoptera: Aleyrodidae). *Biochemical Genetics* 33: 205-214.

Brown JK, Fletcher D, Bird J (1993). First report of *Passiflora* leaf mottle caused by a whitefly-transmitted geminivirus in Puerto Rico. *Plant Disease* 77: 1264.

Brown JK, Frohlich DR, Rosell RC (1995). The sweetpotato or silverleaf white-flies: Biotypes of *Bemisia tabaci* or a species complex? *Annual Review of Entomology* 40: 511-534.

Brown JK, Orstrow KM, Idris AM, Stenger DC (2000). Chino del tomaté virus: Relationships to other begomoviruses and the identification of A component variants that affect symptom expression. *Phytopathology* 90: 546-552.

Brown JK, Torres-Jerez I, Idris AM, Banks GK, Wyatt SD (2001). The core region of the coat protein gene is highly useful for establishing the provisional identification and classification of begomoviruses. *Archives of Virology* (in press).

Burban C, Fishpool LDC, Fauquet C, Fargette D, Thouvenel J-C (1992). Host-associated biotypes within West African populations of the whitefly *Bemisia tabaci* (Genn.), Homoptera: Aleyrodidae. *Journal of Applied Entomology* 113: 416-423.

Byrne DN, Bellows Jr. TS, Parrella MP (1990). Whiteflies in agricultural systems. In Gerling D (ed.) *Whiteflies: Their Bionomics, Pest Status, and Management*, Intercept Ltd., Andover, Hants, UK, pp. 227-261.

Cicero JM, Hiebert E, Webb SE (1995). The alimentary canal of *Bemisia tabaci* and *Trialeurodes abutilonea* (Homoptera, Sternorrhynchi): Histology, ultrastructure and correlation to function. *Zoomorphology* 115: 31-39.

Coats SA, Brown JK, Hendrix DL (1994). Biochemical characterization of biotype-specific esterases in the whitefly *Bemisia tabaci* Genn. (Homoptera: Aleyrodidae). *Insect Biochemistry and Molecular Biology* 24: 723-728.

Cock MJW (1993). *Bemisia tabaci, an update 1986-1992 on the cotton whitefly with an annotated bibliography*. Commonwealth Agricultural Bureau IIBC, Silwood Park, UK, 78 pp.

Cohen S, Kern J, Harpaz I, Ben-Joseph R (1988). Epidemiological studies of the tomato yellow leaf curl virus (TYLCV) in the Jordan Valley, Israel. *Phytoparasitica* 16: 259-270.

Colvin J, Otim-Nape GW, Holt J, Omongo C, Seal S, Stevenson P, Gibson G, Cooter RJ, Thresh JM (1999). Factors driving the current epidemic of severe cassava mosaic disease in East Africa. In *Proceedings of VIIth International Plant Virus Epidemiology Symposium, Plant Virus Epidemiology: Current Status and Future Prospects*, April 11-16, 1999. Plant Virus Epidemiology Committee of the International Society of Plant Pathology, Aguadulce (Almeria), Spain. p. 76.

Costa AS (1976). Whitefly-transmitted plant diseases. *Annual Review of Phytopathology* 14: 429-449.

Costa HS, Brown JK (1991). Variation in biological characteristics and esterase patterns among populations of *Bemisia tabaci*, and the association of one population with silver leaf symptom induction. *Entomologia experimentalis et applicata* 61: 211-219.

Costa AS, Brown JK, Sivasubramaniam S, Bird J (1993). Regional distribution, insecticide resistance, and reciprocal crosses between the "A" and "B" biotypes of *Bemisia tabaci*. *Insect Science and Application* 14: 127-138.

Czosnek H, Laterrot H (1997). A worldwide survey of tomato yellow leaf curl viruses. *Archives of Virology* 142: 1391-1406.

Dawson WO, Hilf ME (1992). Host-range determinants of plant viruses. *Annual Review of Plant Physiology and Plant Molecular Biology* 43: 527-555.

De Barro PJ, Driver F (1997). Use of RAPD PCR to distinguish the B biotype from other biotypes of *B. tabaci* (Gennadius) (Hemiptera: Aleyrodidae). *Australian Journal of Entomology* 36: 149-152.

Duan YP, Powell CA, Webb SE, Purcifull DE, Hiebert E (1997). Geminivirus resistance in transgenic tobacco expressing mutated BC1 protein. *Molecular Plant-Microbe Interactions* 10: 617-623.

Faria JC, Gilbertson RL, Hanson SF, Morales FJ, Ahlquist P, Loniello AO, Maxwell DP (1994). Bean golden mosaic geminivirus type II isolates from the Dominican Republic and Guatemala: Nucleotide sequences, infectious pseudorecombinants, and phylogenetic relationships. *Phytopathology* 84: 321-329.

Fauquet CM, Pita J, Deng D, Torres-Jerez I, Otim-Nape WG, Ogwal S, Sangare A, Beachy RN, Brown JK (1998). A study in Africa: The East African cassava mosaic virus epidemic in Uganda. *Proceedings of the Second International Whitefly and Geminivirus Workshop*, San Juan, Puerto Rico, June 8-12.

Fondong V (1999). Viruses associated with the cassava mosaic disease in Cameroon: Spread and molecular characterization. PhD dissertation, University of Witwatersrand, Johannesburg, South Africa, 152 pp.

Frischmuth T, Roberts S, Von Arnim A, Stanley J (1993). Specificity of bipartite geminivirus movement proteins. *Virology* 196: 666-673.

Frohlich D, Torres-Jerez I, Bedford ID, Markham PG, Brown JK (1999). A phylogeographic analysis of the *Bemisia tabaci* species complex based on mitochondrial DNA markers. *Molecular Ecology* 8: 1593-1602.

Fuchs MA, Buta C (1997). The role of peptide modules in protein evolution. *Biophysical Chemistry* 66: 203-210.

Ghanim M, Rosell RC, Campbell LR, Czosnek H, Brown JK, Ullman DE (2000). Microscopic analysis of the digestive, salivary, and reproductive organs of *Bemisia tabaci* (Gennadius) (Hemiptera: Aleyrodidae) B type. *Journal of Morphology* (in press).

Gilbertson RL, Hidayat SH, Paplomatas EJ, Rojas MR, Hou YM, Maxwell DP (1993). Pseudorecombination between infectious cloned DNA components of tomato mottle and bean dwarf mosaic geminiviruses. *Journal of General Virology* 74: 23-31.

Gildow F (1982). Coated-vesicle transport of luteoviruses through salivary glands of *Myzus persicae* 72: 1289-1296.

Gildow F (1993). Evidence for receptor-mediated endocytosis regulating luteovirus acquisition by aphids. *Phytopathology* 83: 270-277.

Gill RJ (1990). The morphology of whiteflies. In Gerling P (ed.), *Whiteflies: Their Bionomics, Pest Status, and Management*, Intercept Ltd., Andover, Hants, UK, pp. 13-46.

Gill RJ (1992). A review of the sweet potato whitefly in southern California. *Pan-Pacific Entomologist* 68: 144-152.

Gillette WK, Meade TJ, Jeffrey JL, Petty ITD (1998). Genetic determinants of host-specificity in bipartite geminivirus DNA A components. *Virology* 251: 361-369.

Guirao P, Beitia F, Cenis JL (1997). Biotype determination of Spanish populations of *Bemisia tabaci* (Hemiptera: Aleyrodidae). *Bulletin of Entomological Research* 87: 587-593.

Hanley-Bowdoin L, Settlage SB, Orozco BM, Nagar S, Robertson D (1999). Geminiviruses: Models for plant DNA replication, transcription, and cell cycle regulation. *Critical Review in Plant Science* 18: 71-106.

Harrison BD (1985). Advances in geminivirus research. *Annual Review of Phytopathology* 23: 55-82.

Harrison BD, Liu YL, Khalid S, Hameed S, Otim-Nape GW, Robinson DJ (1997). Detection and relationships of cotton leaf curl virus and allied whitefly-transmitted geminiviruses occurring in Pakistan. *Annals of Applied Biology* 130: 61-75.

Harrison BD, Liu YL, Zhou X, Robinson DJ, Calvert L, Munoz C, Otim Nape GW (1997). Properties, differentiation and geographical distribution of geminiviruses that cause cassava mosaic disease. *African Journal of Root and Tuber Crops* 2: 19-22.

Harrison BD, Robinson DJ (1999). Natural genomic and antigenic variation in whitefly-transmitted geminiviruses (Begomoviruses). *Annual Review of Phytopathology* 37: 369-398.

Harrison BD, Zhou X, Otim-Nape GW, Liu YL, Robinson DJ (1997). Role of a novel type of double infection in the geminivirus-induced epidemic of severe cassava mosaic in Uganda. *Annals of Applied Biology* 131: 437-448.

Hill JE, Strandberg JO, Hiebert E, Lazarowitz SG (1999). Asymmetric infectivity of pseudorecombinants of cabbage leaf curl virus and squash leaf curl virus: Implications for bipartite geminivirus evolution and movement. *Virology* 250: 283-292.

Höfer P, Engel M, Jeske H, Frischmuth T (1997). Host range limitation of a pseudorecombinant virus produced by two distinct bipartite geminiviruses. *Molecular Plant-Microbe Interactions* 10: 1019-1022.

Hou YM, Gilbertson RL (1996). Increased pathogenicity in a pseudorecombinant bipartite geminivirus correlates with intermolecular recombination. *Journal of Virology* 70: 5430-5436.

Hou YM, Paplomatas EJ, Gilbertson RL (1998). Host adaptation and replication properties of two bipartite geminiviruses and their pseudorecombinants. *Molecular Plant-Microbe Interactions* 11: 208-217.

Hunter W, Hiebert E, Webb SE, Polston JE, Tsai JH (1996). Precibarial and cibarial chemosensilla in the whitefly, *Bemisia tabaci* (Gennadius) (Homoptera: Aleyrodidae). *International Journal of Insect Morphology and Embryology* 25: 295-304.

Hunter WB, Hiebert E, Webb S, Tsai JH, Polston JE (1998). Location of geminiviruses in the whitefly *Bemisia tabaci* (Homoptera: Aleyrodidae). *Plant Disease* 82: 1147-1151.

Idris AM (1997). Biological and molecular differentiation of subgroup III geminiviruses. PhD dissertation, The University of Arizona, Tucson, AZ, 155 pp.

Idris AM, Bird J, Brown JK (1999). First report of a bean-infecting begomovirus from *Macroptilium lathyroides* in Puerto Rico. *Plant Disease* 83: 1071.

Idris AM, Brown JK (2000). Molecular characterization of full-length begomovirus DNAs from cotton, hollyhock and okra in Sudan in relation to the epidemiology of cotton leaf curl disease-Sudan. *Phytopathology* 90: 537 (abstr.).

Ingham DJ, Lazarowitz SG (1993). A single missense mutation in the BR1 movement protein alters the host range of the squash leaf curl geminivirus. *Virology* 196: 694-702.

Ingham DJ, Pascal E, Lazarowitz SG (1995). Both bipartite geminivirus movement proteins define viral host range, but only BL1 determines viral pathogenicity. *Virology* 207: 191-204.

Kirk AA, Lacey LA, Brown JK, Ciomperlik MA, Goolsby JA, Vacek DC, Wendel LE, Napompeth B (2000). Variation within the *Bemisia tabaci* s.l. species complex (Hemiptera:Aleyrodidae) and its natural enemies leading to successful biological control of *Bemisia* biotype B in the USA. *Bulletin of Entomological Research* 90: 317-327.

Krake LR, Rezaian MA, Dry IB (1998). Expression of the tomato leaf curl geminivirus C4 gene produces virus-like symptoms in transgenic plants. *Molecular Plant-Microbe Interactions* 11: 413-417.

Lazarowitz SG (1991). Molecular characterization of two bipartite geminiviruses causing squash leaf curl disease: Role of viral replication and movement functions determining host range. *Virology* 180: 70-80.

Lazarowitz SG (1992). Geminiviruses: Genome structure and gene function. *Critical Review in Plant Sciences* 11: 327-349.

Legg JP (1996). Host-associated strains within Ugandan populations of the whitefly *Bemisia tabaci* (Genn.) (Homoptera, Aleyrodidae). *Journal of Applied Entomology* 120: 523-527.

Legg JP, Ogwal S (1998). Changes in the incidence of African cassava mosaic virus disease and the abundance of its whitefly vector along south-north transects in Uganda. *Journal of Applied Entomology* 122: 169-178.

Liu Y, Robinson D, Harrison BD (1998). Defective forms of cotton leaf curl virus DNA-A that have different combinations of sequence deletion, duplication, inversion and rearrangement. *Journal of General Virology* 79: 1501-1508.

Mansoor S, Khan S, Bashir A, Saeed M, Zafar Y, Malik K, Briddon RW, Stanley J, Markham PG (1999). Identification of a novel circular single-stranded DNA associated with cotton leaf curl disease in Pakistan. *Virology* 259: 190-199.

Mayo MA, Pringle CR (1998). Virus taxonomy—1997. *Journal of General Virology* 79: 649-657.

McGovern RJ, Polston JE, Danyluk GM, Abouzid AM (1994). Identification of a natural weed host of tomato mottle geminivirus in Florida. *Plant Disease* 78: 1102-1106.

McGrath PF, Harrison BD (1995). Transmission of tomato leaf curl geminiviruses by *Bemisia tabaci*: Effects of virus isolate and vector biotype. *Annals of Applied Biology* 126: 307-316.

Morin S, Ghanim M, Zeidan M, Czosnek H, Verbeek M, Van den Heuvel JF (1999). A GroEL homologue from endosymbiotic bacteria of the whitefly *Bemisia tabaci* is implicated in the circulative transmission of tomato yellow leaf curl virus. *Virology* 30: 75-84.

Mound LA (1983). Biology and identity of whitefly vectors of plant pathogens. In Plumb RT, Thresh JM (eds.), *Plant Virus Epidemiology: The Spread and Control of Insect-Borne Diseases*, Blackwell Scientific, Oxford, UK, pp. 305-313.

Mound LA. Halsey SH (eds.) (1978). *Whitefly of the world: A systematic catalogue of the Aleyrodidae (Homoptera) with Host Plant and Natural Enemy Data.* British Museum (Natural History), London, UK, and John Wiley & Sons, Chichester.

Muniyappa V (1980). Whiteflies. In Harris KF, Maramorosch K (eds.), *Vectors of Plant Pathogens.* Academic Press, New York, pp. 39-85.

Nateshan H, Muniyappa V, Swanson M, Harrison, BD (1996). Host range, vector relations and serological relationships of cotton leaf curl virus from southern India. *Annals of Applied Biology* 128: 233-244.

Noris E, Vaira AM, Caciagli P, Masenga V, Gronenborn B, Accotto GP (1998). Amino acids in the capsid protein of tomato yellow leaf curl virus that are crucial for systemic infection, particle formation, and insect transmission. *Journal of Virology* 72: 10050-10057.

Padidam M, Beachy RN, Fauquet CM (1995). Classification and identification of geminiviruses using sequence comparisons. *Journal of General Virology* 76: 249-263.

Padidam M, Sawyer S, Fauquet CM (1999). Possible emergence of new geminiviruses by frequent recombination. *Virology* 265: 218-225.

Pascal E, Sanderfoot AA, Ward BM, Medville R, Turgeon R, Lazarowitz SG (1994). The geminivirus BR1 movement protein binds single-stranded DNA and localizes to the cell nucleus. *The Plant Cell* 6: 995-1006.

Paximadis M, Idris AM, Torres-Jerez I, Villarreal A, Rey MEC, Brown JK (1999). Characterization of geminiviruses of tobacco in the Old and New World. *Archives of Virology* 144: 703-717.

Pessoni LA, De Zimmermann MJ, Faria JC (1997). Genetic control of characters associated with bean golden mosaic geminivirus resistance in *Phaseolus vulgaris* L. *Brazilian Journal of Genetics* 20: 51-58.

Petty ITD, Miller CG, Meade-Hash TJ, Schaffer RL (1995). Complementable and noncomplementable host adaptation defects in bipartite geminiviruses. *Virology* 212: 263-267.

Polston JE, Anderson PK (1997). The emergence of whitefly-transmitted geminiviruses in tomato in the Western hemisphere. *Plant Disease* 81: 1358-1369.

Pooma W, Gillette WK, Jeffrey JL, Petty ITD (1996). Host and viral factors determine the dispensability of coat protein for bipartite geminivirus systemic movement. *Virology* 21: 264-268.

Qin S, Ward BM, Lazarowitz SG (1998). The bipartite geminivirus coat protein aids BR1 function in viral movement by affecting the accumulation of viral single-stranded DNA. *Journal of Virology* 79: 9247-9256.

Rey MEC, Berry S, Banks GK, Markham PG, Brown JK (2000). A study of whitefly populations on cassava in Southern Africa. In *Proceedings of the International Symposium on Tropical Tuber Crops,* Thiruvananthapuram, India, January 19-22 (abstr.).

Rojas MR, Noueiry AO, Lucas WJ, Gilbertson RL (1998). Bean dwarf mosaic geminivirus movement proteins recognize DNA in a form- and size-specific manner. *Cell* 95: 105-113.

Rosell RC, Bedford ID, Frohlich DR, Gill RJ, Brown JK, Markham PG (1997). Analysis of morphological variation in distinct populations of *Bemisia tabaci*

(Homoptera: Aleyrodidae). *Annals of Entomological Society of America* 90: 575-589.

Rosell RC, Torres-Jerez I, Brown JK (1999). Tracing the geminivirus-whitefly transmission pathway by polymerase chain reaction in whitefly extracts, saliva, hemolymph, and honeydew. *Phytopathology* 89: 239-246.

Rybicki EP (1994). A phylogenetic and evolutionary justification for three genera of Geminiviridae. *Archives of Virology* 139: 49-77.

Rybicki EP (1998). A proposal for naming geminiviruses: A reply by the *Geminiviridae* Study Group Chair. *Archives of Virology* 143: 421-424.

Sanderfoot AA, Lazarowitz SG (1996). Getting it together in plant virus movement: Cooperative interactions between bipartite geminivirus movement proteins. *Trends in Cell Biology* 6: 353-358.

Traboulsi R (1994). *Bemisia tabaci:* A report on its pest status with particular reference to the Near East. *FAO Plant Protection Bulletin* 42: 33-58.

Varma PM (1963). Transmission of plant viruses by whiteflies. *Bulletin of National Institute of Science* 24: 11-33.

Velez JJ, Bassett MJ (1998). Inheritance of resistance to bean golden mosaic virus in common bean. *Journal of the American Society of Horticultural Science* 123: 628-631.

Ward BM, Medville R, Lazarowitz SG, Turgeon R (1997). The geminivirus BL1 movement protein is associated with endoplasmic reticulum-derived tubules in developing phloem cells. *Journal of Virology* 71: 3726-3733.

Wool D, Gerling D, Belloti AC, Morales FJ (1993). Esterase electrophoretic variation in *Bemisia tabaci* (Genn.) (Homoptera: Aleyrodidae) among host plants and localities in Israel. *Journal of Applied Entomology* 115: 185-196.

Zhou X, Liu Y, Calvert L, Munoz C, Otim-Nape GW, Robinson D, Harrison BD (1997). Evidence that DNA-A of a geminivirus associated with severe cassava mosaic disease in Uganda has arisen by interspecific recombination. *Journal of General Virology* 78: 2101-2111.

Zhou X, Liu Y, Robinson D, Harrison BD (1998). Four DNA-A variants among Pakistani isolates of cotton leaf curl virus and their affinities to DNA-A of geminivirus isolates from okra. *Journal of General Virology* 79: 915-923.

Zhou X, Robinson D, Harrison BD (1998). Types of variation in DNA-A among isolates of East African cassava mosaic virus from Kenya, Malawi and Tanzania. *Journal of General Virology* 79: 2835-2840.

Chapter 14

Translational Strategies in Members of the Family *Caulimoviridae*

Mikhail M. Pooggin
Lyubov A. Ryabova
Thomas Hohn

INTRODUCTION

Caulimoviruses are spherical or bacilliform plant viruses containing a circular double-stranded (ds) DNA genome of 7.1 to 8.2 kb. Depending on the gene arrangement, the following six genera in the family *Caulimoviridae* are distinguished (Pringle, 1999; see Chapter 1 of this book; see Figure 14.1). Viruses with icosahedral particles are classified in genera *Caulimovirus* (type species, *Cauliflower mosaic virus*, CaMV), "SbCMV-like viruses" (type species, *Soybean chlorotic mottle virus*, SbCMV), "CsVMV-like viruses" (type species, *Cassava vein mosaic virus*, CsVMV), and "PVCV-like" viruses (type species, *Petunia vein clearing virus*, PVCV). Viruses with bacilliform particles (formerly badnaviruses) are in genera *Badnavirus* (type species, *Commelina yellow mottle virus*, ComYMV) and "RTBV-like viruses" (type species, *Rice tungro bacilliform virus*, RTBV).

Replication of caulimoviruses proceeds via a pregenomic RNA, which is produced in the nucleus by host RNA polymerase II and reverse transcribed in the cytoplasm using viral reverse transcriptase and, probably, a capsid or capsid protein as a cofactor (Rothnie et al., 1994; Mesnard and Carriere, 1995). Viruses in the family *Retroviridae*, composed of animal retroviruses, and those in the families *Hepadnaviridae* and *Caulimoviridae*, with pararetroviruses of animals and plants, respectively, use this replication cycle (Hull and Covey, 1995). Retroviruses encapsidate an RNA genome and replicate through a

The authors wish to thank Johannes Fütterer, Etienne Herzog, and Helen Rothnie for critical reading of the manuscript. M. M. Pooggin is currently supported by a FEBS long-term fellowship.

FIGURE 14.1. Genome Organization of Viruses in Six Distinct Genera of Family *Caulimoviridae*

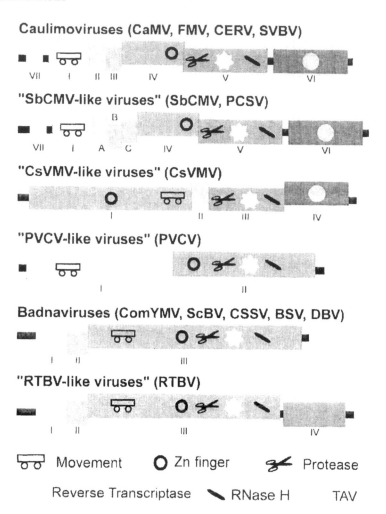

Note: Common motifs are shown with symbols and named at the bottom. CaMV = *Cauliflower mosaic virus;* FMV = *Figwort mosaic virus;* CERV = *Carnation etched ring virus;* SVBV = *Strawberry vein banding virus;* SbCMV = *Soybean chlorotic mottle virus;* PCSV = *Peanut chlorotic streak virus;* CsVMV = *Cassava vein mosaic virus;* PVCV = *Petunia vein clearing virus;* RTBV = *Rice tungro bacilliform virus;* ComYMV = *Commelina yellow mottle virus;* ScVB = *Sugarcane bacilliform virus;* CSSV = *Cacao swollen shoot virus;* BSV = *Banana streak virus;* DBV = *Dioscorea bacilliform virus.*

dsDNA provirus integrated into the host chromosomes, whereas pararetro-viruses replicate through the RNA intermediate and their DNA generally remains as episomes in infected nuclei (Rothnie et al., 1994). However, the distinction might not be rigorous, since the foamy retroviruses (viruses belonging to the genus *Spumavirus* that cause "foamy" cytopathology in cell cultures) also exist as episomes (Linial, 1999), and the PVCV, a pararetro-virus, might encode an integrase (Richert-Pöggeler and Shepherd, 1997).

In several caulimoviruses, fragmented and defective DNAs have been found to occur as integrates (Harper et al., 1999; Jakowitsch et al., 1999; Ndowora et al., 1999). All six genera of family *Caulimoviridae* have coding regions for capsid protein, protease, and reverse transcriptase with sequence motifs related to animal retroviruses and also to plant and animal retrotransposons (Hull and Covey, 1995). Furthermore, caulimoviruses possess genes that have no equivalent in animal viruses but do correspond to functionally related genes of RNA plant viruses, i.e., genes for cell-to-cell movement via tubular structures spanning the cell wall (Perbal et al., 1993; see Chapter 7 in this book), and to those for insect transmissibility (see Chapter 3). Icosahedral cauli-moviruses are generally transmitted by aphids (Pirone, 1991), and many badnaviruses by mealybugs and leafhoppers (Lockhart and Olszewski, 1994). RTBV might have lost a functional gene for insect transmission and depends on *Rice tungro spherical virus* (genus *Waikavirus*, family *Sequi-viridae*) for its transmissibility (Hull, 1996). In addition, genera *Caulimovirus* and SbCMV-like viruses encode a multifunctional protein (called trans-activator or TAV) involved in translation control, inclusion body formation, virus assembly, virus protein stabilization, and determination of host range and symptom severity (Hohn and Fütterer, 1997; Kobayashi et al., 1998).

The coding regions of caulimoviruses can be part of polyproteins, which are processed before or during assembly. In fact, the capsid protein (CP) is always cleaved out from a capsid preprotein, and protease and reverse transcriptase are always processed from a common Pol precursor.

Viruses, especially plant viruses, have very compact genomes. Overlapping open reading frames (ORFs), polyproteins (as mentioned earlier), multiple use of RNA (e.g., as messenger (m) RNA and replicative intermediate), and polycistronic translation are some of the strategies used to keep the genome as small as possible, while retaining all necessary functions.

Viruses in some families have evolved highly specialized translation control mechanisms (Fütterer and Hohn, 1996), e.g., RNA-containing luteoviruses (Miller et al., 1995; see Chapter 9 of this book) and caulimoviruses, which are the subject of this chapter.

ARCHITECTURE OF VIRAL RNA

A basic rule of eukaryotic translation states that one monocistronic RNA is required for every protein being translated (Kozak, 1999). However, this rule seems to be violated in most caulimoviruses. For instance, CaMV has seven ORFs, but probably only three types of mRNA: the pregenomic 35S RNA derived from the 35S promoter (Guilley et al., 1982), spliced derivatives thereof (Kiss-László et al., 1995), and a subgenomic RNA derived from the 19S promoter (Odell et al., 1981). Notably, only the 19S RNA is monocistronic, coding for the transactivator protein (see Figure 14.2).

FIGURE 14.2. Genome Map of *Cauliflower mosaic virus* (CaMV)

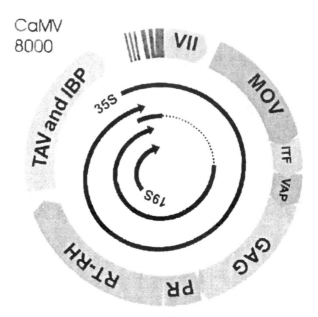

Note: The successive open reading frames encode for movement protein (MOV); insect transmission factor (ITF), virion-associated protein (VAP); structural proteins (GAG); enzymatic polyprotein comprising protease (PR), reverse transcriptase, and RNase H (RT-RH) domains; and a multifunctional protein (TAV and IBP) involved in transactivation of polycistronic translation and formation of viral inclusion bodies. The pregenomic 35S RNA, its spliced derivatives, and the subgenomic 19S RNA are shown with inner circular arrows.

On the pregenomic and spliced RNAs, the ORFs are arranged in tight successive order, either with very short intercistronic distances or overlapping each other for a few nucleotides, a configuration resembling a prokaryotic operon and implying polycistronic translation. The occurrence of polar mutations and corresponding second site revertants led to the proposal that polycistronic translation occurs on CaMV RNA (Dixon and Hohn, 1984). Later, at least two distinct mechanisms of polycistronic translation were recognized in different genera of family *Caulimoviridae:* transactivated reinitiation (e.g., in CaMV [Bonneville et al., 1989] and *Figwort mosaic virus* (FMV) [Gowda et al., 1989]) and leaky scanning due to a natural paucity of translation start codons in the 5'-proximal ORFs (e.g., in RTBV [Fütterer et al., 1997]).

The pregenomic RNA of caulimoviruses is further characterized by a leader sequence of 400 to 700 nucleotides (nt) preceding the first large ORF (in the following referred to as "leader"). These leaders are highly structured in all cases. In the case of CaMV, computer-folding (Fütterer et al., 1988) and chemical and enzymatic analyses (Hemmings-Mieszczak et al., 1997) revealed that the leader consists of a strong and extended stem structure, flanked by less-structured regions (see Figure 14.3). The CaMV stem structure exists in at least two polymorphic varieties, one of which involves a long-range pseudoknot (Hemmings-Mieszczak et al., 1997). The central extended stem structures can be folded in the leaders of all members of family *Caulimoviridae* tested so far (Pooggin et al., 1999). Furthermore, the leaders contain several short ORFs (sORFs). In all cases, with the possible exception of SbCMV, one of these sORFs terminates a few nucleotides upstream of the extended stem structure and is conserved with respect to its 5'-proximal location and short distance from the downstream structure, but not in sequence (Pooggin et al., 1999). It was shown for CaMV that mutations removing the start or stop codon of this sORF (sORF A) led to first and second site revertants, revealing the importance of this three-codon sORF for virus replication (Pooggin et al., 1998, 2001).

GENERAL ENHANCEMENT OF EXPRESSION

The first 60 nt of the CaMV leader preceding sORF A (supporting sequence 1 [S1]) are unstructured and poor in G-residues and were found generally to enhance translation (Fütterer, Gordon, et al., 1990; Hemmings-Mieszczak and Hohn, 1999). In this respect, S1 resembles the Ω-leader sequence from *Tobacco mosaic virus*, which acts as a translational enhancer (Gallie et al., 1987).

FIGURE 14.3. Structural Organization of the Pregenomic RNA Leader of *Cauliflower mosaic virus* (CaMV)

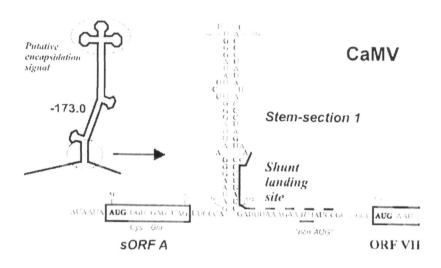

Note: The large stem-loop structure formed in the leader is schematically drawn with a thick line. Its stability in kcal/mol is given. A putative encapsidation signal in the upper part of the structure is encircled. Stem-section 1, the most stable structural element at the base of the structure, is encircled and enlarged alongside, together with adjacent unstructured regions. sORF A is boxed; its amino acids are indicated below. AUG start codons are in bold; a potential non-AUG initiation codon is underlined within the shunt landing site (defined by Fütterer et al., 1993). The nucleotide numbering is from the leader 5'-end.

Interestingly, in the DNA context, S1 also acts as a transcriptional enhancer element (Hohn et al., 1996). Likewise, the corresponding region from RTBV enhances expression at both transcriptional and translational levels (Chen et al., 1994, 1996; Schmidt-Puchta et al., 1997), and the same may also be true for *Mirabilis mosaic virus*, a caulimovirus (Dey and Maiti, 1999).

SHUNTING

The Concept

The default mechanism of eukaryotic translation is described by the scanning model (Kozak, 1989b; Pestova et al., 1998), according to which eukaryotic ribosomes are first recruited to the capped 5'-end of mRNA with the help of initiation factors, and then scanned along the RNA to reach the initiation codon. Accordingly, strong secondary structure in the RNA leader is inhibitory to translation, due to the energy required to melt it before or during scanning. Upstream start codons are also inhibitory, since translation of the corresponding sORFs leads to ribosome release before the main ORF is reached. Situations allowing a reasonable level of translation despite the presence of sORFs could be explained by leaky-scanning or reinitiation events (Kozak, 1991) or a combination thereof (e.g., in the yeast GCN4 system; Hinnebusch, 1997). Leaky scanning occurs when an upstream AUG start codon is in a suboptimal nucleotide context (lacking purine at position −3 or guanosine at position +4, with respect to the A of the start codon [Kozak, 1986]). Furthermore, cap-independent translation depending on an internal ribosome entry site (IRES) has been described for a number of specific viral and cellular RNAs (Jackson and Kaminski, 1995; Skulachev et al., 1999; see Chapter 9 in this book).

One would predict extremely inefficient translation from RNAs of caulimoviruses due to the general organization of the leader with its strong secondary structure, presence of several sORFs, and, for example, in the case of CaMV, an sORF (sORF F) overlapping the first long ORF (ORF VII). Despite this prediction, a reasonable level of translation from CaMV RNA was observed in transfected plant protoplasts (Fütterer et al., 1989; Fütterer, Bonneville, et al., 1990) in vitro (Schmidt-Puchta et al., 1997), and also in transgenic plants (Schärer-Hernández and Hohn, 1998). To elucidate this phenomenon, a deletion analysis of the CaMV leader was performed, leading to the identification of regions either stimulating or inhibiting translation and to the proposal of the "shunt" model, according to which some of the stimulatory regions are required to bypass (shunt) the inhibitory regions (Fütterer, Gordon, et al., 1990). The shunt model was later verified by showing that (1) in contrast to picornavirus IRES, shunting in CaMV requires capped RNA; (2) the center of the leader tolerates additional inhibitory elements, such as very strong hairpins and long reporter ORFs; and (3) a shunt structure can be assembled from two separate RNAs ("trans-shunt") (Fütterer et al., 1993). Later, shunting was also shown for RTBV (Fütterer et al., 1996), and theoretical considerations make it the likely mode of translation of the pregenomic RNA in most members of *Caulimoviridae* family (Pooggin et al., 1999). Shunting has also been observed for a number of animal viruses (Curran and Kolakofsky, 1988; Li, 1996; Yueh and Schneider, 1996; Latorre et al., 1998; Remm et al., 1999) and proposed for the human c-myc proto-oncogene (Carter et al., 1999).

The Details and the Mechanism

In vitro translation systems provide the means to analyze details of the shunt mechanism. Shunting functions both in wheat germ extract (Schmidt-Puchta et al., 1997) and in reticulocyte lysate (Ryabova and Hohn, 2000); thus, it is not restricted to viral host plants. In both in vitro systems and plant protoplasts, three elements are required for CaMV shunting: (1) a cap (Schmidt-Puchta et al., 1997), (2) an sORF (Dominguez et al., 1998; Hemmings-Mieszczak and Hohn, 1999; Pooggin et al., 2000), and (3) a stem structure (Dominguez et al., 1998; Hemmings-Mieszczak et al., 1998). Furthermore, the spacing between the sORF and the stem modulates shunting efficiency (Dominguez et al., 1998). Shunting is optimal if the stop codon of the sORF is located about 6 nt upstream of the base of the stem structure (Pooggin et al., 2000). The strength of the stem structure provides another means of modulating shunt efficiency. The original CaMV stem structure is strong enough for near optimal shunting; reduction in the number of base pairs reduces shunting efficiency (Dominguez et al., 1998; Hemmings-Mieszczak et al., 1998). The primary sequence of the stem structure is of only minor importance. Notably, the bottom section of the stem structure (stem-section 1) contains three repeats of oligo-G sequences paired with oligo-C sequences (see Figure 14.3). Mutation of these motifs (Cs to Gs *or* Gs to Cs) impaired shunting, while the compensatory mutations (Cs to Gs *and* Gs to Cs) restored it (Dominguez et al., 1998; Hemmings-Mieszczak et al., 1998). Interestingly, the original stem structure can be exchanged for a completely different one (e.g., the very strong stem used as scanning inhibitor by Kozak [1989a]) without affecting shunting (Hemmings-Mieszczak and Hohn, 1999). The size of the sORF plays a role: sORFs of between two and ten codons allow efficient shunting, whereas further extending their length leads to substantial reduction of shunt-mediated translation. On the other hand, one-codon (start-stop) ORFs cannot induce shunting (Pooggin et al., 2000). Also, the primary sequence of the sORF has little influence. A notable exception is an sORF with the sequence "MAGDIS" derived from the leader of the mRNA for AdoMetDC, an enzyme involved in polyamine biosynthesis. In the presence of polyamines, ribosomes translating the MAGDIS sORF stall near its stop codon and reinitiation at a downstream ORF is inhibited (Mize et al., 1998). Replacement of sORF A in the CaMV leader with the MAGDIS sORF abolishes shunt-mediated translation (Ryabova and Hohn, 2000), thus confirming that proper translation and termination at sORF A is required for shunting.

A remarkable property of shunting ribosomes is their reduced fidelity of AUG recognition. Non-AUG start codons at the shunt landing site are readily recognized, e.g., in RTBV (Fütterer et al., 1996) and in CaMV (Ryabova and Hohn, 2000).

According to these findings, we visualize shunting in CaMV to occur as follows (see Figure 14.4): (1) Ribosomes enter the 35S RNA at the capped 5'-end and start scanning in a 3'-direction until they reach the sORF A start codon. (2) sORF A is translated and properly terminated. (3) Some of the ribosomes released from sORF A still carry a set of initiation factors sufficient to make them shunt and reinitiation competent, but insufficient to melt the downstream stem structure. (4) Shunting leads to a bypass of the struc-

FIGURE 14.4. Model for Ribosome Shunt in Family *Caulimoviridae*

sORF and Stem-Mediated Shunting

Scanning   Translation   Shunting   Reinitiation
            of sORF

Note: The 40S ribosomal subunit (smaller oval) with a set of initiation factors (small circles) migrates (scans) from the capped 5'-end (shown with a cap symbol) along the pregenomic RNA leader (a thick line) until it reaches the 5'-proximal short ORF (a closed box) located in front of the extended stem structure (the bottom part of which is shown). The 60S ribosomal subunit (bigger oval) binds and translation of the sORF proceeds. Upon translation the stem structure is partially melted and some of the initiation factors are lost. The 40S subunit released at the sORF stop codon cannot penetrate further into the base-paired region, probably due to lack of certain factors with helicase activity, but instead anchors to the unstructured sequence downstream of the stem (i.e., shunts to the landing site). In the landing site, the shunted ribosome can either recognize a non-AUG start codon or resume scanning and reach the AUG start codon of the main ORF (an open box).

tured region and to reinitiation at an AUG or a non-AUG start codon close to the shunt landing site (see Figure 14.3).

The efficient recognition of non-AUGs probably reflects a transient lack of initiation "fidelity" factors (e.g., eIF1/1A, Pestova et al., 1998; or eIF5, Huang et al., 1997) on the shunting ribosome, which might be lost during translation of the sORF. Identification of the set of initiation factors associated with the shunting ribosome will be required to clarify further the molecular mechanism of ribosome shunt.

LEAKY SCANNING
(BACILLIFORM CAULIMOVIRUSES)

The RTBV genome has four ORFs (see Figure 14.1). The products of ORFs I, II, and III are translated from the pregenomic RNA (Fütterer et al., 1997), with the splicing event between the first sORF and ORF IV, allowing production of the ORF IV protein (Fütterer et al., 1994). As in CaMV, translation downstream of the leader on the RTBV pregenomic RNA is initiated by the shunt mechanism (Fütterer et al., 1996) that requires similar conserved structural elements (Pooggin, Hohn, and Fütterer, in preparation). After shunting, a fraction of the landing ribosomes translates ORF I from a non-AUG start codon (AUU) located at the shunt landing site (Fütterer et al., 1996). The others resume scanning and reach ORF II. Since the ORF II AUG is in a weak initiation context, about half of the scanning ribosomes ignore this start codon and reach ORF III, which has a strong AUG (Fütterer et al., 1997). Strikingly, other than the start codon of ORF II, no AUG is present in any reading phase within the RTBV ORF I and II regions to intercept scanning ribosomes and, therefore, preclude downstream translation. The apparent ability of ribosomes to scan over a distance of about 1,000 nt is consistent with the high processivity of the scanning ribosome observed in vitro (Kozak, 1998).

Apart from the absence of ORF IV, the gene organization of badnaviruses is highly related to that of RTBV: the long region between the leader and ORF III is almost devoid of AUGs in all cases, thus suggesting that they also use the leaky-scanning strategy for polycistronic translation (Fütterer et al., 1997, Pooggin et al., 1999). In *Cacao Swollen shoot virus* and *Dioscorea bacilliform virus*, the ORF II AUG in a weak context is the only AUG in this region. Three other badnaviruses do contain additional AUGs within the ORF I/ORF II region, creating either an in-frame sORF with the C-terminus of ORF I (in ComYMV, *Sugarcane bacilliform virus* and *Banana streak virus* [BSV]) or an out-of-frame sORF (also in BSV); both slightly overlap ORF II. However, these AUGs are in suboptimal sequence context, and if translated, the sORFs would probably allow reinitiation (Fütterer and Hohn, 1992; Hinnebusch, 1997). Due to the overlap, these sORFs may specifically downregulate production of the ORF II protein, the function of which is still unknown (Cheng et al., 1996).

ACTIVATED POLYCISTRONIC TRANSLATION (ICOSAHEDRAL CAULIMOVIRUSES)

Unlike bacilliform caulimoviruses, the icosahedral plant pararetroviruses seem to use a reinitiation strategy for polycistronic translation. In general, reinitiation of translation in eukaryotes is very inefficient unless the upstream ORF is short and/or the distance between ORFs is long (Kozak, 1987; Hinnebusch, 1997). However, in the presence of the CaMV-encoded TAV, the reinitiation capacity of the ribosome that has already translated one ORF appears to be activated strongly (Bonneville et al., 1989; Fütterer and Hohn, 1991). TAV function has been also reported in FMV (Gowda et al., 1989) and *Peanut chlorotic streak virus* (PCSV) (Edskes et al., 1996; Maiti et al., 1998). Homology in the functional domain of TAV (miniTAV; DeTapia et al., 1993) is shared by ORF VI of all other caulimoviruses and SbCMV-like viruses and, to a lesser extent, by ORF IV of CsVMV (DeKochko et al., 1998). RTBV has a distinct gene at a similar genome position of yet unknown function, whereas a corresponding gene does not appear to exist in the badnaviruses or PVCV (see Figure 14.1).

In the presence of TAV, a reporter gene fused to any internal ORF of the pregenomic RNA of CaMV (Fütterer, Bonneville, et al., 1990) or FMV (Scholthof, Gowda, et al., 1992) could be efficiently translated, thus supporting the hypothesis that polycistronic translation occurs on this RNA. TAV itself is produced from a monocistronic mRNA transcribed from the subgenomic 19S promoter (Odell et al., 1981). Interestingly, TAV can also activate its own translation (Driesen et al., 1993), probably from the pregenomic RNA or its spliced version. It has been postulated that all internal ORFs (from I to VI) are coupled in a relay race translation, in which ribosomes do not dissociate after translation of one ORF but reach the next AUG codon (by forward or short backward scanning) and reinitiate translation there. Although, in the case of CaMV, some internal ORFs (e.g., ORFs III and IV) can also be translated from spliced versions of the pregenomic RNA, the spliced RNAs are still polycistronic and require TAV for internal cistron translation (Kiss-László et al., 1995).

Transactivation by CaMV TAV does not require any specific viral sequences and can operate also on artificial polycistronic mRNA. However, certain features of this RNA are necessary for effective TAV action, especially the absence of long overlaps between the upstream and downstream ORFs (Fütterer and Hohn, 1991). Furthermore, a short ORF in the leader can enhance TAV-mediated translation from the second cistron in a length-dependent manner (Fütterer and Hohn, 1991, 1992). In FMV, transactivation requires the first 70 nt of the FMV leader (Edskes et al., 1997). Interestingly, this region contains an sORF. The role of the 5'-proximal sORF in transactivated translation of distal cistrons is unknown. It is noteworthy that proper translation of the corresponding sORF in the CaMV leader is a prerequisite not only for efficient shunting but also for shunt enhancement by TAV (Pooggin et al., 2000).

In the FMV pregenomic RNA, other *cis*-elements, including ORF VII and certain regions within ORF I and VI, have also been implicated in the transactivation process, based on deletion analyses (Gowda et al., 1991; Scholthof, Wu, et al., 1992; Edskes, et al., 1997). However, the role of these putative elements remains to be elucidated. In particular, the reported deletion of ORF VII that abolished downstream translation (Gowda et al., 1991) spanned a 3'-part of the FMV leader containing the descending arm of the leader hairpin and the shunt landing site (see Pooggin et al., 1999). According to our shunt model (discussed earlier), such a deletion would abolish ribosome shunting and therefore preclude translation initiation on the resulting RNA that still contains inhibitory sORFs in the remaining part of the leader.

Arabidopsis plants transgenic for a dicistronic reporter gene, with the GUS-coding region in second position, produce GUS efficiently only when infected with CaMV, or when crossed with plants transgenic for TAV (Zijlstra and Hohn, 1992). In both cases, TAV activates translation. Similar experiments with transgenic tobacco plants that express a PCSV-derived dicistronic construct only upon infection by PCSV (Maiti et al., 1998) further confirm the role of TAV in polycistronic translation. Interestingly, CaMV TAV transgenic plants exhibit phenotypic abnormalities, such as low fertility, late flowering, leaf yellowing, and senescence, indicating that TAV interferes with plant gene regulation (Baughman et al., 1988; Balázs, 1990; Zijlstra et al., 1996).

The mechanism of transactivation of polycistronic translation still remains a mystery. One of many possible scenarios is that TAV helps the posttermination ribosome to regain initiation factors that were lost during the translation event. Consistent with this scenario is the finding that TAV can physically interact with a subunit of eIF3 and a ribosomal protein from

Arabidopsis, as revealed by a yeast two-hybrid screen (Park, Hohn, and Ryabova, in preparation).

TRANSACTIVATION AND SHUNTING

Besides its function in polycistronic translation, TAV can enhance the efficiency of ribosome shunt on the CaMV leader (Fütterer et al., 1993). Recently, we have discovered that shunt enhancement by TAV depends on proper translation of sORF A (Pooggin et al., 2000), thus supporting the reinitiation step of the shunt model (see Figure 14.4). Most likely, TAV enhances the reinitiation capacity of the shunting ribosome. Interestingly, certain modifications of the sORF A-coding content could selectively reduce the degree of shunt transactivation without affecting the basal shunt efficiency (Pooggin et al., 2000, 2001). This suggests that the competence of the ribosome for transactivated reinitiation depends on a translational event at an sORF. It also parallels previous findings of our laboratory showing the positive effect of a leader sORF on TAV-activated translation of an internal ORF (Fütterer and Hohn, 1991, 1992). We therefore assume that proper translation of the 5'-proximal sORF is required not only for efficient shunt but also for further downstream translation from polycistronic RNA in those caulimoviruses that encode TAV.

RTBV lacks TAV, but translation downstream of the leader is mediated by shunting (Fütterer et al., 1996) and also depends on the consensus shunt elements of a 5'-proximal sORF followed by a stable stem structure (Pooggin, Hohn, and Fütterer, in preparation). However, the RTBV shunt cannot be significantly enhanced by CaMV TAV. This suggests that TAV may require a specific binding site (missing in the RTBV leader) to exert its effect on shunting. Although RNA-binding domains have been recognized in TAV (DeTapia et al., 1993), no specific target has been identified so far. Interestingly, our preliminary results reveal that the basal level of ribosome shunting in RTBV is two to three times higher than that in CaMV, exactly the factor that is compensated in CaMV by TAV. We assume that RTBV has evolved to possess intrinsically more efficient shunt elements. To study this, shuffling experiments between the CaMV and RTBV leaders are now under way for identification of those elements and a possible TAV-binding site.

BIOLOGICAL SIGNIFICANCE OF RIBOSOME SHUNT

Our recent results with CaMV demonstrate the high degree of correlation between the efficiency of sORF A-mediated shunt and the infectivity of the virus (Pooggin et al., 2001), thus indicating for the first time a biological significance of the shunt mechanism. What could be the role of ribosome shunting in the viral infection cycle? The strategy of pararetroviruses dictates a multiple use of the pregenomic RNA in the cytoplasm of infected cells. First, it must be translated to produce viral proteins and then, most likely, encapsidated into previrions for replication via reverse transcription to produce viral DNA (reviewed by Mesnard and Carriere, 1995). The sorting of viral RNA for translation or encapsidation could be regulated by some *cis*-acting RNA elements. A conserved purine-rich sequence in the leader of CaMV and some other caulimoviruses has been discussed as a putative encapsidation signal (Fütterer et al., 1988; Pooggin et al., 1999). It resides in the upper part of the leader structure (see Figure 14.3) and, in the case of CaMV, binds specifically to the viral CP (Guerra-Peraza et al., 2000). Since this region of the leader is bypassed by shunting ribosomes, its secondary structure is maintained for the interaction with the CP, even in the state of a polysome. Thus, in late infection, accumulation of viral CP could lead to sequestering of the RNA from polysomes into previrions.

Foamy retroviruses may use a replication strategy similar to plant pararetroviruses (reviewed by Linial, 1999). A putative packaging signal probably resides within the RNA leader. Computer-aided analysis has revealed that this leader contains multiple sORFs and folds into an extended stem structure (Pooggin, unpublished data). A structural organization similar to that of caulimoviruses was evident: an sORF terminates a few nucleotides upstream of the stable base of the structure, thus suggesting that, also in this case, a shunt mechanism regulates the sorting of viral RNA for translation or encapsidation.

GAG-POL TRANSLATION

In most animal retroviruses, translation past the 3'-end of the Gag-coding sequence results in the synthesis of Gag-Pol fusion proteins. This occurs either by frameshifting or stop codon suppression. The coassembly of these fusion proteins with CP provides an elegant mechanism to produce Pol-containing virus particles. In CaMV (Penswick et al., 1988; Schultze et al., 1990), hepadnaviruses (Schlicht et al., 1989), and foamy viruses (Enssle et al., 1996), Gag and Pol translation are separate events, and for *hepatitis B virus*, it was shown that Pol is packed as Pol-RNA complex into the viral capsid (Bartenschlager et al., 1990). In contrast, the genome organization of bacilliform caulimoviruses and PVCV suggests that, in these cases, Gag and Pol are produced as obligate fusion proteins. How the surplus of Pol proteins is handled in these cases is not known.

MAKING USE OF TRANSLATIONAL CONTROL

Shunting as described previously is not tight. Not all ribosomes shunt: a subpopulation continues scanning toward the center of the leader (Fütterer et al., 1993; Pooggin et al., 2000; Ryabova et al., 2000). Gene expression units can be constructed based on the CaMV leader, which has one ORF in its center and one at its 3'-end (Fütterer et al., 1993). Depending on whether ribosomes scan or shunt, either the first or the second ORF is translated. The expression ratio of the two ORFs could be manipulated by the strength of the stem structure and the position and length of the upstream sORF.

The potential of directing polycistronic gene expression by exploiting a leaky-scanning mechanism was assessed by employing ORFs I and II of RTBV, or a modified AUG-free reporter gene, as upstream ORFs in polycistronic constructs. With certain modifications in these constructs to increase RNA stability, polycistronic translation was achieved both in protoplasts and in transgenic rice plants (J Fütterer, personal communication).

CONCLUDING REMARKS

Plant pararetroviruses have evolved several sophisticated mechanisms to regulate multiple RNA usage and to modify the default pathway of translation in plant cells and thereby express viral proteins from compact polycistronic RNAs. These mechanisms include ribosome shunting, leaky scanning, and activated reinitiation. The shunting process keeps a central region of the RNA leader, where a putative RNA encapsidation signal resides, free of ribosomes. Therefore, modulation of shunt efficiency either by conformational changes in the leader secondary structure or by transacting factors (e.g., TAV) might provide a means to control the onset and rate of translation versus those of encapsidation of the pregenomic RNA. The viral protein TAV required for polycistronic translation in at least two genera of caulimoviruses appears to be a specialized initiation factor that "restores" a prokaryotic-like capacity to reinitiate translation at internal ORFs in eukaryotic ribosomes. Those caulimoviruses that lack TAV function have elaborated the leaky-scanning strategy of polycistronic translation (as an alternative to activated reinitiation) by clearing a large region of their pregenomic RNAs of AUG start codons. Further investigation of translation control in plant pararetroviruses will certainly reveal new features of the eukaryotic translation machinery and, in particular, new aspects in the behavior of scanning ribosomes.

REFERENCES

Balázs E (1990). Disease symptoms in transgenic tobacco induced by integrated gene VI of cauliflower mosaic virus. *Virus Genes* 3: 205-211.

Bartenschlager R, Junker-Niepmann M, Schaller H (1990). The P gene product of hepatitis B virus is required as a structural component for genomic RNA encapsidation. *Journal of Virology* 64: 5324-5332.

Baughman GA, Jacobs JD, Howell SH (1988). Cauliflower mosaic virus gene VI produces a symptomatic phenotype in transgenic tobacco plants. *Proceedings of the National Academy of Sciences, USA* 85: 7333-7337.

Bonneville J, Sanfaçon H, Fütterer J, Hohn T (1989). Posttranscriptional trans-activation in cauliflower mosaic virus. *Cell* 59: 1135-1143.

Carter PS, Jarquin-Pardo M, DeBenedetti A (1999). Differential expression of Myc1 and Myc2 isoforms in cells transformed by eIF4E: evidence for internal ribosome repositioning in the human c-myc 5'-UTR. *Oncogene* 18: 4326-4335.

Chen G, Müller M, Potrykus I, Hohn T, Fütterer J (1994). Rice tungro bacilliform virus: Transcription and translation in protoplasts. *Virology* 204: 91-100.

Chen G, Rothnie HM, He X, Hohn T, Fütterer J (1996). Efficient transcription from the rice tungro bacilliform virus promoter requires elements downstream of the transcription start site. *Journal of Virology* 70: 8411-8421.

Cheng C-P, Lockhart BEL, Olszewski NE (1996). The ORF I and II proteins of Commelina yellow mottle virus are virion-associated. *Virology* 223: 263-271.

Curran J, Kolakofsky D (1988). Scanning independent ribosomal initiation of the Sendai virus X protein. *The EMBO Journal* 7: 2869-2877.

DeKochko A, Verdaguer B, Taylor N, Carcamo R, Beachy RN, Fauquet C (1998). Cassava vein mosaic virus (CsVMV), type species for a new genus of plant double stranded DNA viruses? *Archives of Virology* 143: 945-962.

DeTapia M, Himmelbach A, Hohn T (1993). Molecular dissection of the cauliflower mosaic virus translational transactivator. *The EMBO Journal* 12: 3305-3314.

Dey N, Maiti IB (1999). Structure and promoter/leader deletion analysis of mirabilis mosaic virus (MMV) full-length transcript promoter in transgenic plants. *Plant Molecular Biology* 40: 771-782.

Dixon LK, Hohn T (1984). Initiation of translation of the cauliflower mosaic virus genome from a polycistronic mRNA: Evidence from deletion mutagenesis. *The EMBO Journal* 3: 2731-2736.

Dominguez DI, Ryabova LA, Pooggin MM, Schmidt-Puchta W, Fütterer J, Hohn T (1998). Ribosome shunting in cauliflower mosaic virus: Identification of an essential and sufficient structural element. *Journal of Biological Chemistry* 273: 3669-3678.

Driesen M, Benito-Moreno RM, Hohn T, Fütterer J (1993). Transcription from the CaMV 19S promoter and autocatalysis of translation from CaMV RNA. *Virology* 195: 203-210.

Edskes HK, Kiernan JM, Shepherd RJ (1996). Efficient translation of distal cistrons of a polycistronic mRNA of a plant pararetrovirus requires a compatible interaction between the mRNA 3' end and the proteinaceous trans-activator. *Virology* 224: 564-567.

Edskes HK, Kiernan JM, Shepherd RJ (1997). Multiple widely spaced elements determine the efficiency with which a distal cistron is expressed from the polycistronic pregenomic RNA of figwort mosaic caulimovirus. *Journal of Virology* 71: 1567-1575.

Enssle J, Jordan I, Mauer B, Rethwilm A (1996). Foamy virus reverse transcriptase is expressed independently from the Gag protein. *Proceedings of the National Academy of Sciences, USA* 93: 4137-4141.

Fütterer J, Bonneville J, Gordon K, DeTapia M, Karlsson S, Hohn T (1990). Expression from polycistronic cauliflower mosaic virus pregenomic RNA. In McCarthy JEG, Tuite MF (eds.), *Posttranscriptional control of gene expression,* Springer-Verlag, Berlin, pp. 349-357.

Fütterer J, Gordon K, Bonneville J, Sanfaçon H, Pisan B, Penswick JR, Hohn T (1988). The leading sequence of caulimovirus large RNA can be folded into a large stem-loop structure. *Nucleic Acids Research* 16: 8377-8390.

Fütterer J, Gordon K, Pfeiffer P, Sanfaçon H, Pisan B, Bonneville J, Hohn T (1989). Differential inhibition of downstream gene expression by the CaMV 35S RNA leader. *Virus Genes* 3: 45-55.

Fütterer J, Gordon K, Sanfaçon H, Bonneville JM, Hohn T (1990). Positive and negative control of translation by the leader sequence of cauliflower mosaic virus pregenomic 35S RNA. *The EMBO Journal* 9: 1697-1707.

Fütterer J, Hohn T (1991). Translation of a polycistronic mRNA in presence of the cauliflower mosaic virus transactivator protein. *The EMBO Journal* 10: 3887-3896.

Fütterer J, Hohn T (1992). Role of an upstream open reading frame in the translation of polycistronic mRNA in plant cells. *Nucleic Acids Research* 20: 3851-3857.

Fütterer J, Hohn T (1996). Translation in plants—rules and exceptions. *Plant Molecular Biology* 32: 159-189.

Fütterer J, Kiss-László Z, Hohn T (1993). Nonlinear ribosome migration on cauliflower mosaic virus 35S RNA. *Cell* 73: 789-802.

Fütterer J, Potrykus I, Bao Y, Li L, Burns TM, Hull R, Hohn T (1996). Position-dependent ATT translation initiation in gene expression of the plant pararetrovirus rice tungro bacilliform virus. *Journal of Virology* 70: 2999-3010.

Fütterer J, Potrykus I, Valles Brau MP, Dasgupta I, Hull R, Hohn T (1994). Splicing in a plant pararetrovirus. *Virology* 198: 663-670.

Fütterer J, Rothnie HM, Hohn T, Potrykus I (1997). Rice tungro bacilliform virus open reading frames II and III are translated from the polycistronic pregenomic RNA by leaky scanning. *Journal of Virology* 71: 7984-7989.

Gallie DR, Sleat TE, Watts JW, Turner PC, Wilson TMA (1987). The 5' leader sequence of TMV RNA enhances the expression of foreign gene transcripts in vitro and in vivo. *Nucleic Acids Research* 15: 3257-3273.

Gowda S, Scholthof HB, Wu FC, Shepherd RJ (1991). Requirement of gene VII in cis for the expression of downstream genes on the major transcript of figwort mosaic virus. *Virology* 185: 867-871.

Gowda S, Wu FC, Scholthof HB, Shepherd RJ (1989). Gene VI of figwort mosaic virus (caulimovirus group) functions in posttranscriptional expression of genes on the full-length RNA transcript. *Proceedings of the National Academy of Sciences. USA* 86: 9203-9207.

Guerra-Peraza O, DeTapia M, Hohn T, Hemmings-Mieszczak M (2000). Interaction of the cauliflower mosaic virus coat protein with the pregenomic RNA leader. *Journal of Virology* 74: 2067-2072.

Guilley H, Dudley RK, Jonard G, Balázs E, Richards KE (1982). Transcription of cauliflower mosaic virus DNA: Detection of promoter sequences, and characterization of transcripts. *Cell* 30: 763-773.

Harper G, Osuji JO, Heslop-Harrison JS, Hull R (1999). Integration of banana streak badnavirus into the *Musa* genome: Molecular and cytogenetic evidence. *Virology* 255: 207-213.

Hemmings-Mieszczak M, Hohn T (1999). A stable hairpin preceded by a short ORF promotes nonlinear ribosome migration on a synthetic mRNA leader. *RNA* 5: 8-16.

Hemmings-Mieszczak M, Steger G, Hohn T (1997). Evidence for alternative structures of the CaMV 35S RNA leader: Implications for viral expression and replication. *Journal of Molecular Biology* 267: 1075-1088.

Hemmings-Mieszczak M, Steger G, Hohn T (1998). Regulation of CaMV 35S RNA translation is mediated by a stable hairpin in the leader. *RNA* 4: 101-111.

Hinnebusch AG (1997). Translational regulation of yeast GCN4. A window on factors that control initiator-tRNA binding to the ribosome. *Journal of Biological Chemistry* 272: 21661-21664.

Hohn T, Corsten S, Rieke S, Müller M, Rothnie HM (1996). Methylation of coding region alone inhibits gene expression in plant protoplasts. *Proceedings of the National Academy of Sciences, USA* 93: 8334-8339.

Hohn T, Fütterer J (1997). The proteins and functions of plant pararetroviruses: Knowns and unknowns. *Critical Reviews in Plant Sciences* 16: 133-161.

Huang H, Yoon H, Hannig EM, Donahue TF (1997). GTP hydrolysis controls stringent selection of the AUG start codon during translation initiation in *Saccharomyces cerevisiae*. *Genes and Development* 11: 2396-2413.

Hull R (1996). Molecular biology of rice tungro viruses. *Annual Reviews in Phytopathology* 34: 275-279.

Hull R, Covey SN (1995). Retroelements: Propagation and adaptation. *Virus Genes* 11: 105-118.

Jackson RJ, Kaminski A (1995). Internal initiation of translation in eukaryotes: The picornavirus paradigm and beyond. *RNA* 1: 985-1000.

Jakowitsch J, Mette MF, Van der Winden J, Matzke MA, Matzke AJ (1999). Integrated pararetroviral sequences define a unique class of dispersed repetitive DNA in plants. *Proceedings of the National Academy of Sciences, USA* 96: 13241-13246.

Kiss-László Z, Blanc S, Hohn T (1995). Splicing of cauliflower mosaic virus 35S RNA is essential for viral infectivity. *The EMBO Journal* 14: 3552-3562.

Kobayashi K. Tsuge S. Nakayashiki H. Mise K. Furusawa I (1998). Requirement of cauliflower mosaic virus open reading frame VI product for viral gene expression and multiplication in turnip protoplasts. *Microbiology and Immunology* 42: 377-386.

Kozak M (1986). Point mutations define a sequence flanking the AUG initiator codon that modulates translation by eukaryotic ribosomes. *Cell* 44: 283-292.

Kozak M (1987). Effects of intercistronic length on the efficiency of reinitiation by eucaryotic ribosomes. *Molecular and Cellular Biology* 7: 3438-3445.

Kozak M (1989a). Circumstances and mechanisms of inhibition of translation by secondary structure in eucaryotic mRNAs. *Molecular and Cellular Biology* 9: 5134-5142.

Kozak M (1989b). The scanning model for translation—an update. *Journal of Cell Biology* 108: 229-241.

Kozak M (1991). Structural features in eukaryotic mRNAs that modulate the initiation of translation. *Journal of Biological Chemistry* 266: 19867-19870.

Kozak M (1998). Primer extension analysis of eucaryotic ribosome-mRNA complexes. *Nucleic Acids Research* 26: 4853-4859.

Kozak M (1999). Initiation of translation in prokaryotes and eukaryotes. *Gene* 234: 187-208.

Latorre P. Kolakofsky D. Curran J (1998). Sendai virus Y proteins are initiated by a ribosomal shunt. *Molecular and Cellular Biology* 18: 5021-5031.

Li J (1996). Molecular analysis of late gene expression in budgerigar fledgling disease virus. PhD thesis. University of Giessen. Germany. 108 pp.

Linial ML (1999). Foamy viruses are unconventional retroviruses. *Journal of Virology* 73: 1747-1755.

Lockhart BEL. Olszewski NE (1994). Badnaviruses. In Webster RG. Granoff A (eds.). *Encyclopedia of Virology*. London. San Diego. Academic Press. Volume 1. pp. 139-143.

Maiti IB. Richins RD. Shepherd RJ (1998). Gene expression regulated by gene VI of caulimovirus: Transactivation of downstream genes of transcripts by gene VI of peanut chlorotic streak virus in transgenic tobacco. *Virus Research* 57: 113-124.

Mesnard JM. Carriere C (1995). Comparison of packaging strategy of retroviruses and pararetroviruses. *Virology* 213: 1-6.

Miller WA. Dinesh-Kumar SP. Paul CP (1995). Luteovirus gene expression. *Critical Reviews in Plant Sciences* 14: 179-211.

Mize GJ. Ruan H. Low JJ. Morris DR (1998). The inhibitory upstream open reading frame from mammalian S-adenosylmethionine decarboxylase mRNA has a strict sequence specificity in critical positions. *Journal of Biological Chemistry* 273: 32500-32505.

Ndowora T. Dahal G. LaFleur D. Harper G. Hull R. Olszewski NE. Lockhart B (1999). Evidence that badnavirus infection in *Musa* can originate from integrated pararetroviral sequences. *Virology* 255: 214-220.

Odell JT, Dudley RK, Howell SH (1981). Structure of the 19S RNA transcript encoded by the cauliflower mosaic virus genome. *Virology* 111: 377-385.

Penswick JR, Hübler R, Hohn T (1988). A viable mutation in cauliflower mosaic virus, a retroviruslike plant virus, separates its capsid protein and polymerase genes. *Journal of Virology* 62: 1460-1463.

Perbal MC, Thomas CL, Maule AJ (1993). CaMV gene I product (pI) forms tubular structures which extend from the surface of infected protoplasts. *Virology* 195: 281-285.

Pestova TV, Borukhov SI, Hellen CU (1998). Eukaryotic ribosomes require initiation factors 1 and 1A to locate initiation codons. *Nature* 394: 854-859.

Pirone TP (1991). Viral genes and gene products that determine insect transmissibility. *Seminars in Virology* 2: 81-87.

Pooggin MM, Fütterer J, Skryabin KG, Hohn T (1999). A short open reading frame terminating in front of a stable hairpin is the conserved feature in pregenomic RNA leaders of plant pararetroviruses. *Journal of General Virology* 80: 2217-2228.

Pooggin MM, Fütterer J, Skryabin KG, Hohn T (2001). Ribosome shunt is essential for infectivity of cauliflower mosaic virus. *Proceedings of the National Academy of Sciences, USA* 98: 886-891.

Pooggin MM, Hohn T, Fütterer J (1998). Forced evolution reveals the importance of short open reading frame A and secondary structure in the cauliflower mosaic virus 35S RNA leader. *Journal of Virology* 72: 4157-4169.

Pooggin MM, Hohn T, Fütterer J (2000). Role of a short open reading frame in ribosome shunt on the cauliflower mosaic virus RNA leader. *Journal of Biological Chemistry* 275: 17288-17296.

Pringle CR (1999). Virus taxonomy. The Universal System of Virus Taxonomy, updated to include the new proposals ratified by the International Committee on Taxonomy of Viruses during 1998. *Archives of Virology* 144: 421-429.

Remm M, Remm A, Ustav M (1999). Human papillomavirus type 18 E1 protein is translated from polycistronic mRNA by a discontinuous scanning mechanism. *Journal of Virology* 73: 3062-3070.

Richert-Pöggeler KR, Shepherd RJ (1997). Petunia vein-clearing virus: A plant pararetrovirus with the core sequences for an integrase function. *Virology* 236: 137-146.

Rothnie HM, Chapdelaine Y, Hohn T (1994). Pararetroviruses and retroviruses: A comparative review of viral structure and gene expression strategies. *Advances in Virus Research* 44:1-67.

Ryabova LA, Hohn T (2000). Ribosome shunting in cauliflower mosaic virus 35S RNA leader is a special case of reinitiation of translation functioning in plant and animal systems. *Genes and Development* 14: 817-829.

Ryabova LA, Pooggin MM, Dominguez DI, Hohn T (2000). Continuous and discontinuous ribosome scanning on the cauliflower mosaic virus 35S RNA is controlled by short open reading frames. *Journal of Biological Chemistry* 275: 37278-37284.

Schärer-Hernández NG. Hohn T (1998). Non-linear ribosome migration on cauliflower mosaic virus 35S RNA in transgenic tobacco plants. *Virology* 242: 403-413.

Schlicht H-J. Salfeld J. Schaller H (1989). Synthesis and encapsidation of duck hepatitis B virus reverse transcriptase do not require formation of core-polymerase fusion proteins. *Cell* 56: 85-92.

Schmidt-Puchta W. Dominguez DI. Lewetag D. Hohn T (1997). Ribosome shunting in vitro. *Nucleic Acids Research* 25: 2854-2860.

Scholthof HB. Gowda S. Wu FC. Shepherd RJ (1992). The full-length transcript of a caulimovirus is a polycistronic mRNA whose genes are *trans* activated by the product of gene VI. *Journal of Virology* 66: 3131-3139.

Scholthof HB. Wu FC. Gowda S. Shepherd RJ (1992). Regulation of caulimovirus gene expression and the involvement of *cis*-acting elements on both viral transcripts. *Virology* 190: 403-412.

Schultze M. Hohn T. Jiricny J (1990). The reverse transcriptase gene of CaMV is translated separately from the capsid gene. *The EMBO Journal* 9: 1177-1185.

Skulachev MV. Ivanov PA. Karpova OV. Korpela T. Rodionova NP. Dorokhov YL. Atabekov JG (1999). Internal initiation of translation directed by the 5'-untranslated region of the tobamovirus subgenomic RNA I(2). *Virology* 263: 139-154.

Yueh A. Schneider RJ (1996). Selective translation initiation by ribosome jumping in adenovirus-infected and heat-shocked cells. *Genes and Development* 10: 1557-1567.

Zijlstra C. Hernández N. Gal S. Hohn T (1996). *Arabidopsis* plants harbouring the cauliflower mosaic virus transactivator gene show strongly delayed flowering. *Virus Genes* 13: 5-17.

Zijlstra C. Hohn T (1992). Cauliflower mosaic virus gene VI controls translation from dicistronic expression units in transgenic *Arabidopsis* plants. *Plant Cell* 4: 1471-1484.

Chapter 15

Recombination in Plant DNA Viruses

Thomas Frischmuth

INTRODUCTION

The evolution of plant viruses proceeds by natural mutation and recombination. Natural occurrence of mutations and recombination between plant viruses has been deduced from sequence analysis of virus isolates for RNA as well as DNA viruses. Besides recombination, pseudorecombination between multipartite viruses has also been recognized as a driving force for the evolution of new viruses (see Figure 15.1).

The most closely related virus isolates differ by point mutations, often silent. However, comparison of more and more distantly related isolates reveals increasing evidence of nonsilent mutational differences and insertions/deletions of increasing size. When genes and genomes between families are compared, major intergroup recombinational events are observed. However, recombination can also be observed between closely related virus species. Recent results suggest that recombination between viruses and viral-derived sequences in the plant genome as well as in transgenic plants must also be included in consideration of viral evolution.

Within the plant DNA viruses, two families containing single-stranded or double-stranded DNA can be distinguished.

So far one family *(Geminiviridae)* and one genus *(Nanovirus)* have been identified within the single-stranded DNA (ssDNA) viruses. The six genera of plant virus family *Caulimoviridae* contain double-stranded DNA (dsDNA) genomes. They are the plant pararetroviruses because their replication cycle resembles that of animal retroviruses (Hull, 1992; Rothnie et al., 1994).

The author is most grateful to B. D. Harrison, J. E. Polston, J. L. Dale, S. Covey, R. W. Briddon, and T. Hohn for providing information prior to publication.

FIGURE 15.1. Diagram Indicating the Apparent Relative Importance of Various Evolutionary Processes in the Generation of Genetic Novelty Among Viruses

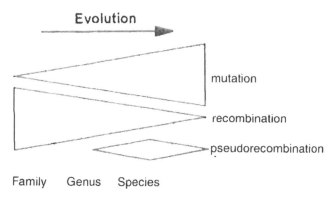

Source: Modified according to Gibbs et al., 1997: 4.

Family Geminiviridae

Geminiviruses are characterized by circular ssDNA genomes of 2.5 to 3.0 kb encapsidated in twinned (geminate) particles of 20 by 30 nm diameter. The genome is organized as monopartite (one circular DNA molecule) or bipartite (two circular DNA molecules) (see Figure 15.2).

Geminiviruses are currently divided into three genera on the basis of their genome organization and biological properties (Palmer and Rybicki, 1997). The overall genomic structure is very similar in all three genera (see Figure 15.2). The intergenic region (IR) contains sequences for viral replication and gene expression. Members of the genus *Mastrevirus* have monopartite genomes transmitted by leafhopper insect vectors to monocotyledonous plants (see Figure 15.2). However, in recent years, a number of geminiviruses belonging to the genus *Mastrevirus,* have been found to infect dicotyledonous plants (Morris et al., 1992; Liu et al., 1997).

Members of the genus *Curtovirus* have a monopartite genome, infect dicotyledonous plant species, and are transmitted by leafhopper or treehopper vectors. *Beet curly top virus* (BCTV; Stanley et al., 1986) represents the type species; its genome organization is outlined in Figure 15.2. So far two other members, *Horseradish curly top virus* (HrCTV; Klute et al., 1996) and *Tomato pseudo-curly top virus* (TPCTV; Briddon et al., 1996), have been identified.

FIGURE 15.2. Genome Organization of Geminiviruses

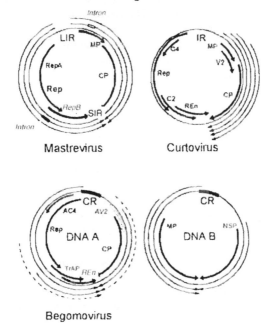

Mastrevirus

Curtovirus

Begomovirus

Note: The diagram shows the consensus genome organization of double-stranded replicative forms of geminiviruses. Intergenic regions (IR, LIR, SIR) are indicated. The part of the IR that is almost sequence identical between DNAs A and B of begomoviruses is indicated by a black box (CR, common region). Open reading frames (ORFs) are indicated by black arrows within the circle, and where a gene's function is known, the name of the gene product is indicated. CP = coat protein; MP = movement protein; NSP = nuclear shuttle protein; Rep = replication associated protein; TrAP = transcription activator protein; REn = replication initiation protein. ORF *AV2* is present only in Old World begomoviruses (open arrow). Arrows outside the circles represent transcripts, whereby transcripts indicated by dotted arrows are not produced by all begomoviruses (arrowheads = 3'-ends). Intron sequences of mastreviruses are indicated by black and open boxes.

The genome organizations of HrCTV and TPCTV differ from that of BCTV, in that HrCTV lacks the replication enhancer gene (REn) and TPCTV, the movement protein gene (MP).

Members of the third genus, *Begomovirus*, have a bipartite genome organization (see Figure 15.2; DNAs A and B), with a few exceptions, e.g., the mediterranean isolates of *Tomato yellow leaf curl virus* (TYLCV; Navot et al., 1991) and *Tomato leaf curl virus* from Australia (TLCV; Dry et al., 1993), which have a monopartite genome organization. The begomoviruses infect dicotyledonous plant species and are transmitted by the whitefly *Bemisia tabaci*. Approximately 180 nucleotides (nt) of the IR of bipartite begomoviruses are almost identical between DNAs A and B (common region [CR]).

The replication initiation protein (Rep) is the only protein that is absolutely necessary for viral replication. Rep of curtoviruses and begomoviruses is produced by a single open reading frame (ORF), whereas splicing of the complementary sense transcript is essential for production of functional RepA-RepB fusion protein (Rep) of mastreviruses (see Figure 15.2).

The product of the transcription activator protein (TrAP) gene of begomoviruses is required for expression of viral coat protein (CP) and nuclear shuttle protein (NSP).

Plants infected with *REn* mutants develop mild symptoms, and viral DNA is greatly reduced in these plants. REn functions as a replication enhancer via interaction with Rep and, by that, modulates the function of Rep, replication versus repression.

The most abundant protein in infected tissues is the CP. The CP of bipartite begomoviruses is not essential for viral replication, systemic spread, or symptom development. In contrast, the CP of mastreviruses and curtoviruses is required for viral movement. The CP protects viral DNA during transmission and determines vector specificity.

Besides the *CP* of mastreviruses, the second virion-sense gene, *MP* (see Figure 15.2), is involved in viral movement, and the gene product has been located in ultra-thin sections to secondary plasmodesmata. The BCTV gene product of virion-sense ORF V3 seems to be involved in viral movement. However, this ORF is not present in TPCTV, a second member of the curtoviruses. For the monopartite members of the begomovirus group, little is known about viral movement, except one report that ORF AC4 seems to be involved in movement of TYLCV.

Mutational analysis showed that both genes encoded by DNA B of bipartite begomoviruses are involved in viral movement (see Figure 15.2). The two proteins have different functions during movement. The NSP moves the viral DNA in and out of the nucleus. In the cytoplasm NSP interacts with MP and is translocated to the cell periphery. At the cytoplasm membrane MP forms endoplasmic reticulum-derived tubules that extend through the cell wall to the next cell. Other data suggest that MP is able to increase the molecular size exclusion limit (SEL) of plasmodesmata. Therefore, two hypotheses for movement of bipartite begomoviruses are currently discussed. Viral DNA moves to the next cell either by endoplasmic reticulum-derived tubules or via MP-modified plasmodesmata (geminiviruses are reviewed in Frischmuth, 1999; for further details, see also Chapters 7 and 17).

Nanoviruses

The nanoviruses were recently recognized as members of an independent virus genus. Nanoviruses are characterized by isometric virions of about 20 nm in diameter and a multicomponent circular ssDNA genome with components of about 1 kb each. So far five members have been identified, *Banana bunchy top virus* (BBTV; Harding et al., 1993), *Subterranean clover stunt virus* (SCSV; Chu et al., 1993), *Faba bean necrotic yellows virus* (FBNYV; Katul et al., 1993), *Milk vetch dwarf virus* (Sano et al., 1993), and a fifth tentative member, Coconut foliar decay virus (CFDV; Rohde et al., 1990).

CFDV has a monopartite genome of 1.3 kb (Rohde et al., 1990) (see Figure 15.3b) and is transmitted by the planthopper *Myndus taffini* (Julia et al.,

FIGURE 15.3. Genome Organization of Nanoviruses

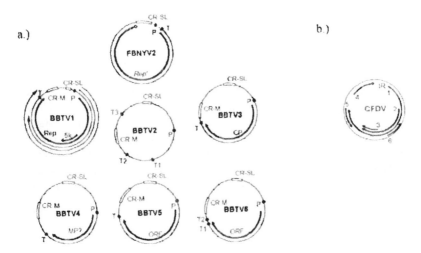

Note: The genome organization of *Banana bunchy top virus* (BBTV) and *Faba bean necrotic yellows virus* (FBNYV) component 2 (a) and of Coconut foliar decay virus (CFDV) (b). The locations of the stem-loop common region (CR-SL) and the major common region (CR-M) are indicated by open boxes, and the putative promoter (P) and termination (T) sequences by black boxes. Position and orientation of open reading frames (ORFs) are indicated by black arrows, and for FBNYV DNA2 by an open arrow. Where a gene function is known, the name of the gene product is indicated (see legend to Figure 15.2). Arrows outside the circle of BBTV1 represent transcripts detected in infected plants (arrowheads = 3'-ends).

1985), whereas other members of this group are aphid transmitted (Grylls and Butler, 1956; Franz, 1997).

The general genome organization of nanoviruses is outlined in Figure 15.3. Until now, six distinguishable components were isolated from plants infected with BBTV (Burns et al., 1995) and FBNYV (Katul et al., 1997) and seven from those infected with SCSV (Boevink et al., 1995). Each component encodes in principle for a single gene in virion-sense orientation (Figure 15.3a). Due to transcript analysis of BBTV DNA1, a second ORF with a coding capacity of 5 kDa was postulated (see Figure 15.3a). However, no product or function has been associated with this putative ORF. Some components, such as FBNYV DNA2, encode putative duplicates of the *Rep* gene normally encoded by DNA1, whereas other components, such as BBTV DNA2, have no coding capacity at all (see Figure 15.3a). The function of such components in the pathogenicity of nanoviruses is unknown. All components possess a noncoding IR of which, in general, two sequence parts have sequence homologies among all components, the major common region (CR-M) and the stem-loop common region (CR-SL) (see Figure 15.3). Both common regions harbor sequences that are involved in viral replication.

As yet very little is known about gene products and their functions in nanoviruses. Genomic component 5 of FBNYV encodes the CP and has a counterpart on component 3 of BBTV and on component 5 of SCSV (see Figure 15.3; Katul et al., 1997; Wanitchakorn et al., 1997). The second characterized gene product on BBTV genomic component 1 is Rep (Figure 15.3a; Harding et al., 1993). Biochemical studies have shown that the Rep protein has cleavage and ligation activities similar to those of the geminivirus Rep protein (Hafner et al., 1997). Functions of other ORFs are not known, except that the gene product of component 4 might be involved in viral movement (Katul et al., 1997).

Family Caulimoviridae

Genus Caulimovirus

Caulimoviruses are named after the type species of the genus, *Cauliflower mosaic virus* (CaMV). CaMV has a circular dsDNA genome 7.8 to 8.2 kb in size with one or more site-specific discontinuities in each DNA strand (see Figure 15.4). The genome is encapsidated in a nonenveloped spherical particle of about 45 to 50 nm in diameter. Most caulimoviruses are aphid transmitted. The coding regions are interrupted by an IR of approximately 600 bp, with various regulatory elements such as the 35S promoter.

FIGURE 15.4. Genome Organization of Caulimoviruses

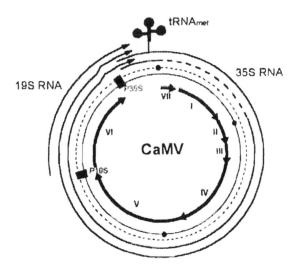

Note: The genome organization of the type-species *Cauliflower mosaic virus* (CaMV) is shown. The double circle represents the genomic DNA (dotted circle = minus-sense strand and continuous circle = plus-sense strand) with the discontinuities shown as spots. Open reading frames (ORFs) are indicated by black arrows within the double circle. The position of the two promoter sequences (P35S and P19S) are indicated by black boxes. Outer arcs and circles show the genomic locations of the two major transcripts, (35S RNA and 19S RNA) and spliced transcripts, whereby the dotted line indicates the positions of intron sequences (arrowheads = 3'-ends). The position where the replication initiator methionine transer (t) RNA (tRNAmet) binds in the pregenomic 35S RNA is indicated.

From this promoter, a transcript spanning the entire genome, plus a duplication approximately 180 nt-long, is produced. A second smaller IR occurs upstream of gene VI, containing the so-called 19S promoter. From this promoter, a transcript spanning gene VI is produced (see Figure 15.4; Rothnie et al., 1994).

ORF I encodes the MP. The ORF II product is involved in the transmission of CaMV by aphids. The gene III product has sequence-nonspecific nucleic acid-binding activity and is required for viral infectivity and may function in virus assembly. ORF IV encodes the CP of CaMV. The gene V product is a multifunctional protein with reverse transcriptase (RT), RNase H, and protease activities and plays a central role in viral replication. The gene VI product is the matrix protein for the major inclusion body and is absolutely necessary for the translation of genes from the polycistronic 35S RNA. Furthermore, the gene VI product is involved in virus assembly, host range determination, and severity of symptoms. The function and role of ORF VII remain obscure, but it is not required for replication, and its product is not detected in infected plants (for more details, see Chapter 14 in this book).

Genera "SbCMV-like viruses," "CsVMV-like viruses," "PVCV-like viruses," Badnavirus and "RTBV-like viruses"

Members of the genera "SbCMV-like viruses" (type species, *Soybean chlorotic mottle virus*), "CsVMV-like viruses" (type species *Cassava vein mosaic virus;* CsVMV) and "PVCV-like viruses" (type species, *Petunia vein clearing virus*) have isometric particles.

Viruses in the genera *Badnavirus* (type species, *Commelina yellow mottle virus;* ComYMV) and "RTBV-like viruses" (type species, *Rice tungro bacilliform virus;* RTBV) have a circular dsDNA genome of 7.5 to 8 kb encapsidated in bacilliform virions of 120 to 150 × 30 nm in size.

In contrast to caulimoviruses, members of these five gemera infect a wide range of economically important crop plants, such as rice, cassava, sugarcane, etc. They are transmitted in a semipersistent manner by mealybugs or leafhoppers.

As in caulimoviruses, both strands have site-specific discontinuities and only the minus-sense strand codes for genes (see Figure 15.4). The coding capacity varies from three ORFs (I-III) in ComYMV (Medberry et al., 1990), to four (I-IV) in RTBV (Hay et al., 1991), to five (I-V) in CsVMV (Calvert et al., 1995) (see Figure 15.4). ORF I of RTBV lacks an AUG initiation codon, and the translation is initiated at an AUU codon (Hay et al., 1991). The coding regions are interrupted by an IR of approximately 700 bp that contains various regulatory elements, such as a promoter similar to the 35S promoter of CaMV (see Figure 15.4). As in CaMV, this promoter produces a transcript spanning the entire viral genome (Figure 15.4). Not much is known about gene functions of these viruses. From sequence comparisons of members of these five genera with CaMV, some enzymatic functions can be assigned to certain ORFs. Most of these functions are located on multifunctional proteins. Although the insect vectors involved in transmission of some of these viruses are known, viral functions have not been identified. In the case of RTBV, transmission by leafhoppers requires coinfection with *Rice tungro spherical virus* (Jones et al., 1991; see also Chapter 3).

RECOMBINATION IN PLANT ssDNA VIRUSES

Nucleotide Sequence Evidence

The overall genome organization deduced from the nucleotide sequences of geminiviruses is very similar among all genera (see Figure 15.2). However, when sequences in a phylogenetic analysis are compared, the three groups differ greatly from one another (see Figure 15.5). Whereas mastreviruses are phylogenetically distinct from one another even if they are from the same geographical region, a close relationship can be observed between members within the genus *Begomovirus* (see Figure 15.5).

When sequences of geminiviruses are compared by computer programs, such as dot matrix programs, no sequence similarities are detected between mastreviruses and begomoviruses, whereas curtoviruses and begomoviruses have similarities in the coding regions for *Rep*, *TrAP*, and *REn*, but none in the *CP*, *V2*, and *V3* genes (see Figure 15.6).

Taking the phylogenetic analysis of single genes into account, clearly, at least *Rep* and *C4* genes of curtoviruses are derived from a begomovirus ancestor (Klute et al., 1996; Briddon et al., 1998). These analyses led to the proposal that curtoviruses originated from a recombination event between a leafhopper-transmitted mastrevirus and a whitefly-transmitted begomovirus (Stanley et al., 1986; Howarth and Vandemark, 1989; Lazarowitz, 1992; Klute et al., 1996; Briddon et al., 1998), by which, according to the "modular evolution" theory (Botstein, 1980), the *Rep*, *C4*, and, with limitations, the *TrAP* and *REn* modules were donated by a begomovirus ancestor and the *CP/V2* module by a mastrevirus. Although mastreviruses are generally found in monocot plants, some exceptions of dicot-infecting mastreviruses have been described (see Figure 15.5; Morris et al., 1992; Liu et al., 1997). Therefore, mixed infection of plants with mastreviruses and begomoviruses is theoretically possible, although this has not been found in nature so far.

Infection of plants with two or more begomoviruses has been observed in nature (Lazarowitz, 1991; Harrison et al., 1997). The severe cassava mosaic disease in Uganda seems to be caused by a recombinant *African cassava mosaic virus* (ACMV)/*East African cassava mosaic virus* (EACMV) virus, designated UgV (Zhou et al., 1997). ACMV and EACMV are considered two virus species (Hong et al., 1993), and both viruses are present in mixed infection of cassava in the field. Interestingly, UgV represents a virus created by interspecific recombination. This is particularly visible in the *CP* gene: The 5'-219 nt are 99 percent identical to EACMV (only 79 percent to ACMV); the following 459 nt are 99 percent identical to ACMV (75 percent to EACMV); and the 3'- 93 nt are again 98 percent identical to EACMV (76 percent

FIGURE 15.5. Phylogenetic Tree obtained from Alignments of DNA A Nucleotide Sequences of *Tomato leaf curl virus* (ToLCV) from Panama, *Sida golden mosaic virus* (SiGMV) from Costa Rica, *Sida golden mosaic virus* (SiGMV) from Honduras, *Bean dwarf mosaic virus* (BDMV) from Colombia, *Abutilon mosaic virus* (AbMV) from the West Indies, *Tomato yellow leaf curl virus* (ToLCV) from Israel, *African cassava mosaic virus* (ACMV) from Kenya, *Beet curly top virus* (BCTV) from the United States, *Tomato pseudo-curly top virus* (TPCTV) from the United States, *Wheat dwarf virus* (WDV) from Europe, *Maize streak virus* (MSV) from Kenya, and *Tobacco yellow dwarf virus* (TYDV) from Australia

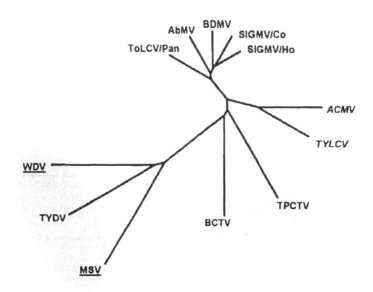

Note: The dendrogram was calculated using the neighbor-joining and bootstrap (1,000 replications) options of CLUSTAL W. Vertical branches are arbitrary; horizontal branches are proportional to calculated mutation distances. Begomoviruses from the Old World are indicated in italics, and monocot-infecting mastreviruses are underlined.

to ACMV) (Zhou et al., 1997). From mixed-infected cassava plants in Cameroon, several genomic components belonging to either the ACMV or EACMV cluster, as well as recombinant molecules, have been isolated (Fondong et al., 2000). Recombination between the two infecting viruses has been observed within both genomic components, DNAs A and B. In contrast to the recombinant UgV strain, recombination has been detected throughout the entire genomic DNA A or B, whereby apparently no recombination has taken place between DNAs A and B (Fondong et al., 2000).

FIGURE 15.6. Dot Matrix Comparison of the Sequences of *Beet curly top virus* (BCTV) and *African cassava mosaic virus* (ACMV)

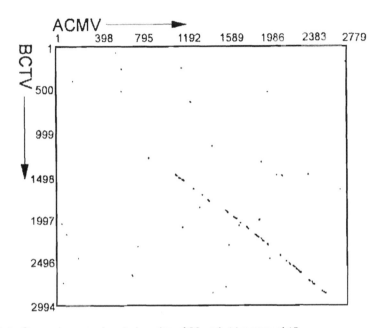

Note: Comparison used a window size of 20 and stringency of 15.

The production of subgenomic DNA, so-called defective interfering DNA (DI DNA), has been observed in all geminivirus genera (Frischmuth and Stanley, 1993). DI DNAs generally have one or more parts of a genome deleted. It is assumed that the production of DI DNA is due to intramolecular recombination, and most DI DNAs have short sequence duplication at the deletion border (see Figure 15.7).

Characteristically, DI DNAs contain only viral-derived sequences. They are dependent for their replication and movement on the parental virus and interfere with viral proliferation in transgenic plants (Frischmuth and Stanley, 1993; Frischmuth, Engel, and Jeske, 1997).

Besides these virus-derived DI DNAs, recently, small DNA molecules, described as satellite and nanovirus-like DNAs, have been found in geminivirus-infected plants. The satellite molecules contain only the viral IR (see Figure 15.7), whereas the rest of the molecule represents nonviral DNA sequences (Dry et al., 1997; Stanley et al., 1997). In *Cotton leaf curl virus* (CLCuV)-

FIGURE 15.7. Genome Map of *Beet curly top virus* (BCTV) (Left) with Short Repeat Sequences at the Deletion Points (Right)

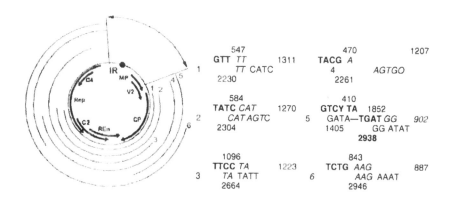

	547			470	1207
	GTT *TT*	1311		**TACG** *A*	
1	*TT* CATC			4	*AGTGO*
	2230			2261	
	584			410	
	TATC *CAT*	1270		**GTCY TA**	1852
2	*CAT AGTC*	5		GATA—**TGAT** *GG*	*902*
	2304			1405 *GG* ATAT	
					2938
	1096			843	
	TTCC *TA*	1223		**TCTG** *AAG*	887
3	*TA* TATT	6		*AAG* AAAT	
	2664			2946	

Note: Left: the extent of the deletions in clones 1 through 6 is indicated. The sequence delimited by the arrows is present in all defective interfering (DI) DNAs. Right: the sizes of the DI DNAs are indicated. For other symbols, see note for Figure 15.2.

infected plants, a small DNA molecule (designated as DNA1) related to nanoviruses has been identified (see Figure 15.8).

As with the DI and satellite DNAs, this nanovirus-like molecule is encapsidated in the *Cotton leaf curl virus* capsid protein (Mansoor et al., 1999). Whether the *Rep* gene of CLCuV DNA1 is functional or whether it is dependent on the Rep protein functions of the coinfecting geminivirus as in the DI and satellite DNAs, is not yet clear. It is known that the *Ageratum yellow vein virus* (AYVV) DNA1 autonomously replicates in leaf disks (Saunders and Stanley, 1999). Replication of gemini- and nanoviruses is very similar, and the loop sequences in their respective stem loops are identical. How the satellite and nanovirus-like DNAs evolved is not clear either. However, all three geminivirus infection-accompanying molecules (DI DNA, satellite DNA, and nanovirus-like DNA) represent a pool of potential DNAs for recombination in mixed-infected plants. One such recombined molecule was isolated from AYVV-infected plants (AYVV Def 19; Saunders and Stanley, 1999). This defective molecule contains sequences of AYVV DNA A, of AYVV DNA1, and of unknown origin (Saunders and Stanley, 1999).

FIGURE 15.8. Genome Organization of DNA1 (Right Molecule) Isolated from *Cotton leaf curl virus* (CLCuV)-Infected Plants Compared with the Genomic Component of *Banana bunchy top virus* (BBTV) Containing the Replication Associated Protein Gene *(Rep)* (Left Molecule)

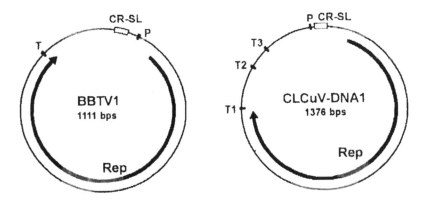

Note: The locations of the stem-loop common region (CR-SL) are indicated by open boxes, and the putative promoters (P) and termination (T) sequences by black boxes.

Recombination Between Begomovirus Genomic Components

The CP of bipartite begomoviruses is not essential for viral replication, systemic spread, or symptom development. After deletion exceeding 200 nt within the *CP* gene of ACMV, a size reversion to almost viral genomic size during systemic infection of plants has been observed (Etessami et al., 1989). Some of the progenies of these mutant viruses were produced by intermolecular recombination with the inoculated DNA B. In other experiments, Roberts and Stanley (1994) demonstrated that lethal mutants within the conserved stem-loop structure of DNA A are also repaired by intermolecular recombination with DNA B, resulting in an infectious virus.

Although bipartite geminiviruses are closely related, the production of viable pseudorecombinants by reassortment of infectious cloned components is generally limited to strains of a particular virus (Stanley et al., 1985; Frischmuth et al., 1993), due to the highly specific nature of the interaction of Rep with the origin of viral replication (Fontes et al., 1994). A second obstacle for the production of viable pseudorecombinants might be the inability of DNA B genes to mediate the movement of heterogenomic DNA A components. For a number of bipartite geminiviruses, complementation of DNA B genes has been demonstrated following coinoculation with both genomic components of one virus and DNA A of another virus (Frischmuth et al., 1993; Sung and Coutts, 1995). So far, four exceptions to viable pseudorecombination between the two distinct geminiviruses have been described (Gilbertson et al., 1993; Frischmuth, Engel, Lauster, et al., 1997; Höfer et al., 1997). The pseudorecombinant virus consisting of *Tomato mottle virus* (ToMoV) DNA A and *Bean dwarf mosaic virus* (BDMV) DNA B was maintained during serial passage and, after several passages, became highly pathogenic. The increased pathogenicity was associated with an increased level of the heterologous DNA B in infected plants. Sequence analysis of the DNA B of the enhanced pathogenic pseudorecombinant virus revealed that intermolecular recombination between the ToMoV DNA A and BDMV DNA B resulted in a new BDMV DNA B molecule in which most of the CR was replaced by that of ToMoV (Hou and Gilbertson, 1996).

These examples show that the prime target for intermolecular recombination between begomovirus DNA A and B components is the CR.

Recombination with the Plant Genome

Recently, geminivirus-related DNA sequences have been discovered in the *Nicotiana tabacum* nuclear genome (Kenton et al., 1995; Bejarano et al., 1996). Sequence analysis of several plant genomic library clones revealed integration of begomovirus DNA A sequences comprising the CR and parts of the *Rep* gene (see Figure 15.9).

The organization of geminivirus-related DNA sequences resembles the patterns that occur by illegitimate recombination of artificially introduced extrachromosomal DNA within eukaryotic chromosomes. Since geminiviruses can be considered extrachromosomal DNA molecules in plants, illegitimate recombination with the plant genome might be responsible for the integration event.

Recombination of geminiviral DNA with the plant genome has been demonstrated experimentally in transgenic plants carrying viral DNA sequences (see Figure 15.10).

When transgenic plants (harboring constructs pCRA1 and pCRA2, containing the *CP*-coding sequences of ACMV) were challenged with an ACMV *CP* mutant, recombination occurred between mutant viral DNA and the integrated construct DNA. This resulted in recombinant virus progeny with "wild-type" characteristics. However, no recombination was observed when transgenic plants harboring construct pJC1 (see Figure 15.10) were challenged (Frischmuth and Stanley, 1998). Notably, CR sequences were present in the random integrated geminivirus-related DNA sequences (see Figure 15.9), as well as in the viral transgenes, when recombination was observed. This suggests that the CR, including the replication origin, facilitates recombination between the plant genome and extrachromosomal viral DNA.

FIGURE 15.9. Schematic Representation of the Arrangement of Nuclear Integrated Geminivirus-Related DNA Sequences

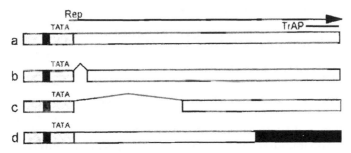

Source: Modified according to Bejarano et al., 1996: 761.

Note: The genome part of the common region (CR = gray box; stem loop = black box) and the replication associated protein *(Rep)*-coding region (blank box, arrow) of a begomovirus (a) are compared with the integrated geminivirus-related DNA copies (b-d). Deleted viral and nonviral sequences (black box) are indicated. *TrAP* = transcription activator protein.

FIGURE 15.10. Schematic Representation of Constructs pCRA1, pCRA2, and pJC1 Used for Transformation of *Nicotiana benthamiana*

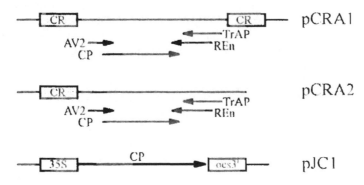

Source: Modified according to Frischmuth and Stanley, 1998: 1266.

Note: The common regions (CR = gray boxes) and genes (arrows) of geminivirus origin are indicated. The coat protein *(CP)* gene in construct pJC1 is under the control of the 35S promoter (35S) and the octopine synthetase terminator (*ocs3'*); *AV 2* = open reading frame; *REn*-replication enhancer; *TrAP*-transcription activator protein.

RECOMINATION IN PLANT dsDNA VIRUSESVIRUSES

Recombination Between Virus Isolates

The CaMV isolates differ by about 5 percent in nucleotide sequence and can often be distinguished by host range, symptom severity, symptom type, etc. This fact has been used to study recombination between CaMV isolates in mixed-infected plants (for example see Dixon et al., 1986; Grimsley et al., 1986; Vaden and Melcher, 1990). Considering all results together, three mechanisms for the generation of recombinant CaMV are possible: (1) ligation of the linear DNAs provided in the inoculum (Geldreich et al., 1986); (2) double-stranded DNA recombination in the nucleus (Choe et al., 1985), and (3) template switching by reverse transcriptase during synthesis of DNA minus strands (Grimsley et al., 1986). The three suggested mechanisms predict different crossover points in the resulting circular virion DNA. Ligation of linear inoculum DNAs produces, in addition to the junction at the ligation site, a junction at the end of the 35S RNA transcript derived from the ligated molecule. This is only the case when the virus inoculum is linearized DNA; therefore, this mechanism probably plays no role in nature. In the case of recombination between two viral dsDNA molecules, crossover points randomly distributed over the viral genome would be expected. In most cases of analyzed recombinant virus progenies in mixed-infected plants, the crossover junctions were close to the 35S RNA start site, and at positions where template switching in the replication cycle of caulimoviruses occurs. This finding favors the third mechanism, whereby intermolecular template switching by the reverse transcriptase occurs during viral replication. A similar mechanism, template switching during replication, has been suggested for RNA virus recombination and is described in detail in Chapter 9 of this book.

Several isolates of RTBV, an RTBV-like virus, were sequenced, and aligned sequences were analyzed by computer program for phylogenetic relationships (Cabauatan et al., 1999; method is described in Figure 15.5). The distribution of sequence clusters suggests recombination between genome segments of various RTBV isolates that resembles recombination illustrated by the phylogenetic analyses for the newly emerged severe UgV variant of ACMV from Uganda. Since the replication cycle of RTBV-like viruses and caulimoviruses is almost identical, the recombination mechanisms suggested for caulimoviruses probably apply for RTBV-like viruses.

Recombination with the Plant Genome

As with geminiviruses, RTBV-like viruse-derived DNA of *Banana streak virus* (BSV) has been discovered in the genome of banana and plantains

(*Musa* spp.; Harper et al., 1999; Ndowora et al., 1999). The majority of integrated BSV-derived DNAs are small pieces of the viral genome. However, one integrated BSV-derived DNA comprises the entire viral genome, interrupted by a cluster of BSV-related and -unrelated sequences (see Figure 15.11).

It has been suggested that this integrated copy might be able to escape from the *Musa* genome through recombination events, thus leading to infection of the plant. The model of BSV genome escape from the *Musa* genome with two steps. In the first step, the interrupting sequence is removed by homologous recombination between the 280 nt-long direct sequence duplication (see Figure 15.11). Second, for the release of the viral genome, two alternative routes are suggested: (1) a homologous recombination between the shorter direct duplication at the ends of the integrated BSV genome or (2) a transcript produced by a plant promoter upstream of the integrated BSV genome. The transcript produces the reverse transcriptase, which uses the transcript for reverse transcription. Intra- or intermolecular template switching then results in a recombinant viral BSV genome (see Figure 15.11).

FIGURE 15.11. Models for the Generation of an Infectious *Banana streak virus* (BSV) Genome by Excision of an Integrated Viral Sequence

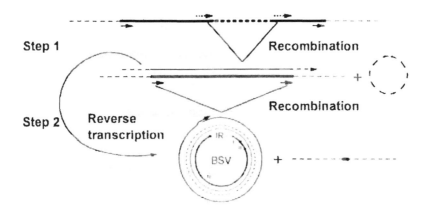

Source: Modified according to Ndowora et al., 1999: 217.

Note: The integrated BSV genome (thick line) is flanked by 98 nt-long direct repeats (arrows below the BSV genome) and is interrupted by a cluster of viral and nonviral sequences (thick dotted line). The interrupting sequences are also flanked by direct repeats (280 nt; dotted line arrows). The flanking *Musa* genome sequences are indicated as dotted lines. The excision models are described in the text.

Similarly, as described previously for the geminivirus ACMV, recombination has been found between infecting CaMV and the viral transgene (Schoelz and Wintermantel, 1993; Wintermantel and Schoelz, 1996; Wintermantel et al., 1997; Király et al., 1998). *Nicotiana bigelovii* was transformed with a construct that expressed gene VI of CaMV strain D4, a gene known to determine systemic infection in solanaceous species, including *N. bigelovii* by CaMV (see Figure 15.12).

Transgenic plants were inoculated with the CaMV strain CM1841, which is unable to infect solanaceous plants species systemically. Recombinants between strain CM1841 and the transgene were isolated from 30 percent of the plants after mechanical inoculation and from 100 percent of the plants after agroinoculation. Simulating conditions of moderate selection pressure, transgenic plants were inoculated with CaMV strain W260 (wild type), which is able to infect solanaceous plant species systemically. Also, in this case, in 3 out of 23 inoculated plants, recombination between W260 and the transgene

FIGURE 15.12. Location of Recombination Junctions Between *Cauliflower mosaic virus* (CaMV) and Transgene Sequences

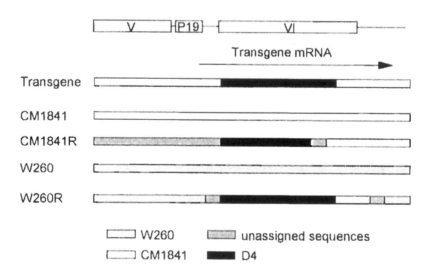

Source: Modified according to Wintermantel et al., 1997: 70.

Note: Transgenic plants contained a segment of CaMV that produces an mRNA of gene VI (CaMV D4 strain) from the 19S promoter. Below the transgene, partial maps of CaMV strains CM1841, W260, and two recombinants of each strain are shown.

was observed. Because W260R (W260 recombinant virus) should not have an advantage over W260, W260R (see Figure 15.12) was inoculated, alone or in mixture with W260, into normal *N. bigelovii*. The W260R (recombinant virus) showed enhanced aggressiveness, and even when W260R was coinoculated with strain W260 in a mixture unfavorable for W260R, the recombinant virus became the predominant virus (Wintermantel et al., 1997).

The recombination between inoculated virus and the transgene occurred within DNA sequences of approximately 100 bp in length, at the 5'- and 3'-ends of the gene VI mRNA produced by the transgenic plant (see Figure 15.12). The mode of recombination consists of an RNA-RNA mechanism, involving a template switch during reverse transcription. The mechanism is therefore similar to the previous mechanism suggested for recombination between CaMV isolates.

CONCLUSIONS

Recombination within plant DNA viruses is certainly one of the driving forces for their evolution. Powerful tools to analyze the history of viral pathogens are computer-based sequence analysis and computerized international databases that provide access to virus sequences. This chapter introduced two computer programs for sequence analyses, phylogenic and dot matrix. However, more programs are available (for example, see Gibbs et al., 1997). Such computer analyses led to the suggestion that the geminivirus BCTV represents a recombinant virus between the two very distinct genera, *Begomovirus* and *Mastrevirus*, of the family *Geminiviridae*.

Intermolecular recombination has been demonstrated between distinct begomoviruses when cloned genomic components of two distinct viruses were used in pseudorecombination experiments. The newly emerged severe ACMV strain from Uganda (UgV) shows clustered sequence similarities to EACMV or ACMV, and since both viruses are detected in the same plant in field isolates, it is assumed that UgV represents an interspecific recombinant virus. In both cases, the experimental approach and the field analysis, the resulting new virus showed enhanced pathogenicity.

Intramolecular recombination seems to be the cause of the production of geminivirus DI DNA because, at the deletion junctions, in most cases, short sequence duplications occur. Besides virus-derived DI DNA, satellite DNAs, consisting of viral- and nonviral-derived sequences, have been found in geminivirus-infected plants. The origin of the nonviral sequences is not yet clear, but they might originate from either the plant genome or other DNA viruses, such as nanoviruses. Nanoviruses have DNA approximately the size of geminivirus DI DNA, and, recently, a nanovirus-like

DNA molecule encapsidated in geminivirus particles was isolated from infected plants. This raises the question of how closely relate geminiviruses and nanoviruses are. The future will show whether the relationship between geminiviruses and nanoviruses has to be reevaluated.

For the recombination of two viruses, it is essential that both viruses are present not only in the same plant but also in the same cell. This problem would be circumvented when viral sequences are present in the plant genome. Gemini-, nano-, as well as caulimoviruses enter the nucleus in their infection cycle; therefore, recombination with the plant genome is easily conceivable. Begomovirus- as well as badnavirus-derived sequences were found in the plant genome. If such plants are infected with a distinct gemini- or badnavirus, recombination with the integrated viral sequences results in exchange of genetic information. That infecting gemini- as well as caulimoviruses acquire genetic information from the plant genome has been demonstrated in transgenic plants carrying viral sequences. Therefore, recombination with the plant genome has to be taken into consideration in the evolution of plant DNA viruses. From the previous discussion, it may be concluded that transgenic plants harboring geminivirus-, nanovirus-, or caulimovirus-derived sequences represent the possibility that artificial recombination will unintentionally cause new viruses to evolve. It must be realized that naturally integrated begomovirus- and badnavirus-derived sequences also have been found in plant genomes.

The involved recombination mechanisms between both DNA virus groups suggest that replication plays an important role in recombination. In most reported cases of geminivirus recombination, between virus genomic components or with the plant genome, the replication origin was involved, and in caulimoviruses, replicative RNA forms were targets for recombination.

REFERENCES

Bejarano ER, Khashoggi A, Witty M, Lichtenstein C (1996). Integration of multiple repeats of geminiviral DNA into the nuclear genome of tobacco during evolution. *Proceedings of the National Academy of Sciences, USA* 93: 759-764.

Boevink P, Chu PW, Keese P (1995). Sequence of subterranean clover stunt virus DNA: Affinities with the geminiviruses. *Virology* 207: 354-361.

Botstein D (1980). A theory of modular evolution for bacteriophages. *Annals of the New York Academy of Sciences* 354: 484-491.

Briddon RW, Bedford ID, Tsai JH, Markham PG (1996). Analysis of the nucleotide sequence of the treehopper-transmitted geminivirus, tomato pseudo-curly top virus, suggests a recombinant origin. *Virology* 219: 387-394.

Briddon RW, Stenger DC, Bedford ID, Stanley J, Izadpanah K, Markham PG (1998). Comparison of a beet curly top virus isolate originating from the old world with those from the new world. *European Journal of Plant Pathology* 104: 77-84.

Burns TM, Harding RM, Dale JL (1995). The genome organization of banana bunchy top virus: Analysis of six ssDNA components. *Journal of General Virology* 76: 1471-1482.

Cabauatan P, Melcher U, Ishikawa K, Omura T, Hibino H, Koganezawa H, Azzam O (1999). Sequence changes in six variants of rice tungro bacilliform virus and their phylogenetic relationships. *Journal of General Virology* 80: 2229-2237.

Calvert LA, Ospina MD, Shepherd RJ (1995). Characterization of cassava vein mosaic virus: A distinct plant pararetrovirus. *Journal of General Virology* 76: 1271-1278.

Choe IS, Melcher U, Richards K, Lebeurier G, Essenberg RC (1985). Recombination between mutant cauliflower mosaic virus DNAs. *Plant Molecular Biology* 5: 281-289.

Chu PW, Keese P, Qiu BS, Waterhouse PM, Gerlach WL (1993). Putative full-length clones of the genomic DNA segments of subterranean clover stunt virus and identification of the segment coding for the viral coat protein. *Virus Research* 27: 161-171.

Dixon L, Nyffenegger T, Delley G, Martinez-Izquierdo J, Hohn T (1986). Evidences for replicative recombination in cauliflower mosaic virus. *Virology* 150: 463-468.

Dry IB, Krake LR, Rigden JE, Rezaian MA (1997). A novel subviral agent associated with a geminivirus: The first report of a DNA satellite. *Proceedings of the National Academy of Sciences, USA* 94: 7088-7093.

Dry IB, Rigden JE, Krake LR, Mullineaux PM, Rezian MA (1993). Nucleotide sequence and genome organization of tomato leaf curl geminivirus. *Journal of General Virology* 74: 147-151.

Etessami P, Watts J, Stanley J (1989). Size reversion of African cassava mosaic virus coat protein gene deletion mutants during infection of *Nicotiana benthamiana*. *Journal of General Virology* 70: 277-289.

Fondong VN, Pita JS, Rey MEC, Kochko A De, Beachy RN, Fauquet CM (2000). Evidence of synergism between African cassava mosaic virus and a new double-recombinant geminivirus infecting cassava in Cameroon. *Journal of General Virology* 81: 287-297.

Fontes EP, Eagle PA, Sipe PS, Luckow VA, Hanley-Bowdoin L (1994). Interaction between a geminivirus replication protein and origin DNA is essential for viral replication. *The Journal of Biological Chemistry* 269: 8459-8465.

Franz A (1997). Diagnosis and biological characterization of faba bean necrotic yellows virus with respect to its epidemiology in West Asia and North Africa [in German]. PhD dissertation, Christian-Albrechts-University, Kiel, Germany, 140 pp.

Frischmuth T (1999). Genome of DNA viruses. In Mandahar CL (ed.), *Molecular Biology of Plant Viruses*. Kluwer Academic Publishers, Boston, Dordrecht, London, pp. 29-46.

Frischmuth T, Engel M. Jeske H (1997). Beet curly top virus DI DNA-mediated resistance is linked to its size. *Molecular Breeding* 3: 213-217.

Frischmuth T, Engel M, Lauster S, Jeske H (1997). Nucleotide sequence evidence for the occurrence of three distinct whitefly-transmitted, *Sida*-infecting bipartite geminiviruses in Central America. *Journal of General Virology* 78: 2675-2682.

Frischmuth T, Roberts S, Von Arnim A, Stanley J (1993). Specificity of bipartite geminivirus movement proteins. *Virology* 196: 666-673.

Frischmuth T, Stanley J (1993). Strategies for the control of geminivirus diseases. *Seminars in Virology* 4: 329-337.

Frischmuth T, Stanley J (1998). Recombination between viral DNA and the transgenic coat protein gene of African cassava mosaic geminivirus. *Journal of General Virology* 79: 1265-1271.

Geldreich A. Lebeurier G, Hirth L (1986). In vivo dimerization of cauliflower mosaic virus DNA can explain recombination. *Gene* 48: 277-286.

Gibbs MJ, Armstrong J, Weiller GF, Gibbs AJ (1997). Virus evolution; the past, and window on the future? In Tepfer M, Balázs E (eds.). *Virus-Resistant Transgenic Plants: Potential Ecological Impact*, Springer-Verlag, Berlin, Heidelberg, New York, pp. 1-19.

Gilbertson RL, Hidayat SH, Paplomatas EJ, Rojas MR, Hou YM, Maxwell DP (1993). Pseudorecombination between infectious cloned DNA components of tomato mottle and bean dwarf mosaic geminiviruses. *Journal of General Virology* 74: 23-31.

Grimsley N, Hohn T, Hohn B (1986). Recombination in a plant virus: Template-switching in cauliflower mosaic virus. *EMBO Journal* 5: 641-646.

Grylls NE, Butler FC (1956). An aphid transmitted virus affecting subterranean clover. *Journal of Australian Institute of Agricultural Sciences* 22: 73-74.

Hafner GJ, Stafford MR, Wolter LC, Harding RM, Dale JL (1997). Nicking and joining activity of banana bunchy top virus replication protein in vitro. *Journal of General Virology* 78: 1795-1799.

Harding RM, Burns TM, Hafner G, Dietzgen RG, Dale JL (1993). Nucleotide sequence of one component of the banana bunchy top virus genome contains a putative replicase gene. *Journal of General Virology* 74: 323-328.

Harper G, Osuji JO, Heslop-Harrison JS, Hull R (1999). Integration of banana streak badnavirus into the *Musa* genome: Molecular and cytogenetic evidence. *Virology* 255: 207-213.

Harrison BD, Zhou X, Otim-Nape GW, Liu Y, Robinson DJ (1997). Role of a novel type of double infection in the geminivirus-induced epidemic of severe cassava mosaic in Uganda. *Annals of Applied Biology* 131: 437-448.

Hay JM, Jones MC, Blakebrough ML, Dasgupta I, Davies JW, Hull R (1991). An analysis of the sequence of an infectious clone of rice tungro bacilliform virus, a plant pararetrovirus. *Nucleic Acids Research* 19: 9993-10012.

Höfer P, Engel M, Jeske H, Frischmuth T (1997). Host range limitation of a pseudorecombinant virus produced by two distinct bipartite geminiviruses. *Molecular Plant-Microbe Interactions* 10: 1019-1022.

Hong YG, Rohinson DJ, Harrison BD (1993). Nucleotide sequence evidence for the occurrence of three distinct whitefly-transmitted geminiviruses in cassava. *Journal of General Virology* 74: 2437-2443.

Hou YM, Gilbertson RL (1996). Increased pathogenicity in a pseudorecombinant bipartite geminivirus correlates with intermolecular recombination. *Journal of Virology* 70: 5430-5436.

Howarth AJ, Vandemark GJ (1989). Phylogeny of geminiviruses. *Journal of General Virology* 70: 2717-2727.

Hull R (1992). Genome organization of retroviruses and retroelements: Evolutionary considerations and implications. *Seminars in Virology* 3: 373-382.

Jones MC, Gough K, Dasgupta I, Rao BL, Cliffe J, Qu R, Shen P, Kaniewska M, Blakebrough M, Davies JW (1991). Rice tungro disease is caused by an RNA and a DNA virus. *Journal of General Virology* 72: 757-761.

Julia JF, Dollet M, Randles J, Calves C (1985). Foliar decay of coconut by *Myndus taffini* (FDMT): New results. *Oléagineux* 40: 19-27.

Katul L, Maiss E, Morozov SY, Vetten HJ (1997). Analysis of six DNA components of the faba bean necrotic yellows virus genome and their structural affinity to related plant virus genomes. *Virology* 233: 247-259.

Katul L, Vetten HJ, Maiss E (1993). Characterisation and serology of virus-like particles associated with faba bean necrotic yellows. *Annals of Applied Biology* 123: 629-647.

Kenton A, Khashoggi A, Parokonny A, Bennett MD, Lichtenstein C (1995). Chromosomal location of endogenous geminivirus-related DNA sequences in *Nicotiana tabacum* L. *Chromosome Research* 3: 346-350.

Király L, Bourque JE, Schoelz JE (1998). Temporal and spatial appearance of recombinant viruses formed between cauliflower mosaic virus (CaMV) and CaMV sequences present in transgenic *Nicotiana bigelovii*. *Molecular Plant-Microbe Interactions* 11: 309-316.

Klute KA, Nadler SA, Stenger DC (1996). Horseradish curly top virus is a distinct subgroup II geminivirus species with *rep* and *C4* genes derived from a subgroup III ancestor. *Journal of General Virology* 77: 1369-1378.

Lazarowitz SG (1991). Molecular characterization of two bipartite geminiviruses causing squash leaf curl disease: Role of viral replication and movement functions in determining host range. *Virology* 180: 70-80.

Lazarowitz SG (1992). Geminiviruses: Genome structure and gene function. *Critical Reviews in Plant Science* 11: 327-349.

Liu L, Van Tonder T, Pietersen G, Davies JW, Stanley J (1997). Molecular characterization of a subgroup I geminivirus from a legume in South Africa. *Journal of General Virology* 78: 2113-2117.

Mansoor S, Khan SH, Bashir A, Saeed M, Zafar Y, Malik KA, Briddon R, Stanley J, Markham PG (1999). Identification of a novel circular single-stranded DNA associated with cotton leaf curl disease in Pakistan. *Virology* 259: 190-199.

Medberry SL, Lockhart BEL, Olszewski NE (1990). Properties of Commelina yellow mottle virus's complete DNA sequence, genomic discontinuities and transcript suggest that it is a pararetrovirus. *Nucleic Acids Research* 18: 5505-5513.

Morris BA, Richardson KA, Haley A, Zhan X, Thomas JE (1992). The nucleotide sequence of the infectious cloned DNA component of tobacco yellow dwarf virus reveals features of geminiviruses infecting monocotyledonous plants. *Virology* 187: 633-642.

Navot N, Pichersky E, Zeidan M, Zamir D, Czosnek H (1991). Tomato yellow leaf curl virus: A whitefly-transmitted geminivirus with a single genomic component. *Virology* 185: 151-161.

Ndowora T, Dahal G, LaFleur D, Harper G, Hull R, Olszewski NE, Lockhart B (1999). Evidence that badnavirus infection in *Musa* can originate from integrated pararetroviral sequences. *Virology* 255: 214-220.

Palmer KE, Rybicki EP (1997). The use of geminiviruses in biotechnology and plant molecular biology, with particular focus on Mastreviruses. *Plant Science* 129: 115-130.

Roberts S, Stanley J (1994). Lethal mutations within the conserved stem-loop of African cassava mosaic virus DNA are rapidly corrected by genomic recombination. *Journal of General Virology* 75: 3203-3209.

Rohde W, Randles JW, Langridge P, Hanold D (1990). Nucleotide sequence of a circular single-stranded DNA associated with coconut foliar decay virus. *Virology* 176: 648-651.

Rothnie HM, Chapdelaine Y, Hohn T (1994). Pararetroviruses and retroviruses: A comparative review of viral structure and gene expression strategies. *Advances in Virus Research* 44: 1-67.

Sano Y, Isogai M, Satoh S, Kojima M (1993). Small virus-like particles containing single-stranded DNAs associated with milk vetch dwarf disease in Japan. Paper presented at the 6th International Congress on Plant Pathology, Montreal, Abstract No. 17.1.27.

Saunders K, Stanley J (1999). A nanovirus-like DNA component associated with yellow vein disease of *Ageratum conyzoides:* Evidence for interfamilial recombination between plant DNA viruses. *Virology* 264: 142-152.

Schoelz JE, Wintermantel WM (1993). Expansion of viral host range through complementation and recombination in transgenic plants. *Plant Cell* 5: 1669-1679.

Stanley J, Markham PG, Callis RJ, Pinner MS (1986). The nucleotide sequence of an infectious clone of the geminivirus beet curly top virus. *EMBO Journal* 5: 1761-1767.

Stanley J, Saunders K, Pinner MS, Wong SM (1997). Novel defective interfering DNAs associated with ageratum yellow vein geminivirus infection of *Ageratum conyzoides. Virology* 239: 87-96.

Stanley J, Townsend R, Curson SJ (1985). Pseudorecombinants between cloned DNAs of two isolates of cassava latent virus. *Journal of General Virology* 66: 1055-1061.

Sung YK, Coutts RH (1995). Pseudorecombination and complementation between potato yellow mosaic geminivirus and tomato golden mosaic geminivirus. *Journal of General Virology* 76: 2809-2815.

Vaden VR, Melcher U (1990). Recombination sites in cauliflower mosaic virus DNAs: Implications for mechanisms of recombination. *Virology* 177: 717-726.

Wanitchakorn R, Harding RM, Dale JL (1997). Banana bunchy top virus DNA-3 encodes the viral coat protein. *Archives of Virology* 142: 1673-1680.

Wintermantel W, Király L, Bourpue J, Schoelz JE (1997). Recombination between cauliflower mosaic virus and transgenic plants that contain CaMV transgenes: Influence of selection pressure on isolation of recombinants. In Tepfer M, Balázs E (eds.), *Virus-Resistant Transgenic Plants: Potential Ecological Impact*, Springer-Verlag, Berlin, Heidelberg, New York, pp. 66-76.

Wintermantel WM, Schoelz JE (1996). Recombination between cauliflower mosaic virus and transgenic plants under conditions of moderate selection pressure. *Virology* 223: 156-164.

Zhou X, Liu Y, Calvert L, Munoz C, Otim-Nape GW, Robinson DJ, Harrison BD (1997). Evidence that DNA-A of a geminivirus associated with severe cassava mosaic disease in Uganda has arisen by interspecific recombination. *Journal of General Virology* 78: 2101-2111.

SECTION V:
RESISTANCE TO
VIRAL INFECTION

Chapter 16

Natural Resistance to Viruses

Jari P. T. Valkonen

INTRODUCTION

A disease is a physiological process. Hence, the study of plant disease is concerned with abnormal plant physiology (Whetzel, 1929). Consequently, the study of resistance examines how a plant can slow down or stop the disease process. Resistant plants are those which do not become infected, which do become infected but at a lower rate, or which develop less severe symptoms than other genotypes. The resistance to viruses may, hypothetically, interfere with any of the steps in the virus infection cycle: entry of the virus or disassembly, replication, movement within and between cells, and transport from the site of initial infection to other parts of the plant.

To reveal the mechanisms of resistance requires examination of the same steps in the virus infection cycle as already explained elsewhere in this book. Because susceptible plant genotypes need to be studied for comparison side by side with the resistant genotypes, the studies on resistance often reveal fundamental, novel interactions and infection mechanisms. Indeed, the resistant plants can serve as "probes" that single out essential mechanisms required for successful virus infection.

The author wishes to warmly thank Dr. Sabina Vidal and Dr. Eugene Savenkov for critically reading the manuscript and for the helpful suggestions for improvement. Professor Kazuo N. Watanabe and Dr. Christiane Gebhardt are gratefully acknowledged for fruitful collaboration in the genetic studies on virus resistance, and Dr. David Baulcombe for making unpublished data available. The author's research on virus resistance is funded by the Forestry and Agriculture Research Foundation (SJFR), Carl Tryggers Stiftelse, Strategisk Forskningsrad, Swedish Royal Academy of Sciences, European Commission, and was in the near past funded also by the Academy of Finland and the University of Helsinki.

DISTINGUISHING A NONHOST FROM A HOST

A virus cannot replicate in a *nonhost*. Non-host plants are therefore *immune* to the virus, and this immunity has to be tested and shown at the level of single cells or protoplasts. The definition for an immune or nonhost plant is therefore quite strict, and the term *nonhost* refers only to plants in which no virus replication takes place. Otherwise, the distinction between a nonhost and a host becomes difficult to draw. If virus replication occurs, the plant is infectible and must be considered a host solely to reflect this fact. Several mechanisms can prevent virus multiplication after initial replication, restrict the infection to the initially infected cells, cause localization of virus infection to restricted areas in the infected leaf, or prevent systemic spread of the virus from the inoculated leaf.

Virus infection can be restricted to the initially infected cells, so-called *subliminal infection* (Sulzinski and Zaitlin, 1982). For example, *Brome mosaic virus* (BMV, genus *Bromovirus*) infects monocot species in the field but is not known to occur in cowpea (a dicot) and does not infect cowpea plants systemically upon mechanical inoculation. However, BMV can infect the protoplasts of cowpea (Fujita et al., 1996). *Tobacco mosaic virus* (TMV, genus *Tobamovirus*) typically infects dicot species in the field, whereas monocots such as barley are not infected. However, the protoplasts of barley can be infected with TMV (Karpova et al., 1997). TMV does not infect orchids systemically, but if the 30 kDa movement protein (MP) gene of TMV is replaced with the MP gene from *Odontoglossum ringspot virus* (a related tobamovirus infecting orchids in nature), the chimeric TMV will systemically infect orchid plants (Fenczik et al., 1995). These examples demonstrate that, although the cells support virus multiplication and synthesis (thousands of new infectious copies of the virus are generated), for which many compatible virus-host interactions are required (Lai, 1998), the cell-to-cell transport functions are not supported.

Natural hosts are plants in which the virus occurs in the field. They are usually, but not always, systemically infected. *Beet necrotic yellow vein virus* (genus *Benyvirus*) and *Potato mop-top virus* (genus *Pomovirus*) are transmitted in soil by protists (previously classified as fungi) and infect the roots, but seldom the aboveground parts, of the sugarbeet and potato plants, respectively (Brunt and Richards, 1989). *Experimental hosts* are plants that can be infected with the virus but in which the virus is not known to occur in the field. They can be systemically infected, or they may only support virus multiplication and cell-to-cell movement in the inoculated leaves and, thus, are not infected systemically. Hosts that develop visible lesions on inoculated leaves are known as *local lesion hosts*.

The term *nonhost* cannot be applied at the plant species level, unless many different genotypes of the species are studied. This is because immunity can be due to a single gene or a few genes that may segregate among genotypes. For example, the recessive gene *pvr1* in *Capsicum chinense* (Murphy et al., 1998) and the dominant gene *Rx* in potato (Bendahmane et al., 1999) provide immunity to *Pepper mottle virus* (PepMoV, genus *Potyvirus*) and *Potato virus X* (PVX, genus *Potexvirus*), respectively. Examples of single genes for virus-specific immunity shared by all genotypes of a species have not been reported, which may be due simply to the difficulty of attributing immunity to a particular gene in the absence of trait segregation.

It is commonly thought that most plant species are nonhosts to most viruses and that susceptibility is an exception to this rule. This conclusion is easily supported by any of the numerous studies that have examined viral host ranges (see, for example, Brunt et al., 1990; Anonymous, 1998). In most species tested with a virus, the inoculated leaves develop no symptoms, no virus is detectable with the standard assays (e.g., enzyme-linked immunosorbent assays [ELISA]), or the plants do not become systemically infected. The picture might turn more complicated if detailed studies at the cellular level are carried out. Anyway, it is difficult to explain these data by specific resistance genes being expressed against most viruses in most plants. Rather, many crucial interactions and steps of the virus infection cycle may fail in most species. In addition, general cellular mechanisms may be conserved in plants and may operate against the early stages of viral infection, in addition to carrying out other control functions to maintain homeostasis. The recently described posttranscriptional gene silencing may provide such an antiviral mechanism (Covey et al., 1997; Ratcliff et al., 1997; Baulcombe, 1999; Selker 1999; discussed later in this chapter and in Chapter 17).

THE TERMS DESCRIBING VIRUS-HOST INTERACTIONS AND RESISTANCE

Pathogenicity and Virulence

Pathogenicity is the ability of an organism to cause a disease (Whetzel, 1929). Plant viruses are *pathogens* because they cause symptoms (disease) at least in some of their hosts. The cryptoviruses form an exception because many of them cause no detectable symptoms in the known hosts (Boccardo et al., 1987). *Virulence* is the measure of pathogenicity (Whetzel, 1929); it can be studied by comparing responses to different strains of the virus in the same host genotype. Highly virulent virus strains (severe strains) induce severe symptoms. Virus strains that cause mild symptoms (mild strains) possess low virulence. An *avirulent* strain causes no symptoms. A *symptomless* virus infection in plants is often referred to as a *latent infection,* whereas in medical virology *latency* usually refers to the time between the infection and the appearance of symptoms (Shaner et al., 1992). A virus strain that accumulates to high titers in the plant has high fitness.

Avirulence and Resistance Genes

When specifically speaking about resistant hosts, virus isolates or strains that fail to infect the resistant plant are avirulent, in contrast to the virulent isolates or strains that overcome the resistance. *Resistance gene* is the host gene that determines the resistance.

Avirulence gene encodes the viral protein *(avirulence determinant)* that shows an incompatible interaction in the resistant host, resulting in a failure at some stage of the infection cycle. In a more strict use, in the context of the gene-for-gene theory (see Recognition of the Virus), the avirulence gene encodes the product *(avirulence factor;* De Wit, 1997) that is recognized by the product of the resistance gene. Consequently, recognition triggers the resistance responses inhibiting viral infection.

Resistance and Susceptibility

Resistance and *susceptibility* are quantitative terms that describe the relative suitability of an infectible genotype to be a host (see Figure 16.1). The use of terms largely follows the recommendations of Cooper and Jones (1983). *Sensitivity* and *tolerance* refer to symptom expression (see Figure 16.1). A sensitive susceptible plant develops severe symptoms, whereas a tolerant susceptible plant contains high virus titers but shows few or no symptoms. A sensitive resistant plant reacts strongly and may die from the resistance expression, for example, due to the development of lethal necrosis following expression of the *hypersensitive response* (HR) (discussed later). A tolerant resistant plant has low virus titers and shows no symptoms. *Immunity* and *extreme resistance* (ER) (discussed later) are the highest degrees of resistance in an infectible host.

Resistance Genetics

The host genes that determine virus resistance are designated as virus resistance genes. Single-gene-mediated resistance to viruses (see Table 16.1) is controlled by dominant or recessive genes. These alternatives are resolved by genetic studies using crosses (Cockerham, 1970; Kyle, 1993).

FIGURE 16.1. Terms Used for Plant Responses to Virus Infections

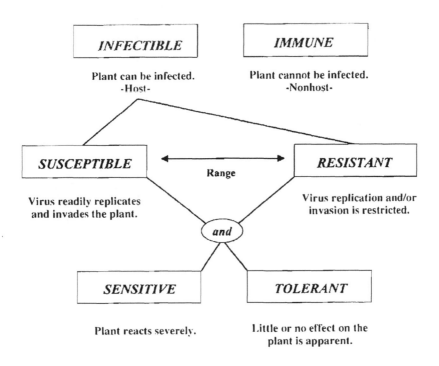

Source: Modified from Cooper and Jones, 1983: 127.

Single-gene-mediated resistance is commonly used in breeding programs due to the simple inheritance. Quantitative resistance is determined by two or several genes that show major effects and by additional minor genes. The genetic loci involved are called major and minor quantitative trait loci (QTL) for virus resistance, respectively (Caranta et al., 1997).

The term *horizontal resistance* is used to describe non-virus-specific resistance conferred by multiple genes and, probably, by several mechanisms with additive effects. It is therefore a form of quantitative resistance. Horizontal resistance can be observed as a generally low incidence of virus infections in field crops. Single-gene-mediated resistance is then called *vertical resistance* (Vanderplank, 1984). Protection against virus infections in the field is often the result of both types of resistance (horizontal and vertical) operating simultaneously.

TABLE 16.1. Examples of Viral Avirulence Genes and the Corresponding Host Resistance Genes

Avirulence Gene	Resistance Gene	Host	Virus	Type of Resistance	Reference
I. Dominant resistance response					
Coat protein	L^2, L^3	pepper	tobamoviruses[1]	HR	Cruz et al. (1997); Gilardi et al. (1998)
	Rx	potato	PVX	ER	Goulden et al. (1993)
	Nx	potato	PVX	HR	Kavanagh et al. (1992)
	N'	tobacco	TMV	HR	Culver and Dawson (1989)
Replicase	N	tobacco	TMV	HR	Padgett et al. (1997); Abbink et al. (1998)
	Tm-1	tomato	ToMV	ER	Hamamoto et al. (1997)
Movement protein	Nb	potato	PVX[2]	HR	Malcuit et al. (1999)
	Tm-2	tomato	ToMV	ER/HR	Weber and Pfitzner (1998)
II. Recessive resistance					
VPg	sbm-1	pea	PSbMV	replication	Keller et al. (1998)
	va	tobacco	potyviruses	movement	Nicolas et al. (1997)

Note: PVX = *Potato virus X* (genus *Potexvirus*); TMV = *Tobacco mosaic virus* (genus *Tobamovirus*); ToMV = *Tomato mosaic virus* (genus *Tobamovirus*); PSbMV = *Pea seed-borne mosaic virus* (genus *Potyvirus*) (Fauquet and Mayo, 1999).

[a] Genes L^2 and L^3 recognize tobamoviruses, except certain strains of *Pepper mild mottle virus* (PMMoV) (Cruz et al., 1997; Gilardi et al., 1998). Gene *va* confers resistance against several potyviruses but no induced resistance response is reported (Nicolas et al., 11997). The chromosomal loci containing these genes in pepper and tobacco, respectively, have not been isolated and characterized. Therefore, it remains to be shown whether the resistance to different viruses is conferred by a single gene or different closely linked virus-specific genes in each of these cases.

[b] *Nb* recognizes the 25 kDa protein that belongs to the so-called triple gene block involved in virus movement.

HOST RESPONSES CONFERRING RESISTANCE TO VIRUSES

The Hypersensitive Response (HR)

HR is a common response to infection with different types of pathogens, but its activation is dependent on the recognition of the pathogen by specific, dominant resistance genes (discussed later). The responses associated with HR include cellular effects at the site of infection, inhibition of virus movement at the edge of the infection lesion, death of the cells at the center of the lesion, and acquired resistance against new infections induced in the tissue around the lesion (*local acquired resistance*, LAR; Ross, 1961a) and in other parts of the plant (*systemic acquired resistance*, SAR; Ross, 1961b).

The features of HR (reviewed by Goodman and Novacky, 1994) in the infected cells include the production of reactive oxygen species, salicylic acid (SA), pathogenesis-related (PR) proteins, and phytoalexins and an increase in lipoxygenase activity. Membrane leakage is observed already a couple of hours after infection of plants showing HR (Westeijn, 1981). Physical changes include callose formation in the plasmodesmata, lignin deposition, and cell wall thickening (Allison and Shalla, 1974). The virus movement is blocked and infection becomes restricted to a relatively small number of cells around the initially infected cell. Due to cell death, a visible necrotic lesion develops at the site of infection a few days after inoculation (see Figure 16.2).

Other genes ("genotypic effect") involved in the full outcome of the resistance response (Ross, 1961b; Baker et al., 1997; Valkonen et al., 1998) and environmental conditions (physiological effects) can reduce the ability of the response to prevent virus movement. Consequently, systemic infection occurs, resulting in necrosis in other parts of the plant. The best-described example of the effect of temperature on HR is the case of the *N* gene-mediated HR to TMV in tobacco. At low temperatures, necrotic local lesions develop in the inoculated leaves and no systemic infection occurs. However, HR is suppressed at temperatures exceeding 30°C, which permits the virus to spread systemically. Consequently, only mosaic symptoms develop in the infected leaves (Ross, 1961a,b; Westeijn, 1981; Dunigan and Madlener, 1995; Padgett et al., 1997).

The necrotic lesion at the infection site is surrounded by a narrow zone of living cells that contains detectable amounts of the virus. Virus movement is blocked within this zone, which shows that cell death is not responsible for blocking the spread of the virus (Ross, 1961a).

FIGURE 16.2. Necrotic Lesions in a Leaf of Potato Cultivar 'A6' Bombarded with Infectious cDNA of *Potato Virus A* (Genus *Potyvirus*) Using a Gene Pistol (Gene Gun, BioRad)

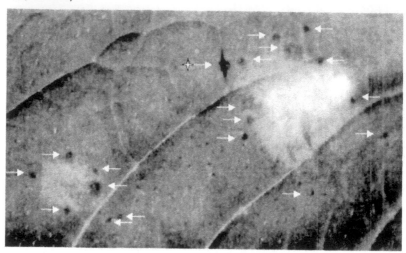

Note: 'A6' carries the gene Na$_{dms}$ conferring HR to PVA (De Bokx, 1972; Valkonen et al., 1996). Bombardment was done at the distance of 2.0 cm (to the left) and 0 cm (to the right) between the leaf and the spacer. The lesions (marked with arrows) became visible 2 to 4 h prior to the photo being taken (four days after inoculation). The lesions appear symmetrically round and sunken. They are initially gray in color but turn dark brown and black within a few hours. One lesion (marked with a star) developed fast on a minor vein. Virus movement is inhibited at the edge of the lesion and no systemic infection will develop. The center of the bombarded area shows phenotypically distinct cell death due to the mechanical damage caused by the microprojectiles and the pressure. Magnification is c. fivefold. (Photo: T Kekarainen and J Valkonen.)

HR requires intercellular communications and cell-to-cell contact. It is expressed in suspension cells (Yano et al., 1998) and calli (Pfitzner and Pfitzner, 1992), but not in protoplasts, nor do protoplasts isolated from plants carrying genes for HR respond to virus infection with cell death (Barker and Harrison, 1984; Adams et al., 1985). Indeed, resistance to viruses associated with HR does not prevent virus multiplication in protoplasts; instead, it prevents virus movement, which is expressed at the tissue level.

Immunity and Extreme Resistance (ER)

Immunity is expressed at infection and no virus amounts exceeding those introduced as an inoculum accumulate in the cells (Cooper and Jones,

1983). For example, the dominant gene *Rx* confers immunity to PVX in potato (Adams et al., 1986; Köhm et al., 1993), while the recessive gene *pvr1* confers immunity to PepMoV in *C. chinense* (Murphy et al., 1998). Extreme resistance (ER) has a phenotype similar to that for immunity in intact plants: no virus is detectable by standard assays such as ELISA and no symptoms develop. However, the difference between ER and immunity is that ER highly reduces but does not completely prevent virus multiplication in inoculated plants and protoplasts (Barker and Harrison, 1984; Karycija et al., 2000). Alternatively, ER results from restricted virus movement, which limits the infection to the initially infected cell or a few cells adjacent to it, and, therefore, the virus titers detected at the tissue level remain extremely low (Sulzinski and Zaitlin, 1982).

The expression of dominant genes for ER or immunity may result in limited necrosis ("pinpoint lesions") in the top leaves following graft inoculation in some genotypes (Ross, 1986). Induction of cell death by *Rx* was recently shown to require a high concentration of the CP elicitor (Bendahmane et al., 1999). *Rx* induces a non-virus-specific resistance at the single-cell level in protoplasts (Köhm et al., 1993). Hence, immunity and ER may be conferred by a mechanism related to HR. Relatedness of ER and HR is also suggested because the genes for immunity and ER are "epistatic" to the genes for HR to the same virus. Comparison of the resistance responses triggered by genes *Rx* and *N* (HR to TMV) in transgenic tobacco plants expressing both the genes has shown that *Rx* induces resistance earlier than *N* (Bendahmane et al., 1999). Therefore, in the presence of the gene for immunity or ER, the virus cannot multiply to titers sufficiently high to trigger the gene for HR, which explains the phenomenon described as epistasis of ER over HR in the literature (Cockerham, 1970; Valkonen et al., 1994; Bendahmane et al., 1999).

Recessive genes also confer immunity or ER. The gene *pvr1* for immunity (Murphy et al., 1998) was already mentioned. The gene *sbm-1* prevents detectable accumulation of certain strains of *Pea seed-borne mosaic virus* (PSbMV, genus *Potyvirus*) in the inoculated leaves of pea (Keller et al., 1998). No induced responses resembling HR have been described for these genes, which does not exclude the possibility that other, as yet unknown, induced responses are involved (Hämäläinen et al., 2000).

Recessive resistance may also result from a weak or incompatible interaction between viral proteins and the host factors involved in virus replication (Lai, 1998). Resistance to viruses based on weak or incompatible interactions between the virus and its host has been described in animals and humans (Brownstein, 1998). A natural deletion mutation in an allele encoding the CC-chemokine receptor 5 makes it nonfunctional as a virus receptor, and, consequently, the individuals homozygous for the mutant allele are highly resistant to *Human immunodeficiency virus* (Dean et al., 1996; Samson et al., 1996). Reduced affinity of *Mouse hepatitis virus* (MHV) to a naturally divergent biliary glycoprotein 1 isoform in an inbred mouse strain makes these mice highly resistant to MHV (Ohtsuka et al., 1996). Reduction of affinity by ten- to hundredfold reduces virus titers in the major target organs by 10^9-fold, which shows that mutations which reduce but do not abolish virus-host interactions are sufficient for creating significant levels of resistance.

Resistance to Virus Translocation

Virus movement in plants was explained in Chapter 7 of this book. In brief, to induce a systemic infection, the virus moves from the initially infected cell to the phloem, is loaded to the sieve elements for long-distance translocation, and is unloaded from the sieve elements in the upper leaves for replication (Santa Cruz, 1999). Numerous examples of virus-host combinations show that virus translocation is blocked at some stage of the movement pathway (Atabekov and Taliansky, 1990; Fenczik et al., 1995; Santa Cruz, 1999; Hämäläinen et al., 2000). In many cases, the block of virus movement occurs in the inoculated leaves at the interface between the mesophyll cells and the bundle sheath cells in the vascular tissue (Santa Cruz, 1999).

Mutations in the viral movement proteins may reduce or inhibit virus movement due to reduced interaction with the host factors required for virus transport (Rajamäki and Valkonen, 1999). Alternatively, mutations in the corresponding host factor involved in virus transport can make the plant more resistant to virus movement, hence making such host genes putative candidates for the recessive resistance genes affecting virus movement (Nicolas et al., 1997; Murphy et al., 1998). However, it is too early to exclude the possibility that recessive resistance to virus movement involves yet unknown responses induced by the virus infection (Hämäläinen et al., 2000). Even in the case of HR, the response that blocks virus movement at the edge of the infection lesion is not yet identified (discussed later).

Non-Virus-Induced Responses Increasing Resistance to Viruses

High or low temperatures and light intensities, salinity, draught, nutritional imbalance or wounding (abiotic stress); plant maturation and senescence; insect feeding; or a disease caused by infection with fungi and bacteria (biotic stress) can alter plant responses to virus infection. Plants respond to many of these stresses by producing PR proteins. Even though they have no known effects on viruses, as mentioned earlier, their induction is strongly correlated with the induction of SAR, which affects viruses as well as other pathogens (discussed later; Murphy et al., 1999).

Plants reaching maturity become less susceptible to virus infections, is called *mature plant resistance*, which is of practical significance in the field (Beemster, 1987). The physiological and molecular bases of mature plant resistance are not well characterized. Cells in older tissues are physiologically less active and may be less supportive to virus replication and movement. Induced responses associated with maturation and senescence may also play a role (Ryals et al., 1996; Murphy et al., 1999).

IDENTIFICATION OF THE GENES INVOLVED IN RESISTANCE

Identification of the Avirulence Gene

Identification of the avirulence gene in the virus is usually approached by comparing the sequences of virus isolates that differ in their virulence (see references in Table 16.1). This can be done without closer knowledge of the corresponding resistance gene. However, it is useful to know whether the resistance is monogenic or polygenic because polygenic resistance may indicate that several viral genes act as avirulence genes simultaneously (Schaad et al., 1997; Hämäläinen et al., 2000). Furthermore, if it is already known whether the resistance restricts virus replication or movement, the studies can focus on seeking mutations in the corresponding viral genes (Rajamäki and Valkonen, 1999; Hämäläinen et al., 2000). Infectious clones of viruses are essential tools, since site-directed mutagenesis or construction of recombinants between different virus isolates is needed to prove which gene is responsible for the avirulence. Mutations that permit an avirulent isolate to overcome resistance are used as further evidence of the positive identification of the avirulence gene (see Table 16.1).

The elicitor of an inducible resistance response such as HR or ER may be identified by transient expression of cloned viral genes in plants using microprojectile bombardment or agroinfiltration (Scofield et al., 1996; Tang et al., 1996; Abbink et al., 1998; Bendahmane et al., 1999). This is an option if no infectious clone of the virus is available. Stable transformation of a susceptible genotype with the putative avirulence gene can also provide useful evidence. The transgenic line is then crossed with a virus-resistant genotype to obtain F_1 progeny. The seedlings germinate and grow initially because they do not respond with HR until they have reached a certain developmental stage, after which lethal necrosis develops in the progeny that contains the avirulence gene (the transgene) and the resistance gene (Honée et al., 1998). Transformation of the resistant genotype with the avirulence gene may result in no transformants due to induction of necrosis at an early stage of regeneration (Pfitzner and Pfitzner, 1992; Gilbert et al., 1998).

Identification of the Resistance Gene

"Genetic dissection" of the resistance mechanism facilitates further molecular study of its function. The different types, mechanisms, or combinations of mechanisms that confer virus resistance are sorted out by making crosses between genotypes differing in their resistance. Resistance expression is tested in progeny by virus inoculation, and progeny is grouped in phenotypic classes. Genetic modeling is carried out to predict the number of genes, their linkage, and their interactions to explain the observed phenotypes and their relative proportions (see Table 16.1; Cockerham, 1970; Kyle, 1993). It is important to note that age, physiological condition of the plants, as well as the environmental conditions (e.g., temperature and light) often affect resistance expression (discussed later; Valkonen et al., 1998; Valkonen and Watanabe, 1999). Consequently, phenotypic expression of resistance may be difficult to distinguish under unfavorable conditions, which may hamper phenotypic grouping of the progeny and genetic modeling.

The studies of inheritance can be continued through mapping the positions of the resistance genes on the chromosomes and through isolation of the genes by "chromosome walking," which requires the availability of a molecular linkage map (Tanksley et al., 1992; Martin et al., 1993; Bendahmane et al., 1999). Alternatively, the resistance gene can be identified using transposon tagging (Johal and Briggs, 1992; Whitham et al., 1994). Direct cloning of resistance genes by polymerase chain reaction (PCR), using degenerate primers designed based on the sequence homology between the conserved motifs in the known gene sequences, has been suggested as a novel strategy (Leister et al., 1996; Sorri et al., 1999). The isolated, putative resistance gene is transferred to a susceptible plant genotype by genetic transformation, and its identity as the resistance gene is subsequently proven by inoculation of the transgenic plants with the virus.

RECOGNITION OF THE VIRUS: THE GENE-FOR-GENE THEORY

According to the *gene-for-gene theory*, an avirulence gene of the pathogen encodes, or results in the production of, a molecule (Avr) that is recognized by the product (R) of the resistance gene. The theory was developed based on genetic evidence revealing that single dominant genes in flax were responsible for the induction of HR following infection with specific strains of the rust fungus *Melampsora lini* (Flor, 1946). The theory concerns only the recognition event and not the following responses that suppress pathogen growth. Later molecular studies on the components involved in pathogen recognition turned Flor's theory into an *elicitor-receptor model*. The signal transduction pathways triggered by pathogen recognition are largely unresolved and currently under study. Comprehensive, detailed reviews on these topics have been published elsewhere (Keen, 1990; Baker et al., 1997; Culver, 1997; Hammond-Kosack and Jones, 1997).

Structures of Viral Avr Products
That Trigger Resistance

The viral Avr gene that is responsible for triggering a dominant resistance gene on the gene-for-gene basis (i.e., the viral gene encoding the avirulence factor) has been described in many virus-resistance gene combinations (Culver, 1997; see Table 16.1).

The gene *N* derived from *Nicotiana glutinosa*, which has been isolated and characterized (discussed elsewhere; Whitham et al., 1994), recognizes the 126 kDa replicase protein of TMV (Padgett et al., 1997). More specifically, the helicase domain acts as the elicitor of the HR and is alone sufficient to trigger the resistance response (Abbink et al., 1998).

The gene *N'* derived from *N. sylvestris* recognizes the coat protein (CP) of TMV (Culver and Dawson, 1989). Expression of CP alone is sufficient to trigger HR in plants carrying *N'* (Pfitzner and Pfitzner, 1992). The right face of the α-helical bundle in the CP forms a recognition surface. It is buried in the virion, and, therefore, intact virions do not elicit *N'*. Mutations in the recognition domain reduce the ability of TMV to form virions and to move systemically in tobacco plants (Taraporewala and Culver, 1996). These data show that the three-dimensional structure of the protein is important for recognition. The resistance gene seems to have evolved to recognize a structure of the viral protein that cannot be mutated without compromising vital functions of the virus. Furthermore, the recognition surface is exposed at virion disassembly, which is the first step in the infection cycle after introduction of the virion into the host cell (Wilson, 1984). Hence, the resistance gene can recognize the virus very early in the infection cycle. The α-helical bundle in the CP is a conserved recognition surface also in other tobamoviruses that trigger the genes L^1 in *C. annuum* and L^3 in *C. chinense* (Dardick et al., 1999; see Table 16.1).

The gene *Rx*, isolated and characterized from potato (Bendahmane et al., 1999), confers immunity to PVX (discussed elsewhere) and is triggered by the CP of PVX. Mutations in the CP within the region of amino acids 121 through 127 (e.g., lysine instead of threonine at position 122) permit PVX to overcome *Rx* but, at the same time, may reduce accumulation of PVX in potato leaves (Goulden et al., 1993). Also, the gene *Nx*, conferring HR to PVX in potato, is elicited by the CP of PVX (Kavanagh et al., 1992), and the amino acids 62 through 78 form a region important for recognition (Santa Cruz and Baulcombe, 1993; 1995). Very few natural isolates of PVX that overcome *Rx* are known (Goulden et al., 1993; Santa Cruz and Baulcombe, 1995), whereas *Nx* recognizes only certain strains of PVX (Santa Cruz and Baulcombe, 1993; 1995). These data may suggest that the recognition domain of *Rx* is more crucial than the recognition domain of *Nx* for the viral functions in susceptible plants. Consequently, the viral mutants born during virus replication in susceptible hosts that would be able to overcome *Rx* may have a greater disadvantage than those overcoming *Nx* and, therefore, be rare among the virus progeny (Goulden et al., 1993).

Structures of Virus Resistance Genes and Their Products

Several resistance genes have been isolated, and the putative structures of their products described (Hammond-Kosack and Jones, 1997). They exist in multiple, highly homologous copies either distributed to different linkage groups (chromosomes) or tightly clustered within a chromosomal region. The highly conserved structures and sequences of the resistance genes isolated from unrelated plant species suggest that a divergent selection and a birth-and-death process may be the primary mechanisms contributing to the set of resistance genes and gene clusters within a species (Michelmore and Meyers, 1998). The gene clusters are syntenic in the related species (Leister et al., 1996), and their existence suggests that the evolution of new resistance specificities occurs by mispairing, recombination, and gene duplication (Hammond-Kosack and Jones, 1997). Single or a few amino acid changes are sufficient to confer pathogen specificity or pathogen strain specificity, consistent with the single amino acid changes in Avr products that permit the pathogen to escape recognition (De Wit, 1997; see references in Table 16.1).

The resistance genes functioning according to the gene-for-gene theory can be grouped into classes based on similar conserved structures and predicted functional domains, irrespective of the type of pathogen (viruses, bacteria, fungi, or nematodes) recognized (see Figure 16.3). Some similarities are also shared between resistance genes and the genes involved in the immune systems of vertebrates and insects.

Two dominant virus resistance genes with functions based on the gene-for-gene theory have been isolated and described: the gene *N*, conferring HR to TMV in tobacco (Whitham et al., 1994; Dinesh-Kumar et al., 1995), and the gene *Rx*, conferring immunity to PVX in potato (Bendahmane et al., 1999). They are rather similar structurally (see Figure 16.3).

The gene *N* contains five exons and encodes a protein (N) of 1.144 amino acids, whereas the gene *Rx* contains three exons and encodes a protein (Rx) of 937 amino acids. Both proteins are putatively cytoplasmic. They contain a nucleotide-binding site (NBS) consisting of a kinase 1A (P-loop) motif, a kinase 2 motif, and a kinase 3a motif, which suggests that binding of ATP motif or GTP is required for protein function. Both proteins also contain a leucine-rich region (LRR) consisting of 14 (N) or 16 (Rx) imperfect repeats. LRRs may play a role in specific protein-protein interactions and cell adhesion or mediate interactions between the protein and the cellular membranes.

The proteins N and Rx differ in their amino-terminal part. The N protein contains a domain similar to the cytoplasmic domains of the *Drosophila* Toll protein, human interleukin receptor IL-1R, and an interleukin 6 (IL-6)-induced primary response protein of mouse myeloid cells (MyD88) (Whitham et al., 1994; Dinesh-Kumar et al., 1995). However, the Rx protein contains a "leucine zipper-like" domain at the amino-terminus, which may play a role in protein dimerization. The carboxy-terminal part of Rx contains three novel domains that have not been described in other resistance gene products (Bendahmane et al., 1999).

A special feature of *N* is that it putatively encodes a truncated protein of 652 amino acids in addition to the full-length protein. The truncated form of N might be a dominant negative regulator of the function of the full-length protein; it might compete for ligand binding, prevent cell death (apoptosis), or be required for signal transduction (Whitham et al., 1994; Dinesh-Kumar et al., 1995).

Recognition and Signal Transduction

Small amounts of the R proteins are expressed also in healthy, non-infected plants, which is probably essential to facilitate rapid recognition and response to pathogen attack at infection (Hammond-Kosack and Jones, 1997).

FIGURE 16.3. Grouping of R Proteins Based on Their Structures Predicted from the Sequences of Cloned Pathogen and Nematode Resistance Genes

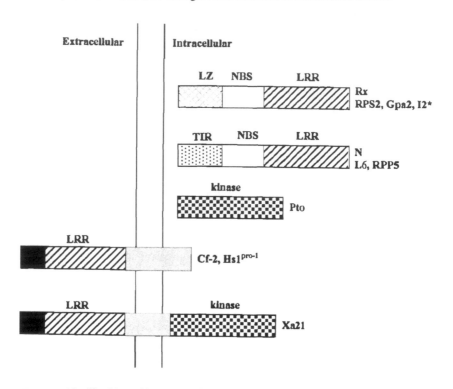

Source: Modified from Hammond-Kosack and Jones, 1997; De Wit, 1997; and Sorri, 1998.

Note: The proteins encoded by the two published virus resistance genes *Rx* (Bendahmane et al., 1999) and *N* (Whitham et al., 1994) are presented using the proteins encoded by some selected genes for resistance to fungi, bacteria, and nematodes for comparison: RPS2 *(Arabidopsis thaliana)—Pseudomonas syringae* pv. *tomato;* Gpa2 (potato)—*Globodera pallida;* I2 (tomato)—*Fusarium oxysporum* forma specialis *lycopersicon;* L6 (flax)—Melampsora lini; RPP5 *(A. thaliana)—Peronospora parasitica;* Pto (tomato)—*Pseudomonas syringae* pv. *tomato;* Cf-2 (tomato)—*Cladosporium fulvum;* Hs1^{pro1} (sugar beet)—*Heterodera* schachtii; Xa21 (rice)—*Xanthomonas oryzae* pv. *oryzae.* The following structures are indicated: LRR = leucine-rich repeat; LZ = leucine zipper; TIR = sequence with homology to the cytoplasmic domain of *Drosophila* Toll and mammalian Il-1R receptors; NBS = nucleotide-binding site.

I2* contains an L2 domain located between the NBS and LRR domains and may therefore belong to a new subclass of resistance genenes.

Even though a protein-protein interaction between the R and the Avr is predicted in the elicitor-receptor model, a positive interaction has been detected in only a single case so far. The kinase Pto encoded by the tomato gene *Pto* (Martin et al., 1993) and AvrPto produced by *Pseudomonas syringae* pathovar *tomato* interact directly, as shown by in vitro studies (Scofield et al., 1996; Tang et al., 1996). Recognition is specific and determined by a single threonine residue in the Pto kinase (Frederick et al., 1998). Although the LRR domain present in all other classes of R proteins (see Figure 16.3) is usually considered the likely determinant of pathogen specificity, Pto does not contain LRR. The lack of physical interaction in the other studied R-Avr combinations may mean that R and Avr do not interact directly, but as components of a protein complex involving other host proteins and possibly also other proteins of the pathogen.

Another dilemma is that no general structural similarities can be found in the avirulence gene products recognized by the structurally similar resistance genes. For example, the 126 kDa replicase of TMV and the CP of PVX recognized by the structurally similar resistance genes *N* and *Rx*, respectively, share apparently no similar domains. The gene *Gpa2*, a homologue of *Rx* from the same gene cluster in potato, recognizes a nematode (Bendahmane et al., 1999). This intriguing discrepancy still awaits an explanation.

The occasionally observed, limited necrotic symptoms induced by *Rx* were long ago proposed as evidence that the dominant genes for immunity or ER to viruses may be similar to the genes for HR that typically induce cell death (Benson and Hooker, 1960). This has now been proved by resistance gene isolation and characterization (Whitham et al., 1994; Bendahmane et al., 1999). The proteins encoded by *N* and *Rx* are probably cytoplasmic, which is consistent with their expected role in virus recognition, since TMV and PVX replicate in the cytoplasm. However, consequently, to activate the defense genes, signal transduction from the cytoplasm to the nucleus is required. Therefore, *N* and *Rx*, similar to several other described dominant resistance genes (Baker et al., 1997; Hammond-Kosack and Jones, 1997), are probably needed for triggering signal transduction, which activates the actual defense response genes. Putative additional host and pathogen proteins participating in a "recognition complex" may provide the means whereby different resistance genes can trigger similar signal transduction pathways and, on the other hand, allow similar resistance genes to trigger partly different pathways.

The genes *N* and *Rx* can trigger similar signal transduction pathways because both can induce cell death (necrosis). However, some differences must also exist, since *Rx* does not usually trigger cell death and HR. Unlike the N protein, the Rx protein does not contain N-terminal domains resembling Toll and IL-R1 involved in signal transduction in animal cells

(Dinesh-Kumar et al., 1995). Instead, Rx contains a "leucine zipper" that may be involved in signal transduction. Both Rx and N contain an NBS domain with three kinase motifs that may be a common mediator of signal transduction for these and many other resistance genes (see Figure 16.3). Furthermore, the kinase domains can provide pathogen specificity (Frederick et al., 1998) and a specific kinase 3a motif has been shown to function as an accurate marker for the gene Ry_{adg} for ER to *Potato virus Y* (genus *Potyvirus*) in potato (Sorri et al., 1999).

The structural conservation of resistance genes that recognize different types of pathogens suggests that a common signal transduction mechanism is involved in the establishment of disease resistance. Existence of a common, conserved network of signal transduction pathways in plants is supported by the ability of the resistance genes to confer pathogen-specific resistance following transfer to another species. For instance, the gene *N* from genus *Nicotiana* confers resistance to TMV in genus *Lycopersicon* (Whitham et al., 1996), and the gene *Rx* from genus *Solanum* confers resistance to PVX in genus *Nicotiana* (Bendahmane et al., 1999).

Production of reactive oxygen intermediates (ROI) occurs within a few minutes after R-Avr recognition and is proposed to function as the primary signal, followed by the induction of SA and nitric oxide (NO). These signals trigger the genes for PR proteins and the other genes involved in HR-associated processes (discussed later). The NO acts in collaboration with ROI and also with a factor that is dependent on SA accumulation to potentiate the defense response and cell death (Delledonne et al., 1998). Many studies indicate that SA is a key signal in the activation of HR and SAR because the endogenous levels of SA are positively correlated with expression levels of resistance and the PR proteins (Ryals et al., 1996). Furthermore, transgenic plants failing to accumulate SA are unable to express SAR and the genes for the PR proteins (Gaffney et al., 1993).

The PR proteins have antifungal or antibacterial activities, but so far none of them has been shown to play a role in virus resistance. It has recently been proposed that the signal transduction pathway leading to PR gene expression and resistance to fungi and bacteria is partly different from the pathway leading to resistance to viruses (Chivasa et al., 1997). The two pathways may be separated into two branches downstream of SA production. Only the branch leading to virus resistance can be inhibited by exogenous application of salicylhydroxamic acid (SHAM), a competitive inhibitor of alternative oxidase (OAX). OAX is involved in the alternative respiratory pathway in plants, suggesting an important role for this pathway in antiviral mechanisms. How OAX functions in defense signal transduction against viruses is not known, but it has been suggested to regulate the expression of host genes that in turn would influence virus movement and replication (Murphy et al., 1999).

Connections Between Programmed Cell Death and Virus Resistance

Cell death associated with HR in plants possesses several features analogous to *programmed cell death (apoptosis)* in animals. Apoptosis in animals differs from passive, "necrotic" cell death because it requires energy and does not cause inflammation. It provides a fundamental mechanism for regulation of development and maintenance of tissue homeostasis and may have a similar role in plants (Pennell and Lamb, 1997; Runeberg-Roos and Saarma, 1998). Common features include activation by similar signals (Ca^{2+}, H_2O_2, growth factor deprivation); similar morphological changes, such as condensation and shrinkage of cells; and fragmentation of nuclear DNA to oligonucleosome-sized pieces, observed as a characteristic DNA ladder in agarose gel electrophoresis (Pennell and Lamb, 1997). In animals and humans, apoptosis is described as a mechanism to remove virus-infected cells (O'Brien, 1998). Cell death associated with HR in plants has a similar function, though it does not remove all of the infected cells and is dispensable from the actual resistance responses blocking the movement of the invading virus (Richael and Gilchrist, 1999).

An intriguing link between HR and programmed cell death is the autonomously developing cell death (ACD) lesions induced in the absence of pathogen infection (Dangl et al., 1996). It is possible that the genes for ACD are mutated forms of the genes that regulate programmed cell death, and that these genes and the genes that trigger HR following pathogen recognition are related, in the sense that all can activate similar pathways (Mittler et al., 1995). ACD lesions are similar to lesions that develop following HR, and the induction of ACD often correlates with induced resistance to pathogens (Dangl et al., 1996). The homozygous condition for the recessive gene *mlo* provides barley with resistance to downy mildew, and ACD is occasionally induced in these resistant barley plants. The dominant allele *Mlo* is proposed to be a negative regulator (suppressor) of programmed cell death (Büschges et al., 1997). Induction of ACD can also be correlated with resistance to viruses (Valkonen and Watanabe, 1999). Indeed, it seems that induction of local and systemic acquired resistance to pathogens is very tightly linked to the induction of cell death (Costet et al., 1999).

VIRAL SUPPRESSOR OF RESISTANCE

Recently, a model was introduced to explain how the success of virus infection is determined at the initial stage of the infection process. It proposes that plants combat virus infection by "gene silencing," a general mechanism normally used for maintaining homeostasis (Covey et al., 1997; Ratcliff et al., 1997). On the other hand, viruses attempt to suppress host gene silencing at an early stage of infection (Anandalakshmi et al., 1998; Brigneti et al., 1998; Jones et al., 1998; Kasschau and Carrington, 1998).

Posttranscriptional Gene Silencing

Studies on transgenic resistance against viruses with an RNA genome using virus-derived genes in antisense orientation revealed that transgene expression could be suppressed or silenced at a posttranscriptional stage, which in turn was correlated with resistance. Hence, posttranscriptional gene silencing (PTGS) was targeted not only against the expression of the transgene but also the RNA virus carrying a homologous sequence. PTGS occurs cytoplasmically and is characterized by a relatively high transcription rate but low to undetectable levels of messenger (m) RNA (De Carvalho et al., 1992; Smith et al., 1994; Ruiz et al., 1998). The PTGS model includes an aberrant transcript of the transgene (or a natural host gene) that is recognized by a specialized host-encoded RNA-dependent RNA polymerase (RdRp) that consequently synthesizes short complementary (c) RNA molecules. These activate a mechanism that degrades all RNA molecules homologous to the double-stranded sequence. Hence, any viral RNA sharing sufficient homology with the transgene sequence will be degraded in the cytoplasm (Sijen et al., 1996; Selker, 1999). Evidence for RdRp enzymes in plants and the existence of homologous genes in fungi and animals suggest that RdRps are conserved in eukaryotic organisms (Schiebel et al., 1998; Cogoni and Macino, 1999). The messenger that mediates a systemic signal for PTGS is believed to be a short RNA molecule of 25 nucleotides (nt) (Hamilton and Baulcombe, 1999; Selker, 1999). Silencing signals can pass through the dividing meristematic cells, but silencing is not activated in such tissues, at least in some cases (Voinnet et al., 1998). PTGS is developmentally regulated, and meiotic resetting takes place (Tanzer et al., 1997).

Viruses induce PTGS upon infection (Baulcombe, 1999). It was first shown in nontransgenic *N. benthamiana* plants inoculated with a cytoplasmically replicating virus (TMV) that carried a *phytoene desaturase (PDS)* gene inserted into the viral genome (Kumagai et al., 1995). Expression of the endogenous *PDS* mRNA was reduced to an undetectable level following infection with the engineered TMV. In some cases, PTGS not only silences host gene expression but also targets the virus carrying the homologous gene insert (Ruiz et al., 1998). This latter phenomenon provides the mechanism whereby the virus-derived sequence inserted into the plant by transformation causes "silencing" of the subsequently infecting virus that contains a homologous sequence. Both homologous RNA sequences will then be degraded in the host cell by the same mechanism. In transgenic plants, PTGS may be preactivated by the transgene transcript and the homologous virus will be subjected to degradation very quickly following infection. Characteristic of PTGS is that only sequences highly homologous to the initiator of PTGS will be targeted. However, the required sequence identities vary considerably case by case, and little is yet known yas to how the threshold value is determined.

A segment of only 60 nt from the MP gene of *Cowpea mosaic virus* (genus *Comovirus*) inserted into a PVX-based gene vector and inoculated to transgenic plants that expressed the homologous, complete viral MP gene was sufficient to activate PTGS (Sijen et al., 1996). Since very short homologous sequences seem sufficient, activation of PTGS upon virus infection in nontransgenic plants could occur by a similar mechanism. though plant genomes are not known to contain large DNA fragments homologous to viral sequences.

PTGS was first associated with natural virus resistance that is expressed as a recovery from infection (Covey et al., 1997; Ratcliff et al., 1997). This is observed in plants of *N. clevelandii* infected with *Tomato black ring virus* (TBRV, genus *Nepovirus;* single-stranded |ss| RNA genome); the new leaves develop no symptoms and are apparently virus free. The new symptomless leaves are resistant to inoculation with isolates of TBRV that share a sufficiently high sequence homology with the initially used isolate (Ratcliff et al., 1997). Plants of kohlrabi *(Brassica oleracea* var. *gongylodes)* recover from infection with *Cauliflower mosaic virus* (genus *Caulimovirus;* double-stranded |ds| DNA genome) by a similar mechanism (Covey et al., 1997).

Suppression of Gene Silencing

The discovery of PTGS as a natural, conserved defense mechanism against viruses in plants (Voinnet et al., 1999) suggested that viruses must have strategies to overcome it. The helper component proteinase (HC-Pro) protein of potyviruses and the 2b protein of *Cucumber mosaic virus* (genus *Cucumovirus*) have been found to suppress PTGS (Anandalakshmi et al., 1998; Brigneti et al., 1998; Kasschau and Carrington, 1998). These proteins were expressed from a heterologous virus genome (PVX-based gene vector) in transgenic plants carrying marker genes that had been silenced by PTGS. Expression of HC-Pro or 2b protein recovered the expression of the marker genes.

Viruses of different families shut off genes in their hosts (Aranda and Maule, 1998), but the reason for preservation of this trait is unknown. Suppression of a general defense mechanism conserved in eukaryotic organisms provides a plausible rationale. Virus strains and isolates differ in virulence, and one reason could be a difference in their ability to suppress PTGS. For instance, amino acid substitutions in HC-Pro affect the titers to which isolates of *Potato virus A* (genus *Potyvirus*) accumulate in tobacco leaves (Andrejeva et al., 1999).

Synergism between two viruses that simultaneously infect a host can lead to increased titers of one of the two viruses and result in more severe symptoms. This may be interpreted as one of the two viruses being more competent in suppression of PTGS, which consequently permits the other virus to multiply and move more efficiently than when infecting alone (Anandalakshmi et al., 1998). Most examples of synergism have been described in hosts that are susceptible to both viruses involved. In plants that are resistant to one of the two viruses, permitting only limited virus multiplication, synergism may overcome resistance. This occurs in sweet potato cultivars (*Ipomoea batatas*) that are extremely resistant to *Sweet potato feathery mottle virus* (SPFMV, genus *Ipomovirus*; Karyeija et al., 2000). Coinfection with SPFMV and an unrelated, phloem-limited *Sweet potato chlorotic stunt virus* (genus *Crinivirus*), to which sweet potato cultivars are susceptible, creates a synergistic interaction between the two viruses. The titers of SPFMV are increased over 600-fold, and very severe symptoms of leaf distortion and stunting develop.

The host genes involved in PTGS are currently being investigated. Many genes known as virus resistance genes may turn out to be involved in PTGS. Using a recombinant TMV carrying the 2b gene from *Tomato aspermy virus* (genus *Cucumovirus*), it was shown that the 2b protein triggers HR in *N. tabacum* that otherwise is susceptible to TMV. However, the same 2b protein expressed from the TMV genome suppresses HR that is triggered by the CP of TMV in *N. benthamiana* (Li et al., 1999). These data indicate that the same viral protein suppresses a resistance response analogous to PTGS in one host but is the specific target for recognition by the resistance gene in another host.

FUTURE USES
OF NATURAL VIRUS RESISTANCE GENES

Resistance to viruses is economically important due to the severe diseases that viruses cause in crop plants, and because viruses cannot be chemically controlled. Control of virus spread by killing the vectors is mostly inefficient or impossible (Peters, 1987). Resistance is particularly important in the vegetatively propagated crops that become virus infected during propagation over several growing seasons in the field. Frequent renewal of planting materials using virus-free stocks is necessary, unless resistant cultivars are available. Production and maintenance of virus-free stocks require specialized institutes and are expensive processes. Also, annual crops can be badly damaged by viruses when the virus vectors occur abundantly and move actively early in the growing season; as a rule, the later the infection occurs, the lower the damage and yield losses.

Isolation of virus resistance genes, such as *N* and *Rx*, allows using the gene sequences directly as markers for resistance in traditional breeding programs. Furthermore, isolated genes can be transferred to susceptible cultivars through genetic transformation. Indeed, *N* and *Rx* confer resistance also when transferred to related plant species (Whitham et al., 1996; Bendahmane et al., 1999), which extends the use of virus resistance genes over natural crossing barriers.

However, problems with the use of single-gene-mediated resistance are not automatically resolved. Many genes for HR are virus specific or virus strain specific (see Table 16.1), which reduces their usefulness in crops infected with several viruses and virus strains in the field. Other genes and the environment (through effects on plant physiology) may modify the outcome of resistance, as explained earlier. Since single amino acid changes in the R proteins are sufficient to determine pathogen specificity (Frederick et al., 1998), it may be possible to engineer and broaden virus specificity of natural resistance genes prior to transfer to the susceptible cultivars.

Small amounts of resistance proteins are expressed also in healthy, noninfected plants (Hammond-Kosack and Jones, 1997), which may mean that the *R* genes or their homologues have other important functions during plant development and in plant physiology. Some resistance genes or similar genes of unknown function have been mapped close to genetic loci controlling physiological traits, such as earliness (early maturation) in potato (Hehl et al., 1999). Thus, perhaps *R* genes originally evolved to control maturation and death (cell death) of the plant, but some of them later evolved to recognize pathogens to trigger quick and localized protection reactions, such as HR and ER. Hence, resistance genes might be engineered and used to modify plant growth and development, for example, the crop maturation time.

REFERENCES

Abbink TEM, Tjernberg PA, Bol JF, Linthorst HJM (1998). Tobacco mosaic virus helicase domain induces necrosis in *N* gene-carrying tobacco in the absence of virus replication. *Molecular Plant-Microbe Interactions* 12: 1242-1246.

Adams SE, Jones RAC, Coutts RHA (1985). Infection of potato protoplasts derived from potato shoot cultures with potato virus X. *Journal of General Virology* 66: 1341-1346.

Adams SE, Jones RAC, Coutts RHA (1986). Expression of potato virus X resistance gene *Rx* in potato leaf protoplasts. *Journal of General Virology* 67: 2341-2345.

Allison AV, Shalla TA (1974). The ultrastructure of local lesions induced by potato virus X: A sequence of cytological events in the course of infection. *Phytopathology* 64: 784-793.

Anandalakshmi R, Pruss GJ, Ge X, Marathe R, Mallory AC, Smith TH, Vance VB (1998). A viral suppressor of gene silencing in plants. *Proceedings of the National Academy of Sciences, USA* 95: 13079-13084.

Andrejeva J, Puurand Ü, Merits A, Rabenstein F, Järvekülg L, Valkonen JPT (1999). Potyvirus HC-Pro and CP proteins have coordinated functions in virus-host interactions and the same CP motif affects virus transmission and accumulation. *Journal of General Virology* 80: 1133-1139.

Anonymous (1998). *Descriptions of Plant Viruses*. [CD-ROM database]. The Association of Applied Biologists, Wellesbourne, UK.

Aranda M, Maule A (1998) .Virus-induced host gene shutoff in animals and plants. *Virology* 243: 261-267.

Atabekov JG, Taliansky ME (1990). Expression of plant virus-coded transport function by different viral genomes. *Advances in Virus Research* 38: 201-248.

Baker B, Zambryski P, Staskawicz B, Dinesh-Kumar SP (1997). Signalling in plant-microbe interactions. *Science* 276: 726-733.

Barker H, Harrison BD (1984). Expression of genes for resistance to potato virus Y in potato plants and protoplasts. *Annals of Applied Biology* 105: 539-545.

Baulcombe DC (1999). Fast forward genetics based on virus-induced gene silencing. *Current Opinions in Plant Biology* 2: 109-113.

Beemster ABR (1987). Virus translocation and mature-plant resistance in potato plants. In De Bokx JA, Van der Want JPH (eds.).*Viruses of Potatoes and Seed-Potato Production*, PUDOC, Wageningen, The Netherlands, pp. 116-125.

Bendahmane A, Kanyuka K, Baulcombe DC (1999). The *Rx* gene from potato controls separate virus resistance and cell death responses. *Plant Cell* 11: 781-791.

Benson AP, Hooker WJ (1960). Isolation of virus X from "immune" varieties of potato. *Solanum tuberosum. Phytopathology* 50: 231-234.

Boccardo G, Lisa V, Lusoni E, Milne RG (1987). Cryptic plant viruses. *Advances in Virus Research* 32: 171-214.

Brigneti G, Voinnet O, Li WX, Ji LH, Ding SW, Baulcombe DC (1998). Viral pathogenicity determinants are suppressors of transgene silencing in *Nicotiana benthamiana. EMBO Journal* 17: 6739-6746.

Brownstein DG (1998). Comparative genetics of resistance to viruses. *American Journal of Human Genetics* 62: 211-214.

Brunt AA, Crabtree K, Gibbs A (1990). *Viruses of Tropical Plants: Description and List from the VIDE Data Base*. CAB International, Wallingford, UK.

Brunt AA, Richards KE (1989). Biology and molecular biology of furoviruses. *Advances in Virus Research* 36: 1-32.

Büschges R, Hollricher K, Panstruga R, Simons G, Wolter M, Frijters A, Van Daelen R, Van der Lee T, Diergaarde P, Groenendijk J, et al. (1997). The barley *Mlo* gene: A novel control element of plant pathogen resistance. *Cell* 88: 695-705.

Caranta C, Lefebvre V, Palloix A (1997). Polygenic resistance of pepper to potyviruses consists of a combination of isolate-specific and broad-spectrum quantitative loci. *Molecular Plant-Microbe Interactions* 10: 872-878.

Chivasa S, Murphy AM, Naylor M, Carr JP (1997). Salicylic acid interferes with tobacco mosaic virus replication via a novel salicylhydroxamic acid-sensitive mechanism. *Plant Cell* 9: 547-557.

Cockerham G (1970). Genetical studies on resistance to potato viruses X and Y. *Heredity* 25: 309-348.

Cogoni C, Macino G (1999). Gene silencing in *Neurospora crassa* requires a protein homologous to RNA-dependent RNA polymerase. *Nature* 399: 166-169.

Cooper JI, Jones AT (1983). Responses of plants to viruses: Proposals for the use of terms. *Phytopathology* 73: 127-128.

Costet L, Cordelier S, Dorey S, Baillieul F, Fritig B, Kauffmann S (1999). Relationship between localized acquired resistance (LAR) and the hypersensitive response (HR): HR is necessary for LAR to occur and salicylic acid is not sufficient to trigger LAR. *Molecular Plant-Microbe Interactions* 12: 655-662.

Covey SN, Al-Kaff VS, Lángara A, Turner DS (1997). Plants combat infection by gene silencing. *Nature* 385: 781-782.

Cruz A de la, López L, Tenllado F, Díaz-Ruíz JR, Sanz AI, Vaquero C, Serra MT, García-Luque I (1997). The coat protein is required for the elicitation of the *Capsicum* L² gene-mediated resistance against the tobamoviruses. *Molecular Plant-Microbe Interactions* 10: 107-113.

Culver JN (1997). Viral avirulence genes. In Stacey G, Keen NT (eds.), *Plant-Microbe Interactions* (Volume 2). Chapman & Hall, New York, pp. 196-219.

Culver JN, Dawson WO (1989). Point mutations in the CP gene of tobacco mosaic virus induce hypersensitivity in *Nicotiana sylvestris*. *Molecular Plant-Microbe Interactions* 2: 209-213.

Dangl JL, Dietrich RA, Richberg MH (1996). Death don't have no mercy: Cell death programs in plant-microbe interactions. *Plant Cell* 8: 1793-1807.

Dardick CD, Taraporewala Z, Lu B, Culver JN (1999). Comparison of tobamovirus coat protein structural features that affect elicitor activity in pepper, eggplant, and tobacco. *Molecular Plant-Microbe Interactions* 12: 247-251.

Dean M, Carrington M, Winkler C, Huttley GA, Smith MW, Allikmets R, Goedert JJ, Buchbinder SP, Vittinghoff E, Gomperts E, et al. (1996). Genetic restriction of HIV-1 infection and progression to AIDS by a deletion allele of the CKR5 structural gene. *Science* 273: 1856-1862.

De Bokx JA (1972). Test plants. In De Bokx JA (ed.), *Viruses of Potatoes and Seed-Potato Production*, PUDOC, Wageningen, The Netherlands, pp. 102-110.

De Carvalho F, Gheysen G, Kushnir S, van Montagu M, Inzé D, Castresana C (1992). Suppression of ß-1,3-glucanase transgene expression in homozygous plants. *EMBO Journal* 11: 2595-2602.

Delledonne M, Cia Y, Dixon RA, Lamb C (1998). Nitric oxide functions as a signal in plant disease resistance. *Nature* 394: 585-588.

De Wit PJGM (1997). Pathogen avirulence and plant resistance: A key role for recognition. *Trends in Plant Science* 2: 452-458.

Dinesh-Kumar SP, Whitham S, Choi D, Hehl R, Corr C, Baker B (1995). Transposon tagging of tobacco mosaic virus resistance gene N: Its possible role in the TMV-N-mediated signal transduction pathway. *Proceedings of the National Academy of Sciences, USA* 92: 4175-4180.

Dunigan DD, Madlener JC (1995). Serine/threonine protein phosphatase is required for tobacco mosaic virus-mediated programmed cell death. *Virology* 207: 460-466.

Fauquet MC, Mayo MA (1999). Abbreviations for plant virus names –1999. *Archives of Virology* 144: 1249-1273.

Fenczik CA, Padgett HS, Holt CA, Casper SJ, Beachy RN (1995). Mutational analysis of the movement protein of odontoglossum ringspot virus to identify a host-range determinant. *Molecular Plant-Microbe Interactions* 8: 666-673.

Flor HH (1946). Genetics of pathogenicity in *Melampsora lini*. *Journal of Agricultural Research* 73: 335-357.

Frederick RD, Thilmony RL, Sessa G, Martin GB (1998). Recognition specificity for the bacterial avirulence protein AvrPto is determined by Thr-204 in the activation loop of the tomato Pto kinase. *Molecular Cell* 2: 241-245.

Fujita Y, Mise K, Okuno T, Ahlquist P, Furusawa I (1996). A single codon in a conserved motif of a bromovirus movement protein gene confers compatibility with a new host. *Virology* 223: 283-291.

Gaffney T, Friedrich L, Vernooij B, Negrotto D, Nye G, Uknes S, Ward E, Kessmann H, Ryals J (1993). Requirement of salicylic acid for the induction of systemic acquired resistance. *Science* 261: 754-756.

Gilardi P, García-Luque I, Serra MT (1998). Pepper mild mottle virus coat protein alone can elicit the *Capsicum* spp. L^3 gene-mediated resistance. *Molecular Plant-Microbe Interactions* 12: 1253-1257.

Gilbert J, Spillane C, Kavanagh TA, Baulcombe DC (1998). Elicitation of *Rx*-mediated resistance to PVX in potato does not require new RNA synthesis and may involve a latent hypersensitive response. *Molecular Plant-Microbe Interactions* 11: 833-835.

Goodman RN, Novacky AJ (1994). *The Hypersensitive Reaction in Plants to Pathogens. A Resistance Phenomenon.* APS Press, St. Paul, Minnesota.

Goulden MG, Köhm BA, Santa Cruz S, Kavanagh T, Baulcombe DC (1993). A feature of the coat protein of potato virus X affects both induced virus resistance in potato and viral fitness. *Virology* 197: 293-302.

Hämäläinen JH, Kekarainen T, Gebhardt C, Watanabe KN, Valkonen JPT (2000). Recessive and dominant genes interfere with the vascular transport of *Potato virus A* in diploid potatoes. *Molecular Plant-Microbe Interactions* 13: 402-412.

Hamamoto H, Watanabe Y, Kamada H, Okada Y (1997). Amino acid changes in the putative replicase of tomato mosaic tobamovirus that overcome resistance in *Tm-1* tomato. *Journal of General Virology* 78: 461-464.

Hamilton AJ, Baulcombe DC (1999). A species of a small antisense RNA in posttranscriptional gene silencing in plants. *Science* 268: 950-952.

Hammond-Kosack KE, Jones JDG (1997) Plant disease resistance genes. *Annual Review of Plant Physiology and Plant Molecular Biology* 48: 575-607.

Hehl R, Faurie E, Hesselbach J, Salamini F, Whitham S, Baker B, Gebhardt C (1999). TMV resistance gene *N* homologues are linked to *Synchytrium endobioticum* resistance in potato. *Theoretical and Applied Genetics* 98: 379-386.

Honée G, Buitink J, Jabs T, De Kloe J, Sijbolts F, Apotheker M, Weide R, Sijen T, Stuiver M, De Wit PJGM (1998). Induction of defense-related responses in Cf9 tomato cells by the AVR9 elicitor peptide of *Cladosporium fulvum* is developmentally regulated. *Plant Physiology* 117: 809-820.

Johal GS, Briggs SP (1992). Reductase activity encoded by the *HM1* disease resistance gene in maize. *Science* 258: 985-987.

Jones AL, Thomas CL, Maule AJ (1998). *De novo* methylation and co-suppression induced by a cytoplasmically replicating plant RNA virus. *EMBO Journal* 17: 6385-6393.

Karpova OV, Ivanov KI, Rodionova NP, Dorokhov YuL, Atabekov JG (1997). Nontransmissibility and dissimilar behaviour in plants and protoplasts of viral RNA and movement protein complexes formed *in vitro*. *Virology* 230: 11-21.

Karyeija RF, Kreuze JF, Gibson RW, Valkonen JPT (2000). Synergistic interactions of a potyvirus and a phloem-limited crinivirus in sweet potato plants. *Virology* 269: 26-36.

Kasschau KD, Carrington JC (1998). A counterdefensive strategy of plant viruses: Suppression of posttranscriptional gene silencing. *Cell* 95: 461-470.

Kavanagh T, Goulden MG, Santa Cruz S, Chapman S, Barker I, Baulcombe DC (1992). Molecular analysis of a resistance-breaking strain of potato virus X. *Virology* 189: 609-617.

Keen NT (1990). Gene-for-gene complementarity in plant-pathogen interactions. *Annual Review of Genetics* 24: 447-463.

Keller KE, Johansen IE, Martin RR, Hampton RO (1998). Potyvirus genome-linked protein (VPg) determines pea seed-borne mosaic virus pathotype-specific virulence in *Pisum sativum*. *Molecular Plant-Microbe Interactions* 11: 124-130.

Köhm BA, Goulden MG, Gilbert JE, Kavanagh TA, Baulcombe DC (1993). A potato virus X resistance gene mediates an induced, nonspecific resistance in protoplasts. *Plant Cell* 5: 913-920.

Kumagai MH, Donson J, Della-Cioppa G, Harvey D, Hanley K, Grill LK (1995). Cytoplasmic inhibition of carotenoid biosynthesis with virus-derived RNA. *Proceedings of the National Academy of Sciences, USA* 92: 1679-1683.

Kyle MM (ed.) (1993). *Resistance to Viral Diseases of Vegetables: Genetics and Breeding*. Timber Press, Oregon.

Lai MMC (1998). Cellular factors in the transcription and replication of viral RNA genomes: A parallel to DNA-dependent RNA transcription. *Virology* 244: 1-12.

Leister D, Ballvora A, Salamini F, Gebhardt C (1996). A PCR-based approach for isolating pathogen resistance genes from potato with potential for wide application in plants. *Nature Genetics* 14: 421-429.

Li H-W, Lucy AP, Guo H-S, Li W-X, Ji L-H, Wong S-M, Ding S-W (1999). Strong host resistance targeted against a viral suppressor of the plant gene silencing defence mechanism. *EMBO Journal* 18: 2683-2691.

Malcuit I, Rosa Marano M, Kavanagh TA, De Jong W, Forsyth A, Baulcombe D (1999). The 24-kDa movement protein of potato virus X elicits Nb-mediated hypersensitive cell death in potato. *Molecular Plant-Microbe Interactions* 12: 536-543.

Martin GB, Brommonschenkel SH, Chunwongse J, Frary A, Ganal MW, Spivey R, Wu T, Earle ED, Tanksley SD (1993). Map-based cloning of protein kinase gene conferring disease resistance in tomato. *Science* 262: 1432-1436.

Michelmore RW, Meyers BC (1998). Clusters of resistance genes in plants evolve by divergent selection and a birth-and-death process. *Genome Research* 8: 1113-1130.

Mittler R, Shulaev V, Lam E (1995). Coordinated activation of programmed cell death and defense mechanisms in transgenic tobacco plants expressing a bacterial proton pump. *Plant Cell* 7: 29-42.

Murphy AM, Chivasa S, Singh DP, Carr JP (1999). Salicylic acid-induced resistance to viruses and other pathogens: A parting of the ways? *Trends in Plant Science* 4: 155-160.

Murphy JF, Blauth JR, Livingstone KD, Lackney VK, Kyle Jahn M (1998). Genetic mapping of the *pvr1* locus in *Capsicum* spp. and evidence that distinct potyvirus resistance loci control responses that differ at the whole plant and cellular levels. *Molecular Plant-Microbe Interactions* 11: 943-951.

Nicolas O, Dunnington SW, Gotow LF, Pirone TP, Hellmann GM (1997). Variations in the VPg protein allow a potyvirus to overcome *va* gene resistance in tobacco. *Virology* 237: 452-459.

O'Brien V (1998). Viruses and apoptosis. *Journal of General Virology* 79: 1833-1845.

Ohtsuka N, Yamada YK, Taguchi F (1996). Difference in virus-binding activity of two distinct receptor proteins for mouse hepatitis virus. *Journal of General Virology* 77: 1683-1692.

Padgett HS, Watanabe Y, Beachy RN (1997). Identification of the TMV replicase sequence that activates the *N* gene-mediated hypersensitive response. *Molecular Plant-Microbe Interactions* 10: 709-715.

Pennell RI, Lamb C (1997). Programmed cell death in plants. *Plant Cell* 9: 1157-1168.

Peters D (1987). Control of virus spread. In De Bokx JA, Van der Want JPH (eds.), *Viruses of Potatoes and Seed-Potato Production,* PUDOC, Wageningen, The Netherlands, pp. 171-174.

Pfitzner UM, Pfitzner AJP (1992). Expression of a viral avirulence gene in transgenic plants is sufficient to induce the hypersensitive defense reaction. *Molecular Plant-Microbe Interactions* 5: 318-321.

Rajamäki M-L. Valkonen JPT (1999). The 6K2 protein and the VPg of *Potato virus A* are determinants of systemic infection in *Nicandra physaloides. Molecular Plant-Microbe Interactions* 12: 1074-1081.

Ratcliff F, Harrison BD, Baulcombe DC (1997). A similarity between viral defense and gene silencing in plants. *Science* 276: 1558-1560.

Richael C, Gilchrist D (1999). The hypersensitive response: A case of hold or fold. *Physiological and Molecular Plant Pathology* 55: 5-12.

Ross AF (1961a). Localized acquired resistance to plant virus infection in hypersensitive hosts. *Virology* 14: 329-339.

Ross AF (1961b). Systemic acquired resistance induced by localized virus infections in plants. *Virology* 14: 340-358.

Ross H (1986). Potato breeding—Problems and perspectives. *Journal of Plant Breeding* (Supp 13), 132 pp.

Ruiz MT, Voinnet O, Baulcombe DC (1998). Initiation and maintenance of virus-induced gene silencing. *Plant Cell* 10: 937-946.

Runeberg-Roos P. Saarma M (1998). Phytepsin, a barley vacuolar aspartic proteinase, is highly expressed during autolysis of developing tracheary elements and sieve cells. *Plant Journal* 15: 139-145.

Ryals JA, Neuenschwander UH, Willits MG, Molina A, Steiner HY, Hunt MD (1996). Systemic acquired resistance. *Plant Cell* 8: 1809-1819.

Samson M, Libert F, Doranz BJ, Rucker J, Liesnard C, Farber CM, Saragosti S, Lapoumeroulie C, Cognaux J, Forceille C, et al. (1996). Resistance to HIV-1 infection in Caucasian individuals bearing mutant alleles of the CCR-5 chemokine receptor gene. *Nature* 383: 722-725.

Santa Cruz S (1999). Perspective: Phloem transport of viruses and macro-molecules—What goes in must come out. *Trends in Microbiology* 7: 237-241.

Santa Cruz S, Baulcombe D (1993). Molecular analysis of potato virus X isolates in relation to the potato hypersensitivity gene *Nx. Molecular Plant-Microbe Interactions* 6: 707-714.

Santa Cruz S, Baulcombe D (1995). Analysis of potato virus X coat protein genes in relation to resistance conferred by the genes *Nx, Nb* and *Rx1. Journal of General Virology* 76: 2057-2061.

Schaad MC, Lellis AD, Carrington JC (1997). VPg of tobacco etch potyvirus is a host genotype-specific determinant for long-distance movement. *Journal of Virology* 71: 8624-8631.

Schiebel W, Pelissier T, Riedel L, Thalmeir S, Schiebel R, Kempe D, Lottspeich F, Sänger HL, Wassenegger M (1998). Isolation of an RNA-directed RNA polymerase-specific cDNA clone from tomato. *Plant Cell* 10: 2087-2101.

Scofield SR, Tobias CM, Rathjen JP, Chang JH, Lavelle DT, Michelmore RW, Staskawicz BJ (1996). Molecular basis of gene-for-gene specificity in bacterial speck disease of tomato. *Science* 274: 2063-2065.

Selker EU (1999). Gene silencing: repeats that count. *Cell* 97: 157-160.

Shaner G, Stromberg EL, Lacy GH, Barker KR, Pirone TP (1992). Nomenclature and concepts of pathogenicity and virulence. *Annual Review of Phytopathology* 30: 47-66.

Sijen T, Wellink J, Hiriart J-B, van Kammen A (1996). RNA-mediated virus resistance: Role of repeated transgenes and delineation of targeted regions. *Plant Cell* 8: 2277-2294.

Smith HA, Swaney SL, Parks TD, Wernsman EA, Dougherty WG (1994). Transgenic plant virus resistance mediated by untranslatable sense RNAs: Expression, regulation, and fate of nonessential RNAs. *Plant Cell* 6: 1441-1453.

Sorri V (1998). Development of a PCR-based breeding marker for resistance to potato Y potyvirus in potato. Licenciate thesis, Department of Plant Biology, Swedish University of Agricultural Sciences, Uppsala, Sweden.

Sorri VA, Watanabe KN, Valkonen JPT (1999). Predicted kinase-3a motif of a resistance gene analogue as a unique marker for virus resistance. *Theoretical and Applied Genetics* 99: 164-170.

Sulzinski MA, Zaitlin M (1982). Tobacco mosaic virus replication in resistant and susceptible plants: In some resistant species virus is confined to a small number of initially infected cells. *Virology* 121: 12-19.

Tang XY, Frederik RD, Zhou JM, Halterman DA, Jia YL, Martin GB (1996). Initiation of plant disease resistance by physical interaction of AvrPto and Pto kinase. *Science* 274: 2060-2063.

Tanksley SD, Ganal MW, Prince JP, de Vicente MC, Bonierbale MW, Broun P, Fulton TM, Giovannoni JJ, Grandillo S, Martin GB, et al. (1992). High-density molecular linkage maps of the tomato and potato genomes. *Genetics* 132: 1141-1160.

Tanzer MM, Thompson WF, Law MD, Wernsman EA, Uknes S (1997). Characterization of post-transcriptionally suppressed transgene expression that confers resistance to tobacco etch virus infection in tobacco. *Plant Cell* 9: 1411-1423.

Taraporewala ZF, Culver JN (1996). Identification of an elicitor active site within the three-dimensional structure of the tobacco mosaic tobamovirus coat protein. *Plant Cell* 8: 169-178.

Valkonen JPT, Jones RAC, Slack SA, Watanabe KN (1996). Resistance specificities to viruses in potato: Standardization of nomenclature. *Plant Breeding* 115: 433-438.

Valkonen JPT, Rokka VM, Watanabe KN (1998). Examination of leaf-drop symptom in virus-infected potato using anther culture-derived haploids. *Phytopathology* 88: 1073-1077.

Valkonen JPT, Slack SA, Plaisted RL, Watanabe KN (1994). Extreme resistance is epistatic to hypersensitive resistance to potato virus Y° in a *Solanum tuberosum* subsp. *andigena*-derived potato genotype. *Plant Disease* 78: 1177-1180.

Valkonen JPT, Watanabe KN (1999). Autonomous cell death, temperature-sensitivity and the genetic control associated with resistance to cucumber mosaic virus (CMV) in diploid potatoes (*Solanum* spp.). *Theoretical and Applied Genetics* 99: 996-1005.

Vanderplank JE (1984). *Disease Resistance in Plants* (Second Edition). Academic Press. Orlando.

Voinnet O, Pinto YM, Baulcombe DC (1999). Suppression of gene silencing: A general strategy used by diverse DNA and RNA viruses in plants. *Proceedings of the National Academy of Sciences, USA,* 96: 14147-14152.

Voinnet O, Vain P. Angell S, Baulcombe DC (1998). Systemic spread of sequence-specific transgene RNA degradation in plants is initiated by localized introduction of ectopic promotorless DNA. *Cell* 95: 177-187.

Weber H. Pfitzner AJP (1998). $Tm\text{-}2^2$ resistance in tomato requires recognition of the carboxy terminus of the movement protein of tomato mosaic virus. *Molecular Plant-Microbe Interactions* 11: 498-503.

Westeijn EA (1981). Lesion growth and virus localization in leaves of *Nicotiana tabacum* cv. Xanthi nc. after inoculation with tobacco mosaic virus and incubation alternatively at 22°C or 32°C. *Physiological Plant Pathology* 18: 357-369.

Whetzel HH (1929). The terminology of phytopathology. *Proceedings of International Congress of Plant Sciences,* Ithaca, NY (1926), pp. 1204-1215.

Whitham S, Dinesh-Kumar SP, Choi D, Hehl R, Corr C, Baker B (1994). The product of the tobacco mosaic virus resistance gene *N:* Similarity to Toll and the interleukin-1 receptor. *Cell* 78: 1101-1115.

Whitham S, McCormick S, Baker B (1996). The *N* gene of tobacco confers resistance to tobacco mosaic virus in transgenic tomato. *Proceedings of the National Academy of Sciences, USA* 93: 8776-8781.

Wilson TMA (1984). Cotranslational disassembly of tobacco mosaic virus *in vitro. Virology* 137: 255-265.

Yano A, Suzuki K, Uchimiya H, Shinshi H (1998). Induction of hypersensitive cell death by a fungal protein in cultures of tobacco cells. *Molecular Plant-Microbe Interactions* 11: 115-123.

Chapter 17

Engineering Virus Resistance in Plants

Erwin Cardol
Jan van Lent
Rob Goldbach
Marcel Prins

INTRODUCTION

The development of recombinant DNA technology and efficient plant cell transformation techniques have allowed plant virologists to design and test a number of biotechnological strategies for increasing virus resistance levels of plants. For these strategies, the concept of "pathogen-derived resistance" (PDR) has been of cardinal importance.

The initial PDR concept (Sanford and Johnston, 1985) suggested the possible broad application of pathogen-originated genes in generating specific host resistance. Deliberate expression of such genes at, e.g., modified expression levels or untimely stages in the pathogen's life cycle, was suggested to be applicable in most host-parasite systems. PDR provides excellent possibilities for all systems involving viruses, considering the relative simplicity of their genomes and the detailed knowledge of many viral gene functions. It is therefore not surprising that most applications of PDR have been reported in viral systems, especially those involving plant viruses. The latter seems true mainly due to the availability of efficient transformation protocols for model plants such as tobacco, which is susceptible to most plant viruses. For successful application of PDR, insight into the molecular biology and genetic makeup of plant viruses is essential. Therefore, the expression strategy of some model plant viruses is illustrated first.

STRUCTURE AND GENETIC ORGANIZATION
OF PLANT VIRAL GENOMES

To date, more than 600 different plant viruses have been identified and described. The vast majority (90 percent) of these viruses have a single-stranded (ss) RNA genome, while relatively few plant viruses have a DNA genome, which can be either double-stranded (ds) (caulimoviruses, badnaviruses) or single-stranded (geminiviruses). In the case of an ssRNA-genome, this can be either of plus polarity (the virus RNA can act directly as messenger RNA) or of minus polarity (the virus contains an RNA polymerase for producing mRNAs upon infection) (see Table 17.1).

Positive-Strand RNA Viruses

Most plant-infecting-viruses have a plus-strand RNA genome. Structurally, positive-strand RNA viruses are rather simple: the RNA genome (or genome parts) is encapsidated in a protein coat (either spherical, baciliform, or rod-shaped), built up by mostly one type (but sometimes two or three) of coat protein (CP).

The genomic RNA's 5'-terminus may consist of a cap structure (similar to cellular messenger [m] RNAs) or a small protein (genome-linked viral protein, or VPg). For many plus-strand RNA viruses, the complete genomic sequence is available, and this basic information provides detailed insight into the genetic organization (gene number and gene order) of these genomes.

Plus-strand RNA viruses may contain genes for the following:

1. An RNA-dependent RNA polymerase, expressed as a single protein or composed out of several subunits (These proteins, also denoted replicases, are involved in the multiplication of the viral RNA.)
2. A coat protein, the basic unit of the protective protein shell, sometimes required for long-distance movement
3. A movement protein, involved in cell-to-cell movement (through plasmodesmata) of the virus particle (e.g., como- and nepoviruses) or the viral RNA genome (e.g., *Tobacco mosaic virus*, TMV) and, in some cases, long-distance movement
4. A transmission factor, aiding the virus in moving from one host plant to another (e.g., poty- and luteoviruses for aphid transmission and nepoviruses for nematode transmission)
5. A protease, when the viral genome is expressed via the production of a polyprotein, from which the functional proteins must be proteolytically released (e.g., como-, nepo-, and potyviruses)
6. A genome-linked protein (VPg) (e.g., como-, nepo-, and potyviruses)
7. Additional proteins of still unknown function, including a putative methyltransferase that may be involved in the capping of viral genomic RNAs

TABLE 17.1. Plant Viral Genome Types

	Number of viruses	Percentage
Plus-stranded RNA	484	77
Minus-stranded RNA	82	13
Double-stranded RNA	27	4
Single-stranded DNA	26	4
Double-stranded DNA	13	2

The following figures give some examples of plant viral genomes, e.g., of TMV (see Figure 17.1), *Brome mosaic virus* (BMV; see Figure 17.2), and *Cowpea mosaic virus* (CPMV; see Figure 17.3). Differences in genome organization between viruses often relate to differences in gene expression strategies. For TMV, this strategy involves a translational "readthrough" and subgenomic mRNAs; for BMV, a divided genome and a subgenomic mRNA; and for CPMV, a divided genome and polyprotein processing.

Negative-Strand RNA Viruses

In contrast to animal viruses, a relatively small number of plant-infecting viruses have a minus-strand (or negative-strand) RNA genome. These include the plant rhabdoviruses (with an unsegmented genome) and tospoviruses (with a tripartite genome). The RNA genome of these viruses is encapsidated by nucleocapsid (N) proteins to form pseudocircular nucleocapsids. Virus particles consist of nucleocapsids enveloped by a lipid membrane. In these envelopes, one or more glycoproteins (G) are found, which form surface projections. For mRNA production (and infectivity), the viral polymerase is included in the virus particle. An example of a plant minus-strand RNA virus is shown in Figure 17.4, which illustrates both the particle morphology and the genomic structure of *Tomato spotted wilt virus* (TSWV), type species of the tospoviruses. Tospoviruses form a distinct genus, *Tospovirus*, within the family *Bunyaviridae*, members of which otherwise infect animals. Whereas the L RNA is completely negative-stranded, with one open reading frame (ORF) in the viral complementary (vc) strand, the M RNA and S RNA are ambisense, with one ORF in the viral (v) strand and a second ORF in the vc strand. L RNA encodes the viral polymerase (denoted L protein); M RNA, the two glycoproteins G1 and G2 and a nonstructural protein (NS_M); and S RNA, the nucleocapsid protein N and a nonstructural protein (NS_S) (see Figure 17.4).

FIGURE 17.1. Genomic Organization and Translational Strategy of *Tobacco mosaic virus* (TMV)

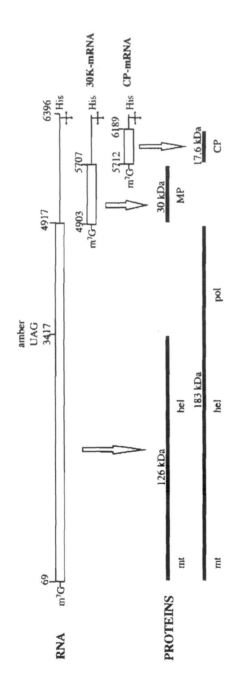

Note: Open reading frames are represented as open bars, with the nucleotide positions of start and stop codons indicated. The 183 kDa protein is an extension of the 126 kDa protein and is produced by readthrough at the amber (UAG) stop codon at position 3417. Both proteins probably represent subunits of the viral replicase. The two other genes encode the (30 kDa) movement protein (MP) and the (17.6 kDa) coat protein (CP) and are expressed from two subgenomic RNAs. Symbols: m⁷G = 5'-cap structure; His = 3'-terminal tRNA-like structure; aa = amino acid; mt, hel, and pol represent the positions of the methyltransferase, helicase, and polymerase domains, respectively.

FIGURE 17.2. Genomic Organization and Expression of the Segmented (Tripartite) RNA Genome of *Brome mosaic virus* (BMV)

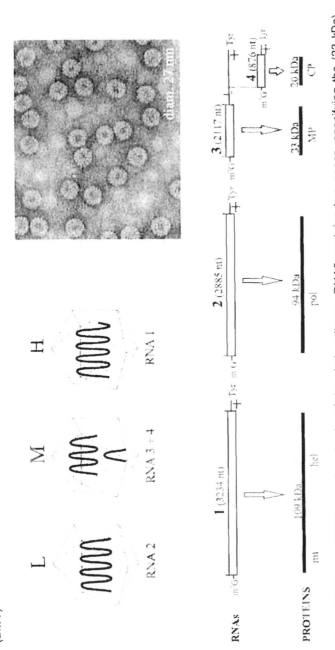

Note: RNA1 and RNA2 encode subunits of the viral replicase; RNA3 contains two genes, specifying the *(33 kDa)* movement protein (MP) and the *(20 kDa)* coat protein (CP), the latter being expressed from a subgenomic mRNA (RNA 4). L, M, and H indicate the low, medium, and high buoyant densities of the respective virus particles. For other symbols, see note to Figure 17.1.

103

FIGURE 17.3. Genomic Organization and Expression of the Bipartite RNA Genome of *Cowpea mosaic virus* (CPMV)

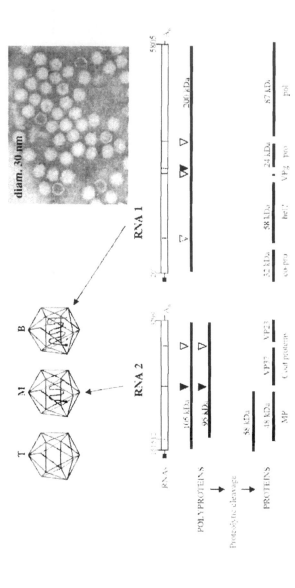

Note: The genomic RNAs are provided with a protein (VPg, black square) at the 5'-end and a poly(A)-tail (A_n) at the 3'-end. Both RNAs contain a single, large open reading frame from which polyproteins are produced that are proteolytically processed to give the mature, functional proteins. The 24 kDa protein represents the viral protease (pro) responsible for the cleavages. For some of the cleavages, the 32 kDa co-pro is needed. T, M, and B represent the relative positions (top, middle, and bottom) of the virus particles in a sucrose gradient. The T component does not contain RNA. For other symbols, see note to Figure 17.1.

FIGURE 17.4. Morphology of the Enveloped Virion of *Tomato spotted wilt virus (TSWV)*

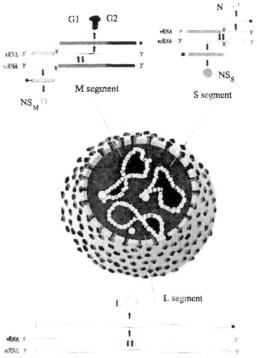

Note: The tripartite RNA genome is wrapped in the N protein to form circular nucleocapsids, which are enwrapped by a lipid envelope containing two types of glycoprotein (G1 and G2). Symbols: v = viral RNA; vc = viral complementary RNA. L, M, and S indicate the relative sizes of the respective viral RNAs (large, medium, and small). N is the viral nucleoprotein; NS_M and NS_S are the nonstructural proteins encoded by the M and S RNAs, respectively. Subgenomic messenger RNAs are capped (indicated by black dots) by snatching these structures from cellular mRNAs.

The Infection Cycle

The infection cycle of plant viruses, as exemplified by TMV in Figure 17.5, includes the following steps (steps 5 and 6 are not included in the figure):

1. Plant cells are penetrated (primary infection), either mechanically (some viruses, such as *Potato virus X* [PVX] and TMV) or with the aid of a biological vector (most viruses), e.g.,
   - insects, especially aphids (many viruses);
   - nematodes (some viruses, such as nepo- and tobraviruses); and
   - fungi (rarely, for instance, *Beet necrotic yellow vein virus* and some potyviruses).

2. Uncoating and translation of the viral genome take place. The viral coat is removed probably by the binding of ribosomes that start translation at the 5'-end ("cotranslational disassembly").
3. Replication of (and concurrent mRNA transcription from) the viral genome occurs. The viral replicase involved in this process is probably built up by both virus-encoded and host-encoded subunits.
4. Further translation of the newly produced viral plus-strand RNAs and mRNAs produces more replicase, coat protein (CP), and movement protein (MP).
5. The nonencapsidated TMV genome cell to cell moves through the plasmodesmata. For this process, the 30 kDa MP is required. The mechanism of cell-to-cell movement is still unknown but should involve modification of the plasmodesmata to allow passage. Some viruses (e.g., CPMV) move as complete viral particles.
6. The virus spreads systemically through the infected plant. This requires, in addition to cell-to-cell movement, long-distance transport through the vascular system (mostly phloem) of the plant.
7. The virus assembles in the infected plant and spreads to other noninfected plants. Since plants do not move, this often requires the aid of a biological vector (see step 1).

ENGINEERED RESISTANCE TO VIRUSES

Plant transformation protocols are available for many plant species, including such important crops as tomato, lettuce, rice, soybean, and potato. As a concept, transformation of crop plants with genetic material of viruses has been successfully exploited for conferring resistance against viruses.

Based upon the PDR theory, several aspects of the plant viral infection process are possible targets for conferring resistance, without interfering with essential host functions (Beachy, 1994). Three types of viral genes are omnipresent in plus-strand RNA viruses, as shown previously, and have thus been most widely used for PDR strategies that involve (altered) protein expression: coat protein genes, replicase genes, and movement protein genes. Other approaches depend on features specific to some viruses, such as the occurrence of interfering satellite or defective viral sequences, or on expression of molecules that do not occur in the natural environment of plant viruses but can be tailored to target viral RNA or proteins, such as specific toxins and antibodies. The most recent and most promising addition to the gamut of possible antiviral strategies is the induction of homology-dependent gene silencing in transgenic plants by viral transgenes, also termed RNA-mediated resistance.

FIGURE 17.5. The Infection Cycle of *Tobacco mosaic virus* (TMV)

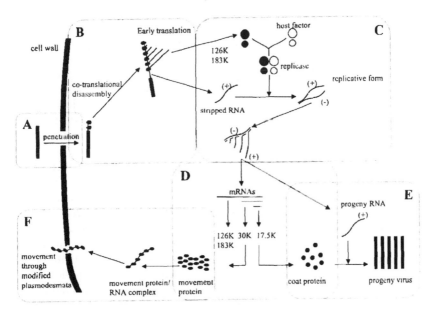

Note: Schematic representation of the TMV replication cycle in plant cells: (A) Penetration of the cell. (B) Dismantling of the virus by cotranslational disassembly after initial binding of ribosomes to the viral RNA. (C) Replication of the viral genome. Parental RNA is duplicated into a minus strand by replicase proteins associated with host components. The viral replicase is subsequently used to produce large numbers of plus-sense viral RNA copies as well as subgenomic messenger RNAs. (D) Translation of viral RNAs into replicases, movement protein, and coat protein. (E) Assembly of progeny virus particles from plus-strand virual sequences and coat protein. (F) Intercellular transport as movement protein-RNA complex.

Coat Protein-Mediated Protection

The strategy now commonly referred to as "coat protein-mediated protection" is based on insights in the conventional phenomenon of "cross-protection." *Cross-protection* is the term used for the phenomenon that a plant, when first inoculated with a mild strain of a given virus, becomes protected against infection with a second, more severe strain of this virus. The mechanisms of cross-protection are not yet fully understood. In a few cases, the coat protein plays a crucial role.

Coat protein-mediated protection refers to the resistance caused by the expression of the viral CP gene in transgenic plants. Accumulation of the CP in transgenic plants has been shown to confer resistance to infection and/or disease development by the virus from which the CP gene is derived and by related viruses.

For CP-mediated protection, complementary (c) DNA that represents the CP gene must be synthesized and cloned. This is relatively simple for viruses with CPs that are encoded by subgenomic (sg) RNAs. However, for viruses in which the CP is part of a polyprotein (e.g., potyviruses and comoviruses), the CP ORF must be artificially provided with an extra AUG start codon.

Because of the genetic structure of most plant (RNA) viruses, which encode their most abundant structural protein (CP) at the 3'-terminal part of the genome, clones of these genes were the first available for genetic studies. Introduction of the CP gene into the plant is mostly done by *Agrobacterium*-mediated gene transfer. Resistance is in all cases recorded as a significant delay in, or an escape from, disease symptom development and a decrease in severity of disease symptoms and virus accumulation. Indeed, CP-mediated resistance is reported for several viruses, as described in numerous reviews dealing with this subject (Beachy et al., 1990; Wilson, 1993; Hackland et al., 1994; Hull, 1994; Kavanagh and Spillane, 1995; Lomonossoff, 1995).

Studies on CP-mediated protection have revealed the following facts:

1. CP-mediated protection works at the protein level. Expression of CP RNA only greatly diminishes protection.
2. In most cases, there is no protection against viral RNA inoculation (exception: PVX).
3. The protection is not absolute; it can be overcome by (very) high virus inoculum concentrations. For TMV/tobacco the resistance level is approximately 10^4, which means that transgenic CP plants become diseased at an inoculum concentration 10,000 times higher than needed for infection of control plants.
4. Protection is rather specific; it works for the corresponding virus or very close relatives that have more than 60 percent amino acid sequence homology in their CPs.
5. CP-mediated resistance is a genuine form of resistance, not tolerance.
6. Resistance segregates as a conventional, single, dominant resistance gene.
7. Resistance works under greenhouse and field conditions.

Although the mechanism on which the observed CP-mediated protection is based still needs to be resolved, it is clear that the presence of CP, not only mRNA (see point 1), is essential. Points (2) and (3) indicate that an early step in the infection cycle is blocked, e.g., disassembly (uncoating) of the challenge virus.

Replicase-Mediated Resistance

Replicase-mediated resistance as a PDR concept was first applied by Golemboski and co-workers (1990). By expressing the 54 kDa readthrough part from the TMV replicase protein, transgenic plants proved highly resistant to the virus. Also, other initial reports were focused on the role of the (mutated) replicase protein, and some evidence was presented for its role in resistance. However, in most cases, no direct correlation could be established between protein expression levels and resistance (Anderson et al., 1992; MacFarlane and Davies, 1992; Donson et al., 1993; Longstaff et al., 1993; Audy et al., 1994; Carr et al., 1994; see also the review on this subject by Baulcombe, 1994).

A number of more recent publications strongly suggest the involvement of replicase gene RNA sequences, rather than the protein. Resistance to *Pepper mild mottle virus* (PMMoV), a tobamovirus, using its 54 kDa protein gene, occurred in two phenotypes (Tenllado et al., 1995), one preestablished phenotype resembling immunity and an induced type of resistance, resulting in highly resistant transgenic plants after initial infection. This resistance was effective against high inoculum doses of PMMV isolates, was not related to transgene expression levels, and was broken by related tobamoviruses, such as TMV. Also, transgenic expression of a truncated 54 kDa protein resulted in PMMV resistance (Tenllado et al., 1996), indicating that the (full-length) protein is not necessary for resistance. Replicase-mediated resistance against *Cymbidium ringspot virus*, a tombusvirus, appeared to correlate with low, rather than high, expression levels and was not functional against related viruses, such as *Artichoke mottled crinkle virus* and *Carnation Italian ringspot virus* (Rubino and Russo, 1995), suggesting the involvement of transgene RNA rather than protein. Replicase-mediated resistance experiments with PVX revealed similar resistance phenotypes (Braun and Hemenway, 1992; Longstaff et al., 1993), and expression of RNA alone was proven sufficient for resistance (Mueller et al., 1995).

Movement Protein-Mediated Resistance

Most cases described so far discuss the use of mutated MP genes to produce dysfunctional MP that might interfere with the movement process and result in an attenuated or delayed virus infection. Reduced TMV accumulation at nonpermissive temperatures was observed in tobacco plants transformed with MP sequences derived from a temperature-sensitive movement mutant (Malyshenko et al., 1993). Inhibition of TMV disease symptom development as well as that of two other tobamoviruses (tobacco mild green mosaic and *sunn-hemp mosaic virus*) in plants expressing a defective MP lacking three amino acids at its N-terminus has been reported by Lapidot and co-workers (1993). Furthermore, these plants showed a delay of several days in the appearance of viral symptoms when inoculated with other nonrelated viruses (Cooper et al., 1995). The expression of a mutated form of the triple gene block (TGB) MPs of *White clover mosaic virus* (WClMV), a potexvirus (Beck et al., 1994), showed broad resistance to WClMV and related viruses, even against *Potato virus S*, a carlavirus, but not against TMV. In conclusion, it seems that the use of defective MPs results in relatively broad resistance, a type of resistance, however, that is broken with increased inoculum doses. Also, use of the TSWV MP gene proved successful for obtaining resistance (Prins et al., 1996), but untranslatable transcripts appeared equally effective, suggesting RNA-mediated resistance.

Satellite and Defective Interfering RNA Protection

Some strains of certain plus-strand RNA viruses have satellites. These small RNA molecules, which are encapsidated in virus particles, do not share sequence homology with viral RNAs. Satellites depend fully on helper viruses for replication and transmission, and most of them can be regarded as molecular parasites.

Most satellite (sat) RNAs are small (300 to 500 bases) and linear. Some satellites encode their own coat proteins and are therefore called *satellite viruses*, whereas others are circular, similar to viroids, and have therefore been originally denoted "virusoids."

In some cases, sat-RNAs suppress symptoms caused by the helper virus, whereas others enhance the symptom severity of virus infection. The former "symptom-attenuating" type of satellites can be used for engineered "satellite protection."

Satellites replicate via a rolling-circle mechanism, which implies the formation of multimeric forms. For some satellites it was shown that these multimeric forms are cleaved to monomers by a "self-cleavage" reaction. Such self-cleavage reactions occur at specific RNA foldings, without the involvement of proteins. The RNA sequence itself possesses catalytic activity, and this autocatalytic sequence is therefore called *ribozyme*.

As discussed earlier, some strains of certain viruses have satellites that coreplicate (and are encapsidated) along with the helper virus genome. Some of these satellites greatly suppress disease symptoms caused by the virus, e.g., satellites of *Cucumber mosaic virus* (CMV) and *Tobacco ringspot virus* (TRSV). The production of such sat-RNAs by transgenic plants has resulted in a reduction of helper virus symptoms when compared to nontransformed plants inoculated with the same virus. This reduction could be correlated with a lower level of helper virus replication in the CMV satellite and TRSV satellite combinations.

It should be noted that the protected plants are not virus free. The transgenic plants have become tolerant, rather than resistant. A point of concern is the risk of conversion of a (plant-produced) symptom-attenuating satellite into a virulent one. Practical application of satellite protection certainly requires the construction of disarmed, nontransmissible satellites. Due to the lack of mild satellites, satellite protection may have only limited economic impact in the future.

For a limited number of plant RNA viruses, e.g., the plus-strand tombus- and carmoviruses and the minus-strand tospoviruses, the occurrence of small defective RNAs, derived from the full-length genomic RNAs by erroneous replication, has been reported. Such defective RNAs, which contain large internal deletions, may sometimes interfere with the replication of the full-length viral genome, leading to symptom attenuation similar to that observed in some satellites. In those cases, the defective RNAs are called defective interfering (DI) RNAs. For TSWV, DI RNAs are always derived from the L RNA (Resende et al., 1991, 1992; Inoue-Nagata et al., 1997; 1998), which encodes the viral polymerase. It is proposed that such RNA molecules, which still must contain all sequences required for replication and encapsidation, have a replicatory advantage, thereby outcompeting the full-length viral RNAs during replication. It can be imagined that transgenic plants expressing full-length cDNA copies of such DI RNAs may become protected against the disease symptoms of virus infection.

Our recent data indicate that this indeed works for TSWV, provided that the transgenically expressed DI RNA molecules are tailored to be the precise replica of the original isolate. This is achieved by supplying the DI cDNA sequence with various types of ribozymes. These molecular scissors can be introduced at the 5'- and 3'-ends of the DI molecule, resulting in accurately copied DI molecules after autocatalytic cleavage. Expression of these molecules in transgenic plants leads to increased tolerance to the virus due to coreplication of the transgenic DI copies. The principle of this mechanism is indicated in Figure 17.6.

FIGURE 17.6. Constructing Replicable Defective Interfering (DI) RNAs in Transgenic Plants

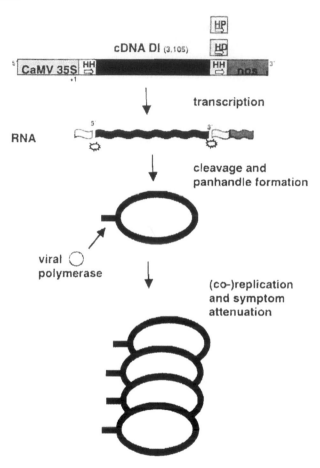

Note: Transgenic RNA expression of *Tomato spotted wilt virus* is hampered by the addition of extra nonviral sequences to both 5'- and 3'-ends. This inhibits the proper formation of a terminal dsRNA region, a so-called panhandle, which is required for efficient replication by the viral polymerase. The use of ribozymes results in self-cleavage of the RNA molecules to perfect DI RNAs. Upon infection with the virus, these RNAs are recognized and coreplicate, ultimately resulting in attenuation of symptoms. HH = hammerhead ribozyme; HP = hairpin ribozyme; HD = hepatitis delta ribozyme. CaMV 35S and nos are promoter and terminator signals, respectively.

RNA-Mediated Resistance

With an increasing number of reports on the use of viral genes for PDR, as described earlier, deviations from the original PDR concept became more frequent. A consistent lack of correlation between expression level of the transgenic protein and levels of resistance was reported, which seemed to be in conflict with the PDR concept (e.g., Stark and Beachy, 1989; Golemboski et al., 1990; Lawson et al., 1990; Gielen et al., 1991; Kawchuk et al., 1991; Van der Wilk et al., 1991). Some of these studies were even unable to show any protein product, suggesting that the expression of the protein was not essential for resistance. In addition, three reports published nearly simultaneously (De Haan et al., 1992; Lindbo and Dougherty, 1992; Van der Vlugt et al., 1992) described untranslatable sequences that were used to confer resistance to TSWV. *Potato virus Y* and *Tobacco etch virus*, the latter two being potyviruses. Moreover, the observed phenotype of the resistance was indistinguishable from that in plants expressing a translatable transgene, yet markedly different from that in reported cases of strictly protein-mediated resistance (Powell et al., 1990). The resistance phenotype in all three cases was independent of the inoculum dose and, as such, resembled immunity, whereas typical CP-mediated resistance levels decreased with increasing virus titers. Another difference was the spectrum of the resistance. RNA-mediated resistance proved to be specific for the virus from which the transgene was derived, while protein-mediated resistance had an effect also on other viruses related to the one used as a source for transformation. As for protein expression levels, no direct correlation could be made between RNA expression levels of the transgene and the levels of resistance against virus infection. For RNA-mediated resistance against TSWV, a clearly negative correlation was reported (Pang et al., 1993). Previous work of De Haan and co-workers (1992) using the same transgene did not show such a strict correlation between RNA expression levels and resistance, although, here, a tendency also observed was that resistant plants generally had lower expression levels. This was also found in resistant plants expressing TSWV NS_M sequences (Prins et al., 1996). These observations, together with others that will be discussed in more detail later, suggested that RNA-mediated resistance is not a form of antisense resistance directed against the replicative strand of the virus; it must operate in a different manner. To date, many more cases of RNA-mediated resistance have been reported, whereas for many (established) cases of PDR, expressed RNA's contribution to resistance has not been examined.

RNA-Mediated Resistance Is Similar to Cosuppression

Various potential mechanisms of cosuppression (or posttranscriptional gene silencing) have been discussed in a number of reviews on the subject (Flavell, 1994; Jorgensen, 1995; Kooter and Mol, 1993; Matzke and Matzke, 1995; Wassenegger and Pelissier, 1998).

Since transgenic RNA-mediated resistance against viruses and cosuppression of endogenous genes share so many similarities, it is conceivable that they are both (induced) manifestations of a basic mechanism residing in plants (and perhaps other organisms) that is involved in the regulation of gene (over)expression. Based on current knowledge, we will construct a model explaining (most of the) observed phenomena (see Figure 17.7). The core of the model is formed by a cellular mechanism that can be induced by the expression of transgenes and subsequently leads to sequence-specific RNA degradation. For both RNA-mediated resistance and cosuppression, passing a threshold level of transgenic (nuclear) expression seems an adequate explanation for the induction of the silencing mechanism. Some reports, however, present strong evidence against this assumption (Van Blokland et al., 1994). Therefore, quantitative expression of a transgene may not be the only requirement for silencing; the quality of the transcript may also play a role (Baulcombe et al., 1996; Metzlaff et al., 1996). Methylations of transgenic loci related to cosuppression may be induced initially by high expression levels of the transgene RNA sequences, either while already in the nucleus (Wassenegger et al., 1994) or after redirecting cytoplasmic RNAs to the nucleus (Lindbo et al., 1993). Extensive methylations of the transgene may cause aberrations in the transcribed mRNAs; these subsequently trigger a resident RNA-dependent RNA polymerase present in the cytoplasm (Dorssers et al., 1982, 1983; Schiebel et al., 1998) to synthesize (short) antisense RNA molecules. These could than form the core of the highly specific RNA-degrading complex that can target specific cellular or viral RNA molecules in the cytoplasm, thereby explaining why the mechanism is so versatile in targeting different sequences yet operates in a very sequence-specific manner. Ribonucleic acids have been described as essential parts of enzymes involved in RNA cleavage and sequence-specific recognition, e.g., for RNase P and for small nuclear ribonucleoproteins (snRNPs) (Baserga and Steitz, 1993; Kirsebom and Svard, 1994; Altman, 1995; Kirsebom, 1995).

Antisense transgenes have been shown to be capable of downregulating expression of endogenous genes as efficiently as sense genes (e.g., Van der Krol et al., 1988, 1990). The original rationale behind the use of antisense transgenes aimed at the expression of stoichiometric amounts of antisense transcripts that can anneal to sense transcripts, thereby making them unstable.

FIGURE 17.7. Model for RNA-Mediated Resistance

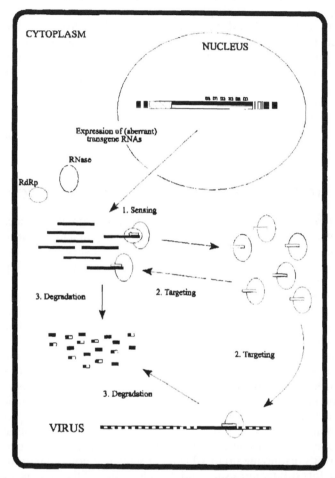

Note: Expression of transgenes in the nucleus leads to an unacceptable level of (aberrant) transcripts. This is sensed (1) by a cytoplasmic factor that includes an RNA-dependent RNA polymerase (RdRp) activity that transcribes short antisense RNAs. These RNAs form the core of a sequence-specific ribonuclease (RNase) that targets (2) and degrades (3) sequences identical or complementary to the transgene, resulting in low steady state transgene RNA levels. Feedback to the nuclear transgene resulting in transgene methylation (m) may cause increased aberration frequencies, thereby reinforcing the silencing. Upon entry of the virus, the viral RNAs, which have the same sequence as the transgene, are also targeted and degraded, resulting in resistance to the virus.

Even though successful as a concept, expression levels of antisense RNA often appeared to be low, displaying features highly resembling sense suppression, such as reduced steady state levels (Van Blokland, 1994). This suggests that suppression of sense as well as antisense sequences can lead to posttranscriptional degradation. If silencing of an endogenous gene can be achieved by posttranscriptionally silencing its antisense transgene, then this requires the ability of the silencing mechanism to operate on both strands. To accomplish this, part of the targeted RNA molecule may not be degraded; instead, it would be retained in the RNA degradation complex, making it capable of targeting both sense and antisense RNAs.

ADVANTAGES OF RNA-MEDIATED RESISTANCE

Even though silencing of transgenes can impose great drawbacks in the transgenic expression of proteins in plants (Finnegan and McElroy, 1994), for pursuing virus resistance, it has proven to be a very powerful strategy. Applying cosuppression-like RNA-mediated resistance against plant viruses is a generally straightforward strategy, one that remains to be explored for many viruses, even for those in which protein expression was initially designed as a means of PDR, but in which protein expression and resistance levels did not directly correlate. Several distinctive properties characterize RNA-mediated virus resistance. First, the observed phenotype resembles immunity in that it is not broken by increased doses of virus or application of viral RNA instead of virus. Second, it has a narrow scope of operation and can be broken by sufficiently (>10 percent at the nucleotide level) heterologous viruses. This limitation can be overcome by simultaneous expression of multiple RNA sequences (Prins et al., 1995). A third property of this resistance based on RNA is that it is expected to be more durable than protein-mediated resistance. RNA-mediated resistance can be broken by related viruses displaying up to 90 percent sequence homology in the target gene, but this still requires substantial modification of the genome of the homologous virus. In contrast, even a single point mutation in a protein has been shown to have major consequences on virus resistance, both in transgenic and natural resistance gene sources (Tumer et al., 1991; Santa Cruz and Baulcombe, 1994). A fourth, and beneficial, aspect of RNA-mediated virus resistance is the advantage of deliberate use of untranslatable RNAs, possibly with extra precautions, such as introduced stop codons. This further decreases the chance of possible unwanted recombinations between transgenic transcripts and RNAs of invading viruses, resulting in novel viruses (Greene and Allison, 1994). Furthermore, transcapsidation (Bourdin and Lecoq, 1991; Candelier-Harvey and Hull, 1993; Lecoq et al., 1993) cannot occur in plants expressing untranslatable RNA of viral coat protein genes. Expression of a viral sequence (not encoding any protein) that induces its own breakdown, resulting in (undetectable) low steady state RNA expression levels due to the induction of a resident plant response capable of very specifically destroying viral sequences, is biosafe and therefore very suitable with respect to public acceptance of genetically modified crops.

CONCLUDING REMARKS

Posttranscriptional silencing of transgenes homologous to viral sequences represents a newly discovered phenomenon that can be successfully applied for developing novel forms of resistance against plant viruses. Accessibility of the viral RNA to be targeted by the silencing mechanism obviously plays an important role. The genomes of plus-strand RNA viruses are relatively easily accessed. The genomes of negative-strand RNA viruses are associated with nucleoproteins throughout their replication cycle and may therefore be less susceptible to RNA degradation. However, viral mRNAs are not encapsidated and are thus accessible for sequence-specific degradation, as shown for TSWV. Potentially, this form of resistance could also operate against DNA viruses, such as geminiviruses (Hong and Stanley, 1996; Kjemtrup et al., 1998). Even though replication of these viruses takes place in the nucleus, mRNAs still end up in the cytoplasm. Targeting of mRNAs by the silencing mechanism and thereby preventing the formation of essential proteins (e.g., those involved in replication and movement) could inhibit virus multiplication or spread.

For the resistance mediated by some viral transgenes an effect of the protein was reported, while plants expressing untranslatable RNAs were all sensitive. Although often biased by preselecting for plants expressing high levels of transgenic protein or transcripts, these incidents may indicate that not all viral sequences are capable of conferring RNA-mediated resistance. By analogy, some transgenes are apparently unable to confer cosuppression to endogenous genes (Elomaa et al., 1995). Possibly, a requirement for specific primary or secondary structure elements in the expressed sequence needs to be met, for which the chance can be augmented by increasing transgene size.

REFERENCES

Altman S (1995). RNase P in research and therapy. *Bio/Technology* 13: 327-329.
Anderson JM, Palukaitis P, Zaitlin M (1992). A defective replicase gene induces resistance to cucumber mosaic virus in transgenic tobacco plants. *Proceedings of the National Academy of Sciences, USA* 89: 8759-8763.

Audy P. Palukaitis P, Slack SA, Zaitlin M (1994). Replicase mediated resistance to potato virus Y in transgenic plants. *Molecular Plant-Microbe Interactions* 7: 15-22.

Baserga SJ, Steitz JA (1993). The diverse world of small ribonucleoproteins. In Gersteland RF, Atkins JF (eds.), *The RNA World*. Cold Spring Harbor Press, Cold Spring Harbor, NY, pp. 359-381.

Baulcombe D (1994). Replicase-mediated resistance: A novel type of virus resistance in transgenic plants. *Trends in Microbiology* 2: 60-63.

Baulcombe DC, English J, Mueller E, Davenport G (1996). Gene silencing and virus resistance in transgenic plants. In Grierson D, Lycett GW, Tucker GA (eds.), *Mechanisms and Applications of Gene Silencing*. Nottingham University Press, Nottingham, England, pp. 127-138.

Beachy RN (1994). Introduction: Transgenic resistance to plant viruses. *Seminars in Virology* 4: 327-328.

Beachy RN, Loesch-Fries S, Tumer NE (1990). Coat protein-mediated resistance against virus infection. *Annual Review of Phytopathology* 28: 451-474.

Beck DL, Vandolleweerd CJ, Lough TJ, Balmori E, Voot DM, Andersen MT, O'Brien IEW, Forster RLS (1994). Disruption of virus movement confers broad-spectrum resistance against systemic infection by plant viruses with a triple gene block. *Proceedings of the National Academy of Sciences, USA* 91: 10310-10314.

Bourdin D, Lecoq H (1991). Evidence that heteroencapsidation between two potyviruses is involved in aphid transmission of a non-aphid transmissible isolate from mixed infections. *Phytopathology* 81: 1459-1464.

Braun CJ, Hemenway CL (1992). Expression of amino-terminal portions or full-length viral replicase genes in transgenic plants confers resistance to potato virus X infection. *Plant Cell* 4: 735-744.

Candelier-Harvey P, Hull R (1993). Cucumber mosaic virus genome is encapsidated in alfalfa mosaic virus coat protein expressed in transgenic tobacco plants. *Transgenic Research* 2: 277-285.

Carr JP, Gal-On A, Palukaitis P, Zaitlin M (1994). Replicase-mediated resistance to cucumber mosaic virus in transgenic plants involves suppression of both virus replication in the inoculated leaves and long distance movement. *Virology* 199: 439-447.

Cooper B, Lapidot M, Heick JA, Dodds JA, Beachy RN (1995). A defective movement protein of TMV in transgenic plants confers resistance to multiple viruses whereas the functional analogue increases susceptibility. *Virology* 206: 307-313.

De Haan P, Gielen JJL, Prins M, Wijkamp IG, Van Schepen A, Peters D, Van Grinsven MQJM, Goldbach R (1992). Characterization of RNA-mediated resistance to tomato spotted wilt virus in transgenic tobacco. *Bio/Technology* 10: 1133-1137.

Donson J, Kearney CM, Turpen TH, Khan IA, Kurath G, Turpen AM, Jones GE, Dawson WO, Lewandowski DJ (1993). Broad resistance to tobamoviruses is mediated by a modified tobacco mosaic virus replicase transgene. *Molecular Plant-Microbe Interactions* 6: 635-642.

Dorssers L, Van der Meer J, Van Kammen A, Zabel P (1983). The cowpea mosaic virus RNA replication complex and the host-encoded RNA-dependent RNA polymerase-template complex are functionally different. *Virology* 125: 155-174.

Dorssers L, Zabel P, Van der Meer J, Van Kammen A (1982). Purification of a host-encoded RNA-dependent RNA polymerase from cowpea mosaic virus-infected cowpea leaves. *Virology* 116: 236-249.

Elomaa P, Helariutta Y, Griesbach RJ, Kotilainen M, Seppanen P, Teeri TH (1995). Transgene inactivation in *Petunia hybrida* is influenced by the properties of the foreign gene. *Molecular General Genetics* 248: 649-656.

Finnegan J, McElroy D (1994). Transgene inactivation: Plants fight back! *Bio/Technology* 12: 883-888.

Flavell RB (1994). Inactivation of gene expression in plants as a consequence of specific sequence duplication. *Proceedings of the National Academy of Sciences, USA* 91: 3490-3496.

Gielen JJL, De Haan P, Kool AJ, Peters D, Van Grinsven MQJM, Goldbach RW (1991). Engineered resistance to tomato spotted wilt virus, a negative-strand RNA virus. *Bio/Technology* 9: 1363-1367.

Golemboski DB, Lomonossoff GP, Zaitlin M (1990). Plants transformed with a tobacco mosaic virus nonstructural gene sequence are resistant to the virus. *Proceedings of the National Academy of Sciences, USA* 87: 6311-6315.

Greene AE, Allison RF (1994). Recombination between viral RNA and transgenic plant transcripts. *Science* 263: 1423-1425.

Hackland AF, Rybicki EP, Thomson JA (1994). Coat protein-mediated resistance in transgenic plants. *Archives of Virology* 139: 1-22.

Hong YG, Stanley J (1996). Virus resistance in *Nicotiana benthamiana* conferred by African cassava mosaic virus replication-associated protein (AC1) transgene. *Molecular Plant-Microbe Interactions* 9: 219-225.

Hull R (1994). Resistance to plant viruses: Obtaining genes by non-conventional approaches. *Euphytica* 75: 195-205.

Inoue-Nagata AK, Kormelink R, Nagata T, Kitajima EW, Goldbach R, Peters D (1997). Effects of temperature and host on the generation of tomato spotted wilt virus defective interfering RNAs. *Phytopathology* 87(11): 1168-1173.

Inoue-Nagata AK, Kormelink R, Sgro JY, Nagata T, Kitajima EW, Goldbach R, Peters D (1998). Molecular characterization of tomato spotted wilt virus defective interfering RNAs and detection of truncated L proteins. *Virology* 248: 342-356.

Jorgensen RA (1995). Co-suppression, flower color patterns, and metastable gene expression states. *Science* 268: 686-691.

Kavanagh TA, Spillane C (1995). Strategies for engineering virus resistance in transgenic plants. *Euphytica* 85: 149-158.

Kawchuk LM, Martin RR, McPherson J (1991). Sense and antisense RNA-mediated resistance to potato leafroll virus in Russet Burbank potato plants. *Molecular Plant-Microbe Interactions* 4: 247-253.

Kirsebom LA (1995). RNase P: A "scarlet pimpernel." *Molecular Microbiology* 17: 411-420.

Kirsebom LA, Svard SG (1994). Base pairing between *Escherichia coli* RNase P RNA and its substrate. *EMBO Journal* 13: 4870-4876.

Kjemtrup S, Sampson KS, Peele CG, Nguyen LV, Conkling MA, Thompson WF, Robertson D (1998). Gene silencing from plant DNA carried by a geminivirus. *Plant Journal* 14: 91-100.

Kooter J, Mol JNM (1993). *Trans*-inactivation of gene expression in plants. *Current Opinions in Biotechnology* 4: 166-171.

Lapidot M, Gafny R, Ding B, Wolf S, Lucas WJ, Beachy RN (1993). A dysfunctional movement protein of tobacco mosaic virus that partially modifies the plasmodesmata and limits virus spread in transgenic plants. *Plant Journal* 4: 959-970.

Lawson C, Kaniewski W, Haley L, Rozmann R, Newell C, Sanders P, Tumer NE (1990). Engineering resistance to mixed virus infection in a commercial potato cultivar: Resistance to potato virus X and potato virus Y in transgenic Russet Burbank. *Bio/Technology* 8: 127-134.

Lecoq H, Ravelonandro M, Wipf-Scheibel C, Monsion M, Raccah B, Dunez J (1993). Aphid transmission of a non-transmissible strain of zucchini yellow potyvirus from transgenic plants expressing the capsid protein of plum pox potyvirus. *Molecular Plant-Microbe Interactions* 6: 403-406.

Lindho JA, Dougherty WG (1992). Pathogen-derived resistance to a potyvirus: Immune and resistant phenotypes in transgenic tobacco expressing altered forms of a potyvirus coat protein nucleotide sequence. *Molecular Plant-Microbe Interactions* 5: 144-153.

Lindbo JA, Silva-Rosales L, Proebsting WM, Dougherty, WG (1993). Induction of a highly specific antiviral state in transgenic plants: Implications for regulation of gene expression and virus resistance. *Plant Cell* 5: 1749-1759.

Lomonossoff GP (1995). Pathogen-derived resistance to plant viruses. *Annual Review of Phytopathology* 33: 323-343.

Longstaff M, Brigneti G, Boccard F, Chapman S, Baulcombe D (1993). Extreme resistance to potato virus X infection in plants expressing a modified component of the putative viral replicase. *EMBO Journal* 12: 379-386.

MacFarlane SA, Davies JW (1992). Plants transformed with a region of the 201-kilodalton replicase gene from pea early browning virus RNA 1 are resistant to virus infection. *Proceedings of the National Academy of Sciences, USA* 89: 5829-5833.

Malyshenko SI, Kondakova OA, Nazarova JuV, Kaplan IB, Atabekov JG (1993). Reduction of tobacco mosaic virus accumulation in transgenic plants producing nonfunctional viral transport proteins. *Journal of General Virology* 74: 1149-1156.

Matzke MA, Matzke AJM (1995). Homology-dependent gene silencing in transgenic plants: What does it really tell us? *Trends in Genetics* 11: 1-3.

Metzlaff M. O'Dell M, Flavell RB (1996). Suppression of chalcone synthase A activities in petunia by the addition of a transgene encoding chalcone synthase A. In Grierson D, Lycett GW, Tucker GA (eds.), *Mechanisms and Applications of Gene Silencing*. Nottingham University Press, Nottingham, England, pp. 21-32.

Mueller E, Gilbert J, Davenport G, Brigneti G, Baulcombe DC (1995). Homology-dependent resistance: Transgenic virus resistance in plants related to homology-dependent gene silencing. *Plant Journal* 7: 1001-1013.

Pang S-Z, Slightom JL, Gonsalves D (1993). Different mechanisms protect tobacco against tomato spotted wilt and impatiens necrotic spot *Tospoviruses*. *Bio/Technology* 11: 819-824.

Powell PA, Sanders PR, Tumer N, Fraley RT, Beachy RN (1990). Protection against tobacco mosaic virus infection in transgenic plants requires accumulation of coat protein rather than coat protein RNA sequences. *Virology* 175: 124-130.

Prins M, De Haan P, Luyten R, Van Veller M, Van Grinsven MQJM, Goldbach R (1995). Broad resistance to tospoviruses in transgenic tobacco plants expressing three tospoviral nucleoprotein gene sequences. *Molecular Plant-Microbe Interactions* 8: 85-91.

Prins M, Resende R de O, Anker C, Van Schepen A, De Haan P, Goldbach R (1996). Engineered resistance to tomato spotted wilt virus is sequence specific. *Molecular Plant-Microbe Interactions* 9: 416-418.

Resende R de O, De Haan P, Avila AC de, Kitajima EW, Kormelink R, Goldbach R, Peters D (1991). Generation of envelope and defective interfering RNA mutants of tomato spotted wilt virus by mechanical passage. *Journal of General Virology* 72: 2375-2383.

Resende R de O, De Haan P, Van de Vossen E, Avila AC de, Goldbach R, Peters D (1992). Defective interfering L RNA segments of tomato spotted wilt virus retain both virus genome termini and have extensive internal deletions. *Journal of General Virology* 73: 2509-2516.

Rubino L, Russo M (1995). Characterization of resistance to cymbidium ringspot virus in transgenic plants expressing a full-length viral replicase gene. *Virology* 212: 240-243.

Sanford JC, Johnson SA (1985). The concept of pathogen derived resistance. *Journal of Theoretical Biology* 113: 395-405.

Santa Cruz S, Baulcombe DC (1994). Molecular analysis of potato virus X isolated in relation to the potato hypersensitivity gene *Nx*. *Molecular Plant-Microbe Interactions* 6: 707-714.

Schiebel W, Pelissier T, Riedel L, Thalmeir S, Schiebel R, Kempe D, Lottspeich F, Sanger HL, Wassenegger M (1998). Isolation of an RNA-dependent RNA polymerase-specific cDNA clone from tomato. *Plant Cell* 10: 2087-2101.

Stark DM, Beachy RN (1989). Protection against potyvirus in transgenic plants: Evidence for broad spectrum resistance. *Bio/Technology* 7: 1257-1262.

Tenllado F, García-Luque I, Serra MT, Díaz-Ruíz JR (1995). *Nicotiana benthamiana* plants transformed with the 54-kDa region of the pepper mild mottle tobamovirus replicase gene exhibit two types of resistance responses against viral infection. *Virology* 211: 170-183.

Tenllado F, García-Luque I, Serra MT, Díaz-Ruíz JR (1996). Resistance to pepper mild mottle tobamovirus conferred by the 54-kDa gene sequence in transgenic plants does not require expression of the wild-type 54-kDa protein. *Virology* 219: 330-335.

Tumer NE, Kaniewski W, Haley L, Gehrke L, Lodge JK, Sanders P (1991). The second amino acid of alfalfa mosaic virus coat protein is critical for coat protein-mediated protection. *Proceedings of the National Academy of Sciences, USA* 88: 2331-2335.

Van Blokland R (1994). Trans-inactivation of flower pigmentation genes in *Petunia hybrida*. PhD dissertation. Free University Amsterdam, The Netherlands.

Van Blokland R, Van der Geest N, Mol JNM, Kooter JM (1994). Transgene-mediated suppression of chalcone synthase expression in *Petunia hybrida* results from an increase in RNA turnover. *Plant Journal* 6: 861-877.

Van der Krol AR, Lenting PE, Veenstra J, Van der Meer IM, Koes RE, Gerats AGM, Mol JNM, Stuitje A (1988). An anti-sense chalcone synthase gene in transgenic plants inhibits flower pigmentation. *Nature* 333: 866-869.

Van der Krol AR, Mur LA, Beld M, Mol JNM, Stuitje A (1990). Flavonoid genes in petunia: Addition of a limited number of gene copies may lead to a suppression of gene expression. *Plant Cell* 2: 291-299.

Van der Vlugt RAA, Ruiter RK, Goldbach R (1992). Evidence for sense RNA-mediated resistance to PVY^N in tobacco plants transformed with the viral coat protein cistron. *Plant Molecular Biology* 20: 631-639.

Van der Wilk F, Willink DPL, Huisman MJ, Huttinga H, Goldbach R (1991). Expression of the potato leafroll luteovirus coat protein gene in transgenic potato plants inhibits viral infection. *Plant Molecular Biology* 17: 431-439.

Wassenegger M, Heimes S, Riedel L, Sänger H (1994). RNA-directed de novo methylation of genomic sequences in plants. *Cell* 76: 567-576.

Wassenegger M, Pelissier T (1998). A model for RNA-mediated gene silencing in higher plants. *Plant Molecular Biology* 37: 349-362.

Wilson TMA (1993). Strategies to protect crop plants against viruses: Pathogen-derived resistance blossoms. *Proceedings of the National Academy of Sciences, USA* 90: 3134-3141

SECTION VI:
METHODS IN
MOLECULAR VIROLOGY

Chapter 18

Antibody Expression in Plants

Thorsten Verch
Dennis Lewandowski
Stefan Schillberg
Vidadi Yusibov
Hilary Koprowski
Rainer Fischer

INTRODUCTION

In the past two decades, advances in plant molecular biology and recombinant antibody engineering have provided new tools for plant biotechnology. Plants are emerging as a safer alternative to microbial or animal cells for the production of pharmaceutical proteins, such as
antibodies. Functional antibody expression in transgenic plants was
demonstrated (Hiatt et al., 1989) and used to create virus-resistant plant
lines (Tavladoraki et al., 1993). The progress and potential of this approach for virus control is reviewed in the first part of this chapter. Furthermore, plant-produced antibodies have been tested in immunotherapy
(Ma and Hein, 1995b). Here, plant viruses as vectors for transient gene
expression offer an alternative to transgenic plants. The development of
this new system for the production of recombinant proteins, such as antibodies, in plants is the focus of the second part of this chapter, followed
by a brief comment on prospects of the presented techniques.

ANTIBODY-MEDIATED VIRAL RESISTANCE
IN TRANSGENIC PLANTS

Plant diseases are a major threat to the world food supply; close to 40 percent of crop yield is still lost to pathogens. Viral diseases are among the

most destructive and the most difficult to manage because of the limited effectiveness of the control measures that have been developed. In this section, we describe the application of novel molecular approaches to generating virus-resistant plants through the expression of antiviral recombinant antibodies. This technology has demonstrated potential in creating plants that are intrinsically resistant to pathogens, and we foresee that it will be of major agronomic importance in the twenty-first century.

In the past, disease control and the generation of resistant plant lines protected against a specific disease, such as a viral infection, were achieved using conventional breeding techniques that exploited the identification of resistant lines, crossing, mutant screenings, and backcrossing. Despite the effort invested in these strategies, many attempts to create a durable virus-resistant plant line either failed or the resistance obtained was rapidly overcome by the pathogen. A major limitation of the conventional breeding strategy is that it is time-consuming.

The constraints of classical approaches for generating resistant plants prompted the exploration of alternative molecular methods, such as the expression of structural or nonstructural viral proteins as well as antivirus antibodies in susceptible hosts. Molecular methods can create plants that are intrinsically resistant to pathogens and have a beneficial impact on the environment, since they lead to significant reductions in the application of agrochemicals, such as the insecticides used to control virus vectors.

Molecular strategies for generating resistant plants have benefited from progress in plant molecular virology that has enhanced our understanding of the molecular basis of plant viral diseases. Fundamental knowledge about the organization of the viral genome, gene functions, and interactions between plants and pathogens has permitted the development of novel strategies for controlling viral diseases of plants (Baulcombe, 1994a,b; Beachy, 1997; see Chapters 16 and 17 in this book). Despite these advances, there are still many gaps in our knowledge of the events leading to viral infection, replication, movement, and vector transmission. Progress toward an understanding of many of these details is essential for developing virus-resistant transgenic plants.

The analysis of viral proteins and enzymes involved in pathogenesis is a difficult task because of low viral expression levels and the difficulty of purifying many plant viruses. However, basic information about the function of and interactions between both structural and nonstructural viral proteins in vivo, and their interaction with host factors or vector components, will aid the design of viral protein derivatives that interfere with pathogenesis.

Transgenic plant technology is a powerful method to produce plants that are resistant to viral pathogens. One strategy for engineering virus resistance is the expression of viral coat protein (CP) or other viral sequences in the host plant (pathogen-derived resistance) (Lomonossoff, 1995; Dawson, 1996). The pathogen-derived resistance approach is based on the expression of viral proteins or sequences that co-opt the viral replication or movement mechanisms (Harrison et al., 1987; Namba et al., 1991; Baulcombe, 1994a; Noris et al., 1996; Beachy, 1997; see Chapter 17 in this book). Other strategies exploit transgenic plants expressing naturally occurring antiviral genes, such as ribosome-inactivating proteins (Taylor et al., 1994; Lam et al., 1996). However, these methods often do not provide complete resistance, are confined to a small spectrum of viral strains, and may create substantial risks to the environment (Wilson, 1993; Baulcombe, 1994a; Falk and Bruening, 1994).

Pathogen-derived resistance strategies are to some extent limited by the available knowledge of viral pathogenesis and the effectiveness of expressing viral sequences to block stages of infection. An alternative strategy is the use of recombinant antibodies (rAbs): they can be tailored to give optimal effects in reducing viral infection processes and can be raised against conserved viral epitopes shared by a wide spectrum of viral pathogens. The use of rAbs to recognize such conserved functional domains of viral CPs, movement proteins (MPs), or replicases is an approach to obtain broad-spectrum resistance and to reduce environmental risk by inactivating the targets inside the plant cell through immunomodulation. The keystone of immunomodulation is the generation of rAbs that recognize essential viral proteins in planta.

The feasibility of using antibodies to increase pathogen resistance has been successfully demonstrated for human viruses (Marasco et al., 1993; Chen et al., 1994; Duan et al., 1994; Gargano and Cattaneo, 1997), and its applications in plant virology are now well established. The potential of rAbs to interfere in plant viral infection was shown in 1993 (Tavladoraki et al., 1993). In this study, expression of a single-chain antibody (scFv) directed against *Artichoke mottled crinkle virus* (genus *Tombusvirus*) CP in the plant cytosol reduced viral infection and delayed symptom development.

Cytosolic expression of the rAb fragments has been the favored strategy, since the majority of processes involved in viral replication and movement take place within this compartment (Wilson, 1993; Baulcombe, 1994a). A disadvantage of using cytosolically expressed antibody is that scFvs do not accumulate to as high a level in the plant cytosol as compared to secreted antibodies (Fecker et al., 1997; Schillberg et al., 1999).

It seems likely that the lack of protein disulfide isomerase and specific chaperones in the cytoplasm results in scFv misfolding and contributes to enhanced proteolysis (Biocca et al., 1995). However, a recent report demonstrated that scFv fragments derived from phage display reached up to 0.3 percent and 1 percent of total soluble protein in the cytoplasm of *Petunia* leaves and petals, respectively, indicating that phage display selection can be used for enrichment of more stable scaffolds that tolerate the absence of disulfide bonds (De Jaeger et al., 1998). Expression levels of recombinant proteins in the cytoplasm of prokaryotes have been enhanced by the addition of a stabilizing fusion protein, and antibody stability in plants has been improved by the fusion of peptide sequences to the antibody C-terminus. Addition of a C-terminal KDEL sequence significantly increased scFv levels in the plant cytoplasm, indicating that fusion to short polypeptides may protect the scFv fragment from proteolytic degradation (Schouten et al., 1996, 1997). Recently, we have shown that fusion of an scFv to a protein, such as the *Tobacco mosaic virus* (TMV, genus *Tobamovirus*) CP, stabilized a cytosolic scFv and raised accumulation levels (Spiegel et al., 1999).

Despite the limitations regarding accumulation levels, cytosolic expression of antibodies is a useful and effective strategy. Zimmermann et al. (1998) showed that even very low levels of a TMV-specific cytosolic scFv (0.00002 percent of total soluble protein) led to a remarkably enhanced, systemic viral resistance in transgenic tobacco plant lines. In contrast, targeting of TMV-specific scFv fragments or full-size antibodies to the intercellular space (apoplast) gave a lower degree of resistance, even though the apoplast-targeted full-size antibody accumulated to a 50,000-fold higher level than the cytosolic scFv (Voss et al., 1995; Zimmermann et al., 1998). These observations demonstrate that selection of the subcellular compartment for antibody expression is critical for engineering viral resistance. However, targeting of virus-specific antibodies to the secretory pathway may be useful to block viral entry and to reduce the number of virions entering the cytosol, where uncoating, replication, and movement take place (Voss et al., 1995; Fecker et al., 1997).

Recently, we developed a novel approach for engineering disease resistant crops by targeting antiviral antibody fragments to the plasma membrane *in planta*. TMV-specific scFv fragments were efficiently targeted to the plasma membrane of tobacco cells by heterologous transmembrane domains of mammalian proteins. Membrane-integrated scFv fusion proteins faced the apoplast and retained both antigen binding and specificity. Transgenic plants expressing membrane-targeted scFv fusion proteins were resistant to TMV infection, demonstrating that membrane-anchored antiviral antibodies were functional in vivo, offering a potentially powerful new method to shield plant cells from invading pathogens (Schillberg et al., submitted).

A disadvantage of the current antibody-mediated resistance approaches may be the choice of viral CP as a target. Plant viral CPs have broad structural diversity; this restricts the effect of the expressed antibodies to a small range of viruses, and, under selective stress, the viral CP sequence may alter without loss of function. Therefore, generation of rAbs directed against conserved functional domains of viral replicases and MPs may provide a better route for obtaining pathogen-resistant plants with broad-spectrum resistance against economically important viruses. In addition, the generation of bivalent bispecific scFv fragments (Fischer, Schumann, et al., 1999), with two specificities, one viral capsid specific and a second specific for either a replicase or an MP, may provide improved tools for antibody-mediated resistance against plant viruses. These rAb approaches can be combined to create transgenic plants in which resistance is pyramided by expressing several specific, targeted rAbs that bind to and neutralize several viral pathogens.

Clearly, a greater understanding of viral pathogenesis will identify key proteins and protein motifs involved in the replication or transmission of viruses. This fundamental research will provide more target antigens and, together with developments in antibody engineering, such as phage display and ribosome display, will increase the number of strategies that can be used to create crop plants that are intrinsically resistant to viral infection in the field.

EXPRESSION OF RECOMBINANT ANTIBODIES USING PLANT VIRUS GENE VECTORS

Recombinant protein production in plants has found a number of applications with the extensive development of different expression techniques, and numerous proteins have been successfully expressed in plants (Whitelam et al., 1993; Goddijn and Pan, 1995; Lomonossoff and Johnson, 1995; Mason and Arntzen, 1995; Turpen and Reinl, 1998). In contrast to bacterial expression systems, plants have posttranslational protein modifications similar to those of mammalian cells, thus allowing the production of correctly folded, functionally active recombinant proteins. This is particularly important for the expression of complex proteins such as antibodies. Unlike animal cell cultures that are currently exploited for antibody expression, plants are not host to human pathogens, such as hepatitis virus or prions. Furthermore, agricultural upscaling is well established and inexpensive, as energy and maintenance costs are negligible. Therefore, plants are attractive for the production of safe, low-cost vaccines and therapeutic proteins for use in humans. Presently, transgenic plants are the most common means for expression of recombinant proteins in plants (Goddijn and Pan, 1995; Knauf, 1995; Hammond, 1999; Richter and Kipp, 1999). However, the creation of transgenic plant lines is often time-consuming and the yields of recombinant proteins vary. Fast, potentially high-yielding alternatives to the stable transformation of plants are plant virus-based gene vectors, which have been used to transiently express recombinant proteins (reviewed by Scholthof and Scholthof, 1996). Plant viruses carrying a foreign gene can spread through the whole plant within one to two weeks postinoculation. Strong viral promoters have already been used in transgenic plants (35S promoter from *Cauliflower mosaic virus*) and are ideal to support high levels of foreign gene expression when used in viral vectors.

Prior to the development of infectious complementary (c) DNA clones of plant RNA viruses, it was proposed that plant viruses could become useful tools to express foreign genes (Hull, 1983; Siegel, 1985). The past decade witnessed the adaptation of a number of plant viruses into transient expression vectors, including TMV (Dawson et al., 1989; Takamatsu et al., 1990), *Potato virus X* (PVX, genus *Potexvirus;* Chapman et al., 1992), *Cowpea mosaic virus* (CPMV, genus *Comovirus;* Porta et al., 1994), *Alfalfa mosaic virus* (AMV, genus *Alfamovirus;* Yusibov et al., 1997), and *Plum pox virus* (PPV, genus *Potyvirus;* Fernandez-Fernandez et al., 1998). Two different strategies were developed: (1) the presentation of proteins or protein domains as a fusion with the CP on the surface of the assembled virus particle and (2) independent expression of complete foreign genes inserted into the viral genome (see Figure 18.1).

Recombinant plant viruses carrying foreign sequences on the particle surface, for example, the jellyfish green fluorescent protein (GFP), were used to study virus movement and viral protein accumulation (Baulcombe et al., 1995). Alternatively, these vectors were used for vaccine development by displaying antigenic determinants of animal and human pathogens on the surface of the virus particles (Usha et al., 1993; Turpen et al., 1995; Yusibov et al., 1997; Modelska et al., 1998; Porta and Lomonossoff, 1998). Most recently, the presentation of scFv on the surface of a particle was demonstrated using PVX (Smolenska et al., 1998). An scFv against the herbicide diuron was fused to the PVX CP with the self-cleaving 2A peptide from foot-and-mouth disease virus as a linker between the two genes. Using this strategy, a mixture of cleaved wild-type-like CP and uncleaved scFv-carrying CP was produced in plants infected with the recombinant virus, and virions consisted of a mixture of either scFv-presenting or wild-type-like CP. The presence of functional scFv on assembled PVX particles was demonstrated by immunosorbent assays. So far, this is one of the largest proteins presented on a virus particle.

FIGURE 18.1. Two *Tobacco mosaic virus* (TMV)-Based Vectors As Examples of the Two Different Strategies of Recombinant Gene Expression from Virus Vectors

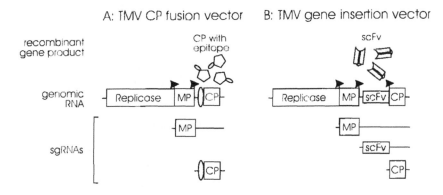

Note: (A) The foreign sequence is fused to the virus coat protein (CP) (here N-terminally) and expressed as a fusion with the CP. (B) The foreign gene is expressed from an additional subgenomic RNA (sgRNA) promoter independent of the CP. MP = movement protein; scFv = single-chain antibody; arrows = sgRNA promoters; boxes = open reading frames.

In addition to CP fusion, viral vectors have been developed to express foreign gene products as separate proteins (see Figure 18.1). These gene insertion vectors avoid the steric constraints that limit the efficiency of CP fusion products to assemble into virions. Expression of a full-size, soluble protein is often desirable for subsequent purification or intracellular targeting. Besides GFP (Baulcombe et al., 1995; Casper and Holt, 1996; Shivprasad et al., 1999) and various enzymes (Dawson et al., 1989; Donson et al., 1991; Chapman et al., 1992; Kumagai et al., 1993), antibodies have been expressed using plant viral vectors (Verch et al., 1998; McCormick et al., 1999). TMV-based vectors were the first to be used for soluble antibody expression from viral vectors in plants.

TMV has a (+)-RNA genome with three open reading frames (ORFs), encoding the replicase that is translated directly from the genomic RNA, and the MP and CP that are expressed from individual subgenomic RNAs (sgRNAs). The CP is one of the most highly expressed viral proteins in plants and can accumulate up to 10 percent of the total soluble protein in infected plants (Matthews, 1991). The foreign protein was produced from the existing CP sgRNA promoter, and an additional CP sgRNA promoter was inserted to express the CP (Dawson et al., 1989). However, the resulting vector was unstable due to recombination between the homologous repeats (Dawson et al., 1989). A considerably more stable vector was subsequently constructed when the CP and CP sgRNA promoter were derived from *Odontoglossum ringspot virus* (genus *Tobamovirus*) (Donson et al., 1991). TMV-based vectors were further improved (Kumagai et al., 1993) and led to a vector 30B (Shivprasad et al., 1999) that was stable upon infection of *Nicotiana benthamiana*, for up to three months, and that systemically infected the plant.

Expression of Anti-TMV Single-Chain Antibodies

The potential of producing functional antibodies in transgenic plants has been reviewed (Hein et al., 1991; Ma and Hein, 1995a,b; Smith and Glick, 1997; Larrick et al., 1998). Antibodies are used in immunotherapy and diagnostics, which require large amounts of inexpensive and safe product, often within a short period of time (Fischer, Liao, et al., 1999). Presently, no established production systems meet all the requirements of antibody production: correct assembly, folding and glycosylation, safety from human pathogens, low cost of production, and ready availability as needed. As described earlier, transgenic plants meet some of these requirements and are presently being tested for commercial antibody production, (Russell, 1999). Virus vectors offer additional advantages, such as ease of genomic manipulation due to the small size of viral genomes, high levels of recombinant protein production, wide host range, and short cycle time. Wide host ranges of some viruses potentially allow rAb production in different host plants, including edible plants using only one vector construct. This way, a number of antibodies could be screened quickly in planta within one to two weeks of inoculation, allowing several rounds of optimization before the first generation of a transgenic plant line, which takes a considerable amount of time. Custom-designed therapeutic antibodies can be produced rapidly. In addition, antibodies are valuable tools. Virus vectors could be used to introduce antibodies interfering specifically with cellular or biochemical pathways at defined developmental stages of the plant to elucidate cellular functions.

To demonstrate the principle of using viral vectors to transiently express antibodies in plants, anti-TMV CP scFv29, which had also been expressed in transgenic plants, was chosen as a model antibody (Schillberg et al., 1999).

To determine the ideal experimental hosts for highest scFv accumulation, *N. tabacum* cv. Xanthi tobacco transgenic for the TMV MP (MP+

'Xanthi' tobacco) and *N. benthamiana* transgenic for the TMV MP (MP+ *N. benthamiana*) were compared. While MP+ 'Xanthi' tobacco was not efficiently infected by the virus vector, MP+ *N. benthamiana* was consistently found to be the best host for highest scFv29 accumulation, as determined by ELISA, and therefore chosen for analysis of a larger number of inoculated plants.

Accumulation levels of scFv in infected leaves were compared between cytosol (scFv29W) and apoplast (scFv29M) and were dependent on the cellular target (see Table 18.1). In the cytosol, scFv29 accumulated up to a mean of 0.047 µg/g fresh weight in inoculated leaves, and in systemically infected leaves, up to 0.17 µg/g fresh weight. Upon apoplastic targeting, scFv29 yielded up to 90 µg/g fresh weight in inoculated and up to 368 µg/g fresh weight in uppermost leaves. Similar compartment-related differences were observed when expressing scFv29 in transgenic plants (S. Zimmerman, personal communication) and were reported for other antibodies (Firek et al., 1993). Possible reasons for these differences, such as misfolding and higher protein turnover, were discussed in the first section.

The ability of plant virus vectors to express functional scFvs that are targeted to various cellular compartments has been shown. We suggest the use of this system as a tool to target structural components or biochemical pathways to elucidate cellular functions. To achieve maximal accumulation of scFv fragments for commercial production, the effects of stabilization and subcellular targeting (compare to first section) need to be considered and tested.

TABLE 18.1. Yields of scFv29 in Leaves of MP+ *Nicotiana benthamiana* Infected with Constructs 30B-scFv29W or 30B-scFv29M of *Tobacco mosaic virus*

Construct	µg scFv / g fresh weight			
	set I, 14 dpi		set II, 13 dpi	
	Inoculated leaf	Upper leaf	Inoculated leaf	Upper leaf
30B-scFv29W	0.0093	0.05	0.047	0.17
cytosol	*± 0.016*	*± 0.09*	*± 0.012*	*± 0.11*
30B-scFv29M	28	194	90	368
apoplast	*± 11*	*± 129*	*± 27*	*± 299*

Note: At 13 or 14 days after inoculation (dpi), scFv accumulation in cytosol or apoplast of inoculated and uppermost leaves of infected *N. benthamiana* was compared to a $F(ab)_2 29$ of known concentration by ELISA. Mean values and the standard deviation (italics) are presented in the table.

Expression of Anticancer Antibodies
Using Plant Virus Vectors

Besides agricultural use, plants are gaining more importance in the production of pharmaceuticals, including therapeutic and diagnostic antibodies (Hein et al., 1991; Ma and Hein, 1995a,b; Smith and Glick, 1997). Custom-designed cancer immunotherapy is one of the possible applications for which high yields of inexpensive recombinant antibody are required within a short time. Here, tissue culture techniques are too costly and production in transgenic plants is not as fast as some alternatives. Therefore, plant virus vectors can be used as a new approach to obtaining potentially higher yields in a relatively short period of time as compared to transgenic plants (Verch et al., 1998; McCormick et al., 1999). McCormick et al. (1999) obtained an scFv derived from the surface immunoglobulin of a mouse B-cell lymphoma (38C13) in *N. benthamiana*. The apoplastic scFv could easily be recovered from the intercellular washing fluid by vacuum infiltration and centrifugation and subsequently purified by affinity chromatography. Correct folding of the plant scFv and the absence of potentially immunogenic monosaccharides were verified and compared to the same scFv produced in bacteria without observing major differences. Most important, mice were protected against tumor challenge after vaccination with the scFv isolated from plants. The authors concluded that time efficiency and high yields of the virus-based transient expression system could be utilized for the production of custom-made antibodies for immunotherapy. For the same reasons, plant virus-based antibody expression could serve to quickly express phage-derived antibodies for assaying antibody function.

However, since scFvs lack constant antibody domains, they are not always suitable for therapy. Humanized full-size antibodies are gaining more importance as reagents to stimulate immune responses through cellular cytotoxicity. However, up to now, production of such full-size antibodies has been faced with serious limitations: animal tissue culture is costly, bacterial overexpression does not result in correctly folded full-size antibodies, and generation of a transgenic plant line is time-consuming (Ma et al., 1995). Therefore, plant virus vectors provide an inexpensive and fast alternative for the production of functional antibodies. The production of scFv fragments, as described previously, requires the introduction of only a single short gene into the viral vector genome. In contrast, full-size antibodies require the coexpression of two genes for heavy chain (HC) and light chain (LC). Three approaches can achieve this goal: (1) insertion of both antibody genes into one virus vector, (2) independent expression of HC and LC from two virus vectors that coinfect the same cell, and (3) complementation of two deficient viral genomes, each of which is carrying one of the antibody genes. Since foreign sequences are recombined out faster with increasing insert length, the efficiency of the first strategy is low and not well suited for the expression of large antibody genes. Therefore, we tested antibody production by coinoculation with two TMV constructs (strategy 2) and, alternatively, through an AMV complementation system (strategy 3).

As a model antibody, we chose the colorectal cancer antibody CO17-1A (Koprowski et al., 1979), which has been extensively studied (Khazaeli et al., 1988; Ragnhammer et al., 1993) and is presently used for immunotherapy in Europe. We cloned the genes encoding HC and LC separately into individual TMV-based vectors, and *N. benthamiana* plants were coinoculated with both constructs (Verch et al., 1998). Coinfection of the cell with the HC- and LC-carrying virus vectors resulted in expression and assembly of functional CO17-1A antibody. Plant-produced rAb CO17-1A bound to the corresponding antigen GA733-2 in ELISA and could be purified using ammonium sulfate precipitation followed by protein A affinity chromatography. However, superinfection of cells with two competing viral constructs was inefficient, and the HC construct showed some degree of recombination and loss of the foreign gene.

As an alternative approach, we used a complementation system based on AMV, a member of the family *Bromoviridae*. AMV is a tripartite RNA virus with a (+)-sense genome (Matthews, 1991). The replicase genes are located on RNA1 and RNA2. RNA3 encodes the cell-to-cell movement protein P3 that is translated directly from genomic RNA3 and the CP that is expressed from subgenomic RNA4. The AMV expression system that we have adapted is based on two RNA3 derivatives and transgenic *N. tabacum* cv. Samsun plants expressing replicase proteins P1 and P2 (Taschner et al., 1991). In the P3-deficient RNA3, the HC gene of antibody CO17-1A was cloned into the position of the MP (Δ=P-17HCK). In the CP-deficient RNA3, the CP gene, which is also required for cell-to-cell movement, was replaced by the LC gene, leaving the CP sgRNA promoter and the 3'-end of the AMV genome intact (Δ=CP-17LC). Upon coinfection, the two movement-deficient RNA3-based vectors (Δ=P-17HCK + Δ=CP-17LC) could complement each other in trans and allow systemic spread of recombinant virus in planta (Verch, unpublished results). Antibody CO17-1A was expressed in systemically infected leaves as detected by ELISA. Recombinant RNA was detected by northern analysis and RT-PCR (reverse transcription–polymerase chain reaction), but a certain degree of vector instability was observed. Better yields of recombinant product and vector stability should be achievable with further optimization of the virus backbone, as shown previously with the development of the TMV-based vector. Despite present limitations, the complementation system is a promising approach to expressing multipartite proteins, such as antibodies, using plant virus-based vectors. Different genes can be more easily and efficiently introduced into the same cell than by superinfection, in which competition between constructs of different fitness and partial resistance of infected cells decrease optimal infection and yields of rAb.

CONCLUSIONS

Antibody expression in plants is becoming increasingly important. Transgenic plants can be engineered with antibody-mediated resistance to pests. Modern tools of crop transformation, together with the broader knowledge of virus structure and functions gained recently, make antibody-mediated resistance feasible for agricultural use. The limited interference of specifically designed antiviral antibodies with cellular functions and plant growth could set competitive standards for safer disease protection, as compared to transgenic plants expressing viral sequences (cross-protection, virus gene-mediated resistance). Advances in tissue- and compartment-specific expression could make antiviral antibodies a powerful tool for studing virus functions during initial and systemic infection of plants through interference with specific virus proteins as an alternative to mutating the virus genome.

Furthermore, antibodies transiently expressed from plant virus vectors represent a powerful tool for basic research. Thereby, antibodies targeting biochemical pathways or cellular signals can be introduced into the plant at specific developmental stages. Such transient expression systems are a valuable alternative to constitutive expression of the foreign gene in transgenic plants. The ease of genetic manipulation of virus vectors saves time and resources in the experimental setup. Here, an alternative to virus vectors is agrofiltration as a transient system, which has been recently used for antibody expression (Vaquero et al., 1999). Full-grown leaves are vacuum infiltrated with recombinant *Agrobacterium tumefaciens,* and protein expression can be detected as soon as two to three days postinoculation. However, this approach is limited to a laboratory scale.

Apart from their potential in plant protection and basic research, antibodies produced either constantly or transiently in plants are becoming more and more important in molecular medicine. The "molecular farming" of antibodies used in immunotherapy or diagnostics is safer and less expensive than tissue culture techniques. Plant virus vectors optimized for antibody expression could be a powerful alternative to transgenic approaches. Possible applications include testing of antibodies designed for expression in transgenic plants, production of custom-designed antibodies for immunotherapy or diagnostics, and rapid, high-yield production of antibodies.

Although antibody expression in plants is a relatively new technology, the potential for basic as well as applied virus research is enormous. Transgenic plants and viral expression vectors open a wide field of applications for the use of antibodies in plants and molecular medicine.

REFERENCES

Baulcombe D (1994a). Novel strategies for engineering virus resistance in plants. *Current Opinion in Biotechnology* 5: 117-124.

Baulcombe D (1994b). Replicase-mediated resistance: A novel type of virus resistance in transgenic plants? *Trends in Microbiology* 2: 60-63.

Baulcombe DC, Chapman S, and Cruz SS (1995). Jellyfish green fluorescent protein as a reporter for virus infections. *The Plant Journal* 7: 1045-1053.

Beachy RN (1997). Mechanisms and applications of pathogen-derived resistance in transgenic plants. *Current Opinion in Biotechnology* 8: 215-220.

Biocca S, Ruberti F, Tafani M, Pierandrei-Amaldi P, Cattaneo A (1995). Redox state of single chain Fv fragments targeted to the endoplasmic reticulum, cytosol and mitochondria. *Biotechnology* 13: 1110-1115.

Casper SJ, Holt CA (1996). Expression of the green fluorescent protein-encoding gene from a tobacco mosaic virus-based vector. *Gene* 173: 69-73.

Chapman S, Kavanagh T, Baulcombe D (1992). Potato virus X as a vector for gene expression in plants. *The Plant Journal* 2: 549-557.

Chen SY, Khouri Y, Bagley J, Marasco WA (1994). Combined intra- and extracellular immunization against human immunodeficiency virus type 1 infection with a human anti-gp 120 antibody. *Proceedings of the National Academy of Sciences, USA* 91: 5932-5936.

Dawson WO (1996). Gene silencing and virus resistance: A common mechanism. *Trends in Plant Science* 1: 107-108.

Dawson WO, Lewandowski DJ, Hilf ME, Bubrick P, Raffo AJ, Shaw JJ, Grantham GL, Desjardins PR (1989). A tobacco mosaic virus-hybrid expresses and loses an added gene. *Virology* 172: 285-292.

De Jaeger G, Buys E, Eeckhout D, De Wilde C, Jacobs A, Kapila J, Angenon G, Van Montagu M, Gerats T, Depicker A (1998). High level accumulation of single-chain variable fragments in the cytosol of transgenic *Petunia hybrida*. *European Journal of Biochemistry* 159: 1-10.

Donson J, Kearney CM, Hilf ME, Dawson WO (1991). Systemic expression of a bacterial gene by a tobacco mosaic virus-based vector. *Proceedings of the National Academy of Sciences, USA* 88: 7204-7208.

Duan L, Bagasra O, Laughlin MA, Oakes JW, Pomerantz RJ (1994). Potent inhibition of human immunodeficiency virus type 1 replication by an intracellular anti-Rev single-chain antibody. *Proceedings of the National Academy of Sciences, USA* 91: 5075-5079.

Falk BW, Bruening G (1994). Will transgenic crops generate new viruses and new diseases? *Science* 263: 1395-1396.

Fecker LF, Koenig R, Obermeier C (1997). *Nicotiana benthamiana* plants expressing beet necrotic yellow vein virus (BNYVV) coat protein-specific scFv are partially protected against the establishment of the virus in the early stages of infection and its pathogenic effects in the late stages of infection. *Archives of Virology* 142: 1857-1863.

Fernandez-Fernandez MR, Martinez-Torrecuadrada JL, Casa, JI, Garcia JA (1998). Development of an antigen presentation system based on plum pox potyvirus. *FEBS Letters* 427: 229-235.

Firek S, Draper J, Owen MR, Gandecha A, Cockburn B, Whitelam GC (1993). Secretion of a functional single-chain Fv protein in transgenic tobacco plants and cell suspension cultures [published erratum appears in *Plant Molecular Biology* 1994, March 24(5):833]. *Plant Molecular Biology* 23: 861-870.

Fischer R, Liao Y-C, Hoffmann K, Schillberg S, Emans N (1999). Molecular farming of recombinant antibodies in plants. *Biological Chemistry* 380: 825-839.

Fischer R, Schumann D, Zimmermann S, Drossard J, Sack M, Schillberg S (1999). Expression and characterization of bispecific single-chain Fv fragments produced in transgenic plants. *European Journal of Biochemistry* 262: 810-816.

Gargano N, Cattaneo A (1997). Inhibition of murine leukaemia virus retrotranscription by the intracellular expression of a phage-derived anti-reverse transcriptase antibody fragment. *Journal of General Virology* 78: 2591-2599.

Goddijn OJM, Pan J (1995). Plants as bioreactors. *Trends in Biotechnology* 13: 379-387.

Hammond J (1999). Overview: The many uses and applications of transgenic plants. In Hammond J, McGarvey P, Yusibov V (eds.), *Plant Biotechnology—New Products and Applications* (Second Edition), Springer-Verlag, Berlin, pp. 1-20.

Harrison BD, Mayo MA, Baulcombe DC (1987). Virus resistance in transgenic plants that express cucumber mosaic virus satellite RNA. *Nature* 328: 799-802.

Hein MB, Tang Y, McLeod DA, Janda KD, Hiatt A (1991). Evaluation of immunoglobulins from plant cells. *Biotechnology Progress* 7: 455-461.

Hiatt A, Cafferkey R, Bowdish K (1989). Production of antibodies in transgenic plants. *Nature* 342: 76-78.

Hull R (1983). Genetic engineering with plant viruses and their potential as vectors. *Advances in Virus Research* 28: 1-33.

Khazaeli MB, Saleh MN, Wheeler RH, Huster WJ, Holden H, Carrano R, LoBuglio AF (1988). Phase I trial of multiple large doses of murine monoclonal antibody CO17-1. II. Pharmacokinetics and immune response. *Journal of the National Cancer Institute* 80: 937-942.

Knauf VC (1995). Transgenic approaches for obtaining new products from plants. *Current Opinion in Biotechnology* 6: 165-170.

Koprowski H. Steplewsk Z. Mitchell K. Herlyn M. Herlyn D, Fuhrer P (1979). Colorectal carcinoma antigens detected by hybridoma antibodies. *Somatic Cell Genetics* 5: 957-971.

Kumagai MH, Turpen TH, Weinzettl N, Della-Cioppa G. Turpen AM, Donson J. Hill ME, Grantham GL. Dawson WO. Chow TP. Piatak M. Grill LK (1993). Rapid, high-level expression of biologically active α-trichosanthin in transfected plants by an RNA viral vector. *Proceedings of the National Academy of Sciences, USA* 90: 427-430.

Lam YH. Wong YS. Wang B, Wong RNS. Yeung HW. Shaw PC (1996). Use of trichosanthin to reduce infection by turnip mosaic virus. *Plant Science* 114: 111-117.

Larrick JW. Yu L. Chen J. Jaiswal S, Wycoff K (1998). Production of antibodies in transgenic plants. *Research in Immunology* 149: 603-608.

Lomonossoff GP (1995). Pathogen-derived resistance to plant viruses. *Annual Reviews in Phytopathology* 33: 323-343.

Lomonossoff G. Johnson JE (1995). Eukaryotic viral expression systems for polypeptides. *Virology* 6: 257-267.

Ma JK, Hein MB (1995a). Immunotherapeutic potential of antibodies produced in plants. *Trends in Biotechnology* 13: 522-527.

Ma JK. Hein MB (1995b). Plant antibodies for immunotherapy. *Plant Physiology* 109: 341-346.

Ma JK, Hiatt A, Hein M, Vine ND, Wang F. Stabila P. Van Dolleweerd C. Mostov K. and Lehner T (1995). Generation and assembly of secretory antibodies in plants. *Science* 268: 716-719.

Marasco WA. Haseltine WA. Chen SY (1993). Design, intracellular expression, and activity of a human anti-human immunodeficiency virus type 1 gp120 single-chain antibody [see comments]. *Proceedings of the National Academy of Sciences, USA* 90: 7889-7893.

Mason HS. Arntzen CJ (1995). Transgenic plants as vaccine production systems. *Trends in Biotechnology* 13: 388-392.

Matthews REF (1991). *Plant Virology* (Third Edition). Academic Press. New York.

McCormick AA. Kumagai MH. Hanley K. Turpen TH. Hakim I. Grill LK. Tuse D. Levy S. Levy R (1999). Rapid production of specific vaccines for lymphoma by expression of the tumor-derived single-chain Fv epitopes in tobacco plants. *Proceedings of the National Academy of Sciences, USA* 96: 703-708.

Modelska A. Dietzschold B. Sleysh N. Fu ZF. Steplewski K. Hooper DC. Koprowski H. Yusibov V (1998). Immunization against rabies with plant-derived antigen. *Proceedings of the National Academy of Sciences, USA* 95: 2481-2485.

Namba S. Ling KS. Gonsalves C. Gonsalves D. Slightom JL (1991). Expression of the gene encoding the coat protein of cucumber mosaic virus (CMV) strain WL appears to provide protection to tobacco plants against infection by several different CMV strains [see comments]. *Gene* 107: 181-188.

Noris E. Accotto GP. Tavazza R, Brunetti A. Crespi S. Tavazza M (1996). Resistance to tomato yellow leaf curl geminivirus in *Nicotiana benthamiana* plants transformed with a truncated viral C1 gene |published erratum appeared in *Virology* 1997, January 20;227(2):519|. *Virology* 224: 130-138.

Porta C, Lomonossoff GP (1998). Scope for using plant viruses to present epitopes from animal pathogens. *Reviews in Medical Virology* 8: 25-41.

Porta C, Spall VE, Loveland J, Johnson JE, Barker PJ, Lomonossoff G (1994). Development of cowpea mosaic virus as a high-yielding system for the presentation of foreign peptides. *Virology* 202: 949-955.

Ragnhammer P, Fagerberg J, Frödin J-E, Hjelm A-L, Lindemalm C, Magnusson I, Masucci G, Mellstedt H (1993). Effect of monoclonal antibody 17-1A and GM-CSF in patients with advanced colorectal carcinoma—Long lasting, complete remissions can be induced. *International Journal of Cancer* 53: 751-758.

Richter L, Kipp PB (1999). Transgenic plants as edible vaccines. In Hammond J, McGarvey P, and Yusibov V (eds.), *Plant Biotechnology—New products and Applications* (Second Edition), Springer-Verlag, Berlin, pp. 159-176.

Russell DA (1999). Feasibility of antibody production in plants for human therapeutic use. In Hammond J, McGarvey P, Yusibov V (eds.), *Plant Biotechnology—New Products and Applications* (Second Edition), Springer-Verlag, Berlin, pp. 119-138.

Schillberg S, Zimmermann S, Voss A, Fischer R (1999). Apoplastic and cytosolic expression of full-size antibodies and Fv fragments in *Nicotiana tabacum*. *Transgenic Research* 8: 255-263.

Scholthof HB, Scholthof K-BG (1996). Plant virus gene vectors for transient expression of foreign proteins in plants. *Annual Reviews of Phytopathology* 34: 299-323.

Schouten A, Roosien J, De Boer JM, Wilmink A, Rosso MN, Bosch D, Stiekema WJ, Gommers FJ, Bakker J, Schots A (1997). Improving scFv antibody expression levels in the plant cytosol. *FEBS Letters* 415: 235-241.

Schouten A, Roosien J, Van Engelen FA, De Jong GAMI, Borst-Vrenssen AWMT, Zilverentant JF, Bosch D, Stiekema WJ, Gommers FJ, Schots A, Bakker J (1996). The C-terminal KDEL sequence increases the expression level of a single-chain antibody designed to be targeted to both the cytosol and the secretory pathway in transgenic tobacco. *Plant Molecular Biology* 30: 781-793.

Shivprasad S, Pogue GP, Lewandowski DJ, Hidalgo J, Donson J, Grill LK, Dawson WO (1999). Heterologous sequences greatly affect foreign gene expression in tobacco mosaic virus-based vectors. *Virology* 255: 312-323.

Siegel A (1985). Plant-virus-based vectors for gene transfer may be of considerable use despite a presumed high error frequency during RNA synthesis. *Plant Molecular Biology* 4: 327-329.

Smith MD, Glick B (1997). Production and applications of plant-produced antibodies. *Food Technology and Biotechnology* 35: 183-191.

Smolenska L, Roberts IM, Learmonth D, Porter AJ, Harris WJ, Wilson TM, Santa Cruz S (1998). Production of a functional single chain antibody attached to the surface of a plant virus. *FEBS Letters* 441: 379-382.

Spiegel H. Schillberg S. Sack M. Holzem A. Nähring J. Monecke M. Liao Y-C. Fischer R (1999). Expression of antibody fusion proteins in the cytoplasm and ER of plant cells. *Plant Science* 149: 63-71.

Takamatsu N, Watanabe Y, Yanagi H, Meshi T, Shiba T, Okada Y (1990). Production of enkephalin in tobacco protoplasts using tobacco mosaic virus RNA vector. *FEBS Letters* 269: 73-76.

Taschner PE, Van der Kuyl AC, Neeleman L. Bol JF (1991). Replication of an incomplete alfalfa mosaic virus genome in plants transformed with viral replicase genes. *Virology* 181: 445-450.

Tavladoraki P. Benvenuto E. Trinea S. DeMartinis D. Galeffi P (1993). Transgenic plants expressing a functional single-chain Fv antibody are specifically protected from virus attack. *Nature* 366: 469-472.

Taylor S. Massiah A. Lomonossoff G. Roberts LM. Lord JM. Hartley M (1994). Correlation between the activities of five ribosome-inactivating proteins in depurination of tobacco ribosomes and inhibition of tobacco mosaic virus infection. *Plant Journal* 5: 827-835.

Turpen TH, Reinl SJ (1998). Tobamovirus vectors for expression of recombinant genes in plants. In Cunnigham C. Porter AJR (eds.), *Recombinant Proteins from Plants: Production and Isolation of Clinically Useful Compounds* (Volume 3), Humana Press, Inc., Totowa, NJ, pp. 89-101.

Turpen TH, Reinl SJ, Charoenvit Y. Hoffman SL. Fallarme V. Grill LK (1995). Malarial epitopes expressed on the surface of recombinant tobacco mosaic virus. *Biotechnology* 13: 53-57.

Usha R. Rohll JB. Spall VE. Shanks M. Maule AJ. Johnson JE. Lomonossoff GP (1993). Expression of an animal virus antigenic site on the surface of a plant virus particle. *Virology* 197: 366-374.

Vaquero C. Sack M. Chandler J. Drossard J. Schuster F. Monecke M. Schillberg S. Fischer R (1999). Transient expression of a tumor-specific single-chain fragment and a chimeric antibody in tobacco leaves. *Proceedings of the National Academy of Sciences, USA* 96: 11128-11133.

Verch T. Yusibov V. Koprowski H (1998). Expression and assembly of a full-length monoclonal antibody in plants using a plant-virus vector. *Journal of Immunological Methods* 220: 69-75.

Voss A. Niersbach M. Hain R. Hirsch HJ. Liao YC. Kreuzaler F. Fischer R (1995). Reduced virus infectivity in *Nicotiana tabacum* secreting a TMV-specific full-size antibody. *Molecular Breeding* 1: 39-50.

Whitelam GC. Cockburn B. Gandecha AR. Owen MRL (1993). Heterologous protein production in transgenic plants. *Biotechnology and Genetic Engineering Reviews* 11: 1-29.

Wilson TM (1993). Strategies to protect crop plants against viruses: Pathogen-derived resistance blossoms. *Proceedings of the National Academy of Sciences, USA* 90: 3134-3141.

Yusibov V. Modelska A. Steplewski K. Agadjanyan M. Weiner D. Hopper DC. Koprowski H (1997). Antigens produced in plants by infection with chimeric plant viruses immunize against rabies virus and HIV-1. *Proceedings of the National Academy of Sciences, USA* 94: 5784-5788.

Zimmermann S, Schillberg S, Liao Y-C, Fischer R (1998). Intracellular expression of TMV-specific single-chain Fv fragments leads to improved virus resistance in *Nicotiana tabacum. Molecular Breeding* 4: 369-379.

Chapter 19

Nucleic Acid Hybridization for Plant Virus and Viroid Detection

Rudra P. Singh
Xianzhou Nie

The discovery of truth and its transmission to others belong together, and their joint exercise can afford satisfactions greater than either one practiced by itself.

Joel Hildebrand

INTRODUCTION

The 1950s was the decade of nucleic acid research. The revolutionary discovery (Watson and Crick, 1953) of DNA structure and its simplistic mode of duplication catalyzed many studies of the chemical and physical behavior of the nucleic acids. The outcome became the foundation for the technology later to be known as *recombinant DNA technology*, or simply *modern biotechnology*. Molecular hybridization is one such by-product of early biotechnology.

Nucleic acid hybridization (NAH) refers to the specific interaction in vitro between two polynucleotide sequences (RNA:RNA, RNA:DNA, or DNA:DNA) by complementary base pairing. The method stems from the fact that the two complementary strands in a nucleic acid duplex or double helix can be denatured and separated by heat and then renatured (annealed, hybridized) under conditions favoring hydrogen bonding of base pairs.

Although nucleic acid hybridization was first demonstrated in 1961, and Gillespie and Spiegelman (1965) introduced a simple filter paper method for it, the method did not achieve wide use until the early 1980s. A series of arti-

The authors wish to thank Ms. Shelley Tucker for typing the various drafts of this manuscript.

cles reviews the early application of NAH in medicine (Meinkoth and Wahl, 1984; Coghlan et al., 1985; Kulski and Norval, 1985; Matthews et al., 1985) and in plant virology (Gould and Symons, 1983; Maule et al., 1983; Karjalainen et al., 1987; McInnes and Symons, 1991) with both solution and filter paper hybridizations. In the mid 1980s, techniques of polymerase chain reaction (PCR) for DNA and reverse transcription-PCR (RT-PCR) for RNA became available for disease diagnosis (Saiki et al., 1985). These provide a relatively higher sensitivity of disease detection than NAH (Van der Wilk et al., 1994; Canning et al., 1996; Singh, 1998a). However, because of the proprietary rights of PCR methodology, commercial exploitation of PCR requires payment of license fees to patent owners. This may be prohibitive to small agricultural disease-diagnosing laboratories dealing with several crops and many diseases of individual crops. Although about 80 items of nucleic acid hybridization technology are also patented (Krika, 1992), it remains freely available. Despite its lower detection sensitivity compared to PCR (Singh, 1998a), there is renewed interest in NAH, and it is becoming the method of choice for small commercial laboratories.

This chapter offers a general appraisal of the various methods using NAH for virus and viroid detection. It aims to provide a methodological framework but leaves the details of the appropriate protocol to be developed and adopted by local researchers in their own regions. NAH has been successfully adopted in a remote field station in the United States without using specialized blotting apparatus, a vacuum for blotting or baking, or an incubator with temperature control, and instead using only periodic agitation by hand (Harper and Creamer, 1995). Several excellent books (Hames and Higgins, 1985; Wilkinson, 1992; Krika, 1992; Kaufman et al., 1995) provide accounts of methodologies, steps, and reagent preparation protocol. Some reviews (Singh, 1989; McInnes and Symons, 1991; Nikolaeva, 1995) provide details of the methods and their application to plant viruses and viroids.

THE PRINCIPLE OF THE NUCLEIC ACID HYBRIDIZATION METHOD

Nucleic acid molecules consist of a sequence of four component nucleotides (A, C, G, and T or U). The nucleotide sequence embodies the specificity to a particular virus or viroid. Therefore, any method that can readily differentiate between sequences should be able to identify a virus or viroid and its strains or isolates.

Molecular hybridization entails a specific binding of two strands of nucleic acids. These can be two RNAs, two DNAs, or one strand of RNA to one strand of DNA. The binding of nucleic acid could take place in solution

(solution hybridization) (Britten and Kohne, 1968), it could use filter paper or membrane *(filter hybridization)* (Gillespie and Spiegelman, 1965), or it could happen on the tissue itself *(in situ hybridization)* (Gall and Pardue, 1971). Depending on the type of nucleic acid separation and their transfer to solid supports, the hybridization method can be termed *Southern transfer hybridization* (Southern, 1975) for DNA, and *northern transfer hybridization* (Alwine et al., 1977) for RNA. Several major steps are involved for widely used solid-support hybridization. These include preparation of probes, sample preparations, sample denaturation application and its immobilization on the membranes, prehybridization and hybridization, washing of the membranes, and the detection of hybridized probes.

Preparation of Probes

About 5 percent of plant viruses have DNA genomes (either single- or double-stranded), 3 percent have double-stranded (ds) RNA, while the remaining 92 percent have single-stranded (ss) RNA genomes (Harrison, 1985). All viroids sequenced so far are single-stranded, covalently closed circular RNA (Singh, 1998b). Therefore, for the detection of plant viruses and viroids, reverse transcription of RNA genomes is needed, which is performed by the RNA-dependent DNA polymerase (reverse transcriptase).

One of the most important aspects in the detection of viruses and viroids is the use of labeled probes (the labels of which are detected easily) that specifically hybridize with their complementary strands. The probes with high specific activity are thus critical in NAH. Various methods for the preparation of DNA and RNA probes are described by Kaufman et al. (1995).

Methods Used for Double-Stranded DNA Probes

Nick Translation

This method was developed for labeling probes with radioisotopes (Rigby et al., 1977). Other labels were later incorporated by this method. The procedure involves using pancreatic DNase I to create single-stranded nicks in the dsDNA, followed by the 5'- to 3'-exonucleolytic action of *Escherichia coli* DNA polymerase I to remove the stretches of single-stranded DNA (ssDNA) starting at the nick and to replace them with new nucleotides. In the presence of Mg^{2+} ions, the DNase generates single-stranded nicks. During the incorporation of new nucleotides, one of the four deoxyribonucleotides (dNTPs) could be radioactively labeled with $\alpha^{32}PdATP$ or a $^{32}PdCTP$ or nonradioactively labeled with biotin or digoxigenin. Examples of the nick translation method used for preparing probes for plant viroids and viruses include *Potato spindle tuber viroid* (PSTVd, genus

Pospiviroid; Pringle, 1999) (Owens and Diener, 1981; Lakshman et al., 1986) and several viruses (Maule et al., 1983).

Random Primers

In a method developed by Feinberg and Vogelstein (1983), labeling is mediated by a Klenow fragment (the large fragment of *E. coli* DNA polymerase I). Random primers (a complete DNase I digest of calf thymus DNA, or a population of random hexanucleotides synthesized by an automatic DNA synthesizer containing all four bases in every position are available from several commercial companies) are used to prime DNA synthesis in vitro from any denatured closed circular or linear dsDNA as a template using the Klenow fragment (Taylor et al., 1976). The DNA product is the exclusive result of primer extension, in the presence of four dNTPs, one of which could be radioactively (^{32}P) or nonradioactively (digoxigenin) labeled. It has been shown that the concentration of primers determines the mean probe size. When primer concentrations are high, the probes are shorter (Hodgson and Fisk, 1987). Therefore, it is possible to manipulate an optimal probe size by varying the primer concentration. Examples of random-priming method usage include *Avocado sunblotch viroid* (ASBVd, genus *Avsunviroid;* Flores, 1986) and *Tomato spotted wilt virus* (TSWV, genus *Tospovirus;* De Haan et al., 1991).

Methods for Single-Stranded DNA and RNA Probes

Single-stranded probes eliminate the potential formation of nonproductive hybrids of reannealed probes because ssDNA or ssRNA probes have only one complementary strand of a given DNA or RNA sequence. The ssDNA probes can be prepared using bacteriophage M13 or the phagemid vector, or cDNA (complementary DNA) reversely transcribed from virus or viroid RNA.

For the M13 universal-sequencing primer system, the DNA samples are cloned in one of the "even" series of M13 vectors, e.g., M13 mp8. The universal probe primer sequence (5'- GAAATTGTTATCC-3') is complementary to the 5'- side of the polycloning site of the family of M13 vectors. It is used to initiate synthesis of the minus-strand from the plus-strand template by the Klenow fragment. Either labeling method (random primer or nick translation) can be used. Thus, the inserted probe sequence remains single stranded and does not require denaturation for the hybridization analysis. Various potato virus cDNA probes have been prepared using this method (Hopp et al., 1991).

An in vitro transcription system is used for labeling RNA. A series of plasmid vectors with polycloning sites are available for the preparation of an ssRNA probe. The polycloning sites are located downstream from the bacteriophage promoters SP6, T3, or T7 in the vector, and the cDNA inserts are cloned between these promoters. The inserts are transcribed in vitro into single-stranded sense or antisense RNA from linear plasmid DNA with the promoters SP6, T3, or T7. During transcription, one of the ribonucleotides (rNTPs) is radioactively or nonradioactively labeled and becomes incorporated into the RNA strand, resulting in a labeled RNA probe. The labeled RNA probe usually has a high specific activity and is much "hotter" than the ssDNA probe (Kaufman et al., 1995).

End Labeling

The 5'-end labeling using bacteriophage T4 polynucleotide kinase is suitable for labeling RNA and DNA, including very small pieces. In this method, the γ phosphate of ATP is transferred to a free 5'-OH group in either DNA or RNA. The enzyme also has phosphatase activity. In the forward reaction, it catalyzes the phosphorylation following removal of 5'-terminal phosphate with alkaline phosphatase, while in the exchange reaction, it catalyzes the exchange of 5'-phosphate with the γ-phosphate of ATP. Both reactions are most efficient with DNA that has the protruding 5'-terminus (Arrand, 1985). Polynucleotide kinase-mediated, ^{32}P-labeled ds-probes have been used for the northern blot hybridization of *Cucumber mosaic virus* (genus *Cucumovirus*) satellite (sat) RNA sequences (Rosner et al., 1983).

The 3'-end labeling is performed by terminal deoxynucleotidyl transferase, which adds deoxyribonucleotides onto the 3'-ends of DNA fragments (Roychoudhury et al., 1979). Both ssDNA and dsDNA are substrates for this enzyme. The PSTVd probes labeled with biotin 7-ATP have been prepared by this method (Roy et al., 1989). In addition, the 3'-end can also be labeled using a Klenow fragment, which has 5' → 3'-polymerase activity. This is particularly useful for filling in the 3'-ends created by the restriction enzymes *Bam* HI, *Hind* III, and *Eco* RI. Depending on the sequence of the protruding 5'-ends of the DNA, nucleotides must be appropriate to fill in the stretches of labeled DNA. If a high specific activity is required, more than one of the α^{32}P-labeled dNTPs can be provided in the reaction mixture.

Polymerase Chain Reaction (PCR) Labeling of Probes

The labeling of probes by the PCR method yields highly specific double-stranded and single-stranded probes. RNA is used in the PCR after reverse transcription. By carrying out asymmetric PCR the target DNA has been shown to yield exclusively single-stranded fragments (Gyllensten and Erlich, 1988). The method requires specific primers for the amplification of the target nucleic acid. Alternatively, the DNA fragment can be cloned into an appropriate vector, using universal primers. Radioactive or nonradioactive labels can be incorporated into the amplified fragments. However, when using digoxigenin, the template that has been found suitable is prepared first by the normal PCR and purified, then used for the incorporation of the label in the second PCR. An RT-PCR-incorporated cDNA probe with digoxigenin has been used for *Potato leaf roll virus* (PLRV, genus *Polerovirus*) (Tautz et al., 1992; Loebenstein et al., 1997).

TYPES OF LABELS

Two types of labels have been used for the preparation of probes. Early studies (Owens and Diener, 1981; Macquaire et al., 1987; Salazar et al., 1988) used radioactive labels consisting of ^{32}P, an isotope of choice because of its high energy that converts into short scintillation counting times and short autoradiographic exposures. All four dNTPs are available in an α-^{32}P-labeled form of high and low specific activity. In addition, γ^{32}P ATP is also available for the 5'-end labeling of RNA and DNA. Similarly, 3'-end labeling of ssDNA can be performed with a single α-^{32}P-cordycepin-5'-triphosphate lacking 3'- OH. However, ^{32}P dNTPs have a short half-life of about 14 days and probes have to be prepared more often. The storage life can be extended in some cases if ^{32}P DNA probes are stored in a buffer containing 5 mM of 2-mercaptoethanol (Barker et al., 1985).

In recent years, the nonradioactive labels biotin and digoxigenin (Forster et al., 1985; McInnes et al., 1989; Roy et al., 1989; Candresse et al., 1990; Hopp et al., 1991; Singh et al., 1994) have also been incorporated into DNA and RNA probes (see Table 19.1). The probes labeled with nonradioactive tags have the following features: (1) they have none of the hazards associated with the use of radioisotopes; (2) the probes are stable for several months; (3) higher probe concentrations can be used without causing high background problems; (4) detection is rapid, i.e., within three hours; and (5) the reagents are inexpensive.

Biotin and digoxigenin are haptens that attach to the nucleic acids. Hapten is detected using a labeled specific binding protein (e.g., antibiotin or antidigoxigenin). Conjugates of avidin or antibiotin and antidigoxigenin antibodies with enzymes or fluorescent groups have been used for the detection of biotin- or digoxigenin-labeled molecules. The principle of biotin-labeled probes is based on the incorporation of biotinylated nucleotides (bio-dUTP and bio-dCTP) into DNA or RNA by nick translation (Rigby et al., 1977) or random priming (Feinberg and Vogelstein, 1983), detection of

TABLE 19.1. Detection of Viruses and Viroids by Nucleic Acid Hybridization, 1990-1997

Virus/Viroids	Probe	Nature of Sample	Sensitivity	References
Viruses				
PVX	Biotin-dsDNA	PVX RNA	50 fg	Eweida et al., 1990
PVY	Biotin-dsDNA*	Nucleic acid, potato	600 pg	Audy et al., 1991
	Digoxigenin-dsDNA*	Nucleic acid, potato	60 pg	Audy et al., 1991
TSWV	³²P-cRNA	TWSV RNA	0.1 pg	De Haan et al., 1991
PVY				
PVX	Biotin-cDNA	Nucleic acid, plants	1.5 pg	Hopp et al., 1991
PLRV				
PVYn	Digoxigenin-dsDNA	PVYᴺ RNA	5 pg	Dhar and Singh, 1994
PVY	Digoxigenin-dsDNA	PVYᴼ RNA	10 pg	Singh and Singh, 1995
SLCV	Digoxigenin-dsDNA*	Nucleic acid, plants	less than ³²P	Harper and Creamer, 1995
BCTV				
LIYV				
ZYMV				
PLRV	Digoxigenin-cRNA	Nucleic acid, plants	1 pg	Loebenstein et al., 1997
Viroids				
CSVd	Biotin-cRNA*	CSVd RNA	5 pg	Candresse et al., 1990
PSTVd	Biotin-cRNA	PSTVd RNA	5 pg	Candresse et al., 1990
PSTVd	Biotin-cDNA	Nucleic acid, plants	1.5 pg	Hopp et al., 1991
PSTVd	Digoxigenin-cDNA. multi.*	PSTVd RNA	2.5 pg	Welnicki and Hiruki, 1992
PSTVd	Digoxigenin-cDNA	PSTVd RNA	2.0 pg	Podleckis et al., 1993
ASSVd	Digoxigenin-cDNA	ASSVd RNA	2.0 pg	Podleckis et al., 1993
PSTVd	Digoxigenin-cDNA. multi.	PSTVd RNA	0.48 pg	Singh et al., 1994

Note: BCTV = *Beet curly top virus;* LIYV = *Lettuce infectious yellows virus;* PLRV = *Potato leaf roll virus;* PVX = *Potato virus X;* PVY = *Potato virus Y,* strain N (PVYᴺ), strain O (PVYᴼ); SLCV = *Squash leaf curl virus;* TSWV = *Tomato spotted wilt virus;* ZYMV = *Zucchini yellow mosaic virus;* ASSVd = *Apple scar skin viroid;* CSVd = *Chrysanthemum stunt viroid;* PSTVd = *Potato spindle tuber viroid.*
• ³²P probes have been used for comparison in these studies.

the label by the streptavidin-alkaline phosphatase reaction, or by use of antibiotin antibodies. A range of biotinylated nucleotides (Bio-11-dUTP, Bio-16-dUTP, Bio-7-dATP, Bio-17-dCTP, Biotin-21-dUTP) are commercially available. A photo-activatable analogue of biotin (Photobiotin VII) has also been used. Probes are prepared from commercially available Photobiotin acetate. Upon brief irradiation with visible light, Photobiotin forms stable linkages with nucleic acids, and the procedure can be used for rapid, small- and large-scale preparation of stable probes (Forster et al., 1985).

The steroid hapten digoxigenin has also been used. In this system, the digoxigenin-labeled dUTP in cDNA or cRNA (complementary RNA) probes is detected by immunoassay, using an antidigoxigenin-alkaline phosphatase conjugate to the digoxigenin moiety and hydrolysis of the particular enzyme substrate, which results in the production of a color (Welnicki and Hiruki, 1992; Podleckis et al., 1993; Dhar and Singh, 1994; Singh et al., 1994; Singh and Singh, 1995; Singh et al., 1999).

SAMPLE PREPARATION

The method of sample preparation depends on several factors and can vary with plant species, cultivars, and plant organs (e.g., leaf, tuber, or corm); the type of viruses and their location in the tissues (e.g., phloem, xylem) and cells (nucleus); the concentration of virus and viroids; and the ease of virus and viroid release from the tissues. Thus, two types of sample preparations have been extensively used. In one case, buffered sap extracts were spotted on a membrane, while in others, nucleic acid extracts were prepared. Sap extracts diluted with buffer have been used for several viruses (Bar-Joseph et al., 1983; Maule et al., 1983; Nikolaeva et al., 1990; LeClerc et al., 1992). Depending upon the virus, the buffer solution may contain antioxidants, ribonuclease inhibitors, virus-stabilizing agents, and enzymes. In other cases, high-ionic-strength buffers with antioxidants and detergents have been used to release the nucleic acids from the nuclei (Owens and Diener, 1981; Barbara et al., 1990). In a few cases, simple squashing or squeezing of the leaf or insect samples onto a membrane has been sufficient to provide good hybridization signals (Baulcombe et al., 1984; Czosnek et al., 1988). However, some plant extracts have been shown to reduce the sensitivity of virus detection up to 20 percent. In the case of DNA viruses, this inhibition could be reversed with dilution of the sap, although this approach may not be enough for some RNA viruses (Maule et al., 1983). Crude sap extracts from healthy and infected sources treated with 1 percent sodium dodecyl sulfate (SDS) have also been shown to respond differently to labeled radioactive or nonradioactive probes. Nonspecific background coloration was observed with biotin-labeled probes but not with those labeled with radioisotopes (Roy et al., 1988).

In the majority of viruses, use of crude sap extracts or clarified sap has been unsatisfactory and has required partial purification and concentration of nucleic acids (Flores, 1986; Candresse et al., 1988; Harper and Creamer, 1995; Singh and Singh, 1995). Several methods for nucleic acid extractions have been described. Some are rapid, with nucleic acids being available for spotting within a few (three to four) hours, i.e., proteinase K treatment followed by extraction with phenol and chloroform and precipitation of nucleic acids by ethanol (Singh and Singh, 1996a). Other methods involve many steps and require up to two days to prepare nucleic acid extracts (Yang et al., 1992; Barker et al., 1993). Methods that do not involve organic solvents are also available for nucleic acid preparation from plant and insect tissues. For example, a method (Spiegel and Martin, 1993) using Tris-HCl buffer, supplemented with lithium dodecyl sulphate, lithium chloride, ethylene diamine tetra-acetic acid (EDTA), sodium deoxycholate, detergent, and antioxidant has been suitable for tuber tissues (Singh and Singh, 1995) and aphid tissues (Singh et al., 1995; Naidu et al., 1998). An ultrasimple procedure for the extraction of nucleic acids from potato plants, tubers, and aphids has recently been described (Singh, 1999). This consists of soaking samples of plant tissue or aphids in a 0.5 percent solution of a detergent (Triton X-405R), incubating the sample at 37° C for 30 min, and centrifuging the content. The supernatant can be used directly for nucleic acid hybridization or RT-PCR.

Many metabolic products occur in plants that interfere with the extraction of intact nucleic acids and mask the true hybridization signals (Newberry and Possingham, 1979; Rowhani et al., 1993; Harper and Creamer, 1995). To remove these interfering compounds, nucleic acid extraction procedures have been modified to involve single- to multistep protocols (Rowhani et al., 1993). Harper and Creamer (1995) found that pigments and other interfering substances from bean, cucumber, tomato, and zucchini plants were eliminated with phenol-chloroform treatment. We have found that the polyphenols, mainly chlorogenic acid or its polymerized products, which occur in most plants and interfere with nucleic acid extraction, can be neutralized by using citric acid in the extraction buffer (Singh et al., 1998).

Sample Support Medium

In the beginning, the nucleic acid hybridization reaction was carried out in liquid medium and was termed *solution hybridization* (Jayasena et al., 1984). The presence of hybridization products in solution hybridization was estimated by S1 nuclease digestion. Although it is suitable for the quantitative measurement of sequence complexity and composition, it is labor intensive and not suitable for large-scale testing. In contrast, filters or membranes, the so-called *solid-support hybridization*, generally referred to as *dot-blot* or *spot hybridization*, are ideally suited to the analysis of multiple samples. These allow the replication of membranes for many sequences, with different probes and under different conditions of hybridization and washing. This process is much more sensitive than solution hybridization of RNA probes. In solid support, denatured DNA and RNA are immobilized on a nitrocellulose or nylon membrane in such a way that self-annealing is prevented, yet bound sequences are available for hybridization with an added nucleic acid probe.

Nitrocellulose efficiently binds DNA and RNA, except for very small fragments of fewer than 300 nucleotides (nt). However, the nitrocellulose membranes are very fragile and require high ionic strength for quantitative binding of both DNA and RNA. The binding efficiency is much reduced at low ionic strength (Southern, 1975). On the other hand, nylon membranes are more pliable, easier to handle, and can be used indefinitely. A pore size of 0.45 μm is suitable for large molecules and of 0.22 μm for molecules of fewer than 500 nt. RNA membranes with pore sizes of 0.1 to 0.22 μm have been found to be more efficient.

Sample Application and Immobilization

For the application, 1 to 5 μl samples of sap or nucleic acids are spotted next to each other in dots of uniform diameter (see Figure 19.1). Occasionally, as much as 25 μl per sample has been used (Harper and Creamer, 1995). For quantitative determination, known amounts of DNA dots are applied in an identical way to the same membrane. Commercial apparatuses are available for the uniform application of multiple samples. Nylon membranes do not require pretreatment. However, nitrocellulose membranes must be presoaked in water for 5 min, then in 20x standard saline citrate |SSC| solution (1x SSC = 0.15M sodium chloride |NaCl| and 0.015M sodium citrate) and dried (Thomas, 1980). After sample application, membranes are air-dried or dried under a lamp and baked in vacuo at 80°C for 2 h to immobilize the target nucleic acid sequence.

Denaturation of Samples

DNA and RNA can be denatured prior to or after application to membranes. DNA and dsRNA are denatured by heating or by alkali treatment before sample spotting onto a membrane. Several chemical denaturants for RNA have been described: dimethyl sulphoxide (Bantle et al., 1976), methyl mercuric hydroxide (Bailey and Davidson, 1976), formaldehyde (Lehrach et al., 1977), and glyoxal (McMaster and Carmichael, 1977). However, since researchers have found variation in the response to hybridization with each denaturant, they should be evaluated for each virus-host combination.

FIGURE 19.1. Dot-Blot Detection of Three Preparations, with a Digoxigenin-Labeled Probe

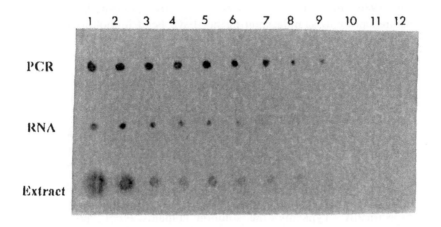

Note: Potato spindle tuber viroid RNA amplified by PCR, hybridized to a twofold dilution; RNA = purified viroid diluted twofold and hybridized; extract = total nucleic acids extracted from the leaves, diluted twofold and hybridized.

Maule et al. (1983) found that encapsidation of viral nucleic acid was not a barrier to either binding to nitrocellulose or hybridization with probes, although, in their studies, it was shown that alkali treatment increased the sensitivity to *Tobacco mosaic virus* (genus *Tobamovirus*) up to tenfold and reduced the sensitivity to *Cowpea mosaic virus* (CPMV, genus *Comovirus*) by 90 percent. When sap was treated with formaldehyde, a slight increase in sensitivity was observed to CPMV and *Pea enation mosaic virus-1* (genus *Enamovirus*). They also showed that the presoaking of nitrocellulose membranes in a high ionic buffer prior to the sample application increased sensitivity to several viruses (Maule et al., 1983). Denaturation with formaldehyde has been shown to be 32 times more effective than formamide for leaf tissues infected with PLRV (Smith et al., 1993). Thus, ample evidence indicates that some sort of denaturation of nucleic acids appears essential for reliable detection of viruses. Similarly, viroids, being single-stranded circular RNA, are thought to be "rodlike" in vivo (Reisner, 1991) and to have benefited from the denaturation step, e.g., heating at 100°C for 3 min (Owens and Diener, 1981) and denaturation with 7.4 percent formaldehyde (Candresse et al., 1988) or 8 percent glyoxal for 5 min at 60°C (Roy et al., 1989). For ASBVd, treatment of the samples with formaldehyde was shown to be better than heating them at 100°C for 3 min in 50 percent formamide followed by rapid cooling (White and Bancroft, 1982; Garger et al., 1983; Flores, 1986). In our laboratory, RNAs for virus and viroid have not been denatured prior to sample spotting but have been immobilized through cross-linking by UV exposure for 30 sec to 3 min over a UV trans-illuminator (Dhar and Singh, 1994; Singh et al., 1994). We have also found that RNA prepared from tuber tissues, denatured by formaldehyde (37 percent) and 20x SSC and incubated at 60°C for 15 min, can be stored at various temperatures (4, -20, and -70°C) and remain suitable for hybridization for up to 15 days (Singh and Singh, 1995).

PREHYBRIDIZATION

In the prehybridization step, the membrane is incubated in a solution that is designed to precoat all the sites on its surface that could bind the probe nonspecifically. Typically, the solution contains Ficoll, polyvinylpyrrolidone, bovine serum albumin (Denhardt's solution; Denhardt, 1966), and heterologous DNA. The use of carrier DNA (e.g., calf thymus DNA or salmon sperm DNA) may not be necessary in all cases, or it could be replaced with other RNAs. For example, Flores (1986) has obtained highly reproducible results with the addition of fractions of nucleic acids, which are insoluble in 2M lithium chloride (LiCl) (mostly ribosomal RNA of high molecular weight) when substituted for calf thymus DNA in ASBVd detection. Prehybridization is mostly carried out at 37 to 42°C in the presence of 50 percent formamide for 16 to 20 h, with some exceptions, which include a prehybridization step of 2 h or elimination of the step (Flores, 1986), or it could be as high as 75°C in the presence of formamide (Barbara et al., 1990) for the separation and detection of *Hop latent viroid* (HLVd) and *Hop stunt viroid* (HSVd).

HYBRIDIZATION

It is necessary to ensure that the added probe is single stranded. This is achieved by boiling or by denaturing the dsDNA probes in alkali. In most cases, hybridization is carried out in the presence of formamide. By including 30 to 50 percent formamide in the hybridization solution, the incubation temperature is reduced to 30 to 42°C. Thus, at this lower temperature the probe is more stable. Non-covalently bound nucleic acids on nitrocellulose membranes are better retained (Anderson and Young, 1985). An inert polymer such as dextran sulphate has been shown to increase hybridization (Wahl et al., 1979). For a single-stranded probe, the rate of hybridization could increase three- to fourfold, but for a double-stranded probe, it could approach 100-fold. Hybrid yields have also increased. In both cases, most of the increase is due to the formation of concatenates, i.e., extensive networks of reassociated probes, which by virtue of single-stranded regions hybridize to membrane-bound nucleic acids and so lead to an overestimation of the extent of hybridization (Anderson and Young, 1985). Solutions containing dextran sulphate are viscous and produce high surface tension on contact with glass surfaces and are thus not suitable for in situ hybridization (Lewis and Wells, 1992). The distantly related sequences can be distinguished by incubation at low temperatures, whereas it should be possible to distinguish closely related sequences by hybridization at high temperatures. For plant viruses and viroids, hybridization temperatures have ranged from 37° (Bar-Joseph et al., 1983) to 75°C (Barbara et al., 1990).

WASHING PROCEDURES

Hybrid stability is directly related to the Tm (melting temperature). The Tm is influenced by base composition, salt concentration, presence of formamide, fragment length, nature of nucleic acid within the hybrid (DNA:DNA, RNA:DNA, RNA:RNA), and mismatch formation. The Tm of RNA:DNA hybrids is 10 to 15°C higher than that of DNA:DNA hybrids, and that of RNA:RNA hybrids is 20 to 25°C higher. To lower the Tm of the hybrids, formamide is used in the hybridization solutions. The specificity of hybrid formation is greatly influenced by the stringency of the final washing steps after hybridization, by increasing the temperature to within 5 to 15°C below Tm, and by lowering the salt concentrations from 2x SSC to 0.1x SSC. The optimal washing temperature has to be experimentally determined for each virus or viroid hybridization. In general, a high-stringency condition consists of 2x SSC and 0.1 percent SDS for 15 min at room temperature, repeated once. This is followed by the same washing solution for 20 min at 65°C with slow shaking, repeated two to four times. A low-stringency condition is similar to the first step of the high-stringency condition but for only 10 min, while the second step is similar but for 15 min at lower temperatures (50 to 55°C).

DETECTION PROCEDURES

After washing, the membranes are processed for the visible result, depending on the nature of the label. Radioactive membranes are visualized through autoradiography at 4°C (Flores, 1986) or generally at −70°C using an XAR film with an intensifying screen (Nikolaeva, 1995). For nonradioactive labels, colorimetric visualization of immunodetected nucleic acid is carried out with nitroblue tetrazolium (NBT) and 5-bromo-4-chloro-3-indolyl-phosphate (BCIP) reagents, following the manufacturer's protocol. Chemiluminescent detection membranes are placed in two sheets of clear acetate while still wet, and the Lumi-phos 530 and, more recently, the CSPD or CDP-star (Bohringer, Mannheim) is added slowly to the membrane, distributed to the whole surface, and incubated overnight for the Lumi-phos 530 or for 10 to 15 min for the CSPD or the CDP-star. Results are visualized and the membrane is photographed (Roy et al., 1988; Singh et al., 1994; Harper and Creamer, 1995). For quantitative detection, the spots can be cut out and counted by scintillation counter for radioactive labels or by densitometer, using various software packages (Candresse et al., 1988; Singh et al., 1996; Singh and Singh, 1997).

SENSITIVITY OF THE DETECTION

Sensitivity of the detection of a virus or viroid hybrid is affected by several factors. Among these are size of the probes, location of the probes on the target molecule, the medium of support, the types of nucleic acids, and their separation protocols.

For potato viruses, the detection sensitivity by NAH had ranged from 600 pg RNA (Audy et al., 1991) to as low as 50 fg RNA (Eweida et al., 1990) (see Table 20.1). We have found that the size of the probe can affect sensitivity immensely. When probe sizes of 0.56, 1.2, 2.5, and 3.25 kb from the 3'-end of strain N of *Potato virus Y* (PVY, genus *Potyvirus*) genome were used, the sensitivity with larger-sized probes of 3.25 kb was high. They detected 5 pg RNA, while the smallest probe of 0.56 kb could detect an RNA concentration of only 1,000 pg or more, a greater than 200-fold difference (Dhar and Singh, 1994). Similarly, a full-sized (359 nt) PSTVd-specific probe was about three times more efficient in dot-blot hybridization when compared with the short (19 to 20 nt) synthetic probe (Skrzeczkowski et al., 1986). The 87 bp insert probe was as sensitive as the full-length PSTVd genome (Welnicki et al., 1989), indicating a minimal effect of probe size for viroid detection. When monomeric, dimeric, tetrameric, and hexameric cRNA probes of PSTVd were compared, a two- to thirtyfold increase in sensitivity over the same size cDNA probes was observed in leaf and dormant tuber extracts; the hexameric probes were more sensitive than the monomeric probes (Singh et al., 1994). Thus, viroid probe size effect has not been as high as for viruses (Dhar and Singh, 1994).

The sensitivity of detection may vary depending on the portion of the genome cloned. Nucleic acid probes from the coat protein gene region of the

MAV isolate of *Barley yellow dwarf virus* (genus *Luteovirus*) hybridized to homologous MAV 27 times greater than the PAV isolate, while probes prepared from other parts of the genome hybridized equally with PAV isolates (Fattouch et al., 1990). Similarly, when different parts of the PVY^O genome were tested as a probe with over 30 isolates of PVY^O, PVY^N, and PVY^{NTN}, probes from certain parts appeared more sensitive than others. The 5'-end containing the P1 gene (1.7 kb) and the middle (the cytoplasmic inclusion, 2.3 kb) showed a one- to fourfold increase in sensitivity (Singh and Singh, unpublished data; see Figures 19.2A, B, and C).

Sensitivity due to hybridization can be improved, depending on the types of nucleic acids and their separation. In some cases, DNA samples are first separated by agarose gel electrophoresis and transferred to either nylon or nitrocellulose membranes, and the immobilized DNA is detected after hybridization with suitably labeled probes. This procedure, the *Southern transfer* (Southern, 1975), has been shown to be very sensitive. For example, when the Southern hybridization method was applied to individual whiteflies carrying *Tomato yellow leaf curl virus* DNA, as low as 0.5 pg DNA or 300,000 copies of viral DNA were detected (Zeiden and Czosnek, 1991). Similarly, the increased sensitivity of viroid RNA was observed when RT-PCR-amplified fragments were subjected to Southern hybridization (see Figures 19.3A and B). Conventional RT-PCR detected PSTVd up to a dilution of 1:100 (see Figure 19.3A), while Southern hybridization detected viroids up to a dilution of 1:1,000 (see Figure 19.3B).

When RNA instead of DNA is separated by electrophoresis on an agarose gel under denaturing conditions, transferred to a nylon or nitrocellulose membrane, and hybridized with a labeled probe to detect RNA species, it is termed *northern transfer* (Alwine et al., 1977). A dramatic increase in sensitivity of RNA detection due to northern hybridization can be demonstrated. For example, no bands were detected with a viroid dilution of 1:100 from the gel containing ethidium bromide (see Figure 19.4A), while after transfer to the nylon membrane and hybridization with a digoxigenin labeled probe, all viroid bands were visible up to a dilution of 1:100 (Figure 19.4B).

The type of label also affects sensitivity. Audy et al. (1991), working with three different potato viruses (*Potato virus S* [PVS, genus *Carlavirus*], *Potato virus X* [PVX, genus *Potexvirus*], *Potato virus Y* [PVY, genus *Potyvirus*]), have shown that colorimetric detection of digoxigenin-labeled DNA gave a level of sensitivity of 60 pg RNA, similar to autoradiography of ^{32}P-labeled probes. However, sulfonated-, biotinylated-, and peroxidase-labeled probes

FIGURE 19.2. (A) Schematic Representation of the RNA of Strain O of *Potato virus Y* (PVYᵒ) with Nucleotides (Upper) and Deduced Amino Acids (Lower) Numbers for Each Gene; (B) Clones Used in Sequencing in Relation to the Representation of the PVYᵒ Genome; (C) The Highest Dilution Factor of the PVYᵒ-Infected Leaf Nucleic Acid Extracts, Detected by Each Clone When Used As Detection Probes in Nucleic Acid Hybridization

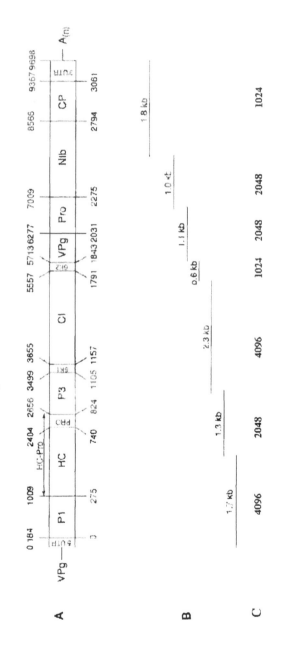

Note: P1-Pro = protease; HC-Pro = helper component proteinase; P3 = P3 protein; 6K1 = 6kDa Protein 1; CI = cylindrical inclusion protein; 6K2 = 6kDa Protein 2; VPg = genome-linked virus protein; NIb = nuclear protein b; CP = coat protein; 3'-UTR = 3'-untranslated region.

Source: Based on Singh and Singh, 1996b.

detected only 600 pg of viral DNA. Similarly, Hopp et al. (1991) have shown that sscDNA probes labeled with biotin detected 1.5 to 6 pg of viral nucleic acid, which is between 20 to 50 times more sensitive than the standard enzyme-linked immunosorbent assay (ELISA) detection. Recently, using an RT-PCR-amplified probe of PLRV labeled with digoxigenin, researchers reported a sensitivity of 2,000 times more than the ELISA (Loebenstein et al., 1997).

Plant species are known to differ in virus concentration. This has been reflected also by an NAH assay of TSWV in various plants. TSWV could be detected in 16 ng of total RNA from tobacco but required 80 ng from tomato and lettuce, and 400 ng from chrysanthemum and pepper plants (Rice et al., 1990).

Variation in sensitivity has been noted also with the type of membrane used for spotting the samples. Nitrocellulose was found more sensitive for ASBVd (Barker et al., 1985; Flores, 1986), while Hybond N+ and Greenscreen-plus were better for TSWV (De Haan et al., 1991). Even the sides of the membrane surface (A or B as defined by manufacturer) could make a difference in sensitivity of detection of hop viroids, namely, HLVd and HSVd (Barbara et al., 1990).

APPLICATION TO VIRUSES AND VIROIDS

Besides detecting and identifying a virus or viroid, NAH technology has been put to other uses. Identification of various DNA and RNA species from the infected sources can be carried out using both Southern and northern hybridization. A few studies have advanced the use of hybridization to establish the presence of viruses and viroids in a particular organ as well as to learn of their distribution and long-distance movement. The techniques have been variously termed *tissue-blot* (Abad and Moyer, 1992; Podleckis et al., 1993) and *tissue-print hybridization* (Más and Pallás, 1995). The detection probes can be radioactive (Abad and Moyer, 1992) or nonradioactive digoxigenin (Podleckis et al., 1993; Más and Pallás, 1995). Both chemiluminescent and colorigenic visualization systems can be applied to nonradioactive labels. This method does not need nucleic acid extraction, which is its greatest advantage. Simply cut ends of leaves, petioles, stems, tubers, or other parts of the plant can be analyzed for their virus distribution. The samples are pressed against pretreated membrane surfaces (nylon or nitrocellulose), and the membranes are processed similar to those of the dot-blot membrane. Researchers have observed, in studies of viroids (*Citrus exocortis viroid, Chrysanthemum stunt viroid*, HSVd, and ASBVd) that the first three viroids can be detected by tissue blot more reliably from stem ends than from leaves, although the

FIGURE 19.3. Increased Sensitivity of *Potato spindle tuber viroid* Detection in Southern Hybridization

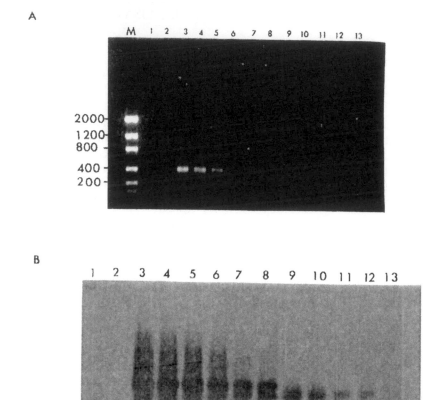

Note: (A) Amplified viroid preparation diluted twofold. Lane M = size markers; lanes 1 and 2 = healthy controls; lane 3 = undiluted; lane 4 = 1:2; lane 5 = 1:4; lane 6 = 1:8; lane 7 = 1:16; lane 8 = 1:32, lane 9 = 1:64; lane 10 = 1:128; lane 11 = 1:200; lane 12 = 1:500, and lane 13 = 1:1000. (B) Same samples as in A, subjected to Southern hybridization.

technique was not reliable for ASBVd (Romero-Durban et al., 1995). Since both PSTVd and *Apple scar skin viroid* have also been detected from various organs of potato, apple, and tomato plants (Podleckis et al., 1993), the failure of ASBVd may reflect differential distribution of viroids in the plant tissues.

Another approach that has not received much attention is the construction of a combination probe for multiple viruses and viroids. This type of probe would be very beneficial for the certification of vegetatively propagated crops or trees for their crop certification or for rapid diagnosis of diseased material at plant quarantine stations. The feasibility of the approach has been demonstrated by constructing a single probe of PSTVd, PVY, and PLRV by using three separate restriction enzyme-incised cDNA fragments and by inserting the three fragments in a vector. Using a nonradioactive digoxigenin label, Welnicki et al. (1994) have shown that one viroid and two viruses (PSTVd, PVY, PLRV) can be successfully detected from potato plants and tubers. Additional work is needed to determine the reliability of this approach with field-grown tubers or plants infected with multiple viruses and also with other virus-plant systems.

CONCLUSIONS

This chapter has reviewed in a simple way the methodology of nucleic acid hybridization and its application to the study of plant viruses and viroids. Attempts were made to provide the essential features of the complete methodology and the various steps involved in NAH. The details of the protocol are not described because several books have been published specifically for that purpose. Examples of the protocols for applications to plant viruses were noted throughout this chapter. It appears that since 1990, the predominant probes are nonradioactive, mainly digoxigenin based. The sensitivity of NAH is in the lower picogram range, and various rapid and reliable methods of sample preparations are available. NAH application to the nucleic acids separated on agarose or polyacrylamide gels and transferred to nitrocellulose or nylon membranes has further improved the sensitivity of detection (see Figures 20.3 and 20.4). The use of tissue-blot or tissue-print hybridization and the combination probe for simultaneous detection of multiple viruses and viroids are areas worth further exploration. However, with the prevailing trend of intellectual property protection, the outflow of information may be curtailed for general usage and may remain concentrated in the hands of some larger disease-testing laboratories.

FIGURE 19.4. Detection of a Northern Blot Following Hybridization with a Digoxigenin-Labeled Probe of *Potato spindle tuber viroid* (PSTVd)

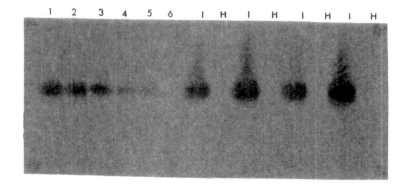

Note: A = RNA detected from a gel containing ethidium bromide; B = same samples as in A; RNA detected after transfer to a nylon membrane and hybridization with a probe; lanes 1-6 = purified PSTVd RNA; lane 1 = undiluted; lane 2 = 1:2; lane 3 = 1:4; lane 4 = 1:10; lane 5 = 1:20; lane 6 = 1:100; I = PSTVd-infected leaf extract; H = healthy leaf extract.

REFERENCES

Abad JA, Moyer JW (1992). Detection and distribution of sweet potato feathery mottle virus in sweet potato by in vitro-transcribed RNA probes (riboprobes), membrane immunobinding assay, and direct blotting. *Phytopathology* 82: 300-305.
Alwine JC, Kemp DJ, Stark GR (1977). Method for detection of specific RNAs in agarose gels by transfer to diazobenzyloxymethyl-paper and hybridization with

DNA probes. *Proceedings of the National Academy of Sciences, USA* 74: 5350-5354.

Anderson MLM, Young BD (1985). Quantitative filter hybridization. In Hames BD, Higgins SJ (eds.), *Nucleic Acid Hybridization a Practical Approach*, IRL Press, Oxford. pp. 73-111.

Arrand JE (1985). Preparation of nucleic acid probes. In Hames BD, Higgins SJ (eds.), *Nucleic Acid Hybridization a Practical Approach*, IRL Press Oxford. pp. 17-45.

Audy P, Parent J, Asselin A (1991). A note on four nonradioactive labelling systems for dot hybridization detection of potato viruses. *Phytoprotection* 72: 81-86.

Bailey JM, Davidson N (1976). Methyl mercury as a reliable denaturing agent for agarose gel eletrophoresis. *Annals of Biochemistry* 70: 75-85.

Bantle JA, Maxwell IH, Hahn WE (1976). Specificity of oligo (dT)-cellulose chromatography in the isolation of polyadenylated RNA. *Annals of Biochemistry* 72: 413-427.

Barbara DJ, Morton A, Adams AN (1990). Assessment of a UK hops for the occurrence of hop latent and hop stunt viroids. *Annals of Applied Biology* 116: 265-272.

Bar-Joseph M, Rosner A, Moscovitz M, Hull R (1983). A simple procedure for the extraction of double-stranded RNA from viral infected plants. *Journal of Virological Methods* 6: 1-18.

Barker H, Webster KD, Reavy B (1993). Detection of potato virus Y in potato tubers: A comparison of polymerase chain reaction and enzyme-linked immunosorbent assay. *Potato Research* 36: 13-20.

Barker JM, McInnes JL, Murphy PJ, Symons RH (1985). Dot-blot procedure with ^{32}P DNA probes for the sensitive detection of avocado sunblotch and other viroids in plants. *Journal of Virological Methods* 10: 87-98.

Baulcombe D, Flavell RB, Boulton RE, Jellis GJ (1984). The sensitivity and specificity of a rapid nucleic acid and hybridization method for the detection of potato virus X in crude sap samples. *Plant Pathology* 33: 361-370.

Britten RJ, Kohne DE (1968). Repeated sequences in DNA. *Science* 161: 529-540.

Candresse T, Macquaire G, Brault V, Monsion M, Dunez J (1990). ^{32}P- and biotin-labelled in vitro transcribed cRNA probes for the detection of potato spindle tuber viroid and chrysanthemum stunt viroid. *Research Virology* 141: 97-107.

Candresse T, Macquaire G, Monsion M, Dunez J (1988). Detection of chrysanthemum stunt viroid (CSV) using nick-translated probes in a dot-blot hybridization assay. *Journal of Virological Methods* 20: 185-193.

Canning ESG, Penrose MJ, Barker I, Coates D (1996). Improved detection of barley yellow dwarf virus in single aphids using RT-PCR. *Journal of Virological Methods* 56: 191-197.

Coghlan JP, Aldred P, Haralambidis J, Niall HD, Penschow JD, Tregar GW (1985). Hybridization histochemistry. *Annals of Biochemistry* 149: 1-28.

Czosnek H, Ber R, Navot N, Zamir D, Antignus Y, Cohen S (1988). Detection of tomato yellow leaf curl in lysates of plants and insects by hybridization with a viral DNA probe. *Plant Disease* 72: 949-951.

De Haan PT, Gielen J, Van Grinsven M, Peters D, Goldbach R (1991). Detection of tomato spotted wilt virus in infected plants by molecular hybridization and PCR. In *Exploring and Exploiting the RNA Genome of Tomato Spotted Wilt Virus*, PhD thesis, Agricultural University, Wageningen, The Netherlands, pp. 93-106.

Denhardt DT (1966). A membrane-filter technique for the detection of complimentary DNA. *Biochemica Biophysica Research Communication* 23: 641-646.

Dhar AK, Singh RP (1994). Improvement in the sensitivity of PVY^N detection by increasing the cDNA probe size. *Journal of Virological Methods* 50: 197-210.

Eweida M, Xu H, Singh RP, Abou Haidar MG (1990). Comparison between ELISA and biotin-labeled probes from cloned cDNA of potato virus X for the detection of virus in crude tuber extracts. *Plant Pathology* 39: 623-628.

Fattouch FA, Ueng PP, Kawata EF, Barbara DJ, Larkins BA, Lister RM (1990). Luteovirus relationships assessed by cDNA clones from barley yellow dwarf viruses. *Phytopathology* 80: 913-920.

Feinberg AP, Vogelstein B (1983). A technique of radio-labelling DNA restriction fragments to high specific activity. *Annals of Biochemistry* 132: 6-15.

Flores R (1986). Detection of citrus exocortis viroid in crude extracts by dot-blot hybridization: Conditions for reducing spurious hybridization results and for enhancing the sensitivity of the technique. *Journal of Virological Methods* 13: 161-169.

Forster AC, McInnes J, Skingle DC, Symons R (1985). Non-radioactive hybridization probes prepared by the chemical labelling of DNA and RNA with a novel reagent: Photobiotin. *Nucleic Acids Research* 13: 745-761.

Gall JG, Pardue ML (1971). In Grossman L, Moldave K (eds.), *Methods in Enzymology* (Volume 21), Academic Press, New York.

Garger SJ, Turpen T, Carrington JC, Morris TJ, Jordan RL, Dodds JA, Grill LK (1983). Rapid detection of plant RNA viruses by dot-blot hybridization. *Plant Molecular Biology Reporter* 1: 21-25.

Gillespie D, Spiegelman S (1965). A quantitative assay for DNA-RNA hybrids with DNA immobilized on a membrane. *Journal of Molecular Biology* 12: 829-842.

Gould AR, Symons RH (1983). A molecular biological approach to relationships among viruses. *Annual Review of Phytopathology* 21: 179-199.

Gyllensten UB, Erlich HA (1988). Generation of single-stranded DNA by the polymerase chain reaction and its application to direct sequencing of the HLA-DQA locus. *Proceedings of the National Academy of Sciences, USA* 85: 7652-7656.

Hames BD, Higgins SJ (eds.) (1985). *Nucleic Acid Hybridization a Practical Approach*. IRL Press, Oxford, England.

Harper K, Creamer R (1995). Hybridization detection of insect-transmitted plant viruses with digoxigenin-labeled probes. *Plant Disease* 79: 563-567.

Harrison BD (1985). Usefulness and limitations of the species concept for plant viruses. *Intervirology* 24: 71-78.

Hodgson CP, Fisk RZ (1987). Hybridization probe size control: Optimized "oligolabelling" *Nucleic Acids Research* 15: 6295.

Hopp HE, Hain L, Bravo-Almonacid F, Tazzini AC, Orman B, Arese AI, Ceriani MF, Saladrigas MV, Celnik R, del Vas M, Mentaberry AN (1991). Development and application of nonradioactive nucleic acid hybridization system for simultaneous detection of potato pathogens. *Journal of Virological Methods* 31: 11-30.

Jayasena KW, Randles JW, Barnett OW (1984). Synthesis of a complementary DNA probe specific for detecting subterranean clover red leaf virus in plants and aphids. *Journal of General Virology* 65: 109-117.

Karjalainen R, Rouhiainen L, Söderlund H (1987). Diagnosis of plant viruses by nucleic acid hybridization. *Journal of Agriculture Science Finland* 59: 179-191.

Kaufman PB, Wu W, Kim D, Cseke LJ (eds.) (1995). *Handbook of Molecular and Cellular Methods in Biology and Medicine.* CRC Press, Boca Raton FL.

Krika LJ (ed.) (1992). *Nonisotopic DNA Probe Techniques.* AP Press, New York.

Kulski JK, Norval M (1985). Nucleic acid probes in diagnosis of viral diseases of man. *Archives of Virology* 83: 3-15.

Lakshman DK, Hiruki C, Wu XN, Lang WC (1986). Use of [^{32}P] RNA probes for the dot-hybridization detection of potato spindle tuber viroid. *Journal of Virological Methods* 14: 309-319.

LeClerc C, Eweida M, Singh RP, AbouHaidar MG (1992). Biotinylated DNA probes for detecting virus Y and aucuba mosaic virus in leaves and dormant tubers of potato. *Potato Research* 35: 173-182.

Lehrach H, Diamond J, Wozney JM, Boedtker H (1977). RNA molecular weight determinations by gel electrophoresis under denaturing conditions, a critical re-examination. *Biochemistry* 16: 4743-4751.

Lewis FA, Wells M (1992). Detection of virus in infected human tissue by in situ hybridization. In Wilkinson DG (ed.), *In Situ Hybridization a Practical Approach,* IRL Press, Oxford, England, pp. 121-136.

Loebenstein G, Akad F, Filatov V, Sadvakasova G, Manadilova A, Bakelman H, Teverovsky E, Lachman O, David A (1997). Improved detection of potato leafroll luteovirus in leaves and tubers with a digoxigenin-labeled cRNA probe. *Plant Disease* 81: 489-491.

Macquaire G, Mansion M, Dunez J (1987). Étude des possibilités de détection précoce du viroide des tubercules fusiformes de la pomme de terre. *Agrinomie* 7: 639-645.

Más P, Pallás V (1995). Non-isotopic tissue-printing hybridization: A new technique to study long-distance plant virus movement. *Journal of Virological Methods* 52: 317-326.

Matthews JA, Batki A, Hynds C, Krika LJ (1985). Enhanced chemiluminescent method for the detection of DNA dot-hybridization assays. *Annals of Biochemistry* 151: 205-209.

Maule AJ, Hull R, Donson J (1983). The application of spot hybridization to the detection of DNA and RNA viruses in plant tissues. *Journal of Virological Methods* 6: 215-224.

McInnes JL, Habili N, Symons RH (1989). Nonradioactive, Photobiotin-labelled DNA probes for routine diagnosis of viroids in plant extracts. *Journal of Virological Methods* 23: 299-312.

McInnes JL, Symons RH (1991). Comparative structure of viroids and their rapid detection using radioactive and nonradioactive nucleic acid probes. In Maramorosch K (ed.), *Viroids and Satellites: Molecular Parasites at the Frontier of Life*, CRC Press, Boca Raton, FL, pp. 21-58.

McMaster GK, Carmichael GG (1977). Analysis of single-stranded nucleic acids on polyacrylamide and agarose gels by using glyoxal and acridine orange. *Proceedings of the National Academy of Sciences, USA* 74: 4835-4838.

Meinkoth J, Wahl G (1984). Hybridization of nucleic acids immobilized on solid supports. *Annals of Biochemistry* 138: 267-284.

Naidu RA, Robinson DJ, Kimmins FM (1998). Detection of each of the causal agents of groundnut rosette disease in plants and vector by RT-PCR. *Journal of Virological Methods* 76: 9-18.

Newberry HJ, Possingham JV (1979). Factors affecting the extraction of intact ribonucleic acid from plant tissues containing interfering phenolic compounds. *Plant Physiology* 60: 543-547.

Nikolaeva OV (1995). Nucleic acid hybridization methods in diagnosis of plant viruses and viroids. In Singh RP, Singh US (eds.), *Molecular Methods in Plant Pathology*, CRC/Lewis Publishers, Boca Raton, FL, pp. 133-144.

Nikolaeva OV, Morozov SY, Zakhariev VM, Skryabin KG (1990). Improved dot-blot hybridization assay for large-scale detection of potato viruses in crude potato tuber extracts. *Journal of Phytopathology* 129: 283-290.

Owens RA, Diener TO (1981). Sensitive and rapid diagnosis of potato spindle tuber viroid disease by nucleic acid spot hybridization. *Science* 213: 670-672.

Podleckis EV, Hammond RW, Hurtt SS, Hadidi A (1993). Chemiluminescent detection of potato spindle tuber and pome fruit viroids by digoxigenin-labelled dot blot and tissue blot hybridization. *Journal of Virological Methods* 43: 147-158.

Pringle CR (1999). Virus taxonomy—1999. The Universal System of Virus Taxonomy, updated to include the new proposals ratified by the International Committee on Taxonomy of Viruses during 1998. *Archives of Virology* 144: 421-429.

Reisner D (1991). Viroids: From thermodynamic to cellular structure and function. *Molecular Plant-Microbe Interactions* 4: 122-131.

Rice DJ, German TL, Mau RFL, Fujimoto FM (1990). Dot-blot detection of tomato spotted wilt virus RNA in plant and thrips tissues by cDNA clones. *Plant Disease* 74: 274-276.

Rigby PWJ, Dieckmann M, Rhodes C, Berg P (1977). Labeling deoxyribonucleic acid to high specificity by nick translation with DNA polymerase I. *Journal of Molecular Biology* 113: 237-257.

Romero-Durban J, Cambra M, Duran-Villa N (1995). A simple imprint-hybridization method for detection of viroids. *Journal of Virological Methods* 55: 37-47.

Rosner A, Bar-Joseph M, Moscovitz M, Mevarech M (1983). Diagnosis of specific viral RNA sequences in plant extracts by hybridization with a polynucleotide kinase-mediated, ^{32}P-labeled, double-stranded RNA probe. *Phytopathology* 73: 699-702.

Rowhani A, Chay C, Golino DA, Falk B (1993). Development of a polymerase chain reaction technique for the detection of grapevine tissue. *Phytopathology* 83: 749-753.

Roy BP, AbouHaidar MG, Alexander A (1988). Construction and use of cloned cDNA biotin and ^{32}P-labeled probes for the detection of papaya mosaic potexvirus RNA in plants. *Phytopathology* 78: 1425-1429.

Roy BP, AbouHaidar MG, Alexander A (1989). Biotinylated probes for the detection of potato spindal tuber viroid (PSTV) inplants. *Journal of Virological Methods* 23: 149-156.

Roychoudhury R, Tu C-PD, Wu R (1979). Influence of nucleotide sequence adjacent to duplex DNA termini on 3'-terminal labeling by terminal transferase. *Nucleic Acids Research* 6: 1323-1333.

Saiki RK, Gelfond DH, Stoffel S, Scharf SJ, Higuchi RG, Horn GT, Mullis KB, Erlich HA (1985). Primer-directed enzymatic amplification of DNA with a thermostable DNA polymerase. *Science* 239: 487-491.

Salazar LF, Balbo J, Owens RA (1988). Comparison of four radioactive probes for the diagnosis of potato spindle tuber viroid by nucleic acid spot hybridization. *Potato Research* 31: 431-442.

Singh M, Singh RP (1995). Digoxigenin-labelled cDNA probes for the detection of potato virus Y in dormant potato tubers. *Journal of Virological Methods* 52: 133-143.

Singh M, Singh RP (1996a). Factors affecting detection of PVY in dormant tubers by reverse transcription polymerase chain reaction and nucleic acid spot hybridization. *Journal of Virological Methods* 60: 47-57.

Singh M, Singh RP (1996b). Nucleotide sequence and genome organization of a Canadian isolate of the common strain of potato virus Y (PVY[n]). *Canadian Journal of Plant Pathology* 18: 209-224.

Singh M, Singh RP (1997). Potato virus Y detection: Sensitivity of RT-PCR depends on the size of fragment amplified. *Canadian Journal of Plant Pathology* 19: 149-155.

Singh M, Singh RP, Moore L (1999). Evaluation of NASH and RT-PCR for the detection of PVY in dormant tubers and its comparison with visual symptoms and ELISA in plants. *American Journal of Potato Research* 75: 61-66.

Singh RP (1989). Molecular hybridization with complementary DNA for plant viruses and viroid detection. In Agnihotri VP, Singh N, Chaube HS, Singh US, Dwivedi (eds.). *Perspectives in Phytopathology*, *Today and Tomorrow's Printers and Publishers*, New Delhi, India, pp. 51-60.

Singh RP (1998a). Reverse-transcription polymerase chain reaction for the detection of viruses from plants and aphids. *Journal of Virological Methods* 74: 125-138.

Singh RP (1998b). Viroids twenty-five years later: Personal reflections. *Topics in Tropical Virology* 1: 131-152.

Singh RP (1999). A solvent-free, rapid and simple virus RNA-release protocol for PLRV detection in aphids and plants by RT-PCR. *Journal of Virological Methods* 83: 27-33.

Singh RP, Boucher A, Lakshman DK, Tavantzis SM (1994). Multimeric non-radio-active cDNA probes improve detection of potato spindle tuber viroid (PSTVd). *Journal of Virological Methods* 49: 221-234.

Singh RP, Boucher A, Sommerville TH, Coleman S (1996). Detection of potato viruses A, M, S, X, Y and leafroll and potato spindle tuber viroid from tissue culture plantlets using single leaf discs. *American Potato Journal* 73: 101-112.

Singh RP, Kurtz J, Boiteau G, Bernard G (1995). Detection of potato leafroll virus in single aphids by the reverse transcription polymerase chain reaction and its potential epidemiological application. *Journal of Virological Methods* 55: 133-143.

Singh RP, Singh M, King RR (1998). Use of citric acid for neutralizing polymerase chain reaction inhibition by chlorogenic acid in potato extracts. *Journal of Virological Methods* 74: 231-235.

Skrzeczkowski LJ, Henning J, Zargorski W, Markiewicz Wkierek R, Welnicki M (1986). The application of synthetic DNA probe for the detection of potato spindle tuber viroid (PSTV). In *Proceedings of International Seminar on Viroids, Plant Viruses and Their Detection*, Warsaw Agricultural University, Warsaw, pp. 107-111.

Smith OP, Damsteegt VD, Keller CJ, Beck RJ (1993). Detection of potato leafroll virus in leaf and aphid extracts by dot-blot hybridization. *Plant Disease* 77: 1098-1102.

Southern EM (1975). Detection of specific sequences among DNA fragments separated by gel electrophoresis. *Journal of Molecular Biology* 98: 503-517.

Spiegel S, Martin RR (1993). Improved detection of potato leafroll virus in dormant potato tubers and microtubers by the polymerase chain reaction and ELISA. *Annals of Applied Biology* 121: 493-500.

Tautz D, Hülskamp M, Sommer RJ (1992). Whole mount in situ hybridization in *Drosophila*. In Wilkinson DG (ed.), *In Situ Hybridization a Practical Approach*, IRL Press, Oxford, England, pp. 61-73.

Taylor JM, Illmensee R, Summers J (1976). Efficient transcription of RNA into DNA by avian sarcoma virus polymerase. *Biochimica and Biophysica Acta* 442: 324-330.

Thomas P (1980). Hybridization of denatured DNA and small RNA fragments transferred to nitrocellulose. *Proceedings of the National Academy of Sciences, USA* 77: 5201-5205.

Van der Wilk F, Korsman M, Zoon F (1994). Detection of tobacco rattle virus in nematodes by reverse transcription and polymerase chain reaction. *European Journal of Plant Pathology* 100: 109-122.

Wahl GM, Stern M, Stark GR (1979). Efficient transfer of large DNA fragments from agarose gels to diazobenzyloxymethyl-paper and rapid hybridization by using dextran sulfate. *Proceedings of the National Academy of Sciences, USA* 76: 3683-3687.

Watson JD, Crick FHC (1953). Molecular structure of nucleic acids: A structure for deoxyribose nucleic acid. *Nature* 171: 737-738.

Welnicki M. Hiruki C (1992). Highly sensitive digoxigenin-labelled DNA probe for the detection of potato spindle tuber viroid. *Journal of Virological Methods* 39: 91-99.

Welnicki M, Skrzeczkowski J, Soltynska A, Jonczyk P, Markiewicz W, Kierzek R, Imiolczyk B. Zagorski W (1989). Characterization of synthetic DNA probe detecting potato spindle tuber viroid. *Journal of Virological Methods* 24: 141-152.

Welnicki M, Zekanowski C, Zagorski W (1994). Digoxigenin-labelled molecular probe for the simultaneous detection of three potato pathogens, potato spindle tuber viroid (PSTVd), potato virus Y (PVY), and potato leafroll virus (PLRV). *Acta Biochim Polonica* 41:473-475.

White BA, Bancroft FC (1982) Cytoplasmic dot hybridization: Simple analysis of relative levels in multiple small cell or tissue samples. *Journal of Biological Chemistry* 257: 8569-8572.

Wilkinson DG (ed.) (1992). *Hybridization a Practical Approach.* IRL Press, Oxford, England.

Yang X, Hadidi A, Garnsey SM (1992). Enzymatic cDNA amplification of citrus exocortis and cachexia viroids from infected citrus hosts. *Phytopathology* 82: 279-285.

Zeiden M, Czosnek H (1991). Acquisition of tomato yellow leaf curl virus by the whitefly *Bemisia tabaci. Journal of General Virology* 72: 2607-2614.

Chapter 20

Application of PCR in Plant Virology

Ralf Georg Dietzgen

INTRODUCTION

Eleven years ago, the journal *Science* named the polymerase chain reaction (PCR) a major scientific development and chose DNA polymerase as its "Molecule of the Year" (Guyer and Koshland, 1989). Since then, this extremely powerful and versatile technique has literally revolutionized research in molecular biology. In plant virology, PCR has been widely adopted and its use in an ever increasing number of novel applications appears to be limited only by the imagination of the researcher. The potential power and sensitivity of PCR is due to the exponential amplification of small amounts of specific target DNA sequences in vitro. Usually three temperature steps are involved to (1) melt the target DNA into single strands, (2) anneal specific oligonucleotide primers, and (3) extend the primers by way of a thermostable DNA polymerase (Saiki et al., 1988). These steps are repeated 30 to 45 times, leading to the production of large amounts of DNA fragments of defined size that are delineated by the two primers. The amplified genetic information can then be identified, analyzed, and utilized in a multitude of downstream applications. Although novel PCR technology development has mostly taken place in medical research, a huge body of published information attests to the many and varied applications of PCR in plant virology. The main topics of this chapter are plant virus diagnosis, virus/strain differentiation, and genome characterization. However, it is beyond the scope of this chapter to give a fully comprehensive and in-depth review of all the scientific literature involving PCR applications in plant virology. Instead, I have attempted to provide an overview of established and emerging PCR applications and some recent technology advances.

The author wishes to thank Dr. D. S. Teakle for editorial comments.

ASPECTS OF PCR TEMPLATE PREPARATION

Many PCR applications, especially plant virus diagnosis, require rapid, easy-to-use, efficient, and reliable methods for isolation of viral nucleic acid template from purified virus preparations or from extracts of plant tissues, soil, or virus vectors. Total plant nucleic acids, total plant DNA or RNA, double-stranded (ds) RNA, poly (A+) RNA, or genomic viral nucleic acids may serve as the template in PCR or reverse transcription (RT) -PCR. The requirements for extraction of nucleic acids suitable for PCR amplification vary with plant species, cultivar, and tissue type, and with the type of viral nucleic acid to be isolated. Due to the high sensitivity of PCR, only pico- or femtogram amounts of target template are required. Most applications do not require high template purity, but removal or attenuation of components that are inhibitory to the reverse transcriptase or DNA polymerase enzymes may be necessary.

Nucleic Acid Extraction and Inhibitors of PCR

PCR amplification can be hampered by the presence in plant extracts of inhibitors of RT and/or PCR, in particular phenolic compounds and polysaccharides. Woody plants, banana leaves, roots and corms, cacao leaves, peanut leaves and seeds, bean leaves and seeds, pepper leaves, sugarbeet roots, gladiolus corms, and potato tubers have been shown to contain PCR inhibitors (Rowhani et al., 1993; De Blas et al., 1994; Saiz et al., 1994; Thomson and Dietzgen, 1995; Singh and Singh, 1996; Hoffmann et al., 1997). The type, location, and concentration of such inhibitors may vary with plant age and cultivar. For example, polysaccharides are increased in young leaves of some peanut cultivars. The presence or absence of inhibitors in a particular tissue can be determined by conducting PCR on samples spiked with uninfected tissue extract. The simplest way to attenuate inhibitors is the dilution of nucleic acid extracts prior to PCR. However, since this may require 100-fold or higher dilutions in some cases (Saiz et al., 1994), this will decrease test sensitivity, and depending on the particular application, sensitivity may then become a limiting factor. Leaves of *Nicotiana* species, tomato, and cucumber do not appear to contain interfering substances in sufficient concentration to inhibit PCR amplification from crude or undiluted extracts (e.g., De Blas et al., 1994).

The most efficient stage at which to remove or reduce inhibitors in plant extracts appears to be during nucleic acid extraction. Woody plant species pose a particular challenge, and various extraction methods have been evaluated and optimized to remove contaminating polyphenolics and polysaccharides (Rowhani et al., 1993; Staub et al., 1995; Wah and Symons, 1997).

Phenolic compounds that are widespread in plants can be removed by chelating to cation exchange resins (Staub et al., 1995) or by complexing with polyvinylpyrrolidone (PVP). Inclusion of citric acid in the extraction buffer prevents darkening of potato extracts and neutralizes chlorogenic acid, the main inhibitory phenolic compound (Singh et al., 1998). Other additives, such as bovine serum albumin, nonionic detergents, ascorbic acid, sodium sulfite, sodium dodecyl sulfate (SDS), and 2-mercaptoethanol, have been added to total nucleic acid extraction buffers at different stages (Rowhani et al., 1993; Wah and Symons, 1997). PCR-amplifiable total plant RNA can be obtained by purification using Sephadex G-50 columns (Vunsh et al., 1991), guanidinium extraction (Seal and Coates, 1998), or LiCl precipitation (Spiegel and Martin, 1993; Rowhani et al., 1993; Halpern and Hillman, 1996). Reliable, but more expensive options for the preparation of total RNA are commercially available spin column matrices, such as RNeasy (Qiagen) (Mackenzie et al., 1997). Zhang et al. (1998) developed a small-scale procedure that uses frozen, ground tissue for subsequent hexadecyl trimethyl-ammonium bromide CTAB-based extraction of DNA viruses and phytoplasmas or for dsRNA extraction of RNA viruses and viroids. Thomson and Dietzgen (1995) have developed a rapid single-step virus release protocol that does not require tissue homogenization while reducing plant inhibitory factors. A small amount of fresh, frozen, or dried tissue is heated in a solution containing 100 mM Tris-HCl, 1 M KCl, and 10 mM EDTA (TPS), centrifuged briefly, and the diluted supernatant is used directly in PCR. This template preparation method has been used to detect several RNA and DNA viruses in diverse plant species and in vector aphids. As suggested by Singh (1998), this simple TPS extraction should be tried before proceeding to more complex extraction procedures, if necessary.

Substances that reduce or prevent inhibition or increase sensitivity and/or specificity may be added directly to the RT and/or PCR reactions. The correct amount of these additives is crucial because they may inhibit PCR at higher concentrations. Dimethyl sulfoxide (DMSO) has been shown to improve amplification of RNA and DNA templates with complex, secondary 5 to 10 percent structure and to increase PCR sensitivity at 5 to 10 percent. However, *Taq* DNA polymerase activity decreases significantly above 10 percent DMSO. Low concentrations of formamide (2 to 5 percent) may increase PCR specificity for GC-rich templates. Addition of 0.1 to 0.5 percent (w/v) skim milk powder may eliminate PCR inhibition due to phenolics and humic acids (De Boer et al., 1995) and has been successfully applied for the detection of viruses in banana leaves and viruliferous aphids. Chemicals including diethyl pyrocarbonate (DEPC), SDS, and urea, which have been shown to inhibit PCR, should be avoided during nucleic acid template preparation or removed prior to PCR. Further details on PCR additives, their benefits, and their disadvantages can be found in general textbooks on PCR technology and in catalogs and newsletters from molecular biology companies.

Plant Material and Soil

PCR amplification of plant viruses and viroids (with a prior RT step for RNA templates) has been achieved from various plant organs, such as leaves, flowers, roots, tree bark, seeds, bulbs, and corms. Preparation of good-quality nucleic acids, free of PCR inhibitors, can be a limiting factor in using PCR for routine virus diagnosis in woody plants such as citrus, apple, cherry, peach, walnut, and grapevine, which are rich in polyphenolics and polysaccharides (Mackenzie et al., 1997; Mathews et al., 1997), but also in polyphenolic-rich banana leaves and corms (Thomson and Dietzgen, 1995). The suitability of Gene Releaser (BioVentures Inc., USA) with intractable tissues has been demonstrated for the subsequent RT-PCR amplification of viruses and viroids from woody and herbaceous plants (Levy et al., 1994). The polymeric matrix is thought to bind inhibitors, and its use does not require organic solvents and may avoid otherwise lengthy and laborious extraction protocols. A simple extraction method for phloem-limited viruses that uses proteinase K and boiling has been used successfully for barley, clover, and beet tissues (French and Robertson, 1994) and may be useful for other tissues with few inhibitor problems. Similarly, a "spot PCR" technique, in which a drop of crude sap is spotted on a small piece of nylon membrane and viral template is released by thermal and chemical treatment, has allowed detection of phloem-limited grapevine viruses (La Notte et al., 1997).

RT-PCR amplification of RNA from soil-borne viruses directly from soil samples requires the removal or neutralization of humic acids and metal ions that may inhibit DNA polymerases. This may be achieved by precipitation of humic acids with 2 percent (w/v) $CaCl_2$ and subsequent 50-fold dilution of the supernatant (Ernst et al., 1996), inclusion of PVP in extraction buffers, use of organic solvents (Skotnicki and Mo, 1996), or use of skim milk (De Boer et al., 1995).

Virus Vectors

Persistently (circulative) and nonpersistently (stylet-borne) aphid-transmitted DNA and RNA viruses have been detected in individual aphids by PCR. Total DNA/RNA extracts (Lopez-Moya et al., 1992; Hu et al., 1996; Stevens et al., 1997) and the commercial Gene Releaser (BioVentures Inc.,

USA) nucleic acid extracts (Mehta et al., 1997) have been used succesfully for template preparation. For details and other animal vectors, readers are referred to Chapter 21 of this book.

PLANT VIRUS DETECTION AND DIFFERENTIATION

The advantages of PCR as a diagnostic tool include exceptional sensitivity, speed, and versatility. PCR is generally 10^2 to 10^5 times more sensitive than enzyme-linked immunosorbent assay (ELISA), the widely used serological diagnostic benchmark. Sensitivity is of particular importance when viruses occur at low concentration (dormant plant tissues, woody tissue) or are unevenly distributed. For example, PCR-based diagnosis of *Citrus tristeza virus* (CTV, genus *Closterovirus*) makes testing of citrus trees possible throughout the year and at virus levels too low for ELISA (Mathews et al., 1997). Diagnostic PCR allows the early detection of virus infection prior to symptom development or in asymptomatic plants, the pooling of samples for large-scale testing, and the checking of ambiguous ELISA results. PCR tests have the potential to detect more than one virus in the one reaction (multiplexing), and diagnosis is amenable to automation. The application of PCR for plant disease diagnosis has been reviewed by Henson and French (1993), and some technical protocols for template extraction, detection, and quantitation of plant viruses can be found in Seal and Coates (1998). A recent review deals specifically with PCR-based detection of viruses infecting banana and sugarcane (Dietzgen et al., 1999).

Applications of PCR-based plant virus diagnosis include the following:

1. Germplasm screening
2. Field surveys to determine virus incidence and geographic distribution
3. Provision of virus-free planting material
4. Domestic and international plant quarantine
5. Detection of mixed virus infections
6. Analysis of virus distribution in different plant tissues
7. Identification of alternative host plants
8. Evaluation of virus-resistant or -tolerant cultivars
9. Analysis of virus transmission by insect, nematode, or fungal vectors

Diagnostic PCR can be performed directly on viruses that have a single-stranded (ss) DNA or DNA genome, whereas RNA-containing viruses require an RT step to generate complementary (c) DNA prior to PCR amplification (RT-PCR). The earliest reports of RT-PCR application to phytopathogenic viruses and viroids appeared in 1989 and 1990 (Puchta and Sanger, 1989; Vunsh et al., 1990). Today, specific PCR and RT-PCR tests have been developed for most of the economically important plant viruses and viroids. Indeed, the rapidly increasing information on viral nucleotide sequences can be expected to enable development of additional virus-specific PCR tests that, when combined, will provide diagnostic PCR test kits to detect diverse viruses infecting specific crops. Such a PCR test set is already available for subterranean clover (Bariana et al., 1994), potato (Singh, 1998), and garlic (Takaichi et al., 1998) and could be assembled for viruses infecting grapevine, sugarcane, and banana or for valuable ornamental plants, such as orchids.

Optimization of diagnostic plant virus PCR has led to the incorporation of technical improvements, some of which are presented in the following material. For denaturation of a dsRNA template, it appears best to boil template and primers for 5 min followed by quenching on ice prior to RT-PCR (Minafra and Hadidi, 1994; Braithwaite et al., 1995). It is advisable to perform a temperature gradient PCR to verify correlation of the calculated and real primer annealing temperatures. Sensitivity of PCR can be improved by amplification of smaller rather than larger DNA fragments (Rosner et al., 1997; Singh, 1998). Increased sensitivity and specificity can also be achieved by the use of a second pair of primers nested within the original PCR product (Mutasa et al., 1996) or by touchdown PCR. A wide range of virus concentrations can be covered by use of a three-primer PCR yielding two virus-specific products, a short one that is preferentially amplified at low virus levels and a long product that is preferred at higher template concentration (Rosner et al., 1997). Inclusion of restriction endonuclease recognition sites at the 5'-ends of primers will enable subsequent directional cloning (Gibbs and Mackenzie, 1997).

Detection of Virus Groups with Degenerate Primers

Universal, broad specificity as well as species-, genus-, or family-specific primers can be designed from conserved nucleotide or amino acid sequence motifs shared by all or several members of the same taxonomic group (see Table 20.1). Due to the degeneracy of the genetic code, this often requires the use of degenerate oligonucleotide primers. A "degenerate primer" is really a mixture of primers in which the nucleotides at one or more defined positions vary by design to represent a consensus sequence. For successful application of degenerate primers in PCR, maximum use of conserved nucleotides and

TABLE 20.1. Broad Spectrum PCR Detection of Members of Plant Virus Genera/
Families with Degenerate Primers

Primer specificity	Family	Genome	Highly conserved region	Reference
Badnavirus	*Caulimoviridae*	dsDNA	tRNA^{xo}/RT/RNase H	Lockhart and Olszewski, 1933
			RT/RNase H	Thomson et al., 1996
Mastrevirus (group I)	*Geminiviridae*	ssDNA	C2 ORF	Rybicki and Hughes, 1990
Begomovirus (group III)	*Geminiviridae*		DNA A, DNA B DNA A	Rojas et al., 1993 Deng et al., 1994
			capsid protein "core"	Wyatt and Brown, 1996
Capillovirus	–	ssRNA	putative RNA polymerase	Marinho et al., 1998
Carlavirus	–	ssRNA	11K /oligo dT	Badge et al., 1996
Carmovirus, Dianthovirus Tombusvirus	*Tombusviridae*	ssRNA	RNA polymerase	Morozov et al., 1995
Closterovirus, Crinivirus	*Closteroviridae*	ssRNA	HSP70	Karasev et al., 1994; Tian et al., 1996
Luteovirus, Polerovirus	*Luteoviridae*	ssRNA	coat protein/17K	Robertson et al., 1991
Potexvirus	–	ssRNA	methyltransferase-helicase/RNA polymerase	Gibbs et al., 1998
Potyviridae	*Potyviridae*	ssRNA	NIb replicase/oligo dT	Gibbs and Mackenzie, 1997
Potyvirus	*Potyviridae*	ssRNA	NIb replicase/coat protein	Langeveld et al., 1991
			coat protein/oligo dT	Pappu et al., 1993 Colinet et al., 1998
(sweet potato)			NIb replicase/coat protein	Yang and Mirkov, 1997
(SCMV complex)			NIb replicase/oligo dT	
Tobamovirus	–	ssRNA	RNA polymerase	Gibbs et al., 1998
Trichovirus	–	ssRNA	RNA polymerase polymearse motifs II, V	Marinho et al., 1998 Saldarelli et al., 1998)
Tospovirus	*Bunyaviridae*	ssRNA	nucleocapsid/3' SRNA	Mumford et al., 1996; Dewey et al., 1996
Vitivirus	–	ssRNA	polymerase motifs II, V	Saldarelli et al., 1998

Note: NIb = nuclear inclusion protein b; SCMV = *Sugarcane mosaic virus.*

minimum degeneracy are important. Use of degenerate primers appears to reduce overall PCR sensitivity. However, incorporation of deoxyinosine at nucleotide positions with more than twofold degeneracy reduces primer complexity and increases sensitivity over fully degenerate primers (Langeveld et al., 1991). A potential problem when using degenerate primers is the sometimes fortuitous amplification of plant-derived sequences, as reported for degenerate potyvirus-specific primers (Pappu et al., 1993; Pearson et al.,

1994) and luteovirus-specific primers (Robertson et al., 1991). Specificity of RT-PCR can be increased by cDNA priming with random hexa-nucleotides and the use of degenerate primers only during PCR (Langeveld et al., 1991; Saldarelli et al., 1998). Benefits of degenerate primers include the potential to detect uncharacterized viruses and new strains (Pearson et al., 1994; Mackenzie et al., 1998) or all viruses of a specific virus group with one set of primers (Tian et al., 1996). Individual viruses or subgroups can often be distinguished by restriction fragment length polymorphism (RFLP) analysis of the PCR product (see Virus Strain Discrimination). Addition of sequences corresponding to the SP6 or T7 promoter sequences or the pUC/M13 forward or reverse primer sequences to the 5'-ends of the degenerate primer sequences enables direct sequencing of the PCR product (Gibbs et al., 1998). Sequence information obtained from the cloned PCR product permits the design of virus/strain-specific PCR primers (Bateson and Dale, 1995; Thomson et al., 1996; Xu and Hampton, 1996; Routh et al., 1998).

Some degenerate primer pairs allow the detection of members of all genera of a family (Gibbs and Mackenzie, 1997), whereas others amplify specific sequences from all species of a genus (Langeveld et al., 1991) or from all species of a genus that infect a particular crop (Colinet et al., 1994; Yang and Mirkov, 1997). "Potyvirid" primers, one of which is *Potyviridae*-specific, the other being complementary to the 3'-poly(A)-tail, are particularly useful because they enable amplification of the complete coat protein (CP) gene and 3'-UTR (Gibbs and Mackenzie, 1997). Closteroviruses are the only viruses known to possess a gene encoding a cellular heat shock protein (HSP) 70 homologue. Therefore, degenerate primers have been designed to target highly conserved motifs in HSP70 for universal RT-PCR detection of this group of viruses (see Table 20.1). Primers targeting the phosphate 1 and 2 motifs enable detection of aphid- and whitefly-trans-mitted closteroviruses (Tian et al., 1996), whereas primers targeting the phosphate 1 and connect 2 (less conserved than phosphate 2) motifs appear to be restricted to PCR amplification of aphid-transmitted closteroviruses. The recently developed GPRIME package of computer programs identifies the most homologous regions of aligned sequences. It was used to design degenerate primers that detect potexviruses and tobamoviruses (see Table 20.1) in Australian orchid collections (Gibbs et al., 1998).

Recently, simultaneous detection and discrimination of two orchid viruses that belong to different genera (*Potexvirus* and *Tobamovirus*) using a pair of degenerate primers was described by Seoh et al. (1998). The primers were designed from homologous regions of the RNA-dependent RNA polymerase (RdRp) that are shared by many plant viruses and may be useful to detect simultaneously several unrelated viruses whose sequences are known. Using touchdown PCR (gradual lowering of the annealing temperature to the calculated optimum in subsequent cycles, which reduces mispriming and increases the yield of the desired product), two virus-specific products of different size with similar band intensities were obtained (Seoh et al., 1998).

Detection of Virus Species and Strains

Appropriate primer design determines RT-PCR specificity for one or more virus species, subgroups, strains, isolates, or pathotypes. For detection and differentiation of virus species or strains, predominantly primers in conserved regions of the CP and viral RNA polymerase genes have been used. Primers designed from more variable regions of the genome enable differentiation of species subtypes (Singh, 1998). A region of high nucleotide sequence divergence in the CP gene among isolates of *Plum pox virus* (PPV, genus *Potyvirus*) has enabled the use of a cherry subgroup-specific RT-PCR in certification and quarantine (Nemchinov and Hadidi, 1998). On the other hand, multiple serogroups of grapevine leafroll-associated virus (GLRaV, genus *Closterovirus*) can be detected with a single set of primers (Routh et al., 1998). Following RT-PCR, with primers complementary to conserved sequences amplifying a fragment from members of both subgroups of a virus species, subgroups can be differentiated by restriction enzyme digest of the PCR product (Rizos et al., 1992; Kruse et al., 1994). PCR-based diagnosis of viruses infecting banana and sugarcane has recently been reviewed (Dietzgen et al., 1999). At least seven virus species are known to infect banana and sugarcane, and PCR assays have been developed for most of them. The main problem for PCR detection is posed by viruses belonging to the genus *Badnavirus*, namely, *Banana streak virus* (BSV) and *Sugarcane bacilliform virus*. Owing to extensive sequence diversity among isolates and apparent integration of BSV sequences into the *Musa* genome (Harper et al., 1999), a PCR assay for detection of the species is not yet available.

Viroids are small, single-stranded circular RNA molecules. In some woody crops they occur in minute amounts, which requires high sensitivity in diagnosis. RT-PCR has been applied for detection of viroids in several crops, including hop, fruit trees, and grapevine (Puchta and Sanger, 1989; Hadidi and Yang, 1990; Staub et al., 1995). Using four viroid-specific primer pairs yielding ^{32}P-labeled PCR products of unique size, Rezaian et al. (1992) were able to detect and differentiate grapevine viroids. Wah and Symons (1997) developed an RT-PCR test in capillaries (fast cycle times due to larger surface area) for low–copy number grapevine viroids that was used for indexing in field and tissue culture samples.

Modified PCR formats using three or four primers in the one reaction can differentiate virus species, strains, or pathotypes. For example, the cucurbit-infecting potyviruses, namely, *Zucchini yellow mosaic virus* (ZYMV), *Watermelon mosaic virus 2* (WMV-2), and *Papaya ringspot virus* (PRSV), have been differentiated in a three-primer RT-PCR that amplified a 1,200 bp product for ZYMV and a 300 bp product for ZYMV and WMV-2 but did not amplify PRSV (Thomson et al., 1995). Three-primer RT-PCR similarly yielded specific products to differentiate between two bean-infecting potyviruses (Saiz et al., 1994). Two primer pairs that reflect a 3'-mismatch between a necrotic and a nonnecrotic strain of Peanut stripe virus (PStV) allowed their "nucleotide polymorphism-based" differentiation (Pappu et al., 1998). *Hemi-nested-PCR* allowed simultaneous detection (species-specific primers in the first round of PCR) and typing (external plus serotype-specific internal primers in the second round of PCR) of PPV isolates (Olmos et al., 1997).

Virus Detection in Vector Insects, Nematodes, and Fungi

The high sensitivity of PCR has allowed reliable virus detection in individual viruliferous insect and nematode vectors. This has opened new avenues for epidemiological studies, and for analysis of molecular virus-vector interactions and virus localization in the vector, and may enable the development of novel approaches for the control of virus spread by vectors. PCR can also be used to determine numbers of viruliferous aphids migrating in a crop, on the basis of which data forecasts and spray warning advice can be issued (Stevens et al., 1997). Both persistently and nonpersistently transmitted RNA and DNA viruses have been detected in various aphid species (Hu et al., 1996; Singh et al., 1996; Mehta et al., 1997; Olmos et al., 1997). A one-minute probing by *Myzus persicae* acquired sufficient cauliflower mosaic virus to be detectable by PCR (Lopez-Moya et al., 1992). High sensitivity is crucial, especially for detection of nonpersistently aphid-transmitted viruses, due to the brief acquisition periods and the small amounts of virus taken up. Successful virus detection has also been reported in viruliferous mealybugs (Minafra and Hadidi, 1994; Thomson et al., 1996), whiteflies (with both virus-specific and geminivirus group-specific degenerate primers) (Navot et al., 1992; Deng et al., 1994), and leafhoppers (Takahashi et al., 1993). PCR amplification of geminivirus DNA in whitefly extracts, saliva, hemolymph, and honeydew was recently used to trace the viral transmission pathway through the vector *Bemisia tabaci* (Rosell et al., 1999). For details, readers are referred to Chapter 21 of this book.

Tobra- and nepoviruses have been detected in viruliferous soil-inhabiting nematodes (Esmenjaud et al., 1994; Van der Wilk et al., 1994). *Beet necrotic yellow vein virus* (BNYVV, genus *Benyvirus*), as well as its fungal vector *Polymyxa betae*, can be detected in sugarbeet roots (Henry et al., 1995; Mutasa et al., 1996). PCR considerably shortens soil testing for BNYVV, which takes seven to eight weeks when seedling bait testing followed by ELISA is performed. Detection of *P. betae* helps with the identification of beet cultivars that are resistant to fungal infection and thus control rhizomania.

Virus-Capture PCR

Inhibitors of RT or PCR can be effectively eliminated by capture of virus particles from crude plant tissue or vector extracts by the surface of polypropylene PCR tubes or microtiter plates or by polystyrene ELISA plates. Components of crude plant extracts that would otherwise inhibit RT-PCR are washed away. Virus particles can be captured nonspecifically by direct binding to the plastic surface (Wyatt and Brown, 1996; Stevens et al., 1997), by binding to nonspecific proteins (Olmos et al., 1996), or by binding to specific polyclonal or monoclonal antibodies, so-called *immunocapture PCR* (Wetzel et al., 1992). The associated use of larger sample volumes (50 to 100 µl instead of 1 µl PCR template) has the added benefit of increasing sensitivity. Comparison of direct-binding and immunocapture techniques showed that a >100-fold higher sensitivity was achieved by immunocapture, whereas direct binding of virus particles to plastic may be inhibited by plant proteins, and its utility needs to be checked for each plant, tissue, and virus group (Nolasco et al., 1993; Rowhani et al., 1995; Barry et al., 1996). Immunocapture PCR appears to be the method of choice when specific antisera are available and highest sensitivity is required (10^4 to 10^6 times more sensitive than ELISA; Rowhani et al., 1995; Jacobi et al., 1998), e.g., for certification of virus-free planting material. Immunocapture of serologically diverse virus isolates (e.g., badnaviruses) or multiple strains/species requires broad-spectrum antisera or a mixture of different antibodies. Immunocapture PCR or RT-PCR works reliably for the entire spectrum of plant viruses, including enveloped tospoviruses (Nolasco et al., 1993; Rowhani et al., 1995; Weekes et al., 1996). Monoclonal antibodies with broad-spectrum specificity for dsRNA enabled immunocapture of viral dsRNA, satellite RNA, and, to a lesser extent, viroids (Nolasco et al., 1993).

Antibody-coating efficiency can be increased by activation of polypropylene surfaces with glutaraldehyde (Kokko et al., 1996) or HCl (Jain et al., 1997). The viral RNA contained in intact captured virions appears to be directly accessible for the initial RT stage of RT-PCR, and chemical or thermal disruption of the capsid is usually not necessary (Nolasco et al.,

1993; Rowhani et al., 1995; Stevens et al., 1997). Immunocapture, washes, and RT-PCR can be conveniently carried out sequentially in the same PCR tube or microplate well. Alternatively, RNA or cDNA derived from polystyrene-captured virions can be transferred to tubes for (RT-) PCR. Viral template contained in Triton X-100 extracts of tissue prints or crude plant extracts spotted on filter paper or nylon membrane appears to be captured nonspecifically by proteins in skim milk or bovine serum albumin (BSA), when used to coat PCR tubes (Olmos et al., 1996). Compared to immunocapture, the sensitivity of this *print-* or *spot-capture* PCR appears to be somewhat reduced, but BSA or skim milk may be an inexpensive alternative to capture virus template for PCR, especially when no specific antisera are available.

Multiplex PCR

The combination of several primer pairs in the same PCR reaction (multiplex PCR) allows the simultaneous amplification of more than one specific DNA fragment. *Multiplex PCR* is useful for (1) simultaneous detection and differentiation of several viruses in the same specimen, (2) incorporation of an internal positive control that would give a PCR product irrespective of the presence of the virus (Thomson and Dietzgen, 1995; Canning et al., 1996), or (3) simultaneous amplification of different target regions from the same template (Takahashi et al., 1993). Multiplex PCR also saves time and reagents. Each PCR product has a unique size by design, so it can be differentiated from the others by gel electrophoresis. *Duplex RT-PCR*, with two virus-specific primer pairs, is frequently used to differentiate viruses infecting the same crop, e.g., *Sugarcane mosaic virus* (SCMV, genus *Potyvirus*), *Fiji disease virus* (FDV, genus *Fijivirus;* Smith and Van de Velde, 1994), *Potato virus Y* (PVY, genus *Potyvirus*), *Potato leafroll virus* (genus *Polerovirus;* Singh et al., 1996), *Grapevine virus B* (genus *Vitivirus*), Grapevine leafroll-associated virus 3 (GLRaV-3, genus *Closterovirus;* Minafra and Hadidi, 1994; La Notte et al., 1997), and tobamoviruses in forest trees (Jacobi et al., 1998).

A multiplex RT-PCR that enables detection of five seed-borne legume viruses has been developed by Bariana et al. (1994) using a cocktail of nine primers. Multiplex PCR assays have recently been developed for the detection of three peanut-infecting viruses (R. G. Dietzgen et al., unpublished data) and of three banana-infecting viruses (incorporating immunocapture; Sharman et al., 2000). A suitable common extraction buffer or nucleic acid extraction protocol for all viruses, design of suitable primers to prevent interference with the heterologous (RT-)PCRs, compatible annealing temperatures, and optimized RT are all vital for converting separate PCR assays to a multiplex reaction. The relative concentrations of the primers, the concentration of the PCR buffer, and the balance between magnesium chloride and deoxynucleotide concentrations are also critically important for a successful multiplex PCR assay (Bariana et al., 1994; Henegariu et al., 1997).

Detection of PCR Products

DsDNA fragments produced during PCR can be visualized by (1) agarose or polyacrylamide gel electrophoresis followed by staining with ethidium bromide or silver nitrate, (2) dot blot or Southern blot membrane hybridization, (3) microplate hybridization and colorimetric detection, or (4) liquid hybridization and fluorescent detection.

Gel electrophoresis separates the PCR products according to size and is most commonly used in the research laboratory. However, it is quite laborious and time-consuming when testing many samples in a diagnostic laboratory and does not easily lend itself to automation.

Various formats of radioactive and nonradioactive membrane hybridization can be used to detect PCR products. Biotinylated or digoxigenin (DIG)-labeled PCR product can be spotted onto a membrane and detected with alkaline phosphatase conjugated to streptavidin or anti-DIG IgG, respectively (Korschineck et al., 1991; Van der Wilk et al., 1994). Alternatively, unlabeled PCR product can be spotted onto a membrane or transferred from an agarose gel for hybridization with a ^{32}P- or DIG-labeled probe (Pappu et al., 1993; Thomas et al., 1997). Southern hybridization is also an effective means of confirming the sequence specificity of PCR products.

For *microplate hybridization*, PCR products are hybridized to oligonucleotide probes and immobilized on microtiter wells followed by colorimetric detection. Use of a sequence-specific hybridization probe provides an additional level of specificity to the PCR primers. Many conceptually similar approaches using the specific interactions of biotin/streptavidin, probe/PCR product, and DIG/anti-DIG IgG for solid phase capture and/or detection have been described. *PCR microplate hybridization* formats include direct adsorption of heat-denatured PCR product to polystyrene wells and hybridization with a DIG-labeled probe (Hataya et al., 1994), nested PCR incorporating DIG and a biotinylated primer for immobilization in streptavidin-coated wells (Weekes et al., 1996), and PCR-ELISA (Olmos et al., 1997; Rowhani et al., 1998).

In *PCR-ELISA* (Roche), PCR products are directly labeled during amplification by incorporation of DIG-11dUTP. Following liquid phase hybridization with a specific biotinylated capture oligonucleotide that is complementary to an internal sequence of the PCR product, the hybrid is immobilized to streptavidin-coated microtiter wells. Alternatively, the capture probe can be immobilized prior to hybridization. Nonspecific material is removed by washing steps, and the bound PCR product is detected by an anti-DIG IgG peroxidase conjugate and color development following addition of a soluble substrate. The extent of color development can then be measured in an ELISA plate reader and is proportional to the amount of specific PCR product. PCR-ELISA is at least as sensitive as gel-based detection, has the potential to be quantitative, and allows simultaneous processing of large numbers of samples, which makes it amenable to automation. PCR-ELISA assays have been developed for viruses of fruit and nut trees and grapevine (Olmos et al., 1997; Rowhani et al., 1998) and for viruses infecting pineapple and banana (R. G. Dietzgen et al., unpublished data).

The Captagene-GCN4 kit (AMRAD Pharmaceuticals, Victoria, Australia) uses the dsDNA-binding protein GCN4 to capture PCR products followed by colorimetric detection. Unlike techniques using DNA capture probes, there is no need to denature the PCR product prior to protein capture. However, this system requires the PCR primers to be modified at their 5'- ends to include either a GCN4 recognition site or biotin (Mutasa et al., 1996).

Universal fluorescent detection can be achieved by way of the fluorescent dye Hoechst 33258, which intercalates into the PCR product in solution and can be measured in a fluorometer (Nolasco et al., 1993). Fluorescent dye-labeled oligonucleotide probes designed to hybridize in solution with one strand of the PCR product are used in real-time PCR assays such as TaqMan and Molecular Beacons.

REAL-TIME, QUANTITATIVE PCR

Real-time PCR systems allow the increase of specific PCR amplicons to be monitored quantitatively during each successive cycle. A pair of primers and an internal hybridization oligonucleotide probe provide dual specificity. Such systems (1) are fully automated, with closed tube product detection providing reduced risk of cross-contamination; (2) can assay up to 96 samples at a time; and (3) do not require any post-PCR manipulations. The sensitivity of real-time PCR approaches the detection of single target molecules. Real-time PCR has the potential for multiplexing, through the use of several probes labeled with different fluorescent dyes to detect two or more viruses simultaneously or to detect an internal control (e.g., a host DNA target) to ensure that a negative result is valid, or for relative quantitation. Pitfalls and the potential of quantitative RT-PCR have recently been reviewed by Freeman et al. (1999). Applications of real-time (RT-)PCR include detection and quantitation of virus genomes and mRNAs and quantitation of transgene mRNA levels.

TaqMan (PE Applied Biosystems) utilizes the $5' \rightarrow 3'$ exonuclease activity of *Taq* DNA polymerase to cleave a specific probe that is labeled with reporter and quencher fluorescent dyes during strand elongation of the target sequence and hence generates a detectable fluorescent signal during amplification. TaqMan appears to be particularly useful in large-scale applications for which high sensitivity, reliability, speed, and quantitative data are required, such as seed indexing (Schoen et al., 1996), epidemiological studies, or field surveys. TaqMan is commercially used for the analysis of *Cucumber mosaic virus* (CMV, genus *Cucumovirus*) in lupine seed in Western Australia (Berryman et al., 1997). Recently, single tube TaqMan RT-PCR assays for FDV of sugarcane, SCMV, and *Tomato spotted wilt virus* (genus *Tospovirus*) have been developed (Dietzgen et al., 1999; D. J. Maclean et al., unpublished data). Looking into the future, one can envisage TaqMan (RT-)PCR assays for multiple viruses or strains and internal control standards with dried predispensed primer-probe combinations in ready-to-use microplate formats.

An alternative real-time detection chemistry technique forms the basis of "molecular beacons"—hairpin-shaped probes which emit fluorescence only when they have bound to their target sequence (Tyagi et al., 1998) and which have recently been commercialized by Roche and Stratagene. The use of differently colored fluorophores allows the detection and quantification of multiple targets in the same reaction.

VIRUS STRAIN DISCRIMINATION

Restriction Fragment Length Polymorphism (RFLP)

When PCR products are cleaved with selected restriction endonucleases and the DNA fragments are subsequently separated by gel electrophoresis, characteristic banding patterns are obtained. These restriction enzyme profiles can be used to differentiate virus species following PCR amplification of DNA fragments with universal, group-specific primers, as demonstrated for luteo-, gemini-, and tospoviruses (Robertson et al., 1991; Rojas et al., 1993; Dewey et al., 1996). PCR-RFLP also enables classification of virus isolates into subgroups (Rizos et al., 1992; Kruse et al., 1994) or pathotypes (Tenllado et al., 1994) and may allow differentiation of serologically indistinguishable isolates (Barbara et al., 1995). Pathotyping tobamoviruses by PCR-RFLP appears to be a simple alternative to bioassays on a set of differential *Capsicum* genotypes (Tenllado et al., 1994). One RFLP group derived from the CP gene PCR product of CTV isolates appears to represent "mild" strains and could be used for rapid presumptive discrimination of mild and severe strains (Gillings et al., 1993). Two primer pairs specific for SCMV or *Sorghum mosaic virus*, both belonging to genus *Potyvirus*, were used in PCR-RFLP for rapid discrimination between strains of the two viruses without the need to interpret symptoms on differential hosts (Yang and Mirkov, 1997). Genetic distances of PCR-RFLP of the PVY CP gene differentiated isolates into three strain groups that correlated with the biological strains (Blanco-Urgoiti et al., 1996).

Single-Strand Conformation Polymorphism (SSCP)

When the strands of dsDNA are separated by heat treatment in the presence of formamide, they attain metastable, sequence-specific folded structures that give them characteristic electrophoretic mobilities in nondenaturing polyacrylamide gels. Even single nucleotide changes may be detected on silver-stained gels, and specific SSCP profiles are obtained for most isolates. SSCP of PCR products has been used for the rapid classification of virus isolates into known strain groups, for the differentiation of strains and variants, and for the detection of mixed infections (Koenig et al., 1995; Rubio et al., 1996; Stavolone et al., 1998). PCR product sizes of 250 to 850 bp appear to be suitable for SSCP analysis. SSCP allowed the differentiation of serologically indistinguishable strains of BNYVV (Koenig et al., 1995) and enabled better differentiation of isolates of *Turnip mosaic virus* (TuMV, genus *Potyvirus*) than biological host assays (Stavolone et al., 1998). Furthermore, differentiations of TuMV isolates by SSCP and by a panel of monoclonal antibodies were perfectly correlated.

Base Excision Sequence Scanning (BESS)

BESS T-Scan (Epicentre Technologies, Madison, WI) enables scanning of PCR products containing DNA sequence variation at single base resolution. The technique involves PCR amplification using one labeled primer in the presence of a limiting amount of dUTP. Treatment of the uracil-containing PCR product with an excision enzyme mix (uracyl N-glycosylase and endonuclease IV) cleaves the DNA at the sites of deoxyuridine incorporation, resulting in a set of nested fragments that are then separated on a standard sequencing gel (Hawkins and Hoffmann, 1997). The banding pattern of an individual sample is virtually identical to the "T" lane sequence of the PCR product. Primers can be labeled with radioisotopes or fluorescent dyes for analysis in conjunction with an automated DNA sequencer (Sambuughin, 1998). The applicability of BESS T-scan and fluorescent dye-labeled primers to identify and differentiate plant virus isolates without cloning and sequencing is currently under investigation (Dietzgen and Mitchell, 2000).

VIRAL GENOME CHARACTERIZATION

Genetic Variability and Phylogenetic Studies

The extent of sequence variation among virus isolates can be assessed by PCR using several pairs of primers targeting different regions of the viral genome. The number and size of amplified DNA bands will differ due to genomic sequence variation in or between the primers. Pair-wise comparison of the banding pattern results in a distance matrix from which a phylogenetic tree can be constructed. For example, two pairs of primers gave sufficient DNA bands to differentiate between 12 isolates of a natural virus population of *Cardamine chlorotic fleck virus* (genus *Carmovirus;* Skotnicki and Mo, 1996). Targeting of genomic regions that are suspected of being responsible for particular strain characteristics opens the possibility of locating particular gene functions.

In a broader approach, Glais et al. (1998) mapped the entire genome of PVY isolates that represented different biological groups varying in their ability to cause necrosis of tobacco and potato. The whole PVY genome was amplified in two large >4(kbp) fragments and subjected to RFLP analysis, using several restriction endonucleases and to clustering analysis. The groupings of isolates appeared to indicate that the tobacco necrosis determinants are likely to be located in the 5'-half of the PVY genome, whereas the potato tuber necrosis determinants are more likely located at the 3'- end. An extensive PCR-RFLP study or sequencing of each gene will be required to more precisely locate these pathogenicity determinants.

For more direct and detailed sequence variability analysis, PCR products may be cloned into plasmid vectors, and their nucleotide sequence determined and subjected to phylogenetic analysis. This approach is illustrated by the analysis of the genetic variability of the PStV CP gene and 3'-UTR (Higgins et al., 1998, 1999) to determine the taxonomic status of PStV isolates from geographically diverse locations and the distribution of recombinant isolates. Sequences for the phylogenetic analysis of a virus species or strain require species- or strain-specific primers (Higgins et al., 1998), whereas group-specific degenerate primers can be used to determine the taxonomic status of different species and strains in a virus complex (Yang and Mirkov, 1997), or of an undescribed species or isolate belonging to this group (Pappu et al., 1997). Genetic variability of virus isolates can also be assessed in relation to their biological properties, such as virulence or the ability to overcome host resistance genes. In an attempt to locate putative resistance-breaking determinants in the genome of *Lettuce mosaic virus* (LMV, genus *Potyvirus*), short regions in the 5'-end nuclear inclusion protein/CP and in the CP were amplified by immunocapture RT-PCR from ten isolates that differed in their ability to overcome the lettuce *Mo1* and *Mo2* resistance genes (Revers et al., 1997). PCR products were sequenced directly without prior cloning. Sequence data were subjected to clustering analysis and compared to the biological properties of the isolates. Phylogenetic analysis of the selected LMV genomic regions showed that isolates grouped according to geographic origin, rather than symptoms on different lettuce genotypes. Similarly, sequence analysis of the CP and 3'-UTR of PStV isolates grouped them according to geographic origin rather than symptoms and virulence on peanut (Higgins et al., 1999).

Acquisition of New Sequence Information

Several kilobase-long fragments of ssRNA virus genomes can be successfully amplified by long template PCR. A DNA polymerase with proofreading activity such as *Pfu, Pwo,* or *Vent* should be used in combination with *Taq* DNA polymerase to increase fidelity and yield. Two overlapping fragments of 5.6 kb and 4.3 kb, covering the entire PVY genome, were amplified by immunocapture RT-PCR using optimized buffers and cycle conditions (Romero et al., 1997). Full-length CMV RNAs 1 (3.4 kb), 2 (3.0 kb), and 3 (2.1 kb) were amplified by RT-PCR and cloned into a plasmid vector, and infectious transcripts were obtained (Hayes and Buck, 1990). Similarly, MacFarlane (1996) RT-PCR amplified the complete RNA2 (3.8 kb) of *Tobacco rattle virus* (genus *Tobravirus*) from an uncharacterized isolate by using a degenerate primer and a primer that targeted a highly conserved sequence motif.

Inverse PCR using abutting, outward-extending primers provides the possibility to amplify full-length infectious clones from the circular ssDNA of geminiviruses (Briddon et al., 1993) or to type virus isolates (Navot et al., 1992). Full-length clones of *Banana bunchy top virus* (BBTV, genus *Nanovirus*) DNA components were also obtained, and inverse PCR confirmed that the BBTV genome was indeed circular (Harding et al., 1993). Fragments of highly purified small circular viroid RNA of unknown sequence may be RT-PCR amplified with random primers, cloned, and partially sequenced. In a subsequent inverse PCR using adjacent primers of opposite polarities, full-length cDNA can be obtained (Navarro et al., 1998).

Several variations of PCR, including inverse PCR (Huang, 1997), reverse PCR, runaround PCR, single-primer PCR, *rapid amplification of cDNA ends* (RACE), and ligation-anchored PCR enable the amplification of unknown sequences adjacent to or between known sequences. Whereas RACE is used to determine the 3'- and 5'- sequences of mRNAs (Bertioli, 1997), *ligation-anchored PCR* is used to determine the ends of viral genomes by using a gene sequence-specific primer in combination with a primer that targets a universal adapter ligated to the 3'- or 5'-end (Wetzel et al., 1994).

Automated fluorescent DNA cycle sequencing of PCR products directly or cloned into a plasmid vector is itself performed by asymmetric PCR using a single primer and a dsDNA template (PE Applied Biosystems). Direct sequencing of PCR products has been reviewed by Rao (1994), who lists critical PCR parameters for good-quality sequencing template as

1. highly specific PCR primers,
2. high annealing temperature close to the estimated Tm of the primers,
3. hot-start PCR,
4. a rapid thermal cycling protocol
5. nested PCR, and
6. as few cycles as possible using a high–copy number target DNA.

MOLECULAR PLANT VIROLOGY APPLICATIONS

Gene Mapping

Pathogen resistance genes can be mapped by a range of generic PCR-based approaches such as randomly amplified polymorphic DNA (RAPD). PCR of plant genomic DNA using a single short (nine to ten) nucleotides |nt| primer of arbitrary sequence is used to generate RAPD markers. The nt primer binds to many different loci and amplifies usually two to ten random DNA sequences among them, which are then separated by agarose gel electrophoresis and visualized by ethidium bromide staining. Polymorphisms are due to point mutations in the primer-binding site or rearrangements in the DNA that alter the size of the PCR product. The method is technically simple, quick to perform, and requires only small amounts of DNA. Generation of RAPDs from near-isogenic lines, which are transmitted as dominant markers, can accelerate plant breeding programs for virus resistance or aid the isolation of natural resistance genes. In DNA amplification fingerprinting (DAF), arbitrary primers as short as 5 nt produce complex banding patterns that are resolved by polyacrylamide gel electrophoresis and silver staining (Caetano-Anolles et al., 1991). The larger number of amplified DNA fragments in DAF as compared to RAPD should make this technique also amenable to the fingerprinting of viral DNA genomes.

Many recently isolated genes for resistance against diverse plant pathogens share conserved sequence motifs. This provides the possibility of isolating additional resistance gene candidates by PCR using degenerate primers (Shen et al., 1998). The suspected integration of BSV sequences into the *Musa* genome was recently confirmed by sequence-specific amplification polymorphism (SSAP) analysis (Harper et al., 1999).

DNA Modification

PCR has found widespread application in the genetic engineering of DNA sequences. PCR-based site-directed mutagenesis has been used to introduce translational start codons, restriction enzyme recognition sites, deletions, insertions, and other sequence modifications. Such modified sequences are then used to study viral gene functions using infectious clones. Complex engineering steps, such as translational fusions to construct vectors for gene expression in bacteria (viral proteins for antiserum production) or plants, are made easy by PCR. Such plant expression vectors are used to express, viral transgenes to induce resistance or reporter proteins, such as green fluorescent protein, to study promoter activity. Engineered DNA constructs should be sequenced to ensure that no errors were incorporated during PCR or vector ligation.

Transgenic Plants

PCR enables early initial screening during regeneration of putative transgenic callus, shoots, or plantlets for the presence of the viral transgene (Worrall, 1998). When gene transfer has been accomplished with *Agrobacterium*, a control PCR using primers targeting the binary vector sequence outside the border repeats of the T-DNA can be used to ensure genuine PCR positive results. However, the results may not be clear-cut because DNA beyond the border repeats has been found integrated into the genome of some transgenic plants (see Worrall, 1998). In a multiplex format, PCR primers for the selectable reporter gene and an endogenous single-copy gene as an internal control can be included. PCR is also useful in identifying transgenic individuals in a segregating population in a breeding program.

Potential risks from field release of plants carrying a viral transgene are based on possible interactions between the transgene sequence and/or its product and another superinfecting virus. *Heteroencapsidation* (i.e., the encapsidation of the genome of the superinfecting virus by the transgenic CP) can be detected by immunocapture RT-PCR using an antiserum to the transgenically expressed CP and specific or broad-spectrum PCR primers to detect any encapsidated viral RNA (Hull, 1998). Transgenes in the environment can also be monitored by PCR and genetically modified foodstuff can be identified for labeling purposes.

Other Applications

The field of plant molecular virology has seen many applications of PCR that are too numerous to discuss in detail in this chapter. Some that have not been discussed include *in situ PCR* to detect target nucleic acids in host tissues, differential display and subtractive hybridization to isolate differentially expressed mRNAs, and colony PCR to identify bacterial colonies carrying recombinant plasmids with the desired insert.

CONCLUSIONS

The exquisite sensitivity and versatility of PCR has made it the method of choice for such applications as (1) diagnosis of plant viruses present at levels below the detection limit of other techniques, (2) establishment of phylogenetic relationships, and (3) identification and characterization of unknown viruses and viroids. The high sensitivity clearly requires special care in setting up and analyzing PCR reactions to avoid contamination. The many variations and applications of PCR attest to its enormous versatility. Beyond doubt, innovative applications will continue to emerge during the second decade of PCR use. Internet resources, such as virus sequence and primer databases, PCR protocols and trouble-shooting guides, and molecular biology discussion groups, will increasingly assist in design, *"in silico"* testing (Gibbs et al., 1998), and ordering of primers, as well as in optimization of PCR conditions in the future (McCann, 1999).

REFERENCES

Badge J, Brunt A, Carson R, Dagless E, Karamagioli M, Phillips S, Seal S, Turner R, Foster GD (1996). A carlavirus-specific PCR primer and partial nucleotide sequence provides further evidence for the recognition of cowpea mild mottle virus as a whitefly-transmitted carlavirus. *European Journal of Plant Pathology* 102: 305-310.

Barbara DJ, Morton A, Spence NJ, Miller A (1995). Rapid differentiation of closely related isolates of two plant viruses by polymerase chain reaction and restriction fragment length polymorphism analysis. *Journal of Virological Methods* 55: 121-131.

Bariana HS, Shannon AL, Chu PWG, Waterhouse PM (1994). Detection of five seedborne legume viruses in one sensitive multiplex polymerase chain reaction test. *Phytopathology* 84: 1201-1205.

Barry K, Hu JS, Kuehnle AR, Sughii N (1996). Sequence analysis and detection using immunocapture-PCR of cymbidium mosaic virus and odontoglossum ringspot virus in Hawaiian orchids. *Journal of Phytopathology* 144: 179-186.

Bateson MF, Dale JL (1995). Banana bract mosaic virus: Characterisation using potyvirus specific degenerate PCR primers. *Archives of Virology* 140: 515-527.

Berryman DI, Selladurai S, Washer S, Jones MGK, Carnegie PR (1997). Applications of the Taqman quantitative PCR system to the analysis of cucumber mosaic virus in lupin seed. In *Proceedings of XIth Biennial APPS Conference*, Perth, Western Australia, p. 210.

Bertioli DJ (1997). Rapid amplification of cDNA ends. In White BA (ed.), *Methods in Molecular Biology* (Volume 67). *PCR Cloning Protocols*, Humana Press, Totowa, NJ, pp. 233-238.

Blanco-Urgoiti B, Sanchez F, Dopazo J, Ponz F (1996). A strain-type clustering of potato virus Y based on the genetic distance between isolates calculated by RFLP analysis of the amplified coat protein gene. *Archives of Virology* 141: 2425-2442.

Braithwaite KS, Egeskov NM, Smith GR (1995). Detection of sugarcane bacilliform virus using the polymerase chain reaction. *Plant Disease* 79: 792-796.

Briddon RW, Prescott AG, Lunness P, Chamberlin LCL, Markham PG (1993). Rapid production of full-length, infectious geminivirus clones by abutting primer PCR (AbP-PCR). *Journal of Virological Methods* 43: 7-20.

Caetano-Anolles G, Bassam BJ, Gresshoff PM (1991). DNA amplification fingerprinting using very short arbitrary oligonucleotide primers. *Bio/Technology* 9: 553-557.

Canning ESG, Penrose MJ, Barker I, Coates D (1996). Improved detection of barley yellow dwarf virus in single aphids using RT-PCR. *Journal of Virological Methods* 56: 191-197.

Colinet D, Kummert J, Lepoivre P, Semal J (1994). Identification of distinct potyviruses in mixedly-infected sweet potato by the polymerase chain reaction with degenerate primers. *Phytopathology* 84: 65-69.

Colinet D, Nguyen M, Kummert J, Lepoivre P, Xia FZ (1998). Differentiation among potyviruses infecting sweet potato based on genus- and virus-specific reverse transcription polymerase chain reaction. *Plant Disease* 82: 223-229.

De Blas C, Borja MJ, Saiz M, Romero J (1994). Broad spectrum detection of cucumber mosaic virus (CMV) using the polymerase chain reaction. *Journal of Phytopathology* 141: 323-329.

De Boer SH, Ward LJ, Li X, Chittaranjan S (1995). Attenuation of PCR inhibition in the presence of plant compounds by addition of BLOTTO. *Nucleic Acids Research* 23: 2567-2568.

Deng D, McGrath PF, Robinson DJ, Harrison BD (1994). Detection and differentiation of whitefly-transmitted geminiviruses in plants and vector insects by the polymerase chain reaction with degenerate primers. *Annals of Applied Biology* 125: 327-336.

Dewey RA, Semorile LC, Grau O (1996). Detection of *Tospovirus* species by RT-PCR of the N-gene and restriction enzyme digestion of the products. *Journal of Virological Methods* 56: 19-26.

Dietzgen RG, Mitchell R (2000). Typing of virus isolates by fluorscent BESS T-Scan technology. *Abstracts of the 15th Australasian Biotechnology Conference*, Brisbane, July 2-6, 2000, p. 75.

Dietzgen RG, Thomas JE, Smith GR, Maclean DJ (1999). PCR-based detection of viruses infecting banana and sugarcane. *Current Trends in Virology* 1: 105-118.

Ernst D, Kiefer E, Drouet A, Sandermann H Jr. (1996). A simple method of DNA extraction from soil for detection of composite transgenic plants by PCR. *Plant Molecular Biology Reporter* 14: 143-148.

Esmenjaud D, Abad P, Pinck L, Walter B (1994). Detection of the region of the coat protein gene of grapevine fanleaf virus by RT-PCR in the nematode vector *Xiphinema index*. *Plant Disease* 78: 1087-1090.

Freeman WM, Walker SJ, Vrana KE (1999). Quantitative RT-PCR: Pitfalls and potential. *BioTechniques* 26: 112-125.

French R, Robertson NL (1994). Simplified sample preparation for detection of wheat streak mosaic virus and barley yellow dwarf virus by PCR. *Journal of Virological Methods* 49: 93-100.

Gibbs A, Armstrong J, Mackenzie AM, Weiller GF (1998). The GPRIME package: Computer programs for identifying the best regions of aligned genes to target in nucleic acid hybridisation-based diagnostic tests, and their use with plant viruses. *Journal of Virological Methods* 74: 67-76.

Gibbs A, Mackenzie A (1997). A primer pair for amplifying part of the genome of all potyvirids by RT-PCR. *Journal of Virological Methods* 63: 9-16.

Gillings M, Broadbent P, Indsto J, Lee R (1993). Characterization of isolates and strains of citrus tristeza closterovirus using restriction analysis of the coat protein gene amplified by the polymerase chain reaction. *Journal of Virological Methods* 44: 305-317.

Glais L, Tribodet M, Gauthier JP, Astier-Manifacier S, Robaglia C, Kerlan C (1998). RFLP mapping of the whole genome of ten viral isolates representative of different biological groups of potato virus Y. *Archives of Virology* 143: 2077-2091.

Guyer RL, Koshland DE Jr. (1989). The molecule of the year. *Science* 246: 1543-1546.

Hadidi A, Yang X (1990). Detection of pome fruit viroids by enzymatic cDNA amplification. *Journal of Virological Methods* 30: 261-270.

Halpern BT, Hillman BI (1996). Detection of blueberry scorch virus strain NJ2 by reverse transcriptase-polymerase chain reaction amplification. *Plant Disease* 80: 219-222.

Harding RM, Burns TM, Hafner G, Dietzgen RG, Dale JL (1993). Nucleotide sequence of one component of the banana bunchy top virus genome contains a putative replicase gene. *Journal of General Virology* 74: 323-328.

Harper G, Osuji JO, Heslop-Harrison JS, Hull R (1999). Integration of banana streak badnavirus into the *Musa* genome: Molecular and cytogenetic evidence. *Virology* 255: 207-213.

Hataya T, Inoue AK, Shikata E (1994). A PCR-microplate hybridization method for plant virus detection. *Journal of Virological Methods* 46: 223-236.

Hawkins GA, Hoffmann LM (1997). Base excision sequence scanning: A new method for rapid sequence scanning and mutation detection. *Nature Biotechnology* 15: 803-804.

Hayes RJ, Buck KW (1990). Infectious cucumber mosaic virus RNA transcribed in vitro from clones obtained from cDNA amplified using the polymerase chain reaction. *Journal of General Virology* 71: 2503-2508.

Henegariu O, Heerema NA, Dlouhy SR, Vance GH, Vogt PH (1997). Multiplex PCR: Critical parameters and step-by-step protocol. *BioTechniques* 23: 504-511.

Henry CM, Barker I, Morris J, Hugo SA (1995). Detection of beet necrotic yellow vein virus using reverse transcription and polymerase chain reaction. *Journal of Virological Methods* 54: 15-28.

Henson JM, French R (1993). The polymerase chain reaction and plant disease diagnosis. *Annual Review of Phytopathology* 31: 81-109.

Higgins CM, Cassidy BG, Teycheney P-Y, Wongkaew S, Dietzgen RG (1998). Sequences of the coat protein gene of five peanut stripe virus (PStV) strains from Thailand and their evolutionary relationship with other bean common mosaic virus sequences. *Archives of Virology* 143: 1655-1667.

Higgins CM, Dietzgen RG, Mat Akin H, Sudarsono, Chen K, Xu Z (1999). Biological and molecular variability of peanut stripe potyvirus. *Current Trends in Virology* 1: 1-26.

Hoffmann K, Sackey ST, Maiss E, Adomako D, Vetten HJ (1997). Immunocapture polymerase chain reaction for the detection and characterization of cacao swollen shoot virus 1 A isolates. *Journal of Phytopathology* 145: 205-212.

Hu JS, Wang M, Sether D, Xie W, Leonhardt KW (1996). Use of polymerase chain reaction (PCR) to study transmission of banana bunchy top virus by the banana aphid (*Pentalonia nigronervosa*). *Annals of Applied Biology* 128: 55-64.

Huang S-H (1997). Inverse PCR: An efficient approach to cloning cDNA ends. In White BA (ed.), *Methods in Molecular Biology* (Volume 67). *PCR Cloning Protocols*, Humana Press, Totowa, NJ, pp. 287-294.

Hull R (1998). Detection of risks associated with coat protein transgenics. In Foster GD, Taylor SC (eds.), *Methods in Molecular Biology* (Volume 81). *Plant Virology Protocols*, Humana Press, Totowa, NJ, pp. 547-555.

Jacobi V, Bachand GD, Hamelin RC, Castello JD (1998). Development of a multiplex immunocapture RT-PCR assay for detection of tomato and tobacco mosaic tobamoviruses. *Journal of Virological Methods* 74: 167-178.

Jain RK, Pappu SS, Pappu HR, Culbreath AK, Todd JW (1997). Detection of tomato spotted wilt tospovirus infection of peanut by immunocapture RT-PCR. *International Arachis Newsletter* 17: 38-39.

Karasev AV, Nikolaeva OV, Koonin EV, Gumpf DJ, Garnsey SM (1994). Screening of the closterovirus genome by degenerate primer-mediated polymerase chain reaction. *Journal of General Virology* 75: 1415-1422.

Koenig R, Luddecke P, Haeberle AM (1995). Detection of beet necrotic yellow vein virus strains, variants and mixed infections by examining single-strand conformation polymorphisms of immunocapture RT-PCR products. *Journal of General Virology* 76: 2051-2055.

Kokko HI, Kivineva M, Karenlampi SO (1996). Single-step immunocapture RT-PCR in the detection of raspberry bushy dwarf virus. *BioTechniques* 20: 842-846.

Korschineck I, Himmler G, Sagl R, Steinkellner H, Katinger HWD (1991). A PCR membrane spot assay for the detection of plum pox virus RNA in bark of infected trees. *Journal of Virological Methods* 31: 139-146.

Kruse M, Koenig R, Hoffmann A, Kaufmann A, Commandeur U, Solovyev AG, Savenkov I, Burgermeister W (1994). Restriction fragment length polymorphism analysis of reverse transcription-PCR products reveals the existence of two major strain groups of beet necrotic yellow vein virus. *Journal of General Virology* 75: 1835-1842.

Langeveld SA, Dore J-M, Memelink J, Derks AFLM, Van der Vlugt CIM, Asjes CJ, Bol JF (1991). Identification of potyviruses using the polymerase chain reaction with degenerate primers. *Journal of General Virology* 72: 1531-1541.

La Notte P, Minafra A, Saldarelli P (1997). A spot-PCR technique for the detection of phloem-limited grapevine viruses. *Journal of Virological Methods* 66: 103-108.

Levy L, Lee I-M, Hadidi A (1994). Simple and rapid preparation of infected plant tissue extracts for PCR amplification of virus, viroid, and MLO nucleic acids. *Journal of Virological Methods* 49: 295-304.

Lockhart BEL, Olszewski NE (1993). Serological and genomic heterogeneity of banana streak badnavirus: Implications for virus detection in *Musa* germplasm. In Ganry J (ed.), *Breeding Banana and Plantain for Resistance to Diseases and Pests*, CIRAD/INIBAP, Montpellier, France, pp. 105-113.

Lopez-Moya JJ, Cubero J, Lopez-Abella D, Diaz-Ruiz JR (1992). Detection of cauliflower mosaic virus (CaMV) in single aphids by the polymerase chain reaction (PCR). *Journal of Virological Methods* 37: 129-138.

MacFarlane SA (1996). Rapid cloning of uncharacterized tobacco rattle virus isolates using long template (LT) PCR. *Journal of Virological Methods* 56: 91-98.

Mackenzie AM, Nolan M, Wei K-J, Clements MA, Gowanlock D, Wallace BJ, Gibbs AJ (1998). Ceratobium mosaic potyvirus: Another virus from orchids. *Archives of Virology* 143: 903-914.

MacKenzie DJ, McLean MA, Mukerji S, Green M (1997). Improved RNA extraction from woody plants for the detection of viral pathogens by reverse transcription-polymerase chain reaction. *Plant Disease* 81: 222-226.

Marinho VLA, Kummert J, Rufflard G, Colinet D, Lepoivre P (1998). Detection of apple stem grooving virus in dormant apple trees with crude extracts as templates for one-step RT-PCR. *Plant Disease* 82: 785-790.

Mathews DM, Riley K, Dodds JA (1997). Comparison of detection methods for citrus tristeza virus in field trees during months of nonoptimal titer. *Plant Disease* 81: 525-529.

McCann S (1999). Web PCR. *Nature Biotechnology* 17: 304.

Mehta P, Brlansky RH, Gowda S, Yokomi RK (1997). Reverse-transcription polymerase chain reaction detection of citrus tristeza virus in aphids. *Plant Disease* 81: 1066-1069.

Minafra A, Hadidi A (1994). Sensitive detection of grapevine virus A, B, or leafroll-associated III from viruliferous mealybugs and infected tissue by cDNA amplification. *Journal of Virological Methods* 47: 175-188.

Morozov SY, Ryabov EV, Leiser R-M, Zavriev SK (1995). Use of highly conserved motifs in plant virus RNA polymerases as the tags for specific detection of carmovirus-related RNA-dependent RNA polymerase genes. *Virology* 207: 213-315.

Mumford RA, Barker I, Wood KR (1996). An improved method for the detection of Tospoviruses using the polymerase chain reaction. *Journal of Virological Methods* 57: 109-115.

Mutasa ES, Chwarszczynska DM, Asker MJC (1996). Single-tube, nested PCR for the diagnosis of *Polymyxa betae* infection in sugar beet roots and colorimetric analysis of amplified products. *Phytopathology* 86: 493-497.

Navarro B, Daros JA, Flores R (1998). Reverse transcription polymersae chain reaction protocols for cloning small circular RNAs. *Journal of Virological Methods* 73: 1-9.

Navot N, Zeidan M, Pichersky E, Zamir D, Czosnek H (1992). Use of the polymeras chain reaction to amplify tomato yellow leaf curl virus DNA from infected plants and viruliferous whiteflies. *Phytopathology* 82: 1199-1202.

Nemchinov L, Hadidi A (1998). Specific oligonucleotide primers for the direct detection of plum pox potyvirus-cherry subgroup. *Journal of Virological Methods* 70: 231-234.

Nolasco G, De Blas C, Torres V, Ponz F (1993). A method combining immunocapture and PCR amplification in a microtiter plate for the detection of plant viruses and subviral pathogens. *Journal of Virological Methods* 45: 201-218.

Olmos A, Cambra M, Dasi MA, Candresse T, Esteban O, Gorris MT, Asensio M (1997). Simultaneous detection and typing of plum pox potyvirus (PPV) isolates by heminested-PCR and PCR-ELISA. *Journal of Virological Methods* 68: 127-137.

Olmos A, Dasi MA, Candresse T, Canbra M (1996). Print-capture PCR: A simple and highly sensitive method for the detection of plum pox virus (PPV) in plant tissue. *Nucleic Acids Research* 24: 2192-2193.

Pappu HR, Pappu SS, Sreenivasulu P (1997). Molecular characterization and interviral homologies of a potyvirus infecting sesame *(Sesamum indicum)* in Georgia. *Archives of Virology* 142: 1919-1927.

Pappu SS, Brand R, Pappu HR, Rybicki EP, Gough KH, Frenkel MJ, Niblett CL (1993). A polymerase chain reaction method adapted for selective amplification and cloning of 3' sequences of potyviral genomes: Application to dasheen mosaic virus. *Journal of Virological Methods* 41: 9-20.

Pappu SS, Pappu HR, Chang CA, Culbreath AK, Todd JW (1998). Differentiation of biologically distinct peanut stripe potyvirus strains by a nucleotide polymorphism-based assay. *Plant Disease* 82: 1121-1125.

Pearson MN, Thomas JE, Randles JW (1994). Detection of an unidentified potyvirus from *Roystonea regia* palm using the polymerase chain reaction and degenerate, potyvirus specific, primers and potential problems arising from the amplification of host plant DNA sequences. *Journal of Virological Methods* 50: 211-218.

Puchta H, Sanger HL (1989). Sequence analysis of minute amounts of viroid RNA using the polymerase chain reaction (PCR). *Archives of Virology* 196: 335-340.

Rao VB (1994). Direct sequencing of polymerase chain reaction-amplified DNA. *Annals of Biochemistry* 216: 1-14.

Revers F, Lot H, Souche S, Le Gall O, Candresse T, Dunez J (1997). Biological and molecular variability of lettuce mosaic virus isolates. *Phytopathology* 87: 397-403.

Rezaian MA, Krake LR, Golino DA (1992). Common identity of grapevine viroids from USA and Australia revealed by PCR analysis. *Intervirology* 34: 38-43.

Rizos H, Gunn LV, Pares RD, Gillings MR (1992). Differentiation of cucumber mosaic virus isolates using the polymerase chain reaction. *Journal of General Virology* 73: 2099-2103.

Robertson NL, French R, Gray SM (1991). Use of group-specific primers and the polymerase chain reaction for the detection and identification of luteoviruses. *Journal of General Virology* 72: 1473-1477.

Rojas MR, Gilbertson RL, Russell DR, Maxwell DP (1993). Use of degenerate primers in the polymerase chain reaction to detect whitefly-transmitted geminiviruses. *Plant Disease* 77: 340-347.

Romero A, Blanco-Urgoiti B, Ponz F (1997). Amplification and cloning of a long RNA virus genome using immunocapture-long RT-PCR. *Journal of Virological Methods* 66: 159-163.

Rosell RC, Torres-Jerez I, Brown JK (1999). Tracing the geminivirus-whitefly transmission pathway by polymerase chain reaction in whitefly extracts, saliva, hemolymph, and honeydew. *Phytopathology* 89: 239-246.

Rosner A, Maslenin L, Spiegel S (1997). The use of short and long PCR products for improved detection of prunus necrotic ringspot virus in woody plants. *Journal of Virological Methods* 67: 135-141.

Routh G, Zhang Y-P, Saldarelli P, Rowhani A (1998). Use of degenerate primers for partial sequencing and RT-PCR-based assays of grapevine leafroll-associated viruses 4 and 5. *Phytopathology* 88: 1238-1243.

Rowhani A, Biardi L, Routh G, Daubert SD, Golino DA (1998). Development of a sensitive colorimetric-PCR assay for detection of viruses in woody plants. *Plant Disease* 82: 880-884.

Rowhani A, Chay C, Golino DA, Falk BW (1993). Development of a polymerase chain reaction technique for the detection of grapevine fanleaf virus in grapevine tissue. *Phytopathology* 83: 749-753.

Rowhani A, Maningas MA, Lile LS, Daubert SD, Golino DA (1995). Development of a detection system for viruses of woody plants based on PCR analysis of immobilized virions. *Phytopathology* 85: 347-352.

Rubio L, Ayllon MA, Guerri J, Pappu H, Niblett C, Moreno P (1996). Differentiation of citrus tristeza closterovirus (CTV) isolates by single-strand conformation polymorphism analysis of the coat protein gene. *Annals of Applied Biology* 129: 479-489.

Rybicki EP, Hughes FL (1990). Detection and typing of maize streak virus and other distantly related geminiviruses of grasses by polymerase chain reaction amplification of a conserved viral sequence. *Journal of General Virology* 71: 2519-2526.

Saiki RK, Gelfand DH, Stoffel S, Scharf SU, Higuchi R, Horn GT, Mullins KB, Ehrlich HA (1988). Primer directed enzymatic amplification of DNA with a thermostable DNA polymerase. *Science* 239: 487-491.

Saiz M, Castro S, De Blas C, Romero J (1994). Serotype-specific detection of bean common mosaic potyvirus in bean leaf and seed tissue by enzymatic amplification. *Journal of Virological Methods* 50: 145-154.

Saldarelli P, Rowhani A, Routh G, Minafra A, Digiaro M (1998). Use of degenerate primers in an RT-PCR assay for the identification and analysis of some filamentous viruses, with special reference to clostero- and vitiviruses of the grapevine. *European Journal of Plant Pathology* 104: 945-950.

Sambuughin N (1998). Analysis of BESS T-Scan mutation data using an ABI automated DNA sequencer. *Epicentre Forum* 5: 7-8.

Schoen CD, Knorr D, Leone G (1996). Detection of potato leafroll virus in dormant potato tubers by immunocapture and fluorogenic 5' nuclease RT-PCR assay. *Phytopathology* 86: 993-999.

Seal S, Coates D (1998). Detection and quantitation of plant viruses by PCR. In Foster GD, Taylor SC (eds.), *Methods in Molecular Biology* (Volume 81), *Plant Virology Protocols*, Humana Press, Totowa NJ, pp. 469-485.

Seoh M-L, Wong S-M, Zhang L (1998). Simultaneous TD/RT-PCR detection of cymbidium mosaic potexvirus and odontoglossum ringspot tobamovirus with a single pair of primers. *Journal of Virological Methods* 72: 197-204.

Sharman M, Thomas JE, Dietzgen RG (2000). Development of a multiplex immunocapture PCR with colorimetric detection for viruses of banana. *Journal of Virological Methods* 89: 75-88.

Shen KA, Meyer BC, Islam-Faridi MN, Chin DB, Stelly DM, Michelmore RW (1998). Resistance gene candidates identified by PCR with degenerate oligonucleotide primers map to clusters of resistance genes in lettuce. *Molecular Plant-Microbe Interactions* 11: 815-823.

Singh M, Singh RP (1996). Factors affecting detection of PVY in dormant tubers by reverse transcription polymerase chain reaction and nucleic acid spot hybridization. *Journal of Virological Methods* 60: 47-57.

Singh RP (1998). Reverse-transcription polymerase chain reaction for the detection of viruses from plants and aphids. *Journal of Virological Methods* 74: 125-128.

Singh RP, Kurz J, Boiteau G (1996). Detection of stylet-borne and circulative potato viruses in aphids by duplex reverse transcription polymerase chain reaction. *Journal of Virological Methods* 59: 189-196.

Singh RP, Singh M, King RR (1998). Use of citric acid for neutralizing polymerase chain reaction inhibition by chlorogenic acid in potato extracts. *Journal of Virological Methods* 74: 231-235.

Skotnicki ML, Mo J-Q (1996). Detection of cardamine chlorotic fleck carmovirus in soil by the polymerase chain reaction. *Australasian Journal of Plant Pathology* 25: 18-23.

Smith GR, Van de Velde R (1994). Detection of sugarcane mosaic virus and Fiji disease virus in diseased sugarcane using the polymerase chain reaction. *Plant Disease* 78: 557-561.

Spiegel S, Martin RR (1993). Improve detection of potato leafroll virus in dormant potato tubers and microtubers by the polymerase chain reaction and ELISA. *Annals of Applied Biology* 122: 493-500.

Staub U, Polivka H, Gross HI (1995). Two rapid microscale procedures for isolation of total RNA from leaves rich in polyphenols and polysaccharides: Application for sensitive detection of grapevine viroids. *Journal of Virological Methods* 52: 209-218.

Stavolone L, Alioto D, Ragozzino A, Laliberte J-F (1998). Variability among turnip mosaic potyvirus isolates. *Phytopathology* 88: 1200-1204.

Stevens M, Hull R, Smith HG (1997). Comparison of ELISA and RT-PCR for the detection of beet yellows closterovirus in plants and aphids. *Journal of Virological Methods* 68: 9-16.

Takahashi Y, Tiongco ER, Cabauatan PQ, Koganezawa H, Hibino H, Omura T (1993). Detection of rice tungro bacilliform virus by polymerase chain reaction for assessing mild infection of plants and viruliferous vector leafhoppers. *Phytopathology* 83: 655-659.

Takaichi M, Yamamoto M, Nagakubo T, Oeda K (1998). Four garlic viruses identified by reverse transcription-polymerase chain reaction and their regional distribution in northern Japan. *Plant Disease* 82: 694-698.

Tenllado F, Garcia-Luque I, Serra MT, Diaz-Ruiz JR (1994). Rapid detection and identification of tobamoviruses infecting L-resistant genotypes of pepper by RT-PCR and restriction analysis. *Journal of Virological Methods* 47: 165-174.

Thomas JE, Geering ADW, Gambley CF, Kessling AF, White M (1997). Purification, properties, and diagnosis of banana bract mosaic potyvirus and its distinction from abaca mosaic potyvirus. *Phytopathology* 87: 698-705.

Thomson D, Dietzgen RG (1995). Detection of DNA and RNA plant viruses by PCR and RT-PCR using a rapid virus release protocol without tissue homogenization. *Journal of Virological Methods* 54: 85-95.

Thomson KG, Dietzgen RG, Gibbs AJ, Tang YC, Liesack W, Teakle DS, Stackebrandt E (1995). Identification of zucchini yellow mosaic potyvirus by RT-PCR and analysis of sequence variability. *Journal of Virological Methods* 55: 83-96.

Thomson KG, Dietzgen RG, Thomas JE, Teakle DS (1996). Detection of pineapple bacilliform virus using the polymerase chain reaction. *Annals of Applied Biology* 129: 57-69.

Tian T, Klaassen VA, Soong J, Wisler G, Duffus JE, Falk BW (1996). Generation of cDNAs specific to lettuce infectious yellows closterovirus and other whitefly-transmitted viruses by RT-PCR and degenerate oligonucleotide primers corresponding to the closterovirus gene encoding the heat shock protein 70 homolog. *Phytopathology* 86: 1167-1173.

Tyagi S, Bratu DP, Kramer FR (1998). Multicolor molecular beacons for allele discrimination. *Nature Biotechnology* 16: 49-53.

Van der Wilk F, Korsman M, Zoon F (1994). Detection of tobacco rattle virus in nematodes by reverse transcription and polymerase chain reaction. *European Journal of Plant Pathology* 100: 109-112.

Vunsh R, Rosner A, Stein A (1990). The use of the polymerase chain reaction (PCR) for the detection of bean yellow mosaic virus in gladiolus. *Annals of Applied Biology* 117: 561-569.

Vunsh R, Rosner A, Stein A (1991). Detection of bean yellow mosaic virus in gladioli corms by the polymerase chain reaction. *Annals of Applied Biology* 119: 289-294.

Wah YFWC, Symons RH (1997). A high sensitivity RT-PCR assay for the diagnosis of grapevine viroids in field and tissue culture samples. *Journal of Virological Methods* 63: 57-69.

Weekes R, Baker I, Wood KR (1996). An RT-PCR test for the detection of tomato spotted wilt tospovirus incorporating immunocapture and colorimetric estimation. *Journal of Phytopathology* 144: 575-580.

Wetzel T, Candresse T, Macquaire G, Ravelonandro M, Dunez J (1992). A highly sensitive immunocapture polymerase chain reaction for plum pox potyvirus detection. *Journal of Virological Methods* 39: 27-37 .

Wetzel T, Dietzgen RG, Dale JL (1994). Genomic organization of lettuce necrotic yellows rhabdovirus. *Virology* 200: 401-412.

Worrall D (1998). PCR analysis of transgenic tobacco plants. In Foster GD, Taylor SC (eds.), Methods in Molecular Biology (Volume 81), *Plant Virology Protocols*, Humana Press,Totowa NJ, pp. 417-423.

Wyatt SD, Brown JK (1996). Detection of subgroup III geminivirus isolates in leaf extracts by degenerate primers and polymerase chain reaction. *Phytopathology* 86: 1288-1293.

Xu L, Hampton RO (1996). Molecular detection of bean common mosaic and bean common mosaic necrosis potyviruses and pathogroups. *Archives of Virology* 141: 1961-1977.

Yang ZN, Mirkov TE (1997). Sequence and relationships of sugarcane mosaic and sorghum mosaic virus strains and development of RT-PCR-based RFLPs for strain discrimination. *Phytopathology* 87: 932-939.

Zhang Y-P, Uyemoto JK, Kirkpatrick BC (1998). A small-scale procedure for extracting nucleic acids from woody plants infected with various phytopathogens for PCR assay. *Journal of Virological Methods* 71: 45-50.

Chapter 21

Plant Virus Detection in Animal Vectors

Rudra P. Singh

INTRODUCTION

Many plant viruses depend on animal vectors for their spread in nature (see reviews by Harris, 1990; Mandahar, 1990; see Chapter 3 in this book). Knowledge of the virus content in vectors of plant viruses is essential to (1) determine the mechanism of transmission and sites of specificity, (2) assess the presence of viruliferous vectors in field populations, and (3) facilitate epidemiological studies and disease management. For example, in some countries, when virus vector populations reach a certain threshold, a warning system for pesticide application is activated (Hull, 1968; Parry, 1987). The objective is to advise growers when to apply pesticides. The advice is based on knowledge of aerial vector populations and their movement into a crop during the growing season. Information on the proportions of viruliferous vectors and the actual threat they present to a crop can then be utilized to identify the areas or regions at greater risk. Such an approach permits pesticide application that is much more discriminating and therefore more environmentally safe. This chapter contains a brief description of vectors and their transmission. The detection of viruses in vectors by earlier methods is reviewed, followed by details of the reverse transcription–polymerase chain reaction (RT-PCR) method.

DETECTION BY EARLIER METHODS

Bioassay

The earliest method of determining whether a suspected virus vector is actually a carrier is by allowing the suspected vector to feed on or probe a

The critical reading and editorial review of the manuscript by Richard Anderson and the typing of the manuscript and technical assistance by Andrea Dilworth of the Potato Research Centre, Fredericton, Canada, are greatly appreciated.

plant susceptible to the virus and determining the presence or absence of the virus in the plant inoculated by the vector. Although this is a time-consuming method, it is still today the most-practiced method used to determine the viruliferous nature of a vector.

Viruses transmitted by leafhoppers have been detected in their vectors by an injection method. Storey (1933) investigated the relationship of *Maize streak virus* (MSV, genus *Mastrevirus*) and the leafhopper *Cicadulina mbila* by extracting the various vector tissues, injecting the extracts into nonviruliferous leafhoppers, and determining whether such vectors become viruliferous. Subsequently, the virus transmissibility was determined by plant transmission tests. A slight modification of this approach was used by Bennett and Wallace (1938) for *Beet curly top virus* (BCTV, genus *Curtovirus*) and the leafhopper *Eutettix tenellus*. They fed the nonviruliferous vector extracts of tissues from infective vectors and then followed the plant transmission test.

In the case of beetles, the macerated vectors of *Cowpea mosaic virus* (genus *Comovirus*) and the regurgitant produced by the vectors were inoculated into indicator plants (Dale, 1953). Thus, the presence of a virus in individual beetles or their tissues was determined by the symptoms in the indicator plants (Scott and Fulton, 1978).

Electron Microscopy

It has been shown that when extracts of aphids *(Myzus persicae)* collected from plants infected with *Potato virus X* (PVX, genus *Potexvirus*), *Tobacco mosaic virus* (TMV, genus *Tobamovirus*), and *Potato virus Y* (PVY, genus *Potyvirus*) were examined by electron microscopy (EM), PVX and TMV, the two viruses nontransmissible by aphids, were detected. PVY, however, which is transmissible, was not detectable (Kikumoto and Matsui, 1962). Electron microscopy was also used to demonstrate the presence of rice dwarf virus (genus *Phytoreovirus*) in the leafhopper *Nephotettix cincticeps* in ultrathin sections (Fukushi and Shikata, 1963). Ultrathin sections of mites *(Eriophyes tulipae)* have been used to demonstrate the presence of *Wheat streak mosaic virus* (genus *Tritimovirus*) in the gut (Paliwal and Slykhuis, 1967). Polyhedral nematode-transmitted viruses (*Xiphinema* spp. and *Longidorus* spp.) have also been demonstrated in nematodes (Roberts and Brown, 1980).

Ultrathin section of aphids showed the presence of *Tobacco severe etch virus* (genus *Potyvirus*) in the maxillary food canal (Taylor and Robertson, 1974), *Anthriscus yellows virus* (genus *Waikavirus*) in the anterior part of the foregut and midgut (Murant et al., 1976), and PVY in the anterior alimentary tract (Lopez-Abella et al., 1981). Similarly, the presence of particles of *Raspberry ringspot virus* and *Tomato black ring virus* (both belonging to genus *Nepovirus*) in the lumen of the buccal cavity and in the space between the stylet and the guiding sheath of the nematode *Longidorus elongatus* (Taylor and Robertson, 1969) has been demonstrated. Immunospecific EM has been utilized to detect *Potato leafroll virus* (PLRV, genus *Polerovirus*) in groups of or individual aphids, such as *Myzus persicae* and *Macrosiphum euphorbiae* (Tamada and Harrison, 1981), and to detect *Barley yellow dwarf virus* (BYDV, genus *Luteovirus*) in *Sitobion avenae* (Paliwal, 1982). Electron microscopical studies using the radioactive label ^{131}I in freeze sections of aphids exposed to PVY and *Tobacco etch virus* (genus *Potyvirus*) have revealed that the label was associated with maxillary stylets and anterior parts of the alimentary canal, as well as with the gut of the vector aphids (Pirone, 1977).

Enzyme-Linked Immunosorbent Assay

The development of a quick and sensitive serological method in the 1970s, i.e., enzyme-linked immunosorbent assay (ELISA), prompted reevaluation of viruses in vectors. Nonpersistently transmitted *Cucumber mosaic virus* (genus *Cucumovirus*) was detected by ELISA in homogenates of single aphids *(Aphis gossypii)* after 90 sec to 20 min of feeding on infected plants (Gera et al., 1978). The virus concentration was estimated to be in the range of 0.01 to 0.1 ng per aphid. Similar studies with another nonpersistently transmitted virus, PVY, were less successful for individual aphids (Carlebach et al., 1982), although the virus could be detected in groups of the aphid *M. persicae*. PVY was not detected in other vector species of aphids *(A. craccivora, A. citricola,* or *A. gossypii)* but was detected in some groups of *Acyrthosiphon pisum,* which is not a vector under their conditions. Similar difficulties have been reported in using ELISA to detect the semipersistently transmitted *Citrus tristeza virus* (CTV, genus *Closterovirus;* Cambra et al., 1981).

The presence of persistently transmitted PLRV in *M. persicae* was detected by ELISA in groups of 30 or more aphids (Clarke et al., 1980). However, by minimizing the nonspecific reactions that often mask the detection of viruses in vectors, PLRV was detected in 67 *M. persicae* aphids after a three-day infection feed. Twenty-eight of these transmitted virus to the plants. The virus concentration was estimated to be 0.01 ng per aphid (Tamada and Harrison, 1981). However, the highly concentrated *Pea enation mosaic virus-1* (PEMV, genus *Enamovirus*) in *Acyrthosiphon pisum* was successfully detected by ELISA in nymphs and adults (Fargette et al., 1982). The virus amounts were 40 ng per ml from nymphs and 200 ng per ml from adults. With prolonged feeding (up to 16 hours), the virus concentration increased further.

Since efforts to detect persistently transmitted luteoviruses in single aphids have been relatively more successful than those aimed at detecting nonpersistent and semipersistent viruses, an attempt has been made to apply this technique for epidemiological purposes to BYDV in Europe (Plumb, 1989). Following the application to detect BYDV in groups of *Rhopalosiphum padi* (Denéchère et al., 1979), cereal aphids caught in suction traps were assessed for their infectivity. When the aphids fed on test plants, a modified ELISA with a flurogenic substrate (MUP-ELISA) showed the presence of the virus (Torrance and Jones, 1982). It was found that the readings had wide differences in both aphid species and assays. For *R. padi* there was a fourfold range, and for *Sitobion avenae*, a fivefold range, for aphids that had access to infected plants (Torrance et al., 1986; Plumb, 1989). Usually readings in MUP-ELISA indicated a much higher number of virus-containing aphids than those obtained from bioassays with test plants.

To improve the ELISA test further, Smith et al. (1991) utilized monoclonal antibodies (Mabs) and a triple-antibody sandwich ELISA (TAS-ELISA) or amplified ELISA and transmission tests from field-collected aphids for *Beet mild yellowing virus* (BMYV, genus *Luteovirus*) and *Beet yellows virus* (BYV, genus *Closterovirus*). They found that Mabs detected the virus in individual aphids from oilseed rape plants but not from sugar beet plants. Differential virus concentration of the host was suspected to be a factor, since amplified ELISA detected virus in single aphids by both Mabs. Moreover, there were wide differences in the number of aphids that contained the virus, as shown by ELISA, and the readings for biological transmission through test plants. Stevens et al. (1995) applied the ELISA and transmission tests to monitor for four years the numbers of winged *M. persicae* and *Macrosiphum euphorbiae* and the presence of BMYV and another member of genus *Luteovirus, Beet western yellows virus* (BWYV), in aphids migrating into spring-sown sugar beet crops. They found that, depending on the previous weather conditions, the number of aphids varied between years. However, the percentage of aphids transmitting viruses and that detected by ELISA were very similar. For example, in one year, 1 percent and 27 percent of *M. persicae* transmitted BMYV and BWYV, respectively, to indicator plants, and ELISA on the same aphids showed that 3 and 33 percent, respectively, contained the two viruses.

Besides in aphids, ELISA has been used to detect viruses in leafhoppers. *Oat blue dwarf virus* (OBDV, genus *Marafivirus*) was detected in *Macrosteles fascifrons* (Azar and Banttari, 1981). However, correlation of virus presence in vectors and the biological transmission to plants was lacking. A better correlation of ELISA and infectivity has been reported with BCTV and its vector, *Circulifer tenellus,* when groups of ten leafhoppers were assessed (Mumford, 1982). Several rice viruses have also been detected in planthoppers by immunological methods. However, a substantial portion, 15 to 30 percent of insects showing positive for viruses in ELISA, did not transmit the viruses to test plants (Hibino and Kimura, 1983; Omura et al., 1984; Caciagli et al., 1985).

Nucleic Acid Hybridization

The use of DNA technology for plant virus detection was soon extended to detect viruses in vectors. Although several examples of radioactive and nonradioactive nucleic acid probes were used to demonstrate presence of viruses in vectors, there is no indication of any attempt to correlate the presence of viruses in vectors with their biological transmission to test plants. In the early 1980s, Jayasena et al. (1984) used radioactively labeled complementary (c) DNA molecular hybridization to detect Subterranean clover red leaf virus (SCRLV, genus *Luteovirus*) in the aphid vectors *Aulacorthum solani* and *M. persicae*. The c DNA detected SCRLV in individuals and groups of *A. solani,* and the average yield of virus was over 170 pg per aphid, with only a trace amount in the nonvector *M. persicae*. Boulton and Markham (1986) applied spot hybridization also with a radioactively labeled nucleic acid probe to detect MSV in several leafhopper species vectors as well as in nonvectors of MSV. The known vector *C. mbila* contained much more virus DNA than the nonvector leafhopper species. However, even in *C. mbila,* the quantities of DNA in individual insects ranged 100-fold. Czosnek et al. (1988) applied the hybridization to detect a DNA virus, *Tomato yellow leaf curl* (TYLCV, genus *Begomovirus*), in individual whiteflies *(Bemisia tabaci)*. An insect lysate was prepared by squashing whiteflies in an Eppendorf tube in a small volume of water containing proteinase K and sodium dodecyl sulphate and incubating at 60°C for two hours. The supernatant obtained by centrifugation (lysate) was used to spot onto nitrocellulose membranes. Distinct spots were observed from DNA of single whiteflies and intensity increased as the number of virus-exposed whiteflies increased in each sample. This simple extraction procedure for DNA viruses appears to be applicable for large-scale testing. Later, Zeiden and Czosnek (1991) further improved the detection sensitivity by using Southern blot hybridization (transfer of viral DNA from gel to nylon membranes). It was shown that as low as 0.5 pg DNA, or 300,000 TYLC viral genome copies, could be detected by this method. By using another simpler method to grind the whiteflies or simply squashing them on membranes for nucleic acid hybridization. Polston et al. (1990) detected *Squash leaf curl virus* (SLCV, another member of genus

Begomovirus) in single whiteflies. However, SLCV was detected more in individual whiteflies ground in buffer and spotted onto membranes than in whiteflies squashed onto membranes. Viral nucleic acids were also detected in *Trialeurodes vaporariorum*, a nonvector whitefly species, by the same protocol (Polston et al., 1990). PLRV has been detected in groups of *M. persicae* by nucleic acid hybridization using a 5'-PLRV cDNA probe consisting of half the PLRV molecule (Smith et al., 1993).

DETECTION BY A NEW METHOD

Reverse Transcription–Polymerase Chain Reaction

In the early 1990s, the newest method of DNA amplification, the polymerase chain reaction (PCR), was introduced for plant pathogen detection. It provides a method of exponentially amplifying specific DNA sequences by in vitro DNA synthesis. Depending on the specificity of the primers, the amplification products can provide both narrow and broad detection capabilities for various isolates of a pathogen (Henson and French, 1993). For the application of viral RNA sequences to PCR, cDNA is synthesized by reverse transcription (RT) and amplified by PCR (RT-PCR). With the availability of nucleotide sequences for many plant viruses and their strains, the development of RT-PCR assays for the detection and diagnosis of viruses in plant tissues and vectors has become feasible (Hadidi et al., 1995; Singh, 1998; see Chapter 20 in this book). RT-PCR methodology has been extensively applied to detect viroids, viruses, bacteria, mycoplasma-like organisms, fungi, and nematodes infecting various plant species (Henson and French, 1993; Hadidi et al., 1995).

RT-PCR for the detection of viruses is severalfold more sensitive than ELISA. Whereas ELISA detects virus concentrations in the lower nanograms or picograms, RT-PCR is capable of detecting viral nucleic acids in femtograms (fg). Amounts as low as 11 fg of BYDV have been detected in the aphid vector *R. padi* (Canning et al., 1996), and 15 fg of *Tobacco rattle virus* (genus *Tobravirus*) has been detected in nematodes (Van der Wilk et al., 1994). Thus, the prospect of detecting plant viruses by RT-PCR has increased, especially those which occur in very low concentrations in their vectors, e.g., nonpersistently transmissible viruses.

General Protocol for RT-PCR

Nucleic acids are the basic requirement for RT-PCR. Various methods have been reported for the extraction of nucleic acids from virus vectors (for details, see Chapter 20). The methods that are found useful for the extraction of RNA from aphids include a nonphenolic method (Singh et al., 1995; Naidu et al., 1998), the guanidine hydrochloride method (Canning et al., 1996), squashing aphids on Whatman paper (Olmos et al., 1997), the commercial kits Flowgen (Lichfield, UK) (Stevens et al., 1997) and Qiagen Neasy plant minikit, (Qiagen) (Naidu et al., 1998), and a proteinase K method (Singh et al., 1996; Singh, 1998). In addition to the quality of nucleic acid extracts, RT-PCR reactions can be inhibited by the polyphenolics and polysaccharides present in extracts (Singh et al., 1998). Dilution of nucleic acid extracts often reduces this inhibition. However, this approach is not always possible because virus nucleic acid concentration may be very low in some insects. Inhibitory substances can be eliminated by modifying nucleic acid extraction procedures by precipitating nucleic acids with isopropanol rather than ethanol (Hänni et al., 1995), using phosphate-buffered saline-Tween-20 in the cDNA mix (Barbara et al., 1995; Singh and Singh, 1996), or using citric acid in the extraction buffer (Singh et al., 1998). Another factor that can affect the sensitivity of virus detection is the size of the amplified fragments. Small-sized fragments are rapidly amplified, and the sensitivity of detection is severalfold greater than for large fragments (Singh and Singh, 1997).

A typical protocol for RT-PCR after extraction of nucleic acids consists of design and synthesis of primers, preparation of reverse transcription (RT) components (buffers, four deoxynucleotide-triphosphates [dNTPs], antisense primer, RT enzyme, ribonuclease inhibitors, dithiothreitol), and cDNA synthesis. For amplification, the protocol involves preparation of PCR reaction components (PCR buffer, four [dNTPs], $MgCl_2$, antisense and sense primers, and *Taq* polymerase), amplification of the target nucleic acids in a thermocycler, and the detection of the amplified fragments by agarose gel electrophoresis. A detailed step-by-step protocol for RT-PCR has been published (Singh, 1998).

DETECTION IN ANIMAL VECTORS

Aphids

A DNA-containing virus, *Cauliflower mosaic virus* (genus *Caulimovirus*) was the first virus to be detected from single aphids *(M. persicae)* using PCR (Lopez-Moya et al., 1992). Hadidi et al. (1993) used RT-PCR to detect PLRV in aphid nucleic acids prepared from groups of 50 to 100 aphids. In their study, an additional band of 1,200 base pairs (bp) was also obtained, along with a specific PLRV band of about 500 bp. Singh et al. (1995) have shown that as short as a five-minute exposure of *M. persicae* to PLRV-infected potato plants can result in the detection of PLRV-specific bands in about 13 percent of individual aphids. With the additional duration (18 to 24 h) of feeding on infected sources, 90 percent of the aphids contained PLRV bands. Mixing a single viruliferous aphid with an increasing number of nonviruliferous aphids and carrying out RT-PCR using nucleic acid extracts from a combined sample showed that as few as 1 in 30 aphids can be detected as PLRV carriers. In addition, it was shown that PLRV can be detected in aphids freshly collected from plants, aphids stored at −70°C, or those stored in 70 percent ethanol at room temperature for several years (Singh et al., 1995). When the RT-PCR method was applied to groups of ten aphids, collected in yellow-pan water traps from potato fields and stored in 70 percent ethanol for seven years (1988 to 1994), PLRV was detected even after such a long time. However, the intensity of the bands from earlier years was weaker than that of bands in recently caught aphids (see Figure 21.1). These findings indicate the possibility of testing stored aphids for the presence of luteoviruses over a longer period, and of determining migrating patterns and the introduction of viruses in a particular region. PLRV-specific bands were detected in the aphids *Aphis nasturtii* and *Macrosiphum euphorbiae* in addition to *M. persicae*, but not in other non-potato-colonizing aphid samples (Singh et al., 1997). RT-PCR analysis of aphids from individual farms showed that not all *M. persicae* aphids caught in yellow-pan traps contained PLRV-specific bands. In Canada, as in other potato-growing countries, *M. persicae* counts are used to warn growers to top-kill potato plants. Thus, virus detection in migrant aphids should facilitate the decision-making process in the management of virus diseases.

BYDV, another member of family *Luteoviridae*, has been detected in single aphids *(R. padi)* by RT-PCR (Canning et al., 1996). In this study, an internal control primer, directed at conserved regions of insect actin, is used along with virus-specific primers. The actin primer serves as a control for each stage of the method. RT-PCR has also been compared to an amplified ELISA method employed to detect BYDV in vector aphids, for use in forecasting systems. The increased sensitivity of RT-PCR was demonstrated when it was found that ELISA-negative aphids were RT-PCR positive. In contrast to members of the family *Luteoviridae*, aphid-transmitted members of the family *Potyviridae* are acquired by aphids in short probes of a few seconds to one minute. Inoculation of virus into plants also occurs in only a few seconds. Nonpersistently transmitted PVY has been used to develop an RT-PCR procedure (Singh et al., 1996). Under controlled conditions, about 40 percent of single *M. persicae* aphids and 15 percent of single *A. nasturtii* aphids acquired PVY, as demonstrated by RT-PCR (Singh et al., 1996).

FIGURE 21.1. Detection of *Potato leafroll virus* in Aphids Collected from Potato Fields from 1988 to 1994

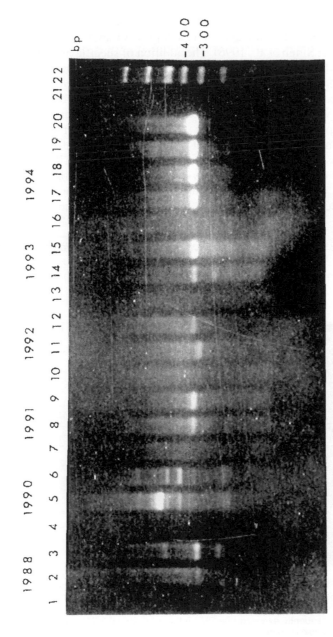

Note: Reverse transcription–polymerase chain reaction products were obtained from batches of ten aphids in 1988 (lanes 1-2), 1990 (lanes 4-6), 1991 (lanes 7-9), 1992 (lanes 10-12), 1993 (lanes 13-15), and 1994 (lanes 16-18). Samples 19 and 20 contained nine field-trapped aphids mixed with a single known viruliferous aphid (fed on infected potato plants in the laboratory). Lane 21 contained no template, and lane 22, the DNA size marker.

509

PVY was detected in aphids collected from plants and tested immediately or after storage in 70 percent ethanol at room temperature. To obtain simultaneous detection of PLRV and PVY, a duplex RT-PCR was developed (see Figure 21.2; Singh et al., 1996). The addition of specific primers for PLRV and PVY detected both viruses in single-aphid preparations containing both viruses under controlled conditions (see Figure 21.2) and in field-collected aphids (see Figure 21.3). Acquisition of PVY by aphids already viruliferous with PLRV was sometimes significantly reduced. However, it is not known whether aphids failed to acquire the second virus or whether the second virus was not amplified to the same extent in duplex RT-PCR. Unfortunately, since PVY is lost after aphids are stored in water, duplex RT-PCR for PLRV and PVY may not be reliable for aphids caught in yellow-pan traps. *Plum poxvirus* (PPV, another economically important member of genus *Potyvirus*) has been detected from individual *A. gossypii* aphids (Olmos et al., 1997)

FIGURE 21.2. Duplex RT-PCR for *Potato virus Y* (PVY) and *Potato leafroll virus (PLRV)* in *Myzus persicae*

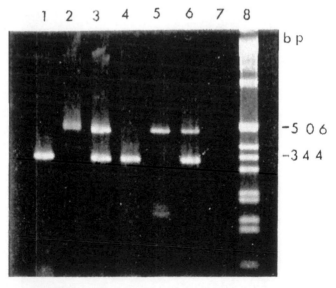

Note: Lane 1 = RT-PCR products of purified PLRV RNA; lane 2 = purified PVY; lane 3 = mixture of purified PLRV and PVY RNAs; lane 4 = aphid fed on a PLRV-infected plant; lane 5 = aphid fed on a PVY-infected plant; lane 6 = aphid fed on a PLRV-infected plant first and subsequently exposed to a PVY-infected plant; lane 7 = nonviruliferous aphid; lane 8 = DNA size markers.

FIGURE 21.3. Application of Duplex RT-PCR for *Potato leafroll virus* (PLRV) and *Potato virus Y* (PVY) Detection

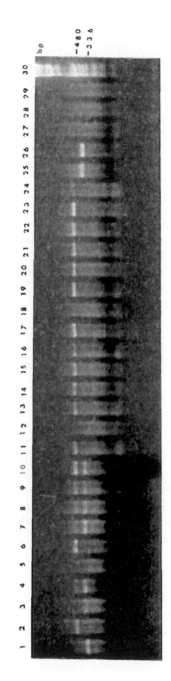

Note: Lanes 1-26 contain aphids collected from field-grown plants and singly analyzed for both viruses by duplex RT-PCR. The bands corresponding to 480 bp are specific to PVY, and those to 336 bp. to PLRV. Lanes 27-29 are known nonviruliferous aphids, and lane 30 contains DNA size markers.

511

by a heminested (where one external primer is common to two PCR reactions) spot-capture PCR (HSC-PCR). In this protocol, aphids exposed to virus sources are individually squashed on Whatman paper and the spots from individual aphids are used for PPV detection by means of heminested PCR. Although the sample preparation procedure of squashing individual aphids appears simple and promising for large-scale application, it has not been evaluated with field catches of aphids.

The semipersistently transmitted BYV has also been detected in aphids *(M. persicae)* (Stevens et al., 1997). It was shown that amplified ELISA and RT-PCR, but not the regular TAS-ELISA (with three antibodies), detected BYV from single-virus-exposed aphids. As with nonpersistently borne PVY (Singh et al., 1996), BYV was not recovered from aphids trapped in water pans (Stevens et al., 1997).

Banana bunchy top virus (BBTV, genus *Nanovirus*) with single-stranded (ss) DNA is transmitted by banana aphid *Pentalonia nigronervosa,* not by mechanical inoculation. With a view toward studying the molecular epidemiology and host range of this devastating virus, a PCR-based detection system was successfully developed to identify BBTV in single viruliferous aphids (Hu et al., 1996).

Similarly, this approach has potential in localizing phloem-limited CTV in its aphid vectors, namely, *Toxoptera citricida, Aphis gossypii,* and *A. spiraecola* (Mehta et al., 1997). The authors were able to localize CTV in mouthparts of the vector and nonvector aphids.

Naidu et al. (1998) have extended the application of the RT-PCR procedures to detect the three causal agents of groundnut rosette disease complex in aphids *(Aphis craccivora). Groundnut rosette assistor virus* (GRAV, genus *Luteovirus*), *Groundnut rosette virus* (GRV, genus *Umbravirus*), and groundnut rosette virus satellite RNA (sat-RNA) were detected by RT-PCR in aphids that had been exposed only to groundnut rosette-diseased plants containing all three agents. Both GRAV and GRV could be detected in total RNA extracted from single aphids that had been exposed to either green- or chlorotic-rosette-infected plants. However, sat-RNA could be amplified only when total RNA extracted from two or more aphids was used. This finding provides an approach for studying the epidemiology of various agents in the development of diseases.

Mealybugs

An immunocapture (IC; trapping virus particles with specific antibodies) RT-PCR method has been used to detect *Grapevine virus A* (genus *Vitivirus*) from mealybugs *(Planococcus ficus)* (Minafra and Hadidi, 1993). Recently, Acheche et al. (1999) have compared the potential of using three methods, i.e., ELISA, RT-PCR, and IC-RT-PCR, to detect *Grapevine leafroll-associated virus 3* (GLRaV-3, genus *Closterovirus*) in the vector mealybugs and grapevine cultures. The IC-RT-PCR was proved to be more sensitive than ELISA and RT-PCR. It provides a specific and most efficient approach to detecting viruliferous mealybug vectors in grapevine plants infected with GLRaV-3. Pineapple plants showing symptoms of mealybug wilt disease have been found to contain two different types of viruses: one is is a closterovirus (Pineapple mealybug wilt-associated virus), and the other a virus with bacilliform particles containing double stranded (ds) DNA (a badnavirus). Interestingly, both viruses are transmitted by mealybugs. A PCR-based sensitive system allowed the detection of *Pineapple bacilliform virus* in its host as well as in potential mealybug vectors (Thomson et al., 1996).

Whiteflies and Leafhoppers

Another application of PCR to detect TYLCV (ssDNA virus) revealed that TYLCV DNA can be amplified from individual whiteflies (Navot et al., 1992). Employing degenerate primers in a PCR-based assay, Deng et al. (1994) successfully amplified whitefly-transmitted geminivirus in plants and vector whiteflies. Furthermore, restriction enzyme analysis of PCR-amplified DNA fragments of a distinct begomovirus being carried in whiteflies allows identification of the virus isolate. This approach is equally important in following the spread of begomovirus in its vector *Bemisia tabaci*, by tracing the presence of viral DNA in different body parts of whiteflies (Atzmon et al., 1998; Rosell et al., 1999; see Chapter 13 in this book).

Nematodes

The first application of RT-PCR in detecting *Tobacco rattle virus* (TRV, genus *Tobravirus*) in groups of 1 to 200 nematodes *(Trichodorus primitivus)* was made by Van der Wilk et al. (1994). Two primer sets from RNA1 of TRV were employed. The first primer set amplified several PCR products, from viruliferous and nonviruliferous nematodes, in addition to TRV-specific 741 bp products. Then with the first-amplification PCR product as a template, a nested PCR was performed using a second internal set of primers, giving only one PCR product with the expected size of 601 bp. However, no PCR products were obtained from nonviruliferous nematode populations. In addition, these researchers developed a nonradioactive dot-blot hybridization procedure that allows the use of a large number of PCR samples for the detection of amplified products. DNA fragments resulting from PCR runs were identified by spotting onto a nylon membrane and hybridizing with a digoxigenin-labeled TRV cDNA fragment as a probe.

Almost at the same time, *Grapevine fanleaf virus* (GFLV, genus *Nepovirus*) was detected in its nematode vector *Xiphinema index*. ELISA was not sensitive enough to detect GFLV in a sample with fewer than ten nematodes. Even modification with biotin-avidin amplification did not yield satisfactory results. Application of RT-PCR, however, allowed detection of GFLV from a single viruliferous *X. index* adult (Esmenjaud and Abad, 1994).

THE PRESENCE OF VIRUS IN A VECTOR
AND BIOLOGICAL TRANSMISSION TO PLANTS

Electron microscopy, serological methods, nucleic acid hybridization, RT-PCR, and combinations thereof have been used to determine the presence of viruses in vectors. These techniques invariably have also detected viruses in nonvector species, although the amount of virus was greater in vector species than in nonvectors (Jayasena et al., 1984), with some exceptions (Polston et al., 1990). Even in the vectors, the amount of virus could vary, up to 10- or 20-fold (Tamada and Harrison, 1981), 100-fold (Boulton and Markham, 1986), and severalfold, depending on the aphid species (Torrance et al., 1986). On the other hand, the problem of presence and biological transmission is more complex in viruses such as TMV, which is not actually insect transmissible but occurs in high concentrations in plants. Aphids from infected plants (Kikumoto and Matsui, 1962) have picked it up, and its transmission to plants by clawing has been demonstrated (Bradley and Harris, 1972). Therefore, it is important to determine the relationship between the virus in a vector and its biological transmission to plants before incorporating information on the presence of virus in any forecasting or warning system.

Recently, the transmission pathway in SLCV-containing whiteflies was studied with PCR. The PCR approach allowed the exact tracking of geminivirus in whiteflies (Rosell et al., 1999; see Chapter 13 in this book).

Examples from various virus-vector studies at present show a variable correlation between the presence of virus in a vector and its transmission to a susceptible host. For aphids, Tamada and Harrison (1981) detected PLRV in 67 aphids by ELISA; however, only 28 of these transmitted the virus. Comparison of virus detection and transmission of OBDV, a marafivirus in leafhoppers, showed that although 81 percent of leafhoppers were viruliferous as determined by ELISA, only 19 to 61 percent transmitted the virus to plants (Azar and Banttari, 1981). A study of whitefly-transmitted virus (Polston et al., 1990) also observed a lack of correlation between the presence of SLCV in whiteflies and virus transmission in the host plant. Viral nucleic acids were detected in 165 out of 219 adult whiteflies that did not transmit the virus. Of the 54 adult whiteflies in which nucleic acids were not detected, one transmitted the virus. However, in 40 of the 41 whiteflies that transmitted the virus, viral nucleic acids were detected. Similarly, when immunosorbent electron microscopy (ISEM) was used to detect nepoviruses in nematodes, the presence of virus did not correlate with infectivity tests (Roberts and Brown, 1980).

Comparative studies using RT-PCR to detect the presence of viruses in vectors and their biological transmission to test plants are increasingly being conducted (Rosell et al., 1999). Singh et al. (unpublished data) have found that 46 out of 60 *M. persicae* were PLRV positive, and of these, 40 transmitted virus to the potato plant. This finding is similar to the relationship observed by Stevens et al. (1995) in the transmission of two luteoviruses to indicator plants and the detection of virus by an improved ELISA. These two observations indicate a closer agreement between the presence of virus and biological transmission and may hold promise (as the sensitivity of virus detection methods increases, more will be known about the relationship between presence and transmission [Singh et al., unpublished data]).

A way to develop more reliable correlations between virus transmissibility and its presence in vectors is to use a group approach. It has been found, for instance, that 220 aphids exposed to doubly infected (PLRV and *Potato spindle tuber viroid* [PSTVd]) plants transmitted PLRV to all the test plants when used in groups of five per plant; when these aphids were tested for PLRV by RT-PCR, 85 to 90 percent of the aphids revealed the presence of PLRV (Singh and Kurz, 1997), whereas PSTVd was present in 7 percent of plants. Similarly, with whiteflies, a good correlation between the presence of virus in a leafhopper vector, as established in ELISA, and its transmission to plants has been observed when leafhoppers were assayed in groups of ten (Mumford, 1982). This approach may be particularly worth testing for viruses that have low transmissibility rates, e.g., BYV (Stevens et al., 1997).

CONCLUSIONS

From the foregoing, it becomes clear that the attempt to determine the proportion of viruliferous vectors from the existing or migrating population has been a continuing process. As the technology of virus detection has evolved, it has been employed to determine the presence of viruses in the vectors. Thus, we see the use of various forms of electron microscopical applications, serological methods, and nucleic acid hybridization protocols.

The results have been encouraging, and this information is available for warning schemes for virus disease management programs. With the discovery of PCR in the mid-1980s, combinations of procedures, such as DNA technology and in vitro amplification, have opened new avenues. Viruses transmissible in a persistent, semipersistent, or nonpersistent manner have been detected with greater sensitivity than ever before. It is possible that PCR and RT-PCR methods of detection could reveal vectors possessing low concentrations of viruses. However, the relationship between the detection of viruses in vectors and the intrinsic capability of vectors to transmit to susceptible plants is still not resolved. Better results may be obtained by using a group approach. Toward this end, experiments should be designed to monitor vectors in the field in large numbers, to determine their spread under natural conditions, and to test the vectors in groups. To achieve such large-scale testing of virus vectors, methods of preparing vector extracts that are simple, rapid, and economical are needed. Efforts in this direction are in progress in the Potato Research Centre laboratory, where aphid extracts prepared with detergents have been found suitable for RT-PCR (Singh, unpublished data).

REFERENCES

Acheche H, Fattouch S, M'Hirsi S, Marzouki N, Marrakchi M (1999). Use of optimised PCR methods for the detection of GLRaV3: A closterovirus associated with grapevine leaf roll in Tunisian grapevine plants. *Plant Molecular Biology Reporter* 17: 31-42.

Atzmon G, Van Oss H, Czosnek H (1998). PCR-amplification of tomato yellow leaf curl virus (TYLCV) DNA from squashes of plants and whitefly vectors: Application to the study of TYLCV acquisition and transmission. *European Journal of Plant Pathology* 104: 189-194.

Azar M, Banttari EE (1981). Enzyme-linked immunosorbent assay versus transmission assay for detection of oat blue dwarf virus in aster leafhoppers. *Phytopathology* 71: 858.

Barbara DJ, Morton A, Spence NJ, Miller A (1995). Rapid differentiation of closely related isolates of two plant viruses by polymerase chain reaction and restriction fragment length polymorphism analysis. *Journal of Virological Methods* 55: 121-131.

Bennett CW, Wallace HE (1938). Relation of the curly top virus to the vector *Eutettix tenellus*. *Journal of Agricultural Research* 56: 31-50.

Boulton MI, Markham PG (1986). The use of squash blotting to detect plant pathogens in insect vectors. In Jones RAC, Torrance L (eds.). *Developments and Applications in Virus Testing*. Association of Applied Biology, Wellesbourne, Warwick. England. pp. 55-69.

Bradley RHE, Harris KF (1972). Aphids can inoculate plants with tobacco mosaic virus by clawing. *Virology* 50: 615-618.

Caciagli P, Roggero P, Luisoni E (1985). Detection of maize rough dwarf virus by enzyme-linked immunosorbent assay in plant hosts and in the planthopper vector. *Annals of Applied Biology* 107: 463-471.

Cambra M, Hermoso de Mendoza H, Moreno P, Navaro L (1981). Detection of citrus tristeza virus (CTV) in aphids by enzyme-linked immunosorbent assay. In *Abstract: Meeting of Virus Disease and Epidemiology*, Association of Applied Biology, Oxford, Wellesbourne, Warwick, England, pp. 71-72.

Canning ESG, Penrose MJ, Barker I, Coates D (1996). Improved detection of barley yellow dwarf virus in single aphids using RT-PCR. *Journal of Virological Methods* 56: 191-197.

Carlebach R, Raccah B, Loebenstein G (1982). Detection of potato virus Y in the aphid *Myzus persicae* by enzyme-linked immunosorbent assays (ELISA). *Annals of Applied Biology* 101: 511-516.

Clarke RG, Converse RH, Kojima M (1980). Enzyme-linked immunosorbent assay to detect potato leafroll virus in potato tubers and viruliferous aphids. *Plant Disease* 64: 43-45.

Czosnek H, Ber R, Navot N, Zamir D, Antignus Y, Cohen S (1988). Detection of tomato yellow leaf curl virus in lysates of plants and insects by hybridization with a viral DNA probe. *Plant Disease* 72: 949-951.

Dale WT (1953). The transmission of plant viruses by biting insects, with particular reference to cowpea mosaic. *Annals of Applied Biology* 40: 384-392.

Denéchère M, Cante F, Lapierre H (1979). Détection immuno-enzymatique du virus de la jaunisse nanisante de lorge dans son vecteur *Rhopalosiphum padi* (L.). *Annals of Phytopathology* 11: 567-574.

Deng D, McGrath PF, Robinson DJ, Harrison BD (1994). Detection and differentiation of whitefly-transmitted geminiviruses in plants and vector insects by the polymerase chain reaction with degenerate primers. *Annals of Applied Biology* 125: 327-336.

Esmenjaud D, Abad P (1994). Detection of a region of the coat protein gene of grapevine fanleaf virus by RT-PCR in the nematode vector *Xiphinema index*. *Plant Disease* 78: 1087-1090.

Fargette D, Jenniskens MJ, Peters D (1982). Acquisition and transmission of pea enation mosaic virus by the individual pea aphid. *Phytopathology* 72: 1386-1390.

Fukushi T, Shikata E (1963). Localization of rice dwarf in its insect vector. *Virology* 21: 503-505.

Gera A, Loebenstein G, Raccah B (1978). Detection of cucumber mosaic virus in viruliferous aphids by enzyme-linked immunosorbent assay. *Virology* 86: 542-545.

Hadidi A, Levy L, Podleckis EV (1995). Polymerase chain reaction technology in plant pathology. In Singh RP, Singh US (eds.). *Molecular Methods in Plant Pathology*, CRC/Lewis, Boca Raton, FL, pp. 167-187.

Hadidi A, Montasser MS, Levy L, Goth RW, Converse RH, Madkour MA, Skrzeckowski LJ (1993). Detection of potato leafroll and strawberry mild yellow-edge luteoviruses by reverse transcription-polymerase chain reaction amplification. *Plant Disease* 77: 595-601.

Hänni C, Brousseau T, Laudet V, Stephelin D (1995). Isopropanol precipitation removes PCR inhibitors from ancient bone extracts. *Nucleic Acids Research* 23: 881-883.

Harris KF (1990). Aphid transmission of plant viruses. In Mandahar CL (ed.). *Plant Viruses*, Volume II, *Pathology*, CRC Press, Boca Raton, FL, pp. 178-204.

Henson JM, French R (1993). The polymerase chain reaction and plant disease diagnosis. *Annual Review of Plant Pathology* 31: 81-109.

Hibino H, Kimura I (1983) Detection of rice ragged stunt virus in insect vectors by enzyme-linked immunosorbent assay. *Phytopathology* 72: 656-659.

Hu JS, Wang M, Sether D, Xie W, Leonhardts KW (1996). Use of polymerase chain reaction (PCR) to study transmission of banana bunchy virus by the banana aphid (*Pentalonia nigronervosa*). *Annals of Applied Biology* 128: 55-64.

Hull R (1968). The spray warning scheme for control of sugar beet yellows in England: Summary of results between 1959-1966. *Plant Pathology* 17: 1-10.

Jayasena KW, Randles JW, Barnett OW (1984). Synthesis of a complementary DNA probe specific for detecting subterranean clover red leaf virus in plants and aphids. *Journal of General Virology* 65: 109-117.

Kikumoto T, Matsui C (1962). Electron microscopy of plant viruses in aphid midguts. *Virology* 16: 509-510.

Lopez-Abella D, Pirone TP, Mernaugh RE, Johnson MC (1981). Effect of fixation and helper component on the detection of potato virus Y in alimentary tract extracts of *Myzus persicae*. *Phytopathology* 71: 807-809.

Lopez-Moya JJ, Cuabero J, Lopez-Abella D, Diaz-Ruiz JR (1992). Detection of cauliflower mosaic virus (CaMV) in single aphids by the polymerase chain reaction (PCR). *Journal of Virological Methods* 37: 129-138.

Mandahar CL (1990). Virus transmission. In Mandahar CL (ed.), *Plant Viruses*, Volume II, *Pathology*, CRC Press, Boca Raton, FL, pp. 206-253.

Mehta P, Brlansky RH, Gowda S (1997). Reverse transcription of polymerase chain reaction of citrus tristeza virus in aphids. *Plant Disease* 81: 1066-1069.

Minafra A, Hadidi A (1993). Detection of grapevine virus A in single mealy bugs by immunocapture reverse transcription-polymerase chain reaction, 11th Meeting of the International Council. In *Study of Viruses and Virus Diseases of the Grapevine* (ICVG), Abstract 139.

Mumford DL (1982). Using enzyme-linked immunosorbent assay to identify beet leafhopper populations carrying beet curly top virus. *Plant Disease* 66: 940-941.

Murant AF, Roberts IM, Elnagar S (1976). Association of virus-like particles with the foregut of the aphid *Cavariella aegopodii* transmitting the semi-persistent viruses anthriscus yellows and parsnip yellow fleck. *Journal of General Virology* 31: 47-58.

Naidu RA, Robinson DJ, Kimmins FM (1998). Detection of each of the causal agents of groundnut rosette disease in plants and vector by RT-PCR. *Journal of Virological Methods* 76: 9-18.

Navot N, Zeidan M, Pichersky E, Zamir D, Czosnek H (1992). Use of the polymerase chain reaction to amplify tomato yellow leaf curl virus DNA from infected plants and viruliferous whiteflies. *Phytopathology* 82: 1199-1202.

Olmos A, Cambra M, Dasi MA, Candresse T, Esteban O, Gorris MT, Asensio M (1997). Simultaneous detection and typing of plum pox potyvirus (PPV) isolates by heminested-PCR and PCR-ELISA. *Journal of Virological Methods* 68: 127-137.

Omura T, Hibino H, Usugi T, Inoue H, Morinaka T, Tzurumachi S, Ong CA, Putta M, Tsuchizaki T, Saito Y (1984). Detection of rice viruses in plants and individual insect vectors by latex flocculation test. *Plant Disease* 68: 374-378.

Paliwal YC (1982). Detection of barley yellow dwarf virus in aphids by serological specific electron microscopy. *Canadian Journal of Botany* 60: 179-185.

Paliwal YC, Slykhuis JT (1967). Localization of wheat streak mosaic virus in the alimentary canal of its vector *Aceria tulipae* Keifer. *Virology* 32: 344-353.

Parry RH (1987). Aphid and virus management in potatoes in eastern Canada. In Boiteau G, Singh RP, Parry RH (eds.), *Potato Pest Management in Canada, Proceedings of Symposium on Improving Potato Pest Protection*, Fredericton NB, Canada, Potato Pest Management in Canada, Canada-New Brunswick Agri-Food Development Agreement, pp. 9-22.

Pirone TP (1977). Accessory factors in nonpersistent virus transmission. In Harris KF, Maramorosch K (eds.), *Aphids As Vectors*, Academic Press, New York, pp. 224-235.

Plumb RT (1989). Detecting plant viruses in their vectors. *Advances in Disease Vector Research* 6: 191-208.

Polston JE, Al-Musa A, Perring TM, Dodds JA (1990). Association of the nucleic acid of squash leaf curl geminivirus with the whitefly *Bemisia tabaci*. *Phytopathology* 80: 850-856.

Roberts IM, Brown DJF (1980). Rayado fino virus: Detection in salivary glands and evidence of increase in virus titre in the leafhopper vector *Dalbulus maidis*. *Turrialba* 31: 78-80.

Rosell RC, Torres-Jerez I, Brown JK (1999). Tracing the geminivirus whitefly transmission pathway by polymerase chain reaction in whitefly extracts, saliva, hemolymph, and honey dew. *Phytopathology* 89: 239-246.

Scott HA, Fulton JP (1978). Comparison of the relationship of southern bean mosaic virus and cowpea strain of tobacco mosaic virus with the bean leaf beetle. *Virology* 84: 207-209.

Singh M, Singh RP (1996). Factors affecting detection of PVY in dormant tubers by reverse transcription polymerase chain reaction and nucleic acid spot hybridization. *Journal of Virological Methods* 60: 47-57.

Singh M, Singh RP (1997). Potato virus Y detection: Sensitivity of RT-PCR depends on the size of fragment amplified. *Canadian Journal of Plant Pathology* 19: 149-155.

Singh RP (1998). Reverse-transcription polymerase chain reaction for the detection of viruses from plants and aphids. *Journal of Virological Methods* 74: 125-138.

Singh RP, Kurz J (1997). RT-PCR analysis of PSTVd aphid transmission in association with PLRV. *Canadian Journal of Plant Pathology* 19: 418-424.

Singh RT, Kurz J, Boiteau G (1996). Detection of stylet-borne and circulative potato viruses in aphids by duplex reverse transcription polymerase chain reaction. *Journal of Virological Methods* 59: 189-196.

Singh RP, Kurz J, Boiteau G, Bernard G (1995). Detection of potato leafroll virus in single aphids by the reverse transcription polymerase reaction and its potential epidemiological application. *Journal of Virological Methods* 55: 133-143.

Singh RP, Kurz J, Boiteau G, Moore LM (1997). Potato leafroll virus detection by RT-PCR in field-collected aphids. *American Potato Journal* 74: 305-313.

Singh RP, Singh M, King RR (1998). Use of citric acid for neutralizing polymerase chain reaction inhibition by chlorogenic acid in potato extracts. *Journal of Virological Methods* 74: 231-235.

Smith HG, Stevens M, Hallsworth PB (1991). The use of monoclonal antibodies to detect beet mild yellowing virus and beet western yellows virus in aphids. *Annals of Applied Biology* 119: 295-302.

Smith OP, Damsteegt VD, Keller CJ, Beck RJ, Hewings AD (1993). Detection of potato leafroll virus in leaf and aphid extracts by dot-blot hybridization. *Plant Disease* 77: 1098-1102.

Stevens M, Hull R, Smith HG (1997). Comparison of ELISA and RT-PCR for the detection of beet yellows closterovirus in plants and aphids. *Journal of Virological Methods* 68: 9-16.

Stevens M, Smith HG, Hallsworth PB (1995). Detection of luteoviruses, beet mild yellowing virus and beet western yellows virus, in aphids caught in sugar beet and oilseed rape crops 1990-1993. *Annals of Applied Biology* 127: 309-320.

Storey HH (1933). Investigation of the mechanism of the transmission of plant viruses by insect vectors. I. *Proceedings of Royal Society B* 113: 463-485.

Tamada T, Harrison BD (1981). Quantitative studies on the uptake and retention of potato leafroll virus by aphids in laboratory and field conditions. *Annals of Applied Biology* 98: 261-276.

Taylor CE, Robertson WM (1969). The location of raspberry ringspot and tomato black ring viruses in the nematode vector *Longidorus elongatus* (de Man). *Annals of Applied Biology* 64: 233-237.

Taylor CE, Robertson WM (1974). Electron microscopy evidence for the association of tobacco severe etch virus with the maxillae in *Myzus persicae* (Sulz). *Phytopathologische Zeitschrift* 80: 257-266.

Thomson KG, Dietzgen RG, Thomas JE, Teakle DS (1996). Detection of pineapple bacilliform virus using the polymerase chain reaction. *Annals of Applied Biology* 129: 57-69.

Torrance L, Jones RAC (1982). Increased sensitivity of detection of plant viruses obtained by using a flurogenic substrate in enzyme-linked immunosorbent assay. *Annals of Applied Biology* 101: 501-509.

Torrance L, Plumb RT, Lennon EA, Gutteridge RA (1986). A comparison of ELISA with transmission tests to detect barley yellow dwarf virus-carrying aphids. In Jones RAC and Torrance L (eds.), *Developments and Applications in Virus Testing*, Association of Applied Biology, Wellesbourne, Warwick, England, pp. 165-176.

Van der Wilk F, Korsman M, Zoon F (1994). Detection of tobacco rattle virus in nematodes by reverse transcription and polymerase chain reaction. *European Journal of Plant Pathology* 100: 109-122.

Zeiden M, Czosnek H (1991). Acquisition of tomato yellow leafcurl virus by the whitefly *Bemisia tabaci*. *Journal of General Virology* 72: 2607-2614.

Index

Page numbers followed by the letter "f" indicate figures; those followed by the letter "t" indicate tables.

Abiotic stress, 376
AbMV. *See Abutilon mosaic virus*
Abortive synthesis. *See* RNA synthesis
Abutilon mosaic virus (AbMV), 46, 348f
ACLSV. *See Apple chlorotic leaf spot virus*
ACMV. *See African cassava mosaic virus*
Acyrthosiphon pisum, 503
Adenoviruses, 269
African cassava mosaic virus (ACMV)
  phylogenetic analysis, 291, 347, 348f
  recombinantion, 347
  taxonomy, 258, 258t
  transmission, 132
  vector specificity, 45
Ageratum yellow vein virus (AYVV), 282t
Agrobacterium tumefaciens, 436
Agrobacterium-mediated (gene transfer), 408
Agroinoculation (agroinfiltration), 356, 377
Agropyron mosaic virus, 226f
Alfalfa mosaic virus (AMV)
  movement, 131
  taxonomy, 10t, 18
  transient expression vector, 430
  transmission, aphid-mediated, 18
  transmission, seed, pollen-mediated, 110
Alfamovirus, alfamoviruses, 18, 131, 136, 180
Allexivirus, allexiviruses, 11t, 20
Alphacryptovirus, 10t, 13

Althaea rosea-infecting virus, 282t
AMV. *See Alfalfa mosaic virus*
Antennapedia, 186
Anthriscus yellows virus (AYV), 40, 502
Antibody expression, 425, 428, 429, 431, 434
Aphis citricola, 503
Aphis craccivora, 503, 512
Aphis gossypii, 503, 512
Aphis nasturtii, 508
Aphis spiraecola, 512
Aphthovirus, 185
Apoptosis, 381, 385
Apple chlorotic leaf spot virus (ACLSV), 11t, 20
Apple scar skin viroid, 12t, 21, 449t, 461
Apple stem grooving virus, 11t, 20
Apple stem pitting virus, 20
Apscaviroid, 12t, 21
Arabidopsis (thaliana), 138, 191, 292, 328, 382f
Arabidopsis thaliana Tal1 virus, 13
Arabis mosaic virus, 106t
Arenaviridae, 178
Arteriviruses, 228
Artichoke mottled crinkle virus, 409, 427
Artichoke yellow ringspot virus, 106t
ASBVd. *See Avocado sunblotch viroid*
Asparagus virus 2, 106t
Assistor luteovirus, 17
Asystasia golden mosaic virus, 297
Augusta disease, 83
Aulacorthum solani, 505
Aureusvirus, 11t, 17

Avenavirus, 11t, 17
Avirulence gene, 370, 372t, 377
Avocado sunblotch viroid (ASBVd),
    12t, 21, 446
Avsunviroid, Avsunviroidae, 12t, 20,
    21, 446
AYV. *See Anthriscus yellows virus*
AYVV. *See Ageratum yellow vein virus*

Bacterial expression system, 429
Bacteriophage promoters, 447
Baculovirus expression system, 39
Badnavirus, badnaviruses, 10t, 12, 317,
    346, 477t
BaMMV. *See Barley mild mosaic virus*
Banana bunchy top virus (BBTV), 343,
    343f, 488, 512
Banana streak virus (BSV), 318, 326,
    355f, 479
Barley mild mosaic virus (BaMMV),
    78t, 91, 92
Barley stripe mosaic virus (BSMV)
    gene expression, 349
    genome organization, 119
    recombinants, 121f
    taxonomy, 11t, 18
    transmission, seed-mediated, 107t,
        109
Barley yellow dwarf virus (BYDV)-
    MAV, PAV, 10t, 16, 179,
    457, 503
Barley yellow mosaic virus (BaYMV),
    11t, 15, 78t, 91
BaYMV. *See Barley yellow mosaic
    virus*
BBTV. *See Banana bunchy top virus*
BCMoV. *See Bean calico mosaic virus*
BCTV. *See Beet curly top virus*
BDMV. *See Bean dwarf mosaic virus*
Bean calico mosaic virus (BCMoV),
    282t, 291
Bean common mosaic virus, 106t, 110
Bean dwarf mosaic virus (BDMV),
    129f, 150, 282t, 348f
Bean golden mosaic virus (BGMV), 9,
    10t, 258, 259f, 282t, 286
Bean yellow mosaic virus, 106t
Beet curly top virus (BCTV). *See also*
    Phylogentic analysis

Beet curly top virus (BCTV) *(continued)*
    detection, 449t
    genomic strucure, 259f, 350f
    movement, 150
    taxonomy, 9, 10t, 258t
    transmission, 45, 502
Beet mild yellowing virus (BMYV),
    504
Beet necrotic yellow vein virus
    (BNYVV). *See also* Genomic
    organization
    detection, 79f, 368, 480-481
    genome, 89, 90f
    RNA recombination, 217
    subgenomic RNA, 216f
    taxonomy, 11t, 18
    transmission, 78t, 79f, 368
Beet western yellows virus (BWYV),
    504
Beet yellows virus, 10t, 19, 40, 181,
    504
Begomovirus, begomoviruses
    coat protein (CP), 284, 294, 305,
        342
    epidemiology, 279, 287, 296, 300
    evolution, 292
    gene expression, 267-274
    replication, 263, 264-267
    taxonomy, 9, 10t, 258, 258t, 340
    transmission, 33t, 45, 96, 293, 294
    variability, 291
Bemisia afer, 301f
Bemisia berbericola, 301f
Bemisia tabaci, 294, 295, 299f, 301f,
    513
Benyvirus, 18, 81, 89, 481
Betacryptovirus, 13
BGMV. *See Bean golden mosaic virus*
Big-vein disease of lettuce, 77
Biotic stress, 376
Blackgram mottle virus, 107t
Blueberry leaf mottle virus, 106t
Blueberry scorch virus, 107t
BMV. *See Brome mosaic virus*
BMYV. *See Beet mild yellowing virus*
BNYVV. *See Beet necrotic yellow vein
    virus*
Broad bean stain virus, 106t
Broad bean true mosaic virus, 106t
Broad bean wilt virus 1, 10t, 15

Brome mosaic virus (BMV), 10t, 134, 147, 203, 386
Brome streak mosaic virus, 226f
Bromoviridae, Bromovirus, bromoviruses, 10t, 18, 46, 134, 206
BSMV. *See Barley stripe mosaic virus*
BSV. *See Banana streak virus*
Buchnera sp., 43
Bunyaviridae, 10t, 14, 34t, 136, 401
BWYV. *See Beet western yellows virus*
BYDV. *See Barley yellow dwarf virus*
Bymovirus, 15, 81, 91, 225

Cabbage leaf curl virus (CaLCV), 282t
Cacao necrosis virus, 106t
Cacao swollen shoot virus, 318f, 326
CaLCV. *See Cabbage leaf curl virus*
CAMV. *See Cauliflower mosaic virus*
Capillovirus, capilloviruses, 11t, 20, 136, 477t
Cap-independent translation, 181, 182, 184, 191f, 323
Cardamine chlorotic fleck virus, 487
Cardiovirus, 186
Carlavirus, carlaviruses, 11t, 20, 41, 106t, 477t
Carmovirus, carmoviruses, 17, 81, 208, 477t
Carnation etched ring virus, 318f
Carnation Italian ringspot virus, 409
Carnation latent virus, 11t, 20
Carnation mottle virus, 11t, 17
Carnation ringspot virus, 11t, 17
Carrot mottle virus, 17
Cassava vein mosaic virus (CsVMV), 10t, 12, 317, 318f, 346
Cauliflower mosaic virus (CAMV, CaMV)
 genomic organization, 318f, 344
 resistance, 387
 taxonomy, 10t, 12, 317
 transmission, 139
Caulimoviridae, Caulimovirus, caulimoviruses, 10t, 39, 134, 318f, 387
CCMV. *See Cowpea chlorotic mottle virus*
CCR. *See* Central conserved region

CdTV. *See Chino del tomate virus*
Cell-to-cell (movement, transport). *See* Movement (transport) of viruses
Central conserved region (CCR), 20
CFDV. *See Coconut foliar decay virus*
Chaperonin, 43, 295
Chayote mosaic virus (ChMV), 282t
Chenopodium quinoa, 113, 114, 115, 138
Chicory yellow mottle virus, 106t
Chimeras, chimeric virus, 36, 45, 94, 111, 215f
Chino del tomate virus (CdTV), 282t, 291
ChMV. *See* Chayote mosaic virus
Chrysanthemum stunt viroid, 449t, 461
Chrysomelidae, 46
Chytridiomycota, 77
Chytrids, 78, 80
Cicadulina mbila, 502, 505
Circoviruses, 9, 264
Circulifer tenellus, 505
Citrus exocortis viroid, 461
Citrus psorosis virus, 14
Citrus tristeza virus (CTV), 40, 475, 503
CLCuV. *See Cotton leaf curl virus*
CLCV. *See Cotton leaf crumple virus*
CLE motifs, 271
CMV. *See Cucumber mosaic virus*
CNV. *See Cucumber necrosis virus*
Coat protein-mediated protection. *See* Resistance
Cocadviroid, 12t, 21
Coccinellidae, 46
Coconut cadang-cadang viroid, 12t, 21
Coconut foliar decay virus (CFDV), 12, 343, 343f
Coleoptera, 46
Coleus blumei viroid 1, 12t, 21
Coleviroid, 12t, 21
Colorectal cancer antibody, 435
Commelina yellow mottle virus (ComYMV), 10t, 12, 317, 318f, 346
Common region (CR), 261, 280, 342, 343f
Comoviridae, Comovirus, comoviruses, 10t, 15, 34t, 46, 133
ComYMV. *See Commelina yellow mottle virus*

Conserved late element, 271, 273f
Control (strategies), 50, 70, 97, 388, 425
Coronavirus(es), 188
Cotranslational disassembly, 406
Cotton leaf crumple virus (CLCV), 282t, 288
Cotton leaf curl virus (CLCuV), 282t, 349, 351f
Cowpea aphid-borne mosaic virus, 106t
Cowpea chlorotic mottle virus (CCMV), 134, 209
Cowpea green vein banding virus, 106t
Cowpea mild mottle virus, 106t
Cowpea mosaic virus (CPMV), 133, 184, 387
Cowpea mottle virus, 107t
Cowpea severe mosaic virus, 106t
CPMV. See Cowpea mosaic virus
CR. See Common region
Crimson clover latent virus, 106t
Crinivirus, 10t, 19, 34t, 388, 477t
Cross-protection, 436
Cryptic viruses, 13
CsVMV. See Cassava vein mosaic virus
CsVMV-like viruses, 12, 317, 318f
CTV. See Citrus tristeza virus
Cucumber mosaic virus (CMV)
    defense mechanism, 387
    detection, 485
    genome structure, 21f
    movement, 131
    RNA recombination, 220f
    satellites, 411
    template-switching mechanism, 206
    transmission, aphid-mediated, 18, 36
    transmission, seed mediated, 109, 116-117
Cucumber necrosis virus (CNV), 81
Cucumovirus, cucumoviruses, 10t, 18, 109, 136, 209
Cucurbit aphid-borne yellows virus, 206, 207f
Curculionidae, 46
Curtovirus, curtoviruses, 9, 10t, 45, 258, 258t
Cymbidium ringspot virus, 409
Cystolic expression, antibodies, 427, 428

Cytoplasmic inclusion protein (bodies), 129f, 227, 227f, 228
Cytorhabdovirus, 10t, 14, 33t

DAG motif, 38, 237
Defective interfering (DI) (DNAs, RNAs), 71, 206, 349, 350f, 412f
Defective interfering RNA protection, 410
Delphacid planthoppers, 49
Desmodium mosaic virus, 106t
Desmotubule, 133
DI (DNAs, RNAs). See Defective interfering
Dianthovirus, dianthoviruses, 11t, 17, 81, 132, 477t
Dioscorea bacilliform virus, 318f, 326
Dorylaimida, 63
Dulcamara mottle virus, 107t

EACMV. See East African cassava mosaic virus
Early viral genes expression, 261
East African cassava mosaic virus, 282t, 304, 347
Eggplant mosaic virus, 107t
Electrical penetration graph technique, 35
Elicitor-receptor model, 379
Elm mottle virus, 106t
Embryo infection, 108-111
EMCV. See Encephalomyocarditis virus
Enamovirus, 11t, 16, 41, 44, 503
Encephalomyocarditis virus (EMCV), 186
Endocytosis, 42
Endosymbiont (bacteria), 43, 45, 50
Enterovirus, 25, 26, 228
Environmental risks (hazards), 38, 70, 98, 427, 490
Epidemiology, 48, 97, 234, 279, 287
Eriophyes tulipae, 502
Escherichia coli, 152, 159, 165, 187, 445
Eutettix tenellus, 502
Externally borne virus, 93, 97, 98

Faba bean necrotic yellows virus
  (FBNYV), 47, 343, 343f
Fabavirus, 10t, 15, 34t, 41
FBNYV. *See Faba bean necrotic
  yellows virus*
Figwort mosaic virus (FMV), 318f, 321
Fiji disease virus, 13, 482
Fijivirus, 10t, 13, 33t, 48, 482
Flaviviridae, 229
FMDV. *See* Foot-and-mouth disease
  virus
FMV. *See Figwort mosaic virus*
Foamy retroviruses, 319, 330
Foot-and-mouth disease virus (FMDV),
  185, 203, 430
Foveavirus, foveaviruses, 11t, 20
Frankliniella occidentalis, 47, 48
Furovirus, furoviruses, 17, 18, 81, 83,
  136
Fusarium oxysporum forma specialis
  lycopersicon, 382f

Geminiviridae, geminiviruses, 9, 150,
  136, 257, 340
Geminivirus
  gene expression, 257, 267, 269
  infection cycle, 259f, 261, 267, 275
  replication, 262, 263, 269
  transcription, 262, 263, 267, 268
  transmission, 44
Gene expression
  early to late, 182
  RNA virus(es), 75, 177f, 182
  units, 331
  virion sense, 260, 274
Gene mapping, 489
Gene-for-gene theory, 370, 378,
  380-381
Genome-linked viral protein (VPg),
  116, 182, 400, 458f
Genomic organization of
  badnaviruses, 318f, 326
  Barley mild mosaic virus, 92f
  Beet necrotic yellow vein virus, 90f
  begomoviruses, 259f
  Brome mosaic virus, 403f
  Cassava vein mosaic virus, 14, 318f
  Caulimoviridae, genera of, 318f

Genomic organization of *(continued)*
  caulimoviruses, 345f
  Cowpea mosaic virus, 404f
  Curtoviruses, 259f
  DNA 1, *Cotton leaf curl virus*, 351f
  geminiviruses, 259f, 341f
  Mastreviruses, 259f
  Nanovirus, 343f, 344
  Peanut clump virus, 88f
  Petunia vein clearing virus, 318f
  Potato mop-top virus, 87f
  Soil-borne wheat mosaic virus, 85f
  Tobacco mosaic virus, 402f
  Tobacco necrosis virus, 84f
  Tobacco rattle virus, 65, 66f
  tombusvirus, 82f
GFLV. *See Grapevine fanleaf virus*
GFP. *See* Green fluorescent protein
GLRaV. *See* Grapevine leafroll-
  associated viruses
Grapevine bulgarian latent virus, 106t
Grapevine fanleaf virus (GFLV), 63,
  69, 106t, 514
Grapevine leafroll-associated virus 3,
  482, 513
Grapevine leafroll-associated viruses
  (GLRaV), 479, 513
Grapevine virus A, 11t, 20, 512
Grapevine virus B, 482
GRAV. *See Groundnut rosette assistor
  virus*
Green fluorescent protein (GFP), 138,
  430, 431, 490
GroEL (homologue), 43, 44, 295
Groundnut rosette assistor virus
  (GRAV), 47, 512
Groundnut rosette disease complex,
  512
Groundnut rosette virus (GRV), 47, 512
Groundnut rosette virus satellite RNA,
  512
GRV. *See Groundnut rosette virus*
Guar symptomless virus, 107t

HAV. *See Hepatitis A virus*
Havana tomato virus (HTV), 281f,
  282t, 291
HC-Pro. *See* Helper component (HC)
  proteinase (Pro)

HCV. *See Hepatitis C virus*
Helper component (HC) proteinase
    (Pro), HC-Pro
  genome organization, PVY", 458f
  movement, 111, 113, 129f, 137, 190
  replication, 111, 113
  transmission, 37, 40, 93, 190
Helper virus, 21, 206, 410, 411
Hepadnaviridae, 317
Heparin, 156, 158f, 159, 162f, 163
Hepatitis A virus (HAV), 185, 186
Hepatitis C virus (HCV), 183, 186
Herpes simplex virus, 269
Heteroencapsidation, 38, 43, 490
Hollyhock leaf crumple virus, 306
Hop latent viroid, 454
Hop stunt viroid, 11t, 20, 454
Hordeivirus, 11t, 18, 107t, 109, 132
Hordeum mosaic virus, 226f
Horseradish curly top virus (HrCTV),
    258t, 340, 341
Host
  gene expression, 386
  range determinants, 290, 319, 346
  resistance genes, 303, 372t, 487
  specific factors (virus movement), 137
Hostuviroid, 11t
HR. *See* Hypersensitive response
HrCTV. *See Horseradish curly top virus*
HTV. *See* Havana tomato virus
Human immunodeficiency virus, 376
Hybridization
  filter, 445
  in situ, 445, 455
  molecular, 443
  northern transfer, 445
  PCR microplate, 483
  solid support, 451
  solution, 445, 451
  southern transfer, 445
  tissue-print, 461, 462
Hypersensitive response (HR), 137,
    370, 372t, 373
Hypoviridae, 228

ICMV. *See Indian cassava mosaic virus*
ICTV. *See* International Committee on
    Taxonomy of Viruses
Idaeovirus, idaeoviruses, 11t, 19, 136

Ilarvirus, ilarviruses, 10t, 18, 136, 180,
    106t
Immunity, 98, 369, 374, 416
Immunocapture PCR. *See* Polymerase
    chain reaction (PCR)
Immunotherapy, 425, 432, 434, 436
Inclusion (bodies, proteins), 136, 227,
    319, 346
Indian cassava mosaic virus (ICMV),
    282t, 291
Indirect invasion (of embryo), 108
Infectious bronchitis virus, 184, 189
Inoculation access period, 70, 293
Intergenic region (IR), 259f, 261, 340,
    341f
Interleukin, 381
Internal ribosome entry sites (IRES),
    176, 183, 323
Internally borne virus, 176, 183, 323
International Committee on Taxonomy
    of Viruses, 3, 4, 225, 284
In-vitro transmission (by fungi), 93
In-vivo transmission (by fungi), 93
Ipomovirus, 11t, 16, 34t, 225
IR. *See* Intergenic region
IRES. *See* Internal ribosome entry sites

Jatropha mosaic virus, 296
Jatropha race, 296

KITC domain, 38
KTER motif, 96

Large intergenic region (LIR), 260, 280
Leaky scanning, 89, 175, 182, 183,
    191f, 326, 331
Lettuce big-vein virus, 10t, 13, 81
Lettuce infectious yellows virus, 10t,
    19, 449t
Lettuce mosaic virus (LMV), 107t, 109,
    230, 231, 488
Lettuce necrotic yellows virus, 10t, 14
Leucine zipper, 381, 382f, 384
Life cycle (of fungal vectors), 78, 79f, 80
Light microscopic autoradiography, 39
Lilium henryi de 1 1 virus, 13

LIR. *See* Large intergenic region
LMV. *See Lettuce mosaic virus*
Local acquired resistance. *See*
  Resistance
Long distance (movement, transport).
  See Movement (transport) of
  viruses
Longidoridae, longidorids, 63, 64
Longidorus, 64, 502
Longidorus elongatus, 67, 503
Longidorus macrosoma, 67
Lucerne australian latent virus, 106t
Luteoviridae, Luteovirus, luteoviruses,
  10t, 16, 179, 191, 192
Lychnis ringspot virus, 107t

Machlomovirus, 11t, 17
Maclura mosaic virus, 11t, 16, 226
Macluravirus, 11t, 16, 34t, 226
Macroptilium golden mosaic virus
  (MGMV), 288, 302, 303
Macrosiphum euphorbiae, 503, 504
Macrosteles fascifrons, 504
Maize chlorotic dwarf virus, 40
Maize chlorotic mottle virus, 11t, 17
Maize dwarf mosaic virus, 226f
Maize rayado fino virus, 11t, 19
Maize stripe virus (MSpV), 10t, 258t,
  274, 285f, 348f
Marafivirus, 19, 49, 504
Mastrevirus, mastreviruses, 258, 477t
Melampsora lini, 379, 382f
Meloidae, 46
Meloidogyne incognita, 232
Melon necrotic spot virus, 108, 107t
Metaviridae, Metavirus, 10t, 13
MGMV. *See Macroptilium golden
  mosaic virus*
MHV. *See Mouse hepatitis virus*
Microprojectile bombardment, 377
Microtubules, 39, 129f
Milk vetch dwarf virus, 343
Minichromosomes, 262
Mirabilis mosaic virus, 322
Mitochondria *COI* gene (sequence)
  298, 299t
Molecular farming, 436
Mouse hepatitis virus (MHV), 376

Movement protein (MP), 128, 129f,
  136, 184, 341f
Movement (transport) of viruses
  cell-to-cell, 128, 129f, 131, 134, 136
  coat protein dependent, 129f,
    133-136
  coat protein independent, 128-133,
    129f
  long-distance, 31, 128
  non-tubule guided, 134-135
  short-distance, 31, 128
  tubule-guided, 133-134
MP. *See* Movement protein
MSpV. *See Maize stripe virus*
MSV. *See Maize streak virus*
Mung bean yellow mosaic virus, 282t
Murine leukemia virus, 188
Murray River encephalitis virus, 25
Myndus taffini, 343
Myzus persicae, 480, 502, 503, 508

Nanovirus, nanoviruses, 9, 10t, 47, 343,
  350
Narcissus latent virus, 226f
Natural resistance. *See* Resistance
Necrovirus, 11t, 17, 81
Nematicides, 70
Nematode-borne virus diseases, 63
Nematostats, 70
Nephotettix cincticeps, 502
Nepovirus, nepoviruses, 64, 69, 133,
  134, 136
Nicotiana benthamiana
  Posttranscriptional gene silencing,
    386
  ScMV accumulation, 432-433
  symptoms induction, 213f
  transformation, 98, 353
  virus movement, 135, 137
Nicotiana bigelovii, 356, 357
Nicotiana clevelandii, 387
Nicotiana debneyi, 96
Nicotiana glutinosa, 379
Nicotiana sylvestris, 379
Nicotiana tabacum, 70, 270f, 352
  cv. Samsun, 232, 435
  cv. Xanthi, 135, 432, 433

Nomenclature of Viruses, International
   Committee on, 3
Nonhost, 298, 368, 369
Nonpersistent transmission (by fungi,
   insects), 33t, 34, 39, 93
Nonstructural (genes, proteins), 65,
   117, 405, 426
Novobiocin, 156, 157f
NSP. *See* Nuclear shuttle protein
Nuclear shuttle protein (NSP), 290,
   341f, 342
Nucleorhabdovirus, 10t, 14, 33t

Oat blue dwarf virus, 504
Oat chlorotic stunt virus, 17
Odontoglossum ringspot virus, 368,
   432
Okra infecting virus, 281f, 282t
Okra leaf-curl virus, 282t
Oleavirus, oleaviruses, 10t, 18, 19
Olive latent virus 2, 10t, 18
Olpidium bornovanus, 78, 78t, 81, 94,
   108
Olpidium brassicae, 77, 78, 78t, 80, 81,
   94
Olpidium cucurbitacearum. *See
   Olpidium bornovanus*
Olpidium radicale. *See Olpidium
   bornovanus*
Olpidium spp., 14, 77-78, 80
Onion yellow dwarf virus, 107t
Ophiovirus, ophioviruses, 10t, 14, 15
Organization and expression of the
   genome. *See* Genomic
   organization
Orthography. *See* Virus classification
Orthomyxoviridae, 178
Oryzavirus, 10t, 13, 33t, 48
Ourmia melon virus, 11t, 19
Ourmiavirus, 11t, 19

Panicovirus, 11t, 17
Panicum mosaic virus, 11t, 17
Papaya ringspot virus, 480
Paralongidorus, 64
Pararetroviruses, 8, 13, 319, 327, 339
Paratrichodorus, 64

Paratrichodorus pachydermus, 67, 68
Parsnip yellow fleck virus, 11t, 15, 40
Partitiviridae, partitiviruses, 10t, 13
Parvoviruses, 264
Passionvine mottle disease, 300
Pathogen-derived resistance, 98, 275,
   308, 399, 406
Pathogenesis-related (PR) proteins, 373
PCR. *See* Polymerase chain reaction
PCSV. *See Peanut chlorotic streak
   virus*
Pea early-browning virus (PEBV), 65,
   107t, 117, 118f, 206
Pea enation mosaic virus (PEMV), 16,
   44, 207f
Pea enation mosaic virus-1, 11t, 16,
   206, 453, 503
Pea enation mosaic virus-2, 17
Pea seed-borne mosaic virus (PSbMV)
   chimeric viruses, 111, 112, 112f,
      113, 113f, 114, 114f
   host resistance gene, 116, 372t
   mutants, 114, 115, 115f
   recessive gene, 375
   transmission, seed-mediated,
      110-116
Peach latent mosaic viroid, 12t, 21
Peach rosette mosaic virus, 106t
Peanut chlorotic streak virus (PCSV),
   318f, 327
Peanut clump virus, 11t, 18, 78t, 86,
   88f
Peanut mottle virus, 107t
Peanut stripe virus, 480
Peanut stunt virus, 106t, 209, 210f
PEBV. *See Pea early-browning virus*
Pecluvirus, 11t, 18, 81, 86, 107t
Pelamoviroid, 12t, 21
PeMoV, PepMoV. *See Pepper mottle
   virus*
PEMV. *See* Pea enation mosaic virus
Pentalonia nigronervosa, 512
PepGMV. *See* Pepper golden mosaic
   virus
Pepper golden mosaic virus (PeGMV),
   283t
Pepper huasteco virus (PHV), 258t,
   263, 270f, 283t
Pepper mild mottle virus (PMMoV),
   372t, 409

Pepper mottle virus (PeMoV,
  PepMoV), 230, 231, 236f,
  239f, 369
Pepper ringspot virus, 65, 107t, 117
Pepper severe mosaic virus, 230
Peregrinus maidis, 49
Peronospora parasitica, 382f
Persistent transmission (by fungi,
  insects), 93
Petunia vein clearing virus (PVCV),
  10t, 12, 317, 318f
Phagemid vector, 446
Pharmaceutical proteins, 425
PHV. *See Pepper huasteco virus*
Phylogenic (analyses, studies, tree)
  begomoviruses, 258, 287, 347, 348f
  cucumovirus, 2a protein, 209, 210f
  potyviridae, 226f
  PVY, 236f, 237, 238f, 239f
Phytoplasmas, 473
Phytoreovirus, 10t, 13, 48, 49, 502
Picornaviridae, picornaviruses, 15, 25,
  181, 182, 228
Pineapple bacilliform virus, 513
Pineapple mealybug wilt-associated
  virus, 513
Pinwheels, 129f, 228
Planococcus ficus, 512
Plasmodiophorids, 78, 80
Plautia stali intestine virus, 186
PLRV. *See Potato leafroll virus*
Plum pox virus (PPV), 182, 479, 510
PMMV. *See Pepper mild mottle virus*
PMTV. *See Potato mop-top virus*
Polerovirus, poleroviruses, 11t, 16, 17,
  44, 477t
Poliovirus, polioviruses, 25, 26, 184,
  185, 186
Polymerase chain reaction (PCR)
  asymmetric, 447, 489
  diagnostic, 475, 476
  duplex RT, 482, 509f, 510, 510f
  ELISA, 483, 484
  heminested, 480, 512
    heminested spot capture, 481, 512
  immunocapture (RT), 488, 490
  in situ, 490
  inhibitors, 472, 474, 481
  inverse, 488
  ligation anchored, 488

Polymerase chain reaction (PCR)
  (continued)
  microplate hybridization, 483
  multiplex, 482
  nested, 483, 489, 513
  real time, 484
  runaround, 488
  spot capture, 482, 512
  touchdown, 476, 479
Polymerase inhibitors, 157f, 476
Polymyxa betae, 78, 89, 96, 97, 481
Polymyxa graminis, 78, 80, 83, 86, 91
Polymyxa spp., 78, 80
Polyprotein, 69, 91, 93, 190, 227f
Polythetic class of viruses, 4
Pomovirus, 11t, 18, 81, 83, 368
Porcine circovirus, 285f
Pospiviroid, Pospiviroidae, 11t, 20,
  446
Posttranscriptional gene degradation,
  416
Posttranscriptional gene silencing
  (PTGS), 190, 369, 386, 414,
  416
Potato aucuba mosaic virus, 37
Potato leafroll virus (PLRV), 11t, 16,
  136, 188, 448
Potato mop-top virus (PMTV), 11, 18,
  78t, 83, 87f, 368
Potato spindle tuber viroid (PSTVd),
  11t, 20, 43, 449t, 453f
Potato stem-mottle, 86
Potato virus A, 374f, 388
Potato virus C, 231
Potato virus M, 132
Potato virus S, 410, 457
Potato virus T, 107t
Potato virus V, 230, 233
Potato virus X (PVX), 20, 150, 182,
  372t, 457
Potato virus Y (PVY)
  genome organization, 458f
  phylogenetic (analysis) tree, 237
  resistance, 413
  signal transduction, 384
  taxonomy, 11t, 15, 16
  translation, IRES, 184
  transmission, 15, 16
  variability, evolution, 225
Potato yellow dwarf virus, 10t, 14

Potato yellow mosaic virus (PYMV),
283t
Potexvirus, potexviruses, 11t, 26, 182,
477t, 478
Pothos latent virus, 17
Potybox, 229
Potyviridae, Potyvirus, potyviruses,
11t, 15, 225, 477t, 478
PPV. *See Plum pox virus*
PR proteins. *See* Pathogenesis-related
(PR) proteins
Programmed cell death. *See* Apoptosis
Prune dwarf virus, 106t
Prunus necrotic ringspot virus, 106t
PSbMV. *See Pea seed-borne mosaic
virus*
Pseudomonas syringae pathovar
tomato, 383
Pseudorecombinants, 65, 116, 210, 213,
351
Pseudoviridae, Pseudovirus, 10t, 13
PSTVd. *See Potato spindle tuber viroid*
PTK domain, 38
PVCV-like viruses, 10t, 317, 318t
PVX. *See Potato virus X*
PVY. *See Potato virus Y*
PYMV. *See Potato yellow mosaic virus*

Quasi-species, 46, 240

Raspberry bushy dwarf virus, 11t, 19
Raspberry ringspot virus, 69, 106t, 117,
502
RB. *See* retinoblastoma-like (Rb)
protein
RCNMV. *See Red clover necrotic
mosaic virus*
Readthrough (RT) (strategy), 43, 66f,
81, 87f, 88f
Receptor-mediated endocytosis, 127,
294
Recombinant
antibodies, 426, 427, 429
DNA technology, 128, 399, 443
proteins, 425, 428, 429
virus(es), 67, 203, 210, 219, 357

Recombination
aberrant homologous, 204, 208, 219,
240
DNA viruses, 339, 347, 351, 354
events, 206, 207f, 219, 225
homologous, 203, 208, 240, 242,
355
illegitimate, 240, 352
non-homologous, 204, 205f, 208,
240
RNA viruses, 203
similarity-assisted, 204, 209, 219
similarity-essential, 204
similarity-nonessential, 204
Red clover necrotic mosaic virus
(RCNMV), 150, 180
Red clover vein mosaic virus, 107t
Ren. *See* Replication enhancer protein
Reoviridae, reoviruses, 10t, 13, 48, 49
Rep. *See* Replication associated
(initiation) protein
Replication associated (initiation)
protein (Rep), 285, 286, 341f,
342
Replication enhancer protein (REn),
265f, 266, 280, 341f
Resistance
antibody-mediated, 429, 436
antisense, 414
breeding for, 70, 98
coat protein-mediated, 407, 408
engineered, 71, 406
extreme, 374
genes for, 382f, 489
horizontal, 371
inducible, 377
local acquired, 373
mature plant, 377
monogenic, 377
movement protein-mediated, 409
multigenic, 377
natural, 98, 367, 389, 489
polygenic, 377
quantitative, 370
replicase-mediated, 409
RNA-mediated, 410, 413, 414, 415f,
416
single-gene-mediated, 317, 371, 389
systemic acquired, 373, 385

Resistance *(continued)*
  transgenic, 307, 386
  vertical, 371
Restriction fragment length
  polymorphism (RFLP), 241,
  243f, 478, 485
Retinoblastoma-like (Rb) protein, 289
Retroid, 8
Retrotransposons, 4, 12, 13
Retroviridae, retroviruses, 8, 13, 317,
  339
RFLP. *See* Restriction fragment length
  polymorphism
Rhabdoviridae, rhabdoviruses, 10t, 14,
  47, 48, 136
Rhizomania, 89, 98, 481
Rhopalosiphum padi, 504, 506
Ribozyme, 410, 412f
Rice dwarf virus (RDV), 49, 502
Rice ragged stunt virus, 10t, 13
Rice stripe virus, 10t, 14
Rice tungro bacilliform virus (RTBV),
  12, 40, 317, 318f
Rice tungro spherical virus, 11t, 15, 40,
  319, 346
Rifampicin, 156, 157f
RNA synthesis
  abortive, 151
  elongated, 156
  initiation of, 148-151, 159, 164, 170
  mechanism, 147
  negative strand, 148, 149f, 153, 159
  termination of, 167, 169, 170
RNA-dependent RNA polymerase, 83,
  147, 176, 203, 386
RTBV. *See Rice tungro bacilliform
  virus*
RTBV-like viruses, 10t, 12, 317, 318t
Ryegrass mosaic virus, 11t, 15, 16,
  226f
Rymovirus, 11t, 15, 16, 225

Saccharomyces cerevisiae, 186
Saccharomyces cerevisiae Ty1 virus,
  10t, 13
Saccharomyces cerevisiae Ty3 virus,
  10t, 13
Satellite, 21, 40
Satellite nucleic acids (DNAs, RNAs),
  21, 208, 241, 349, 350

Satellite (RNA) protection, 410
Satellite viruses, 21, 410
SbCMV. *See Soybean chlorotic mottle
  virus*
SbCMV-like viruses, 10t, 12, 317, 318f
SBWMV. *See Soil-borne wheat mosaic
  virus*
SCMV. *See Sugarcane mosaic virus*
SCRLV. *See* Subterranean clover red
  leaf virus
SCSV. *See Subterranean clover stunt
  virus*
SEL. *See* Size exclusion limit
Semipersistent transmission, 39, 40, 65
Sequiviridae, Sequivirus, 11t, 15, 319
sgRNAs. *See* Subgenomic RNAs
Shallot virus X, 11t, 20
SHMV. *See Sunn-hemp mosaic virus*
Short-distance movement, 31, 128
Sida golden mosaic associated virus,
  283t
Sida golden mosaic virus (SiGMV),
  283t, 348f
Sida race, 296
SiGMV. *See Sida golden mosaic virus*
Signal transduction, 379, 381, 384
Single-chain antibody, 427, 431f
Single-strand conformation
  polymorphism (SSCP),
SIR. *See* Smaller intergenic region
Sitobion avenae, 504
Size exclusion limit (SEL), 130, 131,
  261, 290, 342
SLCV. *See Squash leaf curl virus*
Smaller intergenic region (SIR), 259f,
  260
Sobemovirus, sobemoviruses, 11t, 17,
  46, 108, 182
Soil-borne wheat mosaic virus
  (SBWMV), 11t, 18, 83, 188
Solanum, 384
Solanum melongena, 299t
Solanum nigrum, 289, 299t
Solanum species, 245
Solanum tuberosum, 232
Solanum viarum, 288
Sorghum mosaic virus, 226f, 486
Southern bean mosaic virus, 11t, 17,
  107t, 108, 182
Sowbane mosaic virus, 107t

Soybean chlorotic mottle virus
    (SbCMV). 10t. 12. 317, 318f
Soybean mosaic virus, 107t
SPFMV. See Sweet potato feathery
    mottle virus
Spinach latent virus, 106t
Spongospora spp., 78, 80
Spongospora subterranea, 18, 78, 80,
    86, 96
Spumavirus, 319
SqLCV. See Squash leaf curl virus
Squash leaf curl virus (SLCV,
    SqLCV), 258, 283t, 290, 449t,
    505
Squash mosaic virus, 106t
SSCP. See Single-strand conformation
    polymorphism
SSV. See Sugarcane streak virus
Stipple streak of bean, 83
Strawberry latent ringspot virus, 106t
Strawberry vein banding virus, 318f
Subgenomic (sg) RNA(s), 65, 82f, 175,
    176, 218f
Subliminal infection, 128, 368
Subterranean clover mottle virus, 107t
Subterranean clover red leaf virus
    (SCRLV), 505
Subterranean clover stunt virus
    (SCSV), 9, 10t, 343, 344
Sugarcane bacilliform virus, 318f, 326,
    479
Sugarcane mosaic virus (SCMV), 477t,
    482
Sugarcane streak virus (SSV), 258t
Sunflower mosaic virus, 107t
Sunn-hemp mosaic virus (SHMV),
    107t, 135, 410
Sweet potato chlorotic stunt virus, 388
Sweet potato feathery mottle virus
    (SPFMV), 388
Sweet potato mild mottle virus, 11t, 16,
    226f
Symbionin, 43
Synergism, 290, 388

Tagetitoxin, 156
Taino tomato mottle virus (TTMoV,
    TToMoV), 268, 283t

TAV. See Tomato aspermy virus
Taxonomy of Viruses, International
    Committee on, 3, 284
TBIA. See Tissue-blot immunoassay
TBRV. See Tomato black ring virus
TBSV. See Tomato bushy stunt virus
Tenuivirus, tenuiviruses, 10t, 14, 15,
    49, 136
Ternary complex, 151, 156, 159, 165,
    167
TEV. See Tobacco etch virus,
Texas pepper virus, 258
TGB. See Triple gene block
TGMV. See Tomato golden mosaic
    virus
TLCV, ToLCV. See Tomato leaf curl
    virus
TMV. See Tobacco mosaic virus
TMV-based vector. See Tobacco
    mosaic virus (TMV)
TNV. See Tobacco necrosis virus
Tobacco etch virus (TEV), 37, 136,
    181, 413, 503
Tobacco golden mosaic virus, 132
Tobacco mild green virus, 410
Tobacco mosaic virus (TMV)
    avirulence restance genes, 372t
    -based vectors, 431, 431f
    expression, sg RNA-encoded genes,
        181
    genomic organization, 402f
    host resistance genes, 372t
    infection cycle, 407f
    movement, coat protein-independent,
        128
    taxonomy, 111
    transgenic plants, 432-433
    transient expression vector, 430
    transmission, 108
Tobacco necrosis satellite virus, 21
Tobacco necrosis virus (A)(TNV), 11t,
    17, 81
Tobacco rattle virus (TRV)
    detection, 506, 513
    movement, 136
    recombination, 206
    taxonomy, 11t, 18
    translation, 189
    transmission, 65, 107t, 109

Tobacco ringspot virus (TRSV), 11t, 15, 106t, 109, 411
Tobacco severe etch virus, 502
Tobacco streak virus (TSV), 10t, 18, 19, 106t
Tobacco stunt virus, 13, 81
Tobacco vein mottling virus (TVMV), 37, 38
Tobacco veinal necrosis virus, 234
Tobacco yellow dwarf virus, 285f, 348f
Tobamovirus, tobamoviruses, 18, 26, 136, 181, 477t
Tobravirus, tobraviruses, 18, 64, 65, 69, 70
Tomato aspermy virus (TAV), 106t, 204, 206, 210f, 214f
Tomato black ring virus (TBRV), 69, 106t, 387, 502
Tomato bushy stunt virus (TBSV), 17, 81, 94, 137, 206
Tomato golden mosaic virus (TGMV), 258t, 264, 265f, 283t, 286
Tomato leaf curl virus (TLCV, ToLCV), 283t, 342, 348f, 513
Tomato mosaic virus, 372t
Tomato mottle virus (ToMoV), 283t, 288
Tomato pseudo-curly top virus (TPCTV), 258, 258t, 340, 348f
Tomato ringspot virus, 69, 106t, 133
Tomato spotted wilt virus (TSWV)
  detection, 449t, 485
  genomic structure, particle morphology, 401, 405f
  nonstructural protein, 136, 401, 405f
  taxonomy, 10t, 14
  transcription, ambisense, 178
  transgenic plant, 412f
  transmission, 47
Tomato yellow leaf curl virus (TYLCV), 258t, 259f, 283t, 289, 348f
Tombusviridae, Tombusvirus, tombusviruses, 17, 94, 136, 179, 477t
ToMoV. *See Tomato mottle virus*
Tospovirus, tospoviruses, 14, 47, 136, 401, 477t
Toxoptera citricida, 512

TPCTV. *See Tomato pseudo-curly top virus*
Transcapsidation, 416
Trans-complementation, 291, 292
Transcriptional activation (transactivator) protein (TrAP), 259f, 271, 320, 341f, 353f
Transformation of plants, 98, 353f, 430
Transgenic (plants), 70, 130, 412f, 425, 490
Transient expression system, virus based, 430, 432, 434
Translation (RNA virus)
  initiation of, 181-188
  cap-independent, 181, 182, 184
  elongation of, 188-189
  termination of, 189-190
Transmission of viruses
  by fungi, 43, 77, 78t, 80, 225
  by insects
    aphids, 33t-34t, 39, 40, 47
    beetles, 15, 32, 34t, 46
    leafhoppers, 32, 45, 48, 319, 346
    mealybugs, 32, 33t, 319
    Neuroptera, 34t
    planthoppers, 14, 32, 33t-34t
    thrips, 14, 32, 34t
    treehoppers, 32, 33t, 45
    whiteflies, 45, 46, 225, 285, 293
  mechanical, 44, 46
  by mites, 15, 32, 225
  by nematodes, 63, 64, 65, 67, 475
  through pollen, seed, 105, 106t-107t, 111, 113
  through soil, 77, 474
Transovarial transmission, 45
TrAP. *See* Transcriptional activator protein
Trialeurodes vaporariorum, 301f, 506
Trichodoridae, trichodorids, 63, 64, 68
Trichodorus, 64
Trichodorus primitivus, 513
Trichovirus, trichoviruses, 11t, 20, 107t, 136, 477t
Triple gene block (TGB), 86, 133, 150, 372t, 410
Triplonchida, 63
Tritimovirus, 11t, 16, 226, 502
TRSV. *See Tobacco ringspot virus*

TRV. See Tobacco rattle virus
TSV. See Tobacco streak virus
TSWV. See Tomato spotted wilt virus
TTMoV, TtoMoV. See Taino tomato
     mottle virus
Tubular structures, tubules. 133, 319
Tumor viruses, 183
TuMV. See Turnip mosaic virus
Turnip crinkle virus, 208
Turnip mosaic virus (TuMV), 184, 486
Turnip yellow mosaic virus, 11t, 19,
     107t
TVMV. See Tobacco vein mottling
     virus
TVNV. See Tobacco veinal necrosis
     virus
TYLCV. See Tomato yellow leaf curl
     virus
Tymobox, 19
Tymovirus, tymoviruses 11t, 19, 34t,
     46, 190

Uganda cassava associated virus (UGV,
     UgV), 283t, 304, 347, 348, 354
UGV. See Uganda cassava associated
     virus
Ultrabithorax, 186
Umbravirus, umbraviruses, 11t, 16, 17,
     44, 136
Unassigned plant rhabdoviruses, 14

Varicosavirus, 10t, 13
Vector-mediated seed transmission,
     108
Vegetatively propagated (crops, plants),
     389, 462
Viroids, 4, 20, 443, 444, 449t
Virus. See also Control (strategies);
     Movement (transport)
     of viruses; Transmission of
     viruses
  classification, 3, 8, 10t, 25
  coded proteins, 98
  detection, 443, 449t, 450, 475
  epitopes, 427
  evolution, 203, 206, 219, 225, 339
  nomenclature, 5, 6
  phylogeny, 21

Virus (continued)
  serotypes, 5
  species, 4, 5, 26, 219, 339, 347
  spread of, 229f, 295, 388, 480
Virus classification and nomenclature,
     international code of, 5, 6t, 25
Virus detection in
  aphids, 480, 481, 502, 503, 507
  fungi, 480
  leafhoppers, 502, 504, 513, 514
  mealybugs, 480, 512-513
  nematodes, 480, 502, 506, 513-514
  whiteflies, 457, 480, 505, 513
Virus-free stocks, 389
Virusoids, 410
Virus-specific ribonucleoprotein
     complex (vRNP), 130, 131
Vitivirus, vitiviruses, 11t, 20, 477t, 482,
     512
VPg. See Genome-linked viral protein
vRNP. See Virus-specific
     ribonucleoprotein complex

Waikavirus, waikaviruses, 11t, 15, 34t,
     40, 136, 319
Watermelon mosaic virus 2 (WMV-2),
     480
WCIMV. See White clover mottle virus
WDV. See Wheat dwarf virus
Wheat dwarf virus (WDV), 258t, 348f
Wheat spindle streak mosaic virus,
     226f
Wheat streak mosaic virus (WSMV),
     11t, 16, 107t, 226f, 502
White clover cryptic virus 1, 10t, 13
White clover cryptic virus 2, 10t, 13
White clover mosaic virus, 133, 410
White clover mottle virus (WCIMV),
     410
Whiteflies. See Transmission of viruses
     by; Virus detection in
WMV-2. See Watermelon mosaic
     virus 2
Wound tumor virus, 10t, 13, 48
WSMV. See Wheat streak mosaic virus

Xanthomonas oryzae pv. oryzae, 382f
Xiphinema americanum, 64

Xiphinema index, 514
Xiphinema spp., 64, 67, 70, 502

Yam mosaic virus, 241
Yeast, 186, 187, 188, 191

Zucchini yellow mosaic virus (ZYMV),
   449t, 480
ZYMV. *See Zucchini yellow mosaic
   virus*

Milton Keynes UK
Ingram Content Group UK Ltd.
UKHW022052141024
449569UK00031B/1605